# 6th Process Safety Management Mentoring (PSMM) Forum 2017



Topical Conference at the 2017 AIChE Spring Meeting and 13th Global Congress on Process Safety

San Antonio, Texas, USA
26 - 30 March 2017

ISBN: 978-1-5108-4365-3

**Printed from e-media with permission by:**

Curran Associates, Inc.
57 Morehouse Lane
Red Hook, NY 12571

**Some format issues inherent in the e-media version may also appear in this print version.**



Printed by Curran Associates, Inc. (2017)



AIChE
120 Wall Street, FL 23
New York, NY 10005-4020

Phone: (800) 242-4363
Fax: (203) 775-5177

www.aiche.org

**Additional copies of this publication are available from:**

Curran Associates, Inc.
57 Morehouse Lane
Red Hook, NY 12571 USA
Phone: 845-758-0400
Fax: 845-758-2633
Email: curran@proceedings.com
Web: www.proceedings.com

# TABLE OF CONTENTS

# 482030 AGILE Award Keynote Address

**Abstract:**


Factors in Selecting AGILE Honoree

1. Overall sustained excellent industry leadership in businesses that employ chemical engineers

2. Recognized for efforts in innovation and creating strategies that ensure success

3. Value of the keynote to the meeting attendees and the members of the Institute Schedule

| Event | Time | Location |
|---|---|---|
| Continental Breakfast | 7:00AM - 8:00AM | Henry Gonzalez Convention Center, Lila Cockrell Theatre |
| AGILE Keynote Address | 8:00AM - 9:15AM | Henry Gonzalez Convention Center, Lila Cockrell Theatre |

The 2017 AIChE Spring Meeting and 13th GCPS AGILE Keynote Address will be presented by AGILE award recipient John Y. Televantos, Partner, Arsenal Capital Partners.

John Televantos is a Partner at Arsenal Capital Partners, a middle market private equity firm focused on the specialty industrial, healthcare and specialty financial service sectors. Before joining Arsenal in 2006, Dr. Televantos was President of Aqualon Company, owned by Hercules, Inc.; President and Founder of Helios Chemical Company; and Chief Executive Officer for Foamex International.

Dr. Televantos began his career with Union Carbide in 1977 in silicones research. There he developed two key processes for the manufacture of Methyl Chloride and Siloxane polymers that were commercialized. These processes not only improved efficiency and reduced cost but also dramatically reduced waste. He moved through a variety of management positions there and later at ARCO Chemical, where he became vice president of strategy and development.

At ARCO John co-invented and led a team that developed and commercialized a breakthrough technology in polyol manufacture (a key raw material for polyurethanes) eliminating a whole process step along with waste and emissions, while at the same time opening the door to the creation of new higher-performance polyurethanes. The technology has since been licensed by most of the major global polyol manufacturers with significant waste reduction worldwide.

At Arsenal Capital, Dr. Televantos serves as co-head of the specialty industrial practice. During his tenure, he has been part of several acquisitions and successful exits in the chemicals and materials sectors. In addition to serving on the boards of several Arsenal-owned companies, he is lead director on the board of Balchem Corp.

## 481540 Dan Crowl on Fire - Basics of Flammability *Dan Crowl*


**Abstract:**

Flammable materials are common in industrial facilities - either as a feedstock, solvent, product or fuel. It is important to understand the flammability characteristics of these materials in order to handle them safely. This talk will provide an introduction to flammability. This will include the nature of fires and explosions, the properties of flammable materials, and how to use these properties to handle flammable materials safely

# 481541 How Do Flame Arresters Work?

**Abstract:**

Processes involving the storage, transport, separation, & destruction of flammable vapors are ubiquitous in the chemical & petrochemical industries. These operations present inherent explosion risks that are often addressed through the application & use of Flame Arresters. However, many practicing engineers & recent graduate engineers are unfamiliar with Flame Arresters, which can lead to dangerous misapplications. This introductory presentation details the basic fundamental principles of flame arresters, the industry standards governing these devices, & common flame arrester applications.
Key technical concepts related to flame arresters, such as the MESG, the dynamics of propagating explosions, & arrester element design will be described. The critical regulatory & industry requirements of NFPA 67, NFPA 69, & 33 CFR Part 154 will be cited. A generous cross section of typical flame arrester applications will be provided.

## 481542 BLEVE Basics, A Primer on Boiling Liquid Expanding Vapor Explosions

**Abstract:**

Boiling Liquid Expanding Vapor Explosions (BLEVEs) are a special case of bursting pressure vessels in which the pressure vessel is filled with a liquid that is above its normal atmospheric boiling point. When the vessel fails, the liquid boils and adds significant energy to the bursting vessel event. The thermodynamics of these events can be very complicated and interesting and the consequences of these events can be catastrophic. This session will provide new practitioners with a working knowledge of BLEVE phenomena, potential consequences, prediction methodologies, and some sample event sequences that can lead to a BLEVE. The focus will be on developing fundamentals, discussing common misconceptions, and presenting how to deal with the most common scenarios and the session will include a worked example of a BLEVE calculation to illustrate the techniques.

# 479496 A Proposal for Improving the Efficiency, Effectiveness, Quality, and Usefulness of PHA Identification of Controls

**Abstract:**

Process Hazard Analysis (PHA) is an important discipline for identification and understanding of hazard causes, consequences, consequence severity, controls / safeguards, end consequence likelihood, risk assessment, recommendations for reducing risk, and prioritization of recommendation resolution.

Controls / safeguard discussion and documentation is fraught with potential problems. If controls identified by members of the PHA team are ***not*** adequate and are ***not*** included in the list, team members may reduce their participation as they become increasingly discouraged and disengaged. If controls that are ***not*** adequate ***are*** included in the list, the sheer number of listed, inadequate controls may result in: a false sense of sufficiency of the listed controls; a risk likelihood and risk rank that are less than actual; and reduced recommendation resolution priority, if a recommendation is made at all.

Accordingly, consistency, discipline, and rigor of controls discussion and documentation is recommended for useful, quality PHAs, whatever methodology is employed. Unfortunately, the time required for useful, quality PHAs employing consistency, discipline, and rigor can also result in tedium, reduced participation of the members, and increasing discouragement and disengagement anyway.

This paper outlines a proposal in which leaders of PHA studies list and annotate controls suggested by members of the PHA Team to be reviewed in the study for quick, instructive, and collaborative disposition. The controls listed for each scenario can be sorted and discussed by the annotations which may be modified in the finalizing discussion.

Using such an approach would encourage participation by all members of the PHA Team and reduce occasion for discouragement and disengagement. Application of the approach supports: quicker, improved evaluation and documentation of the effectiveness of controls; improved quickness and quality of evaluating likelihood and risk rank; better evaluation of the need for recommendations; and proper prioritization of recommendation resolution.

# Effective Human Factors Tools to reduce Plant Costs and the Number and Severity of Injuries

**Dennis Attwood, P.Eng., Ph.D.**
**Senior Consultant, Midland Engineering Limited**
**Midland, MI**
**dennis@hf-apps.com**

Prepared for Presentation at
American Institute of Chemical Engineers
2017 Spring Meeting and 13th Global Congress on Process Safety
San Antonio, TX
March 26 – 29, 2017

# Effective Human Factors Tools to reduce Plant Costs and the Number and Severity of Injuries

**Dennis Attwood, P.Eng., Ph.D.**
**Senior Consultant, Midland Engineering Limited**
**Midland, MI**
**dennis@hf-apps.com**

**Human Factors, ergonomics, valve design, risk analysis, HAZOPs, PHA, Critical tasks, procedures**

## Abstract

Human Factors is defined as the "systematic process of designing for human use". This process systematically applies our knowledge of the capabilities, limitations, expectations and needs of people to the equipment they use, the environments they operate in, the tasks they perform and the management systems that guide the safe operation of their plants.

This paper describes easy-to-use and cost-effective Human Factors tools for the solution of major issues in process plants. Considering the breadth of the field of human factors and the limited time available for presentation, the paper limits discussion to the following four applications:

- Assessing the physical risk of operating manual valves to reduce maintenance costs.
- Improving the design of operations procedures.
- A method for applying Human Factors to Process Hazard Analyses.
- Critical task identification and analysis.

Each tool is thoroughly explained and its use is demonstrated with a practical example from the process industries with a summary of Benefits. References for papers explaining each tool in more detail are provided.

## Introduction

The Human Factors discipline is continually evolving. Many in the field are conducting basic research to determine how humans respond to musculoskeletal, physiological and cognitive stressors. Others are developing tools to analyze plant situations and recommend solutions. This paper focuses on some of the many methods developed for analyzing Human Factors in Process Plants. The methods selected herein, can be used to solve a problem that has been identified, for example, to rank the musculoskeletal risk of manual valve operation. They can be used to determine if a Human Factors problem exists, for example, whether procedures are designed to make users make mistakes. Or they can be used to analyze systems during process design to identify opportunities for human error and to provide a proactive solution, for example, to identify critical operator tasks and analyze them to identify appropriate safeguards to mitigate risks.

The methods on which this paper will focus include:

1. Assessing the physical risk of operating manual valves to reduce maintenance costs. The tool ranks each valve in terms of musculoskeletal risk. Plant management can use this ranked list to lower the long term cost of ownership by scheduling maintenance on the highest risk valves first and reducing human injury
2. Improving the design of operations procedures. Statistics indicate that the majority of safety-related incidents in process plants are the results of not following operating procedures and/or not having complete and easy to follow procedures. This set of tools assesses procedures for inherent risk and provides guidance on redesign. Having improved procedures that are easy to follow and understand reduces the probability of human error/injury.
3. A method for applying Human Factors to Process Hazard Analyses. The paper reviews the development of the API Standard designed to include Human Factors in Process Hazard Analyses and provides an improved format for plant use. The benefit of Including Human Factors in the PHA will make sure the key Human Factor risks are addressed and mitigated
4. Critical task identification and analysis. Critical (high risk) process tasks have the potential for personal injury, equipment or facility loss, environmental release, community exposure or business interruption. The tool provides a systematic method of identifying high-risk tasks, analyzing the tasks for inherent risks and identifying appropriate safeguards to mitigate risks. The benefit is to reduce EHS incidents

## Four Human Factors Application Tools

### *1. Assessing the Physical Risk of Operating Manual Valves to reduce Maintenance costs*

Every process plant contains a large number of process valves with a wide diversity of designs that control the flow of feedstock, products and service liquids and gases. The ergonomic issues associated with the design, operation and maintenance of valves include:

- Physical stress to open and close them
- Potential for injuries and subsequent related costs
- Lack of access to process critical valves
- Difficulty to remove and replace them
- Potential for process upsets when valves cannot be operated in the time required

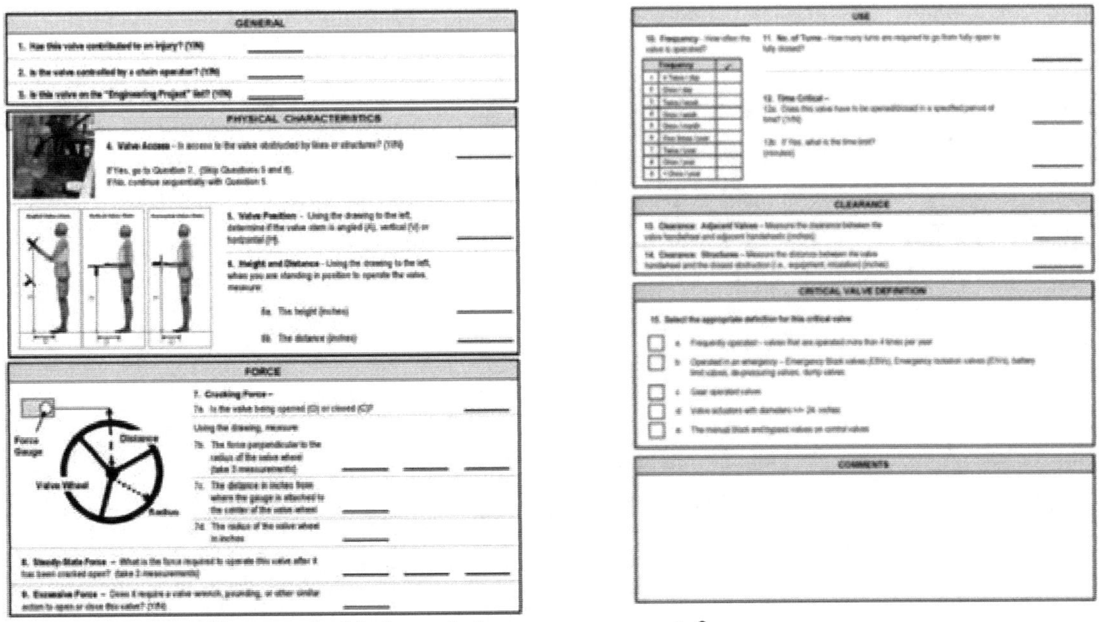


**Figure 1: Valve risk assessment forms**

As a result of injuries suffered while operating valves, a major U.S. petroleum refinery company initiated a program to identify high risk valves and to modify them to reduce the operating risk of injury.

We developed a systematic approach Attwood and Tys (2013) to identify and assess the operations and maintenance issues associated with process significant valves within the refinery. The approach consisted of eight distinct activities:

1. Develop plant standards for human factors requirements in the design, operation and maintenance of valves.
2. Train a group of refinery personnel, consisting of operators, maintenance techs, plant engineers and safety, health and environment staff, in ergonomic issues associated with valve operations and maintenance.
3. Lead the group in developing the definition of "process significant" valves.
4. Identify the valves that meet the definition in each unit of the refinery.
5. Assist refinery personnel to define the characteristics of valves that determine their risk of personnel injury or their potential for causing process upsets.
6. Develop a quantitative scale for each of the characteristics in (5) above.
7. Field assess each of the "process significant" valves against the characteristics.

8. Calculate the sum of the risk characteristics for each valve and develop a risk prioritization ranking list. This will permit the refinery to repair or replace valves, as business conditions permit, according to their position in the ranking list.

The characteristics that were identified to assess the risk of valve operation are shown in Figure 1. We have used this methodology in two different refineries with excellent results. The data were able to distinguish between high and low musculoskeletal risk valves. Figures 2A and 2B, for example, illustrate and describe both a HIGH and a LOW risk valve and recommends practical improvements for each.

**Figure 2A: High risk valve**

High Risk Valve

- Risk score = 79 (out of 112 possible)
- Has contributed to injuries
- Third choice location
- Cracking force > 246 lb.
- Steady state force >149 lb.
- Requires 42 turns to open
- It is recommended to increase the height of the actuator to waist height and to increase the size of the actuator to reduce the steady state rim force required

**Figure 2B: Low risk valve**

Low risk valve

- Risk score = 51 (out of 112 possible)
- Frequently operated
- Cracking force > 181 lb.
- Steady state force > 144 lb.
- Time critical operation (16 turns in 5 minutes)

It is recommended to adjust heights of actuators so pinch points are eliminated and to increase the diameter of the actuator wheels to reduce rim force required

Our valve risk ranking was reviewed by the operations team that provided the data and accepted as a viable risk-ranking technique that can be used to schedule valve maintenance so that the valves with the highest risk are maintained first. Moreover, our recommendations for improving each high risk valve were both practical and economical.

Practical Benefits of this Application

- Reduces the potential for musculoskeletal injuries and days off from work
- Reduces maintenance costs by targeting valves with the highest musculoskeletal risk

## 2. *Improving the design of operations procedures*

The term "procedure" refers to a written description of the steps that a process worker needs to follow to complete a task. The purpose of a procedure is to guide a process worker, step-by-step, through a task. Not every task requires a procedure (Haas, 1999). If the task has few steps, if it is performed frequently and if the consequence of error is minor, it will not likely require a procedure. However, if the task is complex, if it is performed infrequently and if the consequence of making an error is high, a procedure will likely be required. Failure to follow required procedures or failure to follow them correctly accounts for as much as 70% of the error-related incidents made by process workers (HSE, 2005). Incidents include both injuries and errors that affect process safety.

Evaluation Techniques

The effectiveness of a procedure program can be evaluated in a number of different ways. One of the most effective is a site audit to record any gaps between the existing on-site procedure program and the recommended guidelines. The audit might include:

- Observing multiple operators performing the procedure and comparing their performance against the documented procedure to identify deviations
- Evaluating the written procedure with a checklist. The checklist published in Attwood et al, 2003, contains detailed questions that can be used to evaluate written procedures
- Computing a Readability Score which is an open-source index that includes several well-known indices of readability
- Implementing Simplified English. We are proposing to use a powerful new tool to evaluate and develop your procedures. Simply put, simplified English or simplified technical English (ASD, 2005) is a standard set of English words and vocabulary rules that:
    - Reduces ambiguity
    - Improves comprehension for people whose first language is not English
    - Makes translations cheaper and easier
    - Facilitates automatic checking to ensure that procedures and manuals are written according to the rules and vocabulary
    - Improves understanding among users whose first language is not English

We have evaluated the procedure systems in a number of companies and have concluded that the resulting procedures are easier to understand and use and that the risk of making errors is greatly reduced.

Figure 3 illustrates the some common errors that were identified from one company's procedures based on the evaluation checklist identified above.

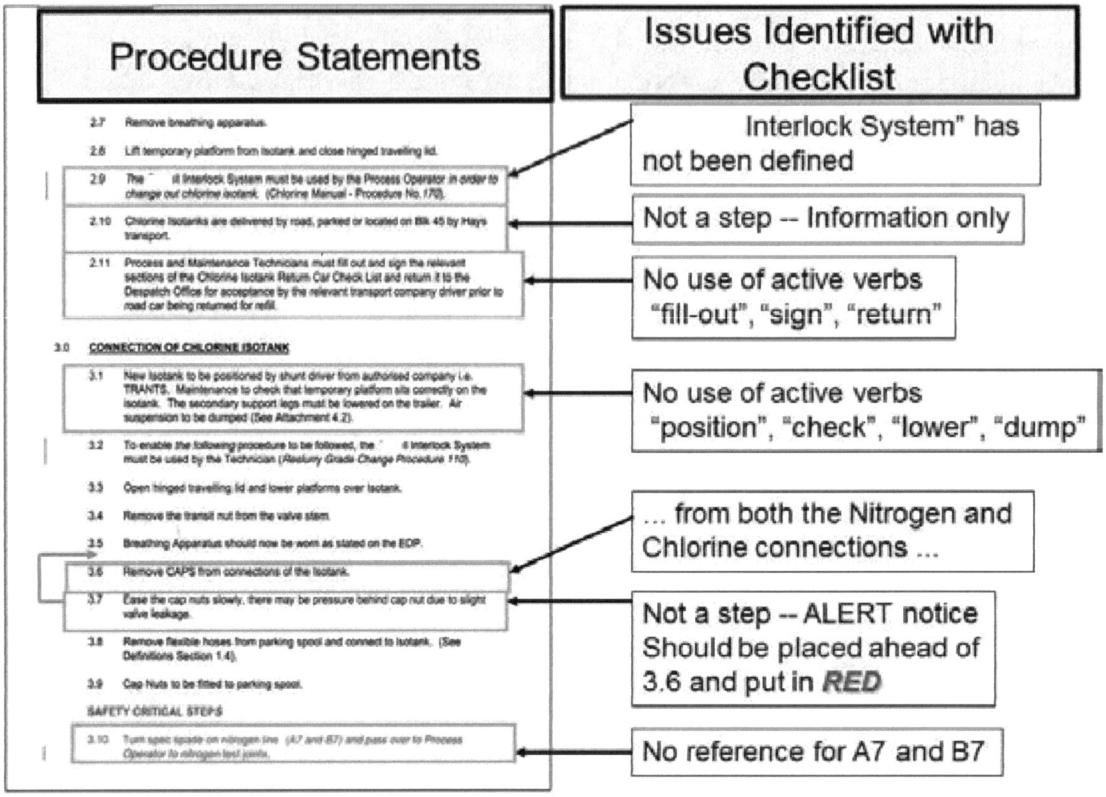


**Figure 3: Procedure design issues**

Practical Benefits of this application:

Our procedures improvement methodology ensures that clearly defined processes are in place to guide the development and implementation of effective procedures. We need to ensure that process considerations such as those below are considered for any procedures improvement program:

- Jobs that require procedures are identified
- Procedures reflect the way work is done and are technically accurate
- Procedures are written at the reading level of the users
- The detail in the procedure is not so complex that novice employees won't understand it, nor so limited that experienced employees will not read it
- Users can quickly locate the correct procedure for the job
- Users are instructed in the application of each procedure
- Procedures are reviewed and updated when the task, equipment or process changes

- Shared knowledge between operations, maintenance and engineering is routinely incorporated into procedures updates
- Users are advised when a procedure changes and instructed on the changes

## *3. A method for applying Human Factors to Process Hazard Analyses*

The principle objective of the HAZOP is to identify process hazards, especially those that affect containment integrity (Attwood and Froehlich, 2003). Some of these hazards, however, are a result of human factors i.e. either with the process equipment itself or with instrumented process control. Many HAZOP protocols are less systematic than they should be for addressing HF concerns such as Human-Machine-Interface (**HMI**). They rely on the use of a checklist without recommending a process for applying it. Since, Human Factors is not the prime reason to perform a HAZOP review, the HAZOP leader will often elect not to review the checklist during the HAZOP proceedings especially during a HAZOP for major capital projects. More likely, the checklist will be reviewed at the end of the HAZOP study resulting in a less critical and systematic approach.

The objective of this tool is to assist the HAZOP team to identify Human Factors (HF) concerns associated with particular pieces of equipment while conducting the HAZOP.

The HAZOP is a sequential process -- a line by line and equipment by equipment review. It requires that each line and/or process system on the Process and Instrumentation Drawing (P&ID) be followed and checked and that the hazards associated with the equipment that is located at each node be identified and discussed. The HF issues need to be highlighted at these nodes as an **integral** part of the HAZOP discussion and not at the end of the HAZOP when the team has completed each P&ID.

The key to this tool is the use of "Screening Questions" for each of 12 pieces of equipment. The screening questions are a simple checklist that identifies the most important HF characteristics of the targeted equipment and /or system so the team can quickly decide whether a potential problem exists and what action to recommend. If a Human Factors concern is identified, it will be recorded on the HAZOP worksheet. It is possible that sufficient information and knowledge is available to resolve the issue during the HAZOP. However, the resolution of the issue will most likely happen outside the HAZOP.

The following equipment and HF issues are included in the screening lists:

1. Valves: sequential; critical; frequent
2. Blind and Blanks, weight, lifting; location; frequency; clearing, draining
3. Pumps and Compressors; maintaining, draining, start-up; shut-down
4. Field displays; critical; start-up; shut-down

5. Firefighting and deluge systems; critical; accessing; escaping; directing
6. Field instrumentation; rodding; clearing; critical
7. Sample points, exposure; PPE
8. Reactors/Driers, critical; switching/regenerating; clearing; entering; catalyst dumping; lifting
9. Vessels (including exchangers); cleaning; exposure; PPE; lifting; entering
10. Furnaces and Fired Heaters; light-off; decoking, flame controlling; switching; PPE; entering
11. Filters; switching; cleaning; clearing; PPE; weight; lifting; entering
12. Loading and Unloading Facilities; height; space; exposure; weight; clearing; disconnecting

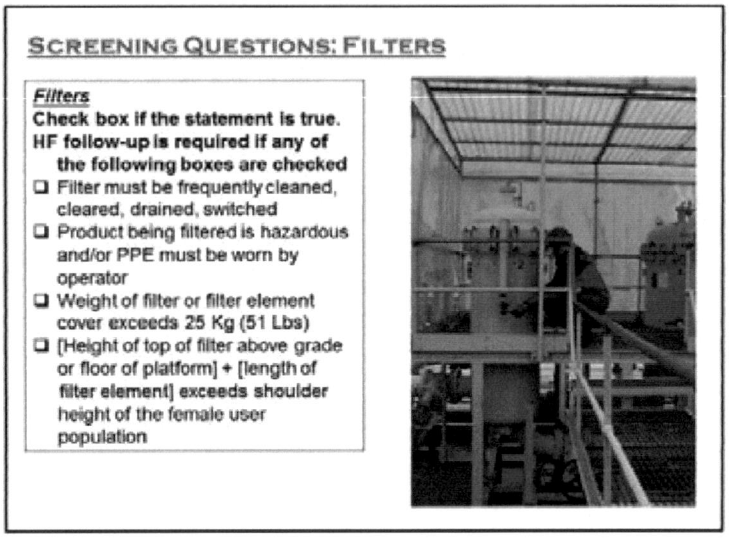


**Figure 4: Screening Questions - Filters**

Screening questions -- Example Format

An example of a "screening list" for Filters is given in Figure 4. This tool was prepared for the API and was published in 2005 (API, 2005). Our company has revised the original tool to make it easier to use. We have also developed a short introductory awareness program to show HAZOP participants how the tool will work in practice

Practical Benefits of this Application

The major benefit of this application is to ensure that Human Factors Issues are identified and discussed as they are encountered during the HAZOP reducing the probability of an incident.

### 4. *Critical task identification and analysis.*

Critical Task Identification Analysis (CTIA) is a systematic method of identifying critical tasks within a process facility, prioritizing their importance, analyzing those tasks that are considered most critical and identifying appropriate interventions to mitigate the risk. Critical in this context means high risk. The tasks that may have the potential for personal injury, equipment or facility loss, environmental release, community exposure, or business interruption are defined as "critical". These tasks represent a hazard and the goal is to identify the hazards and correct them before they turn into incidents.

The phases in the CTIA process are presented in Figure 5. The following paragraphs describe each phase, record examples of the products of each phase and provide insights on the most efficient methods to obtain information.

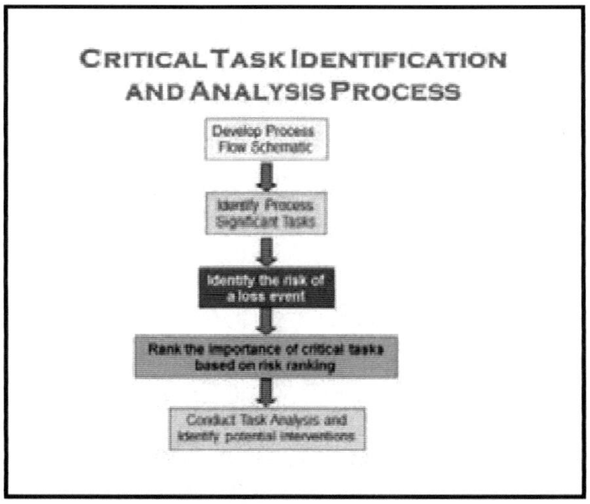


**Figure 5: Phases of a CTIA**

## Develop Process Flow Diagram

The analysis starts by specifying the facility processes and equipment and creating a process flow diagram. A portion of an actual process plant flow diagram is shown in Figure 6.

## Identify process significant tasks

The Critical Task Identification process is a 'brainstorming' technique that is designed to encourage the site team to identify the 'critical' tasks that they could be required to perform in each area of the process facility. A critical task could occur during a process upset or other abnormal situation or it could be a task that is difficult to perform or performed frequently. At each block of the PFS, the team is asked to identify the 'critical' tasks that he or she performs by placing them into one of seven different categories including:

1. Safety-critical tasks, e.g., responding to the failure of a pump flange
2. Quality-critical tasks, e.g., responding to an out-of-limits lab analysis
3. Production-critical tasks, e.g., responding to the loss of an exchanger which reduces output
4. Most frequently performed tasks, e.g., filter changes
5. Difficult to perform or complex tasks
6. Time critical tasks, e.g. responding to tasks that, in a short period of time, will cause a larger upset. An example is a compressor shutdown.
7. Environmentally-critical tasks, e.g., responding to an out-of-limit environmental deviation. For example, a high $SO_2$ reading on the stack of a fired heater.

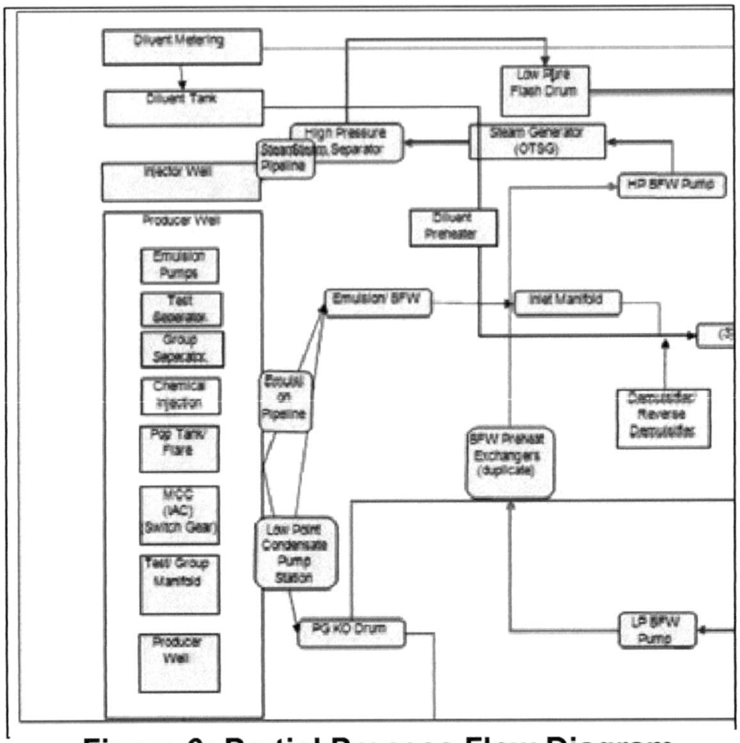


**Figure 6: Partial Process Flow Diagram**

In many cases the same 'critical' tasks can be categorized in different ways. The purpose of the exercise is to identify the major tasks (it is not important which list they appear in).

## Identify the Risk of a Loss Event

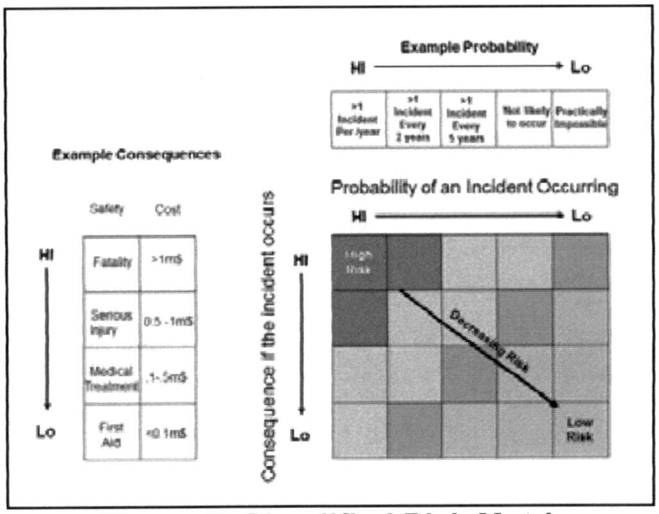


**Figure 7: Simplified Risk Matrix**

No company can afford to spend money and commit resources on issues that have little benefit to the operation. Thus, it is essential to identify the tasks which, when improved, will improve safety, reduce cost or improve the environment. The approach used at this stage is to estimate the risk associated with each task identified in the above exercise. Many companies have developed their own risk matrix. The next step in this exercise is to prioritize each 'critical' task in terms of its risk to the facility.

Figure 7 illustrates a generic risk matrix that is

designed to help workers prioritize tasks. It is presented for illustration purposes only and is not intended for use in formal risk assessments. Experience has shown that a site team can easily categorize tasks into the simplified consequence/probability risk matrix shown in Figure 7. Alternatively, the team can use the risk matrix developed for their company.

## Identify the highest risk tasks

The tasks that most require analysis are those with the highest risk in Figure 7 (shown in RED). The analysis could also be extended to Medium risk tasks shown by the ORANGE sections and the lower risk tasks shown in the GREEN section of Figure 7

## Conduct Task Analysis and Identify Potential Interventions

Finally, a task analysis is conducted on the highest risk tasks. Task analyses can take many forms and are explained in several texts and papers. The one that we use is detailed in CCPS (2010).

Table 1 provides, as an example, the activity analysis of a Feed Water Knockout Drum Sampling Task. Column 1 lists the sequential activities performed by the worker, column 2 lists the Human Factors Concerns for each human activity, column 3 lists the potential errors that can occur for each activity, and column 4 lists the potential solutions that can be used as task design recommendations.

**Table 1: Activity Analysis for Feed Water Knock-out Drum Sampling**

| Activity | Ergonomic / Human Factors Concern | Error Possibility and Consequences | Potential Solution / Mitigation |
|---|---|---|---|
| Check sample drain tank for level by looking at overflow at sample position | Inability to check level | Tank overflows | Add level indication on sample drain tank with level controller in loop for tank pump |
| Approach sample station | Lighting<br>Access | Trips/falls while carrying plastic bottles<br>Poor lighting | Provide adequate lighting<br>Bring sample station to ground<br>Ensure proper head clearance |
| Ensure ventilation is working | Toxicity<br>Air quality<br>Ventilation area | Injury due to exposure (both inside and by ventilation) | Ensure ventilation is installed<br>Provide safe ventilation area outside |

| | | | |
|---|---|---|---|
| Remove cap from sample point on common header | No human factors concern | Poor quality samples | Design issue |
| Displace and purge line | Force<br>Fatigue<br>Proximity | Arm strain<br>Body stress | Ensure sample location and valve are located within visual proximity<br>Ensure valve spring tension is low enough for human use<br>Ensure handle is designed for constant pressure |
| Line up bottle with spout and take sample | Lighting<br>Access<br>Location of sample valves | Poor lighting<br>Take wrong<br>sample | Ensure labeling on valve easily corresponds to bottle (redundant coding with labels/colors)<br>Mount lights outside of |
| Put bottle into carrier | Bending<br>Lifting | Body stress<br>Dropping samples | Provide space on sampling platform to place carrier |
| Select next bottle and repeat sampling process | Bending<br>Lifting | Body stress<br>Dropping samples | Provide space on sampling platform to place carrier |
| Close valve, recap | No new Human Factors | No new hazardous | No recommendations |
| Take samples to lab | No new Human Factors | No new hazardous | No recommendations |

<u>Practical Benefits of this Application</u>

Identifies and analyzes tasks performed by humans that can impact on the overall integrity of the unit, the operating efficiency and reliability of the processes and can influence the safety of humans interacting with the process or facility

Task analysis techniques can be applied to existing tasks or tasks that do not yet exist. The task analysis team is made up of all stakeholders in the plant including engineering, safety and environmental personnel.

## Summary and Conclusions

The effectiveness of a Human Factors program in a process plant is enhanced by the use of applications that can identify and solve human-related issues quickly and increase plant profits or reduce expenses or serve to reduce the number and severity of injuries.

The four applications described in this paper have proven, through previous use, to reduce maintenance costs, reduce the time required to launch projects, and make operations more efficient with lower incident rates. Readers who are interested in learning more about Human Factors analysis techniques are invited to read CCPS (2010) and Attwood et al. (2005).

## References


1. ASD Simplified English Specification ASD-STE100, Aerospace and Defence Industries Association of Europe, 2005.

2. ATTWOOD, D.A., DEEB, J.M., and DANZ-REECE, M.D. "Ergonomic Solutions for the Process Industries". Gulf Professional Publishing (Elsevier), Burlington, MA, December 2005. ISBN 0-7506-7704-X

3. ATTWOOD, D.A. and TYS, A.T. (2013) "A method for ranking the musculoskeletal risk of operating valves in process industries." Proceedings of the Society of Petroleum Engineers, SPE-165690-MS, Pittsburgh, PA, August 2013

4. ATTWOOD, D.A., and FROEHLICH, B., "A Tool to Identify Human Factors Concerns during the HAZOP Process." ASSE Proceedings, Denver, June 2003

5. ATTWOOD, D.A. "Critical Task Analysis" Report prepared for the Surmont II project, Conoco Phillips Canada, Project No. 06-267, November 2007.

6. [1]CCPS "A Practical Approach to Hazard Identification for Operations and Maintenance Workers." Center for Chemical Process Safety Publication, John Wiley and Sons Inc., (Publishers), New York, 2010. ISBN 978-0-470-63524-7.

7. [2]CCPS "Human Factors Methods for Improving Performance in the Process Industries." Center for Chemical Process Safety (CCPS) publication. John Wiley and Sons Inc. (publishers), New York, 2007. ISBN 13 978-0-470-11754-5.

8. Haas, P. M. (1999). "Human Performance Engineering: A Practical Approach to Application of Human Factors." 1999 National Petrochemical Refiners Association (NRPA) National Safety Conference, Paper no. NSC-99-108.

9. HSE (2005) "HSE Human Factors Briefing Note Number 4: Procedures." United Kingdom, Health and Safety Executive.


[1] D. Attwood, contracted co-author
[2] D. Attwood, contracted co-author

# 479344 Selection of Cases for Atmospheric Dispersion Modeling

**Abstract:**

In many industrial facilities, toxic and/or flammable substances are discharged directly to the atmosphere. To ensure compliance with health, safety and environmental regulations, it is important to determine the potential consequences to the public and to personnel working in the facilities. Atmospheric dispersion modeling is a useful tool to provide plant owners the information to make decision in such cases. However, there are numerous potential releases from multiple scenarios, and hence it can be impractical to perform modelling for all cases. Typical industrial standards do not provide sufficient guidance for users about how to effectively choose the representative cases to model. This paper offers the guidance on selecting the relevant cases to perform atmospheric dispersion modeling based on the configuration of relief devices, the nature of discharged fluid, and points of interest in facility.

# Best Practices for Writing Operating Procedures and Trouble-Shooting Guides

**William G. Bridges, President**
**PROCESS IMPROVEMENT INSTITUTE, INC. (PII)**
1321 Waterside Lane, Knoxville, TN 37922
Phone: (865) 675-3458
Fax: (865) 622-6800
**e-mail: wbridges@piii.com**

**Revonda Tew, Principal Engineer & Instructor**
**PROCESS IMPROVEMENT INSTITUTE, INC. (PII)**
1321 Waterside Lane, Knoxville, TN 37922
**e-mail: rtew@piii.com**


Prepared for Presentation at
13th Global Congress on Process Safety
San Antonio, TX
March 27-29, 2017

UNPUBLISHED



13TH GLOBAL CONGRESS ON PROCESS SAFETY

# Best Practices for Writing Operating Procedures and Trouble-Shooting Guides

**William G. Bridges, President**
**PROCESS IMPROVEMENT INSTITUTE, INC. (PII)**

**Revonda Tew, Principal Engineer & Instructor**
**PROCESS IMPROVEMENT INSTITUTE, INC. (PII)**

**Keywords: Procedures, Human Factors, Risk control, process safety management, PSM**

## Abstract

There is no complete, best practice guideline or textbook for writing operating procedures and trouble-shooting guides. This paper presents the proven, best approach for developing accurate operating procedures and for ensuring the page formatting and step writing are optimized to reduce human error rates. This approach and the 32 rules established in this paper are based on the foundations set by Swain and others (in 1970) for control of human error rates, but uses experiences from more than 100 sites where this approach has been successfully followed. The approach and rules for developing operational troubleshooting guides (procedures for responding to process deviations such as those needed for Human IPL) are again the best approaches found and have been applied successfully since the early 1990s. Several case studies are provided that show the gains from following this approach. The guidelines in this paper build upon ones presented in 1999 at CCPS and 2016 at GCPS.

## Background

Operating procedures have always been crucial to the safety, quality, and productivity of process systems. With the advent of new safety and quality standards such as OSHA's process safety management (PSM) regulation, OSHA's personal protective equipment (PPE) regulation, EPA's risk management program (RMP) regulation, and the ISO 9000 quality standard, many companies are facing the daunting task of developing or upgrading their procedures to satisfy varied and sometimes complex and conflicting requirements.

Although there are overlapping characteristics that the various regulations and standards share, each approaches the procedure-writing process from a somewhat different perspective (e.g., quality or safety, protection of workers or protection of the public/environment), and the required level of detail for documenting the procedures differs greatly between these regulations and standards. For instance, the ISO 9000 quality standard advocates documenting procedures and work instructions that impact quality and suggests keeping the procedures as simple as possible, while the OSHA PSM regulation requires detailed procedures that address: all operating modes (startup, shutdown, etc.), operating limits, consequences of deviations, means to avoid hazards, safety and health considerations, and safety systems and their functions. How can companies develop and maintain procedures that ensure productivity and simultaneously satisfy the different regulations and standards? The key is to remember the ultimate goal of the regulations and standards: to reduce human errors that can impact quality, productivity, and/or safety.

Procedure-related errors are errors that occur because some characteristic of the procedure caused task performance to fail. ***This is currently the most critical human factor at most sites since 90% of accidents have at least one root cause related to mistakes within procedures.*** Reducing these procedure deficiencies can reduce human error rates by a factor of 2 to 10, normally.[1, 2]

The concepts and rules presented in this paper are applicable to all industrial settings. Most importantly, this approach **is effective in satisfying both safety and quality concerns.** With this approach, procedures can and have been developed that:

- Provide a sound basis for training
- Provide the users and supervisors with an understanding of what can go wrong and how it can go wrong
- Provides management with an understanding of the inherent risk (to quality, productivity, safety, etc.) of an activity (i.e., the activity addressed by the procedure) and an appreciation of the safeguards that are being relied upon to mitigate this risk.

Three (3) case studies are presented (with data from other sites as well), which illustrate this approach, along with supporting the benefits of the accurate and clear procedures. These case studies' results are representative of the benefits seen at more than 60 sites where the same approach has been followed.

## Fundamentals of Writing Effective Procedures

When developing procedures, or any written material, always remember that the burden of written communication is on the writer, not on the reader. It should never be up to the reader to decipher the meaning of a message.

- Engineers who write procedure manuals "tend to include information that is more relevant to them than it is to the operators actually running the unit" [Sutton, 1992].[3]
- When writing procedures, your guiding principle should always be: if the written procedures are difficult to read, difficult to locate (e.g., if titles do not have a clear meaning), or inconvenient to use, they will seldom be used [Swain, 1983].[1]

***Deficient Procedures are*** the most prevalent problem in process industries since procedures have not traditionally been developed from the perspective of optimizing human factors; instead procedures have been traditionally developed to meet a compliance requirement. Examples of procedure deficiencies (inaccuracies) include:

- Incorrect/incomplete/nonexistent (most procedures we have audited have been only 70-75% accurate – the inaccuracies include missing critical steps, steps as written are not what needs to be done, or the steps are out of sequence)
- No/misplaced/incorrect information in warnings (for example, a warning should **never contain** the action to take; it should instead **emphasize** the action to take) and warnings should be located BEFORE the step applicable to the warning
- Poor format and presentation rules

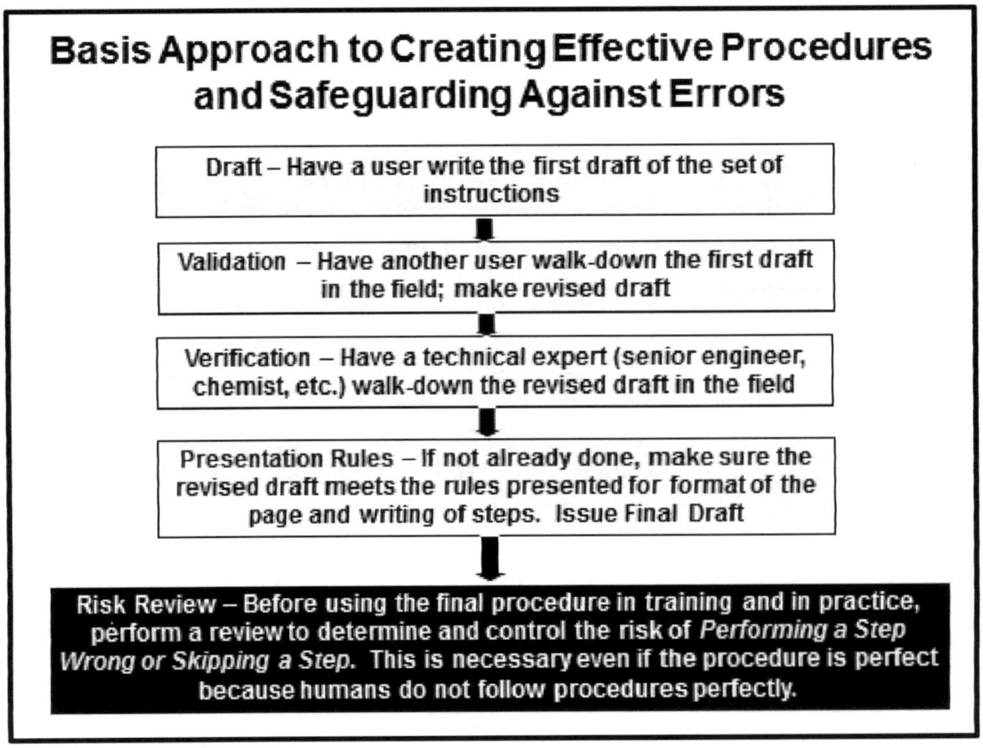


*Figure 1: Best Approach to Developing written Procedures – courtesy of Process Improvement Institute, Inc. (PII), copyright 2003-2017*[4]

The flowchart above describes the basic sequence of steps for writing effective procedures. Note that some basic rules are contained in the flowchart, such as "operators write operating procedures," "maintenance craft-persons write maintenance procedures," and "Lab technicians write lab procedures." Procedures are NOT written by engineers or department superintendents. This is a critical to ensure procedures are written in the user's common language and are written at the right level of detail.

### *Universal Purposes of a Procedure*[4]

1. A written guide to accomplish initial training of employees
2. A quick reference manual for experienced (trained) employees

### *The Right Information to Write – Content is KEY*[4]

The procedure procedural content controls the human error related to "procedure-based" error – if content is wrong, the trainee is very likely to learn the step wrong (trainer uses the procedure as a basis for training or the trainee refers to the procedure step that is wrong when executing the procedure).

Even if the content is correct, not following page formatting best practices can still increase human error rates.

> If the content of the step is wrong it does not matter if the presentation is clear. If the content is perfect, you can further reduce error rates by 50% to 70% by presenting the information properly.

Ask experienced mechanics or operators to describe a simple task that they perform. You will quickly see that even simple tasks can involve many steps. You should consider the following questions when writing each procedure step:

- How is the step performed?
    - Does the action involve interacting with a computer terminal, an automatic controller, or simple devices (gauges, valves, micrometers)?
- Can the actions be performed as written and in the sequence written?
    - Can the equipment be operated as specified?
    - Can the steps be physically performed?
    - Does the user have the training or experience to understand and carry out the action using the information available, or is additional information needed?
- Does the user need to be alerted of any potential hazards (Cautions or Warnings) or need any supporting information (Notes) before performing the action?
- Does the user need to know specific operating ranges or limits to:

- Perform this action
- Recognize the successful completion of the action
- Recognize an actual or potential problem
- Make a decision

*For example, does the user have to verify temperature or pressure visually? Is the user waiting for a computer command or verification? How should the process react? Will exceeding a limit result in an accidental chemical release, personal injury, or quality concern? What is the chance of not operating within safe operating limits, and what would be the consequences? If operation of the component is critical and a deviation will probably cause a failure, state the corrective actions. If failure is unlikely, keep going.*

- Is needed information found on an instrument, panel, or monitor or is it in the procedure or another source such as a graph, table, drawing, or specification sheet? Should this information be included in the procedure or be referenced?
- What is the next logical step? How is the next step affected by what is performed in the current step?
- What are the results of improper task performance? If an action is critical, spell it out in detail. For example, when starting a special pump that may bind or cavitate if not vented and primed properly, the procedure may require more information than simply stating "start the pump."
- Is the action frequently performed? Is it an action that is easily overlooked? Is this a complex piece of critical equipment that has a low frequency of use? These questions can help determine the level of detail in your procedure.
- Is the action performed so infrequently or is it so complicated that the user is not sure how to do it? Is the action so complicated that nobody is ever certain it's done right the first time?
- Is the decision point clearly defined if a decision is required? Unclear decision points can cause arguments and delays in performing actions.

***To address the concerns listed above, a few simple rules are needed for developing content:***

- The procedures should be written at the level of someone who has just completed the basic training for that task. Do not write the procedure for someone just hired or for the 10+ veteran.
- Clearly identify ahead of time what activities need procedures to help ensure error rates are controlled low enough, and identify which activities are considered common "skills" of all of the staff.

  *Example: Starting a pump is typically considered a "skill." This means a procedure step can simply be "Start the benzene recycle pump (P-119)." The procedure does not need to explain how to start the pump. But, starting a pump is a skill that nonetheless must be learned; it requires doing several sub-steps related to positioning of intake and discharge valves, how the pump is throttled at startup, checking local pressure gauge, etc. So, a training module is needed for starting pumps of various types, and the operator needs to learn this necessary skill. Once the skill is learned, it will be applied to so many various cases and so often, that a procedure is no longer needed as a "refresher" on how to "start the pump."*

- The first draft of the procedure must be walked-down in the field by another user. Simply reading through the procedure does not catch enough of the mistakes.
  The revised draft should be reviewed in the field by a technical staff person, such as a process engineer (for operating procedures).
- The procedure needs to be checked to ensure it follow the page format and step writing rules described in the next section. This ensures the accurate steps are clearly presented.

**The target is to reach an accuracy of 95% or better (so, no more than one wrong or missing step out of 20 steps).** The following have been observed in the field by PII staff by direct data collection at more than 110 sites/plants around the world, based on walk-down review of several to dozens of procedures at each unit/plant:

- At 95% accuracy or better and when the same procedures follow 80% of the rules for procedure clarity (presented next), then most users will follow the written procedures and will try to keep the procedures up-to-date
- At about 85% accuracy or less, about half of the users stop using the procedures
- At about 70% accuracy or less, less than 10% of the users will refer to the procedure or will try to keep it up-to-date. ***So, the written procedures are useless***
- Unfortunately, the typical operating procedure walked down by PII staff (accompanied by senior operators, a process engineer, and a shift supervisor) is about 75% accurate (so one step in four is missing or wrong). Usually these inaccurate procedures also follow less than half of the best practices for procedure format, presented next.

### PHA/HAZOP of Procedures

Just as critical as getting the content accurate and presenting the steps clearly, is performing a PHA/HAZOP of the written steps to determine if you have enough independent protection layers (IPLs) against errors that the "imperfect" operators will eventually make when using the "near perfect" written procedures. This analysis is described in detail in Chapter 9 of the *Guidelines for Hazard Evaluation Procedures*, 3rd edition, 2008[5]. **For an update on the methods for such analysis and for a discussion of the extra emphasis on PHA of procedures by US OSHA, refer to the more recent paper (Bridges, Marshall; 2015).**[6, 7, 8]

### Procedure Clarity

The importance of the accuracy of the procedure steps were discussed above. The clarity of the procedure steps (how they are written and how the page is formatted) is also important. Following best practices for step and page format reduces human errors by a multiplying factor of 3 to 5.[1, 9]

**The best practice rules for writing and validating procedures have been published for many years *(see Bridges & Williams, 1997; Madden & Bridges).*[11, 12]** These rules include best practices for formatting of the pages and steps. These have been gradually improved over the past decades and now are incorporated into PII's training materials.[4]

**Table 1 below is a checklist (in auditing format) based on the current set of best practice rules for developing accurate and clear operating, maintenance, and other work-instructions/procedures.**

**TABLE 1 PROCEDURE QUALITY CHECKLIST** *(courtesy PII, copyright 2003-2017)*[4]

| # | Issue | Response |
|---|---|---|
| | **Procedure Content Checklist** | |
| 1 | Is the procedure drafted by a future user of the written procedure? (Engineers should not author procedures to be used by operators or maintenance staff.) | |
| 2 | Is the procedure validated by a walk-down in the field by another future user of the procedures? | |
| 3 | Is the procedure reviewed and commented on by technical staff (engineers or vendors)? | |
| 4 | Is the procedure checked versus the Page and Step format rules below? | |
| 5 | Is a hazard review of step-by-step procedures performed to make sure there are sufficient safeguards (IPLs) against the errors that will occur eventually (when a step is skipped or performed wrong)? | |
| 6 | Is the content measured using "newly trained operators" to judge the % of steps that are missing, steps that are confusing or wrong, and steps that are out-of-sequence? (A score of 95% accuracy of content is good; 98% should be the targeted average.) | |
| | **Page Format Checklist** | |
| 1 | Is the title of the procedure the largest item on the page? | |
| 2 | Is the procedure title clear and consistent with other titles, and does it uniquely describes the topic? | |
| 3 | Are the document control features the smallest items on the page? | |
| 4 | Are temporary procedures clearly identified? | |
| 5 | Is white space used effectively?<br>• Is there one blank line between each step?<br>• Does indentation help the user keep their place?<br>• Are the margins large enough to reduce page congestion? | |
| 6 | Is type size is 12 pt font or larger? | |
| 7 | Is mixed case used for words of steps, with ALL CAPS used only for special cases (such as IF, THEN, AUTO, and WARNING)? | |
| 8 | Is the step number very simple (single level of number used)? Only an integer? | |
| 9 | Have sections or information not necessary to performing the steps been moved to the back or to another part of the manual or training guide? | |
| 10 | Are section titles bold or larger than the text font? Do sections have clear endings? | |
| 11 | Is the decision on electronic presentation versus hard copy correct? Are electronic linkages to procedures clear and accurate and easy to use? If paper is chosen, is the color of the paper appropriate? | |
| 12 | Is the overall page format (such as Outline format or Two-Column [T-Bar] format) appropriate to the use of the procedure? | |
| 13 | Are play script features added for tasks that must be coordinated between two or more users?<br>• Play script is normally used when there are two or more hand-offs of responsibility for steps. | |
| 14 | Are rules followed for formatting of Warnings, Cautions, and Notes? (See annotated rules, such as Warnings are for worker safety and Warnings must clearly standout from rest of page.) | |
| | **Step Writing Checklist** | |
| 1 | Is each step written as a command? | |
| 2 | Is the proper level of detail used throughout? This is judged based on: | |

| | | |
|---|---|---|
| | • *Who* will use the procedures<br>• Same level of detail used in similar procedure steps | |
| 3 | On average, is there only one implied action per instruction? Best practice is to average 1.2. | |
| 4 | Does the procedure indicate when sequence is important?<br>• If sequence matters, each step should be numbered (with an integer or letter)<br>• If sequence does not matter, bullet lists should be used | |
| 5 | Are only common words used? Apply "education" level test (5 grade reading level is best) | |
| 6 | Do all acronyms, abbreviations, and jargon aid understanding?<br>• Develop a list of such terms for use in procedures *and* communication.<br>• Use terms that users use (within reason) | |
| 7 | Is each step specific enough? No room left to guess/interpret:<br>• The meaning of a word or phrase (Check vs. Make sure)<br>• The intent of a step or series of steps<br>• A desired quantity or value<br>• To what equipment the step applies | |
| 8 | Is the procedure free of steps that require in-your-head calculations?<br>• Values expressed as ranges rather than targets with error bands<br>• Conversion tables, worksheets, or graphs provided where needed | |
| 9 | Are graphics to the user's advantage?<br>• No explanatory paragraphs or lengthy instructions that could be replaced by a picture<br>• No impressive graphics that provide no real advantage | |
| 10 | Are references to the user's advantage?<br>• No lengthy explanations or instructions that could be replaced by branching to a reference<br>• No references to a procedure that references still another<br>• No gaps or overlaps between this procedure and a referenced document<br>• If branching, must branch to a procedure, not to a specific step in a procedure | |
| 11 | Are rules followed for writing warnings, cautions, and conditional steps?<br>• No more than 2 per page<br>• No actions within a warning or caution (actions must always be numbered steps)<br>• Warnings and Cautions contain descriptions of potential consequences | |

Each of these rules is explained in detail in the course *Writing Effective Operating and Maintenance Procedures*, by PII[4] (visit www.piii.com for more details). Each rule is backed by experimental or observed data that indicate there is a clear advantage in error reduction by using the rule. Below are a couple of examples from others on procedure writing rules.

***Proper Page Format:*** *In one study, Haney took a procedure presented in narrative format and converted it to column format. He then asked experienced technicians who had been completing the procedure using the narrative format to complete it as presented in the column format. He found that when the technicians used the column format, their errors were reduced by a factor of 3 compared with the narrative format [Haney, 1969].*[10]

***Including Supplemental Checklists:*** *Checklists can be effective aids to performance and often should be used to supplement procedures presented in any format. Checklists can be particularly effective in terms of preventing errors of omission. Some of the literature supports using checklists only for simple procedures that do not need to be performed under time or other constraints [Sutton, 1992].*[3] *However, there is also support for using checklists for emergency procedures in which a core set of steps is used to respond to emergency situations, particularly when the actions follow a simple sequence of steps for shutting down a process [Connelly, 1992].*[13] *Overall, experimental data have shown that the probability of errors of omission and commission can be reduced by a factor of 3 to 10 when checklists are* *used as reminders of key steps in a work instruction [CMA's A Managers Guide to Reducing Human Error, 1990].*[9]

For more data on the history and development of these rules, see Bridges & Williams (1997)[11] and Madden & Bridges (2016)[12] especially refer to originating data and discussion for most of the rules (Swain, 1983).[1] **The procedure on the next page follows the rules** and is an example of the T-Bar format. Several other page formats are also valid.

**Trouble-Shooting Guides (Procedures for Responses to Critical Deviation Alarms)**

Trouble-shooting guides are a special form of operating procedure. They are written for the actions we want the operators to take to recover from a process deviation, *before an emergency situation occurs.* They are called guides since rarely can we predict the process conditions at the time the action is required. Trouble-shooting guides (and necessary training and drills) are required for any action that is considered a Human IPL. The Action Limit is what we show as the Min or Max in a Trouble-Shooting Guide. The action limit triggers the demand to use the trouble-shooting guide.

The starting criteria should be true before going to the effort of developing a procedure for response (before developing a trouble-shooting guide):

- The response is typically still possible, but it is time dependent. The time available is called the process safety time (PST). The operator must complete the diagnosis, make the necessary change(s), and make sure they are out of harm's way by the end of the Maximum Allowable Response Time (MART). [14]
- We usually set an alarm or a pre-alarm to trigger this action. This is usually before the shutdown triggers (ESD occur automatically) or release points (PSV set points) are reached
- The Min and Max shown in the Trouble-shooting Guide table are not the absolute safety limits for a system, but are instead some values that leave us some time to take action to prevent from reaching the absolute limits.
- There is still time to prevent or avoid the final consequence that could occur if we reach the ultimate limits of the process. Usually, we want the MART to be ½ or less of the PST, and we want MART > 10 minutes for trouble shooting in the field/plant and MART > 5 min for trouble-shooting only from the control room. [14]

## Figure 2: Example of Procedure that Follows Most of the Rules in Table 1

PII CHEMICAL COMPANY
Unloading Monomer from Tanker to Storage

Page 1 of 2
Revised 11/4/02, Printed 2/21/16

# Unloading Monomer from Tanker to Storage

| STEPS | DETAILS |
|---|---|
| 1. Wear standard PPE, plus rubber gloves and full-face organic respirator. | Standard PPE includes hard hat, safety glasses with side shields, and steel-toe shoes. |
| 2. Weigh in the tanker. | Record weight on GROSS line of form. |
| 3. Check bill of lading. | ... to verify correct type of material is in tanker. |
| 4. Sign in tanker driver. | Driver must sign in as a visitor and be escorted at all times. |
| 5. Take Certificate of Analysis to QC. | ***CAUTION: To avoid contamination, DO NOT unload until you receive approval from QC.*** |
| 6. Spot tanker. | A) Direct driver to location.<br>B) Verify brake is set.<br>C) Chock the wheels on at least one side between the two rear axles.<br>D) Ground the tanker (attach grounding strap to an unpainted metal surface). |
| 7. Have the storage area operator make sure the storage tank can hold contents of tanker. | ***WARNING:*** *Failure to perform this step can result in an overflow and spill of hazardous Monomer.*<br><br>Perform **Tank Gauging Procedure (SOP-01-804)** or check the storage tank load cell readout. Storage tank operating limits are stenciled on the control panel. |
| 8. Place 5-psig nitrogen pad on tanker. | A) Connect from local 5-psig nitrogen drop to aft bottom nitrogen connection on tanker.<br>B) Open nitrogen supply valve.<br>C) Then open valve at tanker. |
| 9. Connect Monomer unloading hoses. | A) Verify hoses and gaskets are in good condition. (Hoses are normally stored on hose rack at unloading spot.)<br>B) Connect from tanker to air pump. (Use aft bottom connection on tanker. Remove cap.)<br>C) Connect from air pump to storage tank. (Connect to storage tank line labeled "from TW.") |

If the unit has a good PHA/HAZOP, then it is best to extract information from the HAZOP (or What-If) analysis tables to start the development of each guide. The guide is then finished with input from the process experts.[4]

***Table 2 Examples of Creating a Trouble-Shooting Guide from a HAZOP Table***

| **HAZOP Table Entry** | **Trouble-Shooting Guide Entry** |
|---|---|
| Cause: *Bypass valve is open or passing* | Make sure the bypass is tightly closed |
| Safeguard: *Isolation valves for the vessel* | Isolate the vessel, if necessary |
| Safeguard: *Relief valve* | Make sure the relief valve block valves/relief path are open |

The key categories of information needed in a trouble-shooting guide are:[4]

- IMMEDIATE ACTION (by system or by operator)
- DECIDE IF ALARM is REAL
- FINDING and FIXING the CAUSE
- FIX or BYPASS PROBLEM

The example below is of a trouble shooting guide for one critical alarm/action.

## Figure 3: Example of Trouble-Shooting Guide that Follows Best Practices

**Trouble-Shooting Guide**

| Alarm or Indicator: | **PAL 4446 – Low Pressure Alarm for Suction of Organic Feed Pump 40-PM-18.445** |
|---|---|

| | | | |
|---|---|---|---|
| Action Limit: | 5 kPa | | |
| Consequence: | Possible pump seal failure, releasing or spraying organic waste into the berm. | | |
| Process Area: | FB&D Incinerator; Liquid Organic Liquid Feed | **Oper. Mode:** | Normal |
| Drawing #s: | D-400-PI-013 | | |

***IMMEDIATE ACTION (by system or by operator)***

- DCS should shut down the organic feed pump (40-PM-18.445).
- From the DCS display, MAKE SURE the organic feed pump is shutdown.
- HAVE the field operator check for leaks near the organic feed pump.
- IF there is a large leak/release, THEN use the ESD switch to shutdown the unit and then follow/complete the shutdown and isolation procedure, OPS-ESD-117.
- IF there is a minor leak or no leak, THEN:
  - COMPLETE the rest of the trouble-shooting,
  - and DECIDE how to contain the leak for now,

***DECIDE IF ALARM is REAL***

- From the DCS, CHECK the pressure and feed tank level trends. IF the trends indicate the alarm if valid, THEN continue with finding the cause or fixing or bypassing the problem.

***FINDING and FIXING the CAUSE***

- CHECK valves upstream of the organic feed pump to see if any are closed too far, including checking ESD valves.
- CHECK, by feel with hand, if the heat tracing is on; IF Not, then TURN ON or open heat trace valves
- MAKE SURE nitrogen to the pump seal is at the normal operating pressure.
- CHECK if the line if plugged or frozen (skill)

| | | |
|---|---|---|
| ABC Chemical Company<br>Prepared by:<br>*Printed copy of this procedure is good for one job task duration.* | OPS-76-TSG-233 | Incinerator Unit<br>Revised 8/24/2015<br>Printed 2/21/16<br>Page 1 of 2 |

**Figure 3: Example of Trouble-Shooting Guide that Follows Best Practices (continued)**

**Trouble-Shooting Guide**

| Alarm or Indicator: | **PAL 4446 – Low Pressure Alarm for Suction of Organic Feed Pump 40-PM-18.445** |
|---|---|

- CHECK if the line if plugged or frozen (skill)
- CHECK if the level is actually low, use the Organic Feed Tank.
- IF the cause is low level in the feed tank, THEN resolve the problem if necessary based on cause that is found (skill)

***FIX or BYPASS PROBLEM***

- IF necessary, SHUT DOWN the Unit to allow fixing of the problem.
- FOLLOW the proper procedure to resolve problems (repair procedure, line clearing procedure, etc.)
- IF the decision is made to continuing operation without all equipment in normal condition, THEN:
  - FOLLOW Temporary Operating mode, if there is a temporary procedure already written for this possible problem/condition
  - FOLLOW MOC procedures to obtain approval for any non-standard temporary operating procedure or mode

**END**



## Optimal Presentation of Trouble-Shooting Information

Ideally trouble-shooting information should be imbedded in the basic process control system (DCS) so that the operators can access the information on demand, with the click of mouse or key. Using the DCS for display of the steps for response to alarms (trouble-shooting), to be displayed "on demand" is becoming more of the norm each year.

## Comments on Current Industry Consensus Guides for Writing Procedures

None of the consensus textbooks available for procedure writing cover all of the major points of this paper. The best "guidelines" for writing procedures currently exist in the course notebooks used by PII (and likely in the course notebook used by ABS Consulting, as they had similar origins).

Notably, the AIChE/CCPS textbook, *Guidelines for Writing Effective Operating and Maintenance Procedures*, 1996[15]:

- Contains NO description at all on the development, format and use of Trouble-Shooting Guides
- Mentions only about half of the rules listed in this paper (and the related PII course notebook) for formatting of pages and steps
- Emphasizes technical staff or professional procedure writers writing procedures, rather than the Users themselves writing the procedures (it violates the first rule of procedure writing)
- Fails to cover the competency needed by procedure writers
- Recommends and emphasizes having subsections in each work instruction/procedure (between the Title of the task and Step 1) rather than relegating such information either to other sections of the operating manual or rather than turning this information into steps
- Fails to mention the need for a PHA of procedures that focuses on hazard review of deviations from the steps; this is required by best practices and by US OSHA for PHAs under PSM

A better example, though still missing key best practices, is the ASM Consortium *Guidelines for Effective Procedural Practices,* 2010[16]:

- Includes a good discussion on the need and thought processes related to handling abnormal situations (trouble-shooting or response to critical deviations), but there is little description on the format and detailed content of the trouble-shooting guides. The guideline fails to differentiate between routine procedures for startup, shutdown, and online maintenance, and those used for response to critical alarms (trouble-shooting)
- Mentions only about half of the rules listed in this paper (and the related PII course notebook) for formatting of pages and steps
- Mentions that experience users themselves should write their own procedures, but it does not emphasize this rule and leaves the authorship open to engineers or a committee. It specifically mentions using retired senior operators for writing operating procedures. While this is one good way to retain institutional knowledge, we have found that the current users do a better job overall, once they have achieved competency in how to write procedures
- Mentions the competency needed for procedure writers in general
- Mentions the need for a PHA of procedures, but does not reference the method documented in Chapter 9 of *Guidelines for Hazard Evaluation Procedures*, CCPS/AIChE, 3rd edition, 2008,[5] or other related papers on the topic. Also, the description of the PHA is misleading because it focuses on how PHAs should be used in writing procedure content (which is fine for trouble-shooting guides) but fails to emphasize that a step-by-step analysis of deviations from the steps are needed to ensure that the process has enough IPLs against mistakes that can occur when imperfect humans use the procedure.

Though this guideline has deficiencies, it is worth the investment and provides useful insight not found elsewhere. Just be sure to address the deficiencies noted above.

## Implementation of This Approach

PII staff has helped more than 60 sites implement the approach above. The typical approach for successful implementation is:

1. Perform a gap analysis to determine the average accuracy of procedure content, average clarity (as judged versus the procedure format rules), and identify any examples of excellent procedure writing.
2. Convince management of the problem and get their buy-in on the solution.
3. Train (1.5 to 2 days for operators; 1 day for lab or maintenance staff) and coach the users of procedures (normally for 3 to 8 days, to have them achieve full competency) so they can be effective authors. Normally, 20 to 25% of the users (hands-on operators, hands-on maintenance and inspection staff, lab staff) need to be trained as authors. Not all of those trained will make good writers. The organization needs to judge who does the best jobs. About 15 to 20% of the staff should end up being authors.
4. Get organized using project management principles for tracking and scheduling the effort. Use master spreadsheets such as those used at REC Silicon (case study presented later).
5. Use a Traveler form (described later) to make sure each new procedure gets a proper amount of review by other users and by technical staff, such as engineers.
6. Reward procedure writers for their efforts. Especially ensure there are no disincentives for taking the extra effort to write procedures.
7. Track the progress on the effort; fix problems as they arise; enforce schedules (to show management cares).
8. Perhaps use specialized software to make it easy to output the content of procedures into different user-friendly formats such as training tools, training competency records, and field checklists. WordPerfect (formerly from Corel) made this relatively easy with the use of Primary and Secondary merge files; MS Word makes this more difficult. As a result, some specialized software have been developed to help keep formats consistent and to help reformat the content, on demand.
9. Issue procedures through a document control system that does not add extra workload. This will of course make Management of Change (MOC) to procedures easier to track through the steps of change request, to risk review and technical review, and finally to issuing (and replacement of prior versions).
10. Of course, ensure there is JSA performed for procedures that have "personal injury" potential and perform a PHA (or mini-PHA for an MOC) of the steps in the procedures whenever there is "process safety" potential.
11. Train all affected workers in the new procedures.
12. Identify the best writers to become the trainers of new authors, over the long run.

## CASE STUDY 1 – REC Silicon (note, the general approach used for the REC Silicon effort is the same used in nearly all of PII's procedure writing projects, including those described in CASE STUDY 2 and 3; the forms and general approach described in detailed only in CASE STUDY 1 to allowing presenting the approach in the context of actual plant work).

REC Silicon is one of the world's largest manufacturer and suppliers of Silane Gas and specialty silicon gases, Granular Polysilicon from Fluid Bed Reactor technology for solar photovoltaic applications and ultra-pure float zone Polysilicon for the electronic industry.

### SOP Improvement Initiative

REC production facilities in the USA have been actively implementing process safety for many years, with a renewed interest in 2013 to exceed OSHA PSM requirements by meeting industry best practices. REC commissioned gap analyses at both the Butte and Moses Lake production sites to identify areas for PSM best practice implementation. One common finding was that the operating procedures and maintenance procedures needed improvement with respect to accuracy of the procedural steps and presentation clarity (page format and step format). The following are audit results pertaining to procedures:

- Procedure Inaccuracy
- Technical Staff were viewed as owners, not shop floor
- Annual review process ineffective
- Procedure formats needed improvement for reducing human error
- Inconsistent use of Warning, Caution, and Note
- Procedure Naming – not consistent
- Modes of operation contained in one procedure title
- Procedure accessibility was difficult in the legacy database; multiple searches needed to find tool for job
- Procedure Safety Reviews – did not utilize recognized methodologies for procedure safety reviews; i.e., there were no PHA of procedures with process hazards or JSA of tasks with personal hazards

### Step 1 - Procedure Author Training

In early 2014, REC Silicon adopted Process Improvement Institute's (PII's) procedure author training module (*Writing Effective Operating and Maintenance Procedures*[4]) with minor customizations. The training consists of a 2-day class room lecture and about 6 exercises, followed by practical procedure development and field validation steps in the 2-day coaching phase. The classroom course covered all aspects of procedure writing with the focus on techniques that will lower human error rates when using the procedures. Topics included:

- human factors overview
- getting the content right

- page format rules and selection
- step format rules
- addressing operating limits (including creating trouble-shooting guides
- nuances of emergency shutdown and emergency operating procedures
- implementation guides and high level procedures for the task of getting all of the work instructions written
- examples/exercises on each topic

## Step 2 - SOP Management System Revisions

REC Silicon revised its management systems to require SOP writing competency for all personnel developing or changing procedures and to reach full operator certification. The unit trainers have taken over leading the procedure author training. Operators took over procedural ownership from technical staff since these are the people performing the work in the field. Additionally:

- The quality and accuracy of procedures improved and subsequently the quality of operator training and certification program due to developing procedure writing competency
- The quality of the annual reviews and audits improved tremendously with the rewritten procedural format and content changes.

## Step 3 - Developing Troubleshooting Guides Engineering Standard

REC Silicon developed an engineering standard governing the content and format for troubleshooting guides:

- *CCPS Guidelines for Engineering Design for Process Safety*[17] is the basis for determining the safe operating limits.
- Production engineers are the owners of these procedures.
- Engineering competency is required to develop the safe operating limits table based on the available process safety information.
- A consequence of deviation is then developed by the Process Hazard Analysis team.
- Involvement of certified operators is required to develop steps to correct or avoid exceeding the safe limit.
- REC silicon risk ranked its process areas and focused on developing safety troubleshooting guides for high risk process first.
- Over thirty six (36) guides were developed for the Moses Lake Facility.
- These have been developed in conjunction with PHA updates to the safe operating limits.
- Formal training and competency testing of these safety procedures.

## Step 4 – Beginning and Managing The Re-Write Effort:

- Each operating department listed all existing operating procedures into the procedure development tracking form (also called a Traveler, because the form travels with the draft procedure until the procedure is final; see Figure 4 on the next page).

- Each department also developed a list of "skills" that are used across multiple procedures and that operators or maintenance staff are required to attain before using complex procedures.
- Procedures were screened for safety, equipment damage, or quality risk to determine whether 2, 3 or 4 peer field walk-downs were required (see validation traveler workflow in Figure 4).
- The procedure re-write assignments were divided up between all four rotating shifts and the superintendent assigned certified authors (e.g., operators trained/coached as procedure writers) from their pool.
- Each department manages their pool of qualified authors. Cross shift validation was utilized to ensure all operating approaches were considered. The assigned author was responsible for consolidating red line comments.
- A **Validation Traveler** was developed and used to ensure all steps in the SOP re-writing workflow were completed.

**Figure 4** - A **Traveler form** shown to the right accompanies each new or revised procedure through the writing and editing/review process. When complete, the traveler form is then kept in the QA file for the procedure to validate long-term how the initial procedure was developed.

**Procedure Development Traveler - Moses Lake**

| | | | |
|---|---|---|---|
| New Document Number: | | Area: | |
| Procedure Title: | | | |
| Procedure Type: | | | |
| Author: | | | |
| Initial Draft Completion Date: | | | |
| Author Walkdown Date: | | | |
| Author Revision 1 Completion Date: | | | |
| Peer 1: | | Shift: | |
| Peer 1 Walkdown Date: | | | |
| *Optional Peer 2:* | | Shift: | |
| Peer 2 Walkdown Date: | | | |
| *Optional Peer 3:* | | Shift: | |
| Peer 3 Walkdown Date: | | | |
| *Optional Peer 4:* | | Shift: | |
| Peer 4 Walkdown Date: | | | |
| Author Revision 2 Date: | | | |
| Engineer/Verifier: | | | |
| Verification Completion Date: | | | |
| Author Revision 3 Completion Date: | | | |
| QA Person: | | | |
| Author Revision 4 Completion Date: | | | |

*Images and layouts copyrighted by PII, 2008-2017*[4]

- A color coding process was used to flag SOP content that would be deleted, or moved into the unit training manual.
- A total of 184 Moses Lake procedures were tracked through the development, review, and revision process.
- **The overall tracking form below was used to develop KPIs to show accomplishments and to meet set completion dates.**

***Figure 5 Procedure Development Project Status Table***

***Images and layouts copyrighted by PII, 2008-2017*** [4]

**Procedure Development Tracking Form - Moses Lake**

| | | | | | | | Initial Development | | | Validation | | | | |
|---|---|---|---|---|---|---|---|---|---|---|---|---|---|---|
| Item | New Doc # | Old Doc # | Area | Procedure Title | Proc Type | Author(s) | Initial Draft | Author Walkdown | Author Rev 1 | Peer 1 Walkdown | *Opt. Peer 2 Walkdown* | *Opt. Peer 3 Walkdown* | *Opt. Peer 4 Walkdown* | Author Rev 2 |
| 1 | FBP.3.xxx | FBP.3.160 | FBR | Hydrogen Booster Compressor Cleanup | Cleanup | LT/FG | 3/19/14 | 3/19/14 | 3/20/14 | 3/21/14 | 3/21/14 | | | |
| 2 | FBP.3.xxx | FBP.3.160 | FBR | Hydrogen Booster Compressor Operation and Op Restrictions | Operation | LT/FG | 3/20/14 | 3/20/14 | 3/20/14 | 3/21/14 | 3/21/14 | | | |
| 3 | FBP.3.xxx | FBP.3.160 | FBR | Hydrogen Booster Compressor LOTO | LOTO | BCA/OL | 3/19/14 | 3/19/14 | 3/21/14 | | | | | |
| 4 | FBP.3.xxx | FBP.3.160 | FBR | Hydrogen Booster Compressor Standby | Standby | BCA/OL | 3/20/14 | 3/20/14 | 3/21/14 | | | | | |
| 5 | | | 4.0 | STC/TCS from Receiving Tank to STC/TCS Storage Transfer Through Metering Line | Transfer | ME/RR | 3/19/14 | 3/19/14 | 3/20/14 | 3/20/14 | 3/21/14 | 3/21/14 | 3/21/14 | |
| 6 | | | 4.0 | STC/TCS from Receiving Tank to STC/TCS Storage Transfer Through Bulk Line | Transfer | ME/RR | 3/19/14 | 3/19/14 | 3/20/14 | 3/20/14 | 3/21/14 | 3/21/14 | 3/21/14 | |
| 7 | | | 4.0 | H2 Recycle Compressor Switching to Standby | Standby | ME/RR | 3/20/14 | 3/20/14 | 3/20/14 | 3/20/14 | 3/21/14 | 3/21/14 | 3/21/14 | |
| 8 | | | 4.0 | H2 Recycle Compressor Cleanup | Cleanup | ME/RR | 3/20/14 | 3/20/14 | 3/20/14 | 3/20/14 | 3/21/14 | 3/21/14 | 3/21/14 | |

- From PII's procedure author training,[4] the Quality and Content checklist was modified and used as quick reference guide to make sure that all the content development and formatting rules are being followed during the re-write process. We used Microsoft Word and made templates for the T-bar, Flat and Consequence of Deviations.

## Step 5 - Procedure Management of Change

REC Silicon adopted formal safety review methods for procedures and were included in a new procedure management of change procedure. The REC Silicon IT department developed a custom tool to execute these requirements in an electronic workflow. Benefits included improving consistency and quality of MOC procedural safety and health reviews and improving the overall quality of the SOPs with respect to clarity and content; and helped identify human factor deficiencies such as with human machine interface (equipment accessibility, operator feedback, labeling, missing or deficient instruments/controls).

**Overall Summary of the procedure writing initiative:**

For the Moses Lake facility, 398 procedures (operating, maintenance, and trouble-shooting guides) have been developed through this initiative (new and rewrites combined.) REC Silicon has seen great improvements based on the following feedback from workers involved in the initiative:

- Operators are receiving better work instructions
- Sequence of steps is well understood
- Much easier to audit operator compliance to procedure
- **Standardizes how we do work – protects our equipment from damage, protects our people, protects the environment**
- No longer need 20 year operator tribal knowledge and interpretation to complete procedure
- Procedures are not cluttered anymore, they just have the necessary information
- Procedure element now requires involvement of many shop floor workers

- JSA technique is alive on the shop floor
- Procedure can be trusted
- Technical staff no longer doing all procedure work. Freed up to optimize processes and realize other cost saving benefits.

REC Silicon's procedural initiative's success was due to forming a cross-functional project team of key decision makers from all the correct disciplines and having a project sponsor who could remove barriers. REC Silicon held weekly meetings, set short term goals, and tracked KPIs to meet those goals. Leadership was key to support the initiative and the related culture change.

## CASE STUDY 2

As with REC Silicon, PII worked with ROQUETTE in Keokuk, IA (USA) to develop operating procedures for a process that a food additive. There were many procedures for this process and the detailed procedures for each task had never been written. Many of the older staff were being replaced with younger staff and the company decided to take this opportunity to develop detailed procedures to help ensure the consistency of training and coaching of the new workers.

Since there were limited experienced staff, the company trained the newly hired staff (many of whom had years of experience at other companies) in how to write and validate operating procedures. In addition, since ROQUETTE could not spare sufficient workers for the writing effort, PII provided 2 senior staff onsite for 2 weeks. Five (5) workers were trained and coached in procedure writing. These PII senior staff and the newly trained writers (who were also new operators) then worked with the experienced operators to outline the procedure for each task and then develop the best sequence of steps for each task. About 50 tasks were identified that needed procedures. 24 were drafted in the first week onsite and a similar number on the second week onsite. The forms shown in Figure 4 and 5 above were used extensively in this project.

In between the onsite visits by PII, then newly drafted procedures were walked down by the senior operator and by the other new operators. The revised procedures were finally issued for use. Even though the newly hired operators were inexperienced in the process operation, the ones that were trained as procedure writers developed several procedures (with the senior operator input) without help from the PII writers.

The company is considering expanding this effort to the many other process in Keokuk and at their other sites around the world.

## CASE STUDY 3

As with REC Silicon and ROQUETTE in the USA, PII worked with Bahrain National Gas Company (BANAGAS) in Bahrain to develop operating procedures for their two gas process plants and for the gas compression stations upstream of the gas plants. Although the native language of most of the workers is Arabic or Hindi, the official language of the company is English.

This effort was part of a larger project to full implement PSM (and ongoing project that began in 2015 and is scheduled for completion in 2018). There were hundreds of operating procedures for these processes and the detailed procedures for each task had never been written. Similar more than 50 procedures were identified for maintenance activities related to repair, inspection, test, PM, calibration, etc.

The company has many very experience operators and maintenance technicians and many new operators. PII trained 12 operators and 7 maintenance technicians how to write and validate operating procedures. PII provided 2 senior staff onsite for 3 days of coaching following the 2-day course. These PII senior staff and the newly trained writers then worked with the other experienced operators to develop the best sequence of steps for each task and to get the procedures properly formatted. More than a dozen procedures (a combination of operating procedures and maintenance procedures) were written while PII was onsite; the main function of PII staff was to coach newly trained writers as they developed procedures for their departments. About 12 of the 18 BANAGAS writers that were trained and coached reached reasonable competency by the end of the first week (2 days of training and 3 days of coaching). During subsequent visits to the site by PII staff and between site visits via e-mail, PII continue to critique and coach the new minted procedure writers at BANAGAS.

As in the other two companies listed in these case studies, the forms shown in Figure 4 and 5 above were used extensively in this project, and continue to be used as the effort is ongoing. In addition, PII developed a flowchart for the procedure development process (see Figure 6) and developed a procedure writing Guide for use by the BANAGAS procedure writers.

The procedure writing effort is nearing completion, with very minimal input required from PII staff.

## OTHER CASE STUDIES

PII has used the approach above since 2003 at more than 60 plants/sites in North America, South America, Australia, Europe, and the Middle East. The approach has consistently yielded great results; accurate procedures and clear formats. In these cases, the local operators and maintenance staff (the trained and now competent writers) put in 95% to 99% of the procedure writing effort.

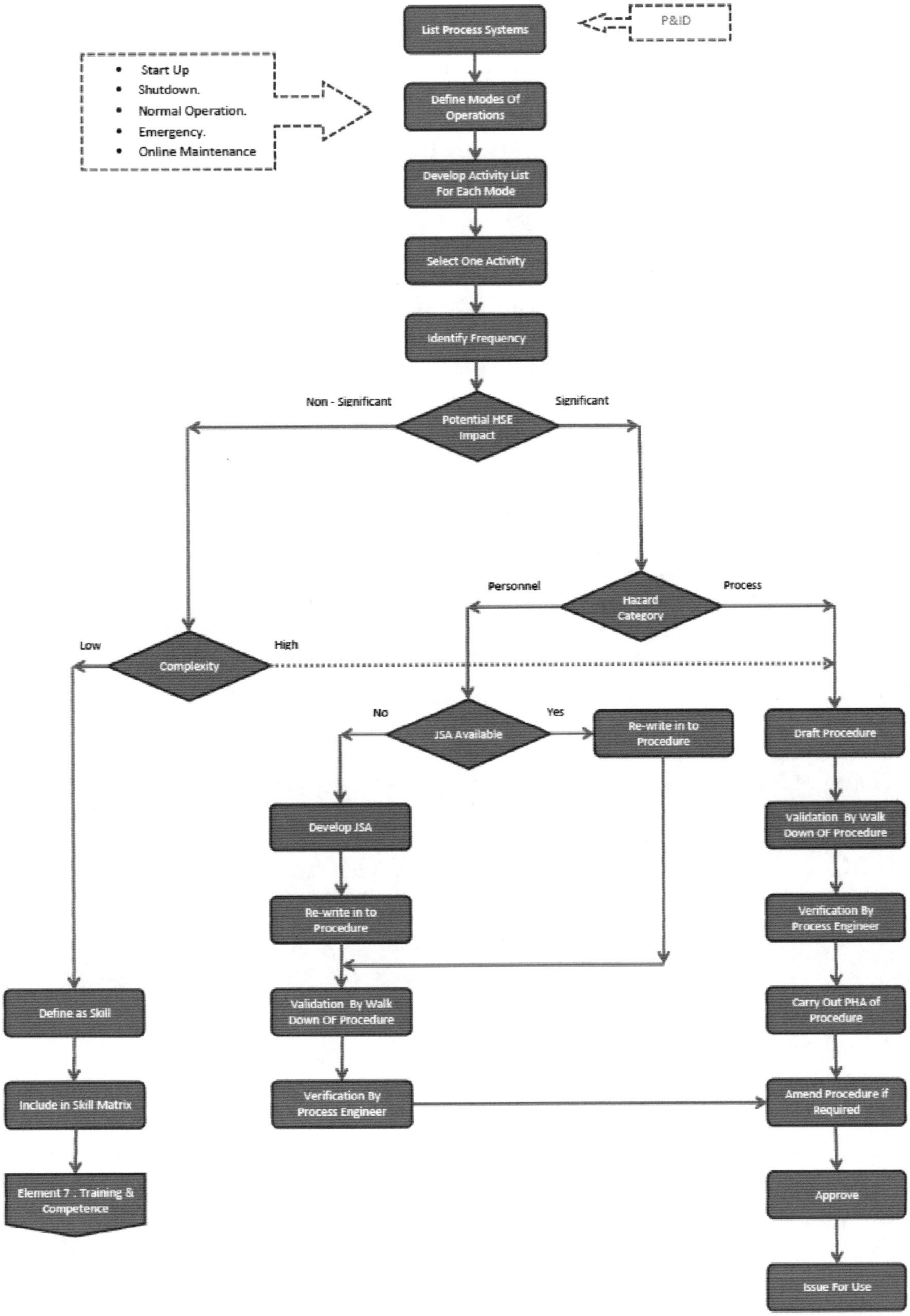


*Figure 6 Procedure Development Flowchart (copyright PII, 2015-17)*

## Conclusions

Having clear and accurate procedures is a must. Poor procedures show up as one of the root causes in 90% of accidents. The approach described in this paper has worked for more than 35 years for one of the authors, and continues to work today. Controlling risk is mostly about controlling human error rates, and having accurate and clear procedures is one key to this control. One other gain from the approach described in this paper is that the users become the owners, and therefore they have a vested interest to keep the procedures up to date and clear.

No doubt, writing all of the procedures for a facility is a large task. But, it can be accomplished by site staff, with support from company process safety or quality control leadership, after enough users become competent authors. Attaining this competency takes less than 2 weeks of work, on average.

## Acronyms Used

**AIChE**– American Institute of Chemical Engineers
**ASM** – Abnormal Situation Management consortium
**BANAGAS** – Bahrain National Gas Company
**CCPS** – Center for Chemical Process Safety (a division of AIChE)
**CMA** – Chemical Manufacturer 's Association, now American Chemical Council (ACC)
**FBR** – Fluidized Bed Reactor
**HAZOP** – Hazard and Operability Analysis
**IPL** - Independent Protection Layer
**JSA** – Job Safety Analysis
**LOPA** – Layer of Protection Analysis
**MART** – Maximum Allowable Response Time
**MOC** – Management of Change
**NUREG** – Nuclear Regulation, US Nuclear Regulatory Commission
**OSHA** – Occupational Safety and Health Administration, US Department of Labor
**PHA** – Process Hazard Analysis
**PII** – Process Improvement Institute, Inc.
**P&ID** – Piping & Instrumentation Diagram
**PSI** – Process Safety Information
**PSM** – Process Safety Management
**PST** – Process Safety Time
**SOP** – Standard Operating Procedure

## References

1. Swain, A. D., Guttmann, H. E., *The Human Reliability Handbook - with Emphasis on Nuclear Power Plant Applications*; NUREG/CR-1278, U.S. Nuclear Regulatory Commission, Division of Facility Operations, 1983.

2. Gertman, D., Blackman, H., Marble, J., Byers, J. and Smith, C., *The SPAR-H Human Reliability Analysis Method,* NUREG/CR-6883, U.S. Nuclear Regulatory Commission, Office of Nuclear Regulatory Research, 2005.
3. Sutton, Ian S., "Writing Operating Procedures," *Proceedings of the International Conference on Hazard Identification and Risk Analysis, Human Factors and Human Reliability in Process Safety*, AIChE, Center for Chemical Process Safety, New York, NY, 1992.
4. *Writing Effective Operating and Maintenance Procedures*, Course Notebook, Process Improvement Institute, Inc., 2003-2017.
5. *Guidelines for Hazard Evaluation Procedures*, 3rd Edition, 2008, CCPS/AIChE.
6. "Necessity of Performing Hazard Evaluations (PHAs) of Non-normal Modes of Operation (Startup, Shutdown, & Online Maintenance)", W. Bridges and Mike Marshall (US OSHA), *18th Annual International Symposium, Mary Kay-O-Connor Process Safety Center, College Station, TX, October 2015. (also updated at 12th GCPS, 2016)*
7. "How to Efficiently Perform the Hazard Evaluation (PHA) Required for Non-Routine Modes of Operation (Startup, Shutdown, Online Maintenance)," W. Bridges and T. Clark, *7th Global Congress on Process Safety*, Chicago, AIChE, March 2011.
8. Bridges, W.G., et. al., "Addressing Human Error During Process Hazard Analyses," *Chemical Engineering Progress*, May 1994
9. *A Manager's Guide to Reducing Human Error*, 1990, Chemical Manufacturers Association (CMA)
10. Haney, R. W., "The Effects of Instructional Format on Functional Testing Performance," *Human Factors*, 11, 1969, pp. 181-187.
11. "Effective Procedure Writing – A Two-Part Series (Part 1: Reducing Risk of Worker Errors; Part 2: Addressing Process Safety Management Requirements for Operating Procedures)," *Chemical Engineering Progress*, 1997, W. Bridges with T. R. Williams (both of JBFA).
12. Madden, L., and Bridges, W., Best Practices for Writing Operating Procedures and Trouble-Shooting Guides, L. Madden and W. Bridges, *12th Global Congress on Process Safety*, Houston, TX, AIChE, March 2016.
13. Connelly, C. S., "Emergency Operating Procedure Writing for Oil Refinery Process Units," *Proceedings of the International Conference on Hazard Identification and Risk Analysis, Human Factors and Human Reliability in Process Safety*, AIChE, Center for Chemical Process Safety, New York, NY, 1992.
14. *Guidelines for Initiating Events and Independent Protection Layers*, 2015, CCPS/AIChE.
15. *Guidelines for Writing Effective Operating and Maintenance Procedures*, 1996, CCPS/AIChE.
16. *ASM Consortium Guidelines – Effective Procedural Practices*, 2010, ASM Joint R&D Consortium.
17. *Guidelines for Engineering Design and Process Safety*, 2nd Edition, 2012, CCPS/AIChE

## 478864 Process Safety Management: It's Not Just for Chemcal Plants!

**Abstract:**

**Process Safety: It's Not Just for Chemical Plants & Refineries!**

**Presented By**

**Richard W. Sarnie, CSP, P.E., ARM-E**

**Director, CCPS Projects**

**Center for Chemical Process Safety**

**American Institute of Chemical Engineers**

In this presentation Mr. Sarnie will discuss lessons learned and relearned over decades in manufacturing and cold storage facilities on preventing and mitigating process safety incidents (LOSS OF CONTAINMENT EVENTS) and how these lessons learned can and should be applied across many types of manufacturing and cold storage operations such as the food, beverage, pharmaceutical and nutraceutical industry. He will:

- Present a rationale as to why a **risk-based** rather than **regulatory-based** approach to managing process safety and other risk exposures is prudent
- Discuss the continuing movement by many organizations to more formal and comprehensive enterprise risk management systems
- A review of the typical steps or phases an organization usually takes in developing and improving management system processes
- Provide an overview of chemical industry best practices for managing process safety (with a focus on AIChE CCPS recommendations as described in the **CCPS Guidelines for Risk Based Process Safety** and many other CCPS books)

- Describe numerous process safety management system deficiencies (root causes) associated with serious chemical process safety incidents that have repeatedly occurred during the last 50+ years
- Discuss how proven process safety management practices and use of multiple layers of protection can help better manage risks in manufacturing and cold storage facilities.

# Hit the Ground Running (Safely):

# Process Safety for the Early Career Engineer

Jerry J. Forest, CCPSC
Celanese
225 E. John Carpenter Freeway Suite 1200, Dallas, TX 75234
Jerry.Forest@Celanese.com

Don Connolley, CSP
BP Corporation
Don.Connolley@bp.com

Amy Gray, Ph.D., PE, CFEI
ESI
aegray@engsys.com

Prepared for Presentation at
American Institute of Chemical Engineers
2017 Spring Meeting and 13th Global Congress on Process Safety
San Antonio, TX
March 26 – 29, 2017

# Hit the Ground Running (Safely):

# Process Safety for the Early Career Engineer

Jerry J. Forest, CCPSC
Celanese
225 E. John Carpenter Freeway Suite 1200, Dallas, TX 75234
Jerry.Forest@Celanese.com

Don Connolley, CSP
BP Corporation
Don.Connolley@bp.com

Amy Gray, Ph.D., PE, CFEI
ESI
aegray@engsys.com

**Keywords:** New Hire, Training, Competency, Mentoring

## Abstract

Engineers entering the workforce are frequently expected to participate in process safety elements such as PHA, MOC, incident investigation, mechanical integrity programs and more. Although there is a recent ABET requirement to include process safety in engineering curricula, many schools do not help students define what to expect in their first year.

This paper will discuss common process safety duties, activities, and responsibilities expected of new engineers. The paper will follow the recently published CCPS book *Process Safety for Undergraduates and Engineers*. It will also introduce the AIChE's Health and Safety Division's Professional Outreach concept on mentoring. The mentoring project will provide a platform to connect experienced process safety engineers with student and local chapters, and individuals who are seeking assistance with specific process safety topics.

## 1. Introduction

When engineers enter industry for the first time they are often surprised that one of their first assignments includes an onboarding process that can include several years of training. The material in these professional development courses is usually not covered in school. Companies may use different onboarding tools and the early career engineer must learn the company specific methodologies for process safety elements.

This paper discusses an overview of how new engineers can hit the ground running with respect to process safety when first entering industry. It includes a brief overview of the Accreditation

Board for Engineering and Technology Inc. (ABET) requirements for university chemical engineering students, typical onboarding training in the first few years of employment, and the type of process safety duties a new hire engineer can expect. The paper also discusses a new pilot process safety mentoring program from the American Institute of Chemical Engineers (AIChE) Safety and Health Division designed to help new hire engineers hit the ground running safely by drawing on those who have years of industrial process safety experience.

## 2. T2 Laboratories and ABET Requirements

In 2007, an explosion occurred at T2 Laboratories in Florida, killing 4 employees and injuring 32 including members of the surrounding community. The U.S. Chemical Safety Board (CSB) investigated the incident and found that the primary cause of the incident was a lack of understanding of the reactive hazards involved with the process. A key recommendation from the incident is addressed to the AIChE:

"Work with the Accreditation Board for Engineering and Technology Inc. to add reactive hazard awareness to baccalaureate chemical engineering curricula requirements." [1]

AIChE responded to this recommendation by working with ABET to require not just reactive chemical hazards in chemical engineering curricula, but all chemical process hazards. [2] Given that ABET accredited schools must address all chemical process hazards in their undergraduate curricula, the first way early career engineers can hit the ground running is by understanding the process safety topics presented in their curricula.

The Center for Chemical Process Safety (CCPS) has gone beyond the original intent of the CSB recommendation with two additional initiatives. The first is publication of *Introduction to Process Safety for Undergraduates and Engineers* (hereafter referred to as *Introduction to Process Safety*) and the second is the undergraduate process safety education initiative.

*Introduction to Process Safety* supplements Crowl and Louvar's *Chemical Process Safety: Fundamentals with Applications* [3] by introducing the student to Risk Based Process Safety. The book is designed to be used as a stand-alone course book or as supplemental material for several chemical engineering courses. *Introduction to Process Safety* was published in 3rd quarter of 2016 and early adopters include Louisiana State University and Kansas State University. Celanese is presently evaluating making the book required reading for all new hire engineers and is considering an onboarding course from the material.

The second initiative in progress from CCPS that goes beyond the original intent of the T2 Laboratories CSB recommendation is an undergraduate process safety initiative. This initiative builds on AIChEs existing Safety and Chemical Engineering Education (SaChE) program and *Introduction to Process Safety* with four key elements:

- Curriculum – enhancing SaChE e-learning modules and translating to multiple languages
- Teaching – Educate professors about process safety with workshops at industrial sites
- Addressing gaps – with industry led "Boot-Camps" for students on campus and
- Maintaining focus – funding key research in process safety [4].

The early career chemical engineer has an opportunity to hit the ground running safely by taking advantage of the material presented by their undergraduate curriculum and by utilizing the resources available by the CCPS. These include the aforementioned SaChE modules, "Boot-Camps", and the resource books *Introduction to Process Safety for Undergraduates and Engineers* and *Chemical Process Safety: Fundamentals with Applications*.

## 3. Onboarding & Training for New Hires

Unfortunately for the new hire engineer, each company will address the various process safety elements in different ways and with differing methodologies. For example, while management of change (MOC) processes may have common elements of risk identification and mitigation, approval, and action item identification and closure, the methodologies used for these elements may differ among companies. While the fundamental principles will look familiar to what was presented in the undergraduate curriculum, the specifics particular to each company may be unfamiliar. Therefore, it is very common for an early career engineer to undergo an onboarding process that might include several years of professional development training.

The final onboarding content is a function of the hiring company and the job position of the early career engineer. The process safety courses given can be mandatory or optional and the material can be delivered by an internal training department or offered by third party vendors. Whatever the case, the onboarding process safety training represents an excellent opportunity for the early career engineer to hit the ground running.

Early career engineers typically are immediately immersed in the following risk based process safety elements: process safety information (PSI), process hazard analysis (PHA), management of change (MOC), pre-startup safety reviews (PSSR), conduct of operations, mechanical integrity (MI), standard operating procedures (SOP), training, metrics, and incident investigation. The extent to which early engineers contribute is usually based on training and experience. Therefore, it is in the engineer's best interest to pursue at a minimum, initial training on these topics.

## 4. What to Expect in the First Year

Although this is a paper on how to hit the ground running with regard to process safety, we would be remiss if we did not mention occupational safety expectations for the early career engineer. Note the majority of this section is taken from *Introduction to Process Safety* [5]. It is safe to say that most U.S. chemical manufacturers place an enormous emphasis on occupational safety – to the extent that they have been criticized by OSHA for placing an emphasis on occupational over process safety. Nevertheless, all engineers should understand their company's expectations for occupational safety and become a model leader by demonstrating behaviors consistent with standards and regulations and exceeding expectations set by the company leadership in this area. Activities that early career engineers can expect include:

- Leading and participating in safety meetings and shift safety meetings
- Unit inspections
- Life critical procedure audits
- 8 – 24 hours of occupational safety training per year

- And more, depending upon an individual company's requirements

*4.1 Social Interactions*

Engineers entering the manufacturing environment are thrust into social interactions for which many are ill prepared. Operators and maintenance crafts have often been part of the same team for years and have unique understanding of their units even if they do not have the vocabulary to express the physical processes in the same way as the new engineers. The main thing for an early career engineer to understand is to approach these groups with great respect. They have a deep understanding of process that is only gained by experience. To become a mentee of these experienced individuals, one must first respect their knowledge, abilities, and attitudes gained by experience.

This section comes with a warning. The early career engineer who adopts an attitude of knowing more than those hourly employees that have run the manufacturing units for years will most certainly fail. The reason simply is the early career engineer does not know as much as these experienced hands. The engineer may possess an understanding of first principals, but must also gain an understanding of the physical process in order to apply first principals.

Learning from operators and crafts will accelerate the learning curve for a new engineer and provide a way to apply technical skills learned at university in a practical way that helps the company achieve results. Interacting with the operators and crafts is also a way to eventually gain their respect and trust, and learn how to function well as a team.

*4.2 New Skills*

Many assignments in the university curriculum include group projects, formal reports, and working with teams. They are designed to prepare students in doing the same activities in their new engineering jobs. New skills in industry related to process safety include assembling process safety information, leading project teams, training operators, writing standard operating procedures, scribing or leading hazard reviews and management of change hazard reviews, delivering presentations and writing formal technical reports (just to name a few skills).

One of the most important skills to acquire is the ability to concisely document reasoning and thinking related to solution of items. Findings and action items are generated from process hazard analysis, management of change, incident investigation, audits, and other team meeting activities. The early career engineer is often assigned much of this important workload and must find solutions to the problems and document with proof that concern is addressed. In many cases, this documentation is regulatory requirement and becomes part of the operating record for the life of the process.

*4.3 Conduct of Operations & Operating Discipline*

One process safety element that has been overlooked in university engineering curricula is conduct of operations. Conduct of operations is a term that "describes human factors and tools that are necessary to produce repeatability in performance and consistency in results" [5]. While much of process safety is concerned with maintaining the integrity of engineered controls, conduct of operations is concerned with maintaining the integrity of administrative and human controls.

The day in a life of an operator can be described by discrete activities performed. These include: reporting to work outfitted, making relief in the field, attending shift change meetings, informal and formal evaluation rounds, issuing permits, preparing equipment for maintenance or operation, operating equipment, and documenting each activity. Early career engineers are often asked to participate in several or all of these aspects of operation.

For example, operators usually document their activities in a log book. Engineers should monitor the operator logs to gain a deeper understanding of unit activities. When problems are documented, engineers should take action. Failure to do so leads to an acceptance of problems and a normalization of deviance. Furthermore, engineers might be asked to create logs and periodically train operators on how to use them.

Other examples of an engineer's role in conduct of operation include participation in equipment evaluation. An engineer might create evaluation sheets by mapping out how an operator should travel through the unit, and define exactly what equipment should be evaluated and how it is to be evaluated. Then, by periodic monitoring of the evaluation rounds, engineers better understand normal operation and are better equipped to solve problems.

Early career engineers are often called to participate in all aspects of conduct of operations and should be familiar with the operating discipline tools to be effective. This familiarity is only gained by experience as it is rarely formally taught in training.

### *4.4 Engineering Discipline*

Engineering discipline is closely related to operating discipline in that it describes activities that engineers and technical personnel do to monitor key operating parameters in the plant to keep the unit running at peak efficiency and normal safe operation. This new skill involves the selection and monitoring of key performance indicators and determination of, monitoring, and evaluating safe operating limits. Engineers monitor trends and take action to prevent process safety incidents based on their understanding of the hazards and risks gained from process safety information and hazard identification and risk assessment processes of their company.

### *4.5 Process Safety Duties*

Early career engineers will be active in most of the 20 elements of risk based process safety. In addition to the company specific training described earlier, participation in these elements will be required. A few common areas are shown in table 1, summarized from *Introduction to Process Safety* [5].

**Table 1: Process Safety Activities for New Engineers**

| Element | Activity |
|---|---|
| Process Knowledge Management | Ensure accuracy of PSI<br>Be familiar with Safety Data Sheets<br>Develop reactivity matrix<br>Develop and update design basis calculations<br>Develop and update P&IDs<br>Verify electrical classification<br>Maintain equipment files |
| Hazard Identification and Risk Assessment | Participate in HIRA, & PHA<br>Assemble required PSI<br>Review applicable RAGAGEPs<br>Compile industry incidents<br>Scribe, document reviews<br>Resolve findings |
| Operating Procedures | Develop and update operating procedures |
| Safe Work Practices | Write and approve permits |
| Asset Integrity and Reliability | Identify equipment covered by inspection, testing, and preventative maintenance (ITPM) programs<br>Write ITPM procedures<br>Analyze inspection results & maintain records |
| Contractor Management | Approve and train contractors<br>Write contractor procedures |
| Management of Change | Identify changes<br>Participate in MOCs<br>Assemble required PSI<br>Resolve items |
| Operational Readiness | Participate in readiness reviews<br>Resolve items |
| Emergency Management | Learn emergency response plans<br>Participate in drills |
| Incident Management | Participate in investigations<br>Resolve items |
| Metrics | Record and maintain data<br>Analyze data |
| Audits | Assemble requested information for audits<br>Become a certified auditor |

The main point of this section is that there is much more to process safety for early career engineers than is generally presented in the university curricula even with the new ABET requirement. While many companies will have onboarding and training to prepare the engineers, many of the skills illustrated above are gained by experience. This is just a sampling of what early career engineers can expect in their first year.

To help engineers prepare for these activities in their first few years, the Safety and Health Division of AIChE is piloting a mentoring program.

## 5. AIChEs Safety and Health Division Pilot Mentoring Program

The American Institute of Chemical Engineers (AIChE) Safety and Health Division's mission is:

"To share, promote, and shepherd process and personal safety for individuals in the field of process safety."

In support of the mission the Division has formed several committees including Professional Outreach, Student Outreach, Communication, and Membership. The Professional Outreach Committee has identified a potential avenue to help the early career engineer hit the ground running safely by creation of a process safety mentoring program.

The objective of the mentoring program is to connect AIChE members experienced in process safety with members new to the area of process safety for the purpose of providing guidance and a resource point. It is envisioned that eligible mentors will be engineers who have at least 10 years of experience in process safety. This experience could have been gained through a variety of pathways. Reviews of the volunteer mentors' experience will be performed by the Safety and Health Division Mentoring Committee.

Eligible mentees will be early or transitional career professionals with less than 5 years of experience in the area of process safety. This career can be in a variety of industries. Reviews of potential mentees' eligibility will be performed by the Mentoring Committee.

The mentoring program is in a pilot stage with plans to focus on Louisiana to solicit mentors and mentees. It is expected that applications for mentors and mentees will be available on the Safety and Health Division website and will include information on where to return information. The mentor application will include name and company info, time commitment desires, areas of expertise, and a brief statement on why the applicant wants to be a part of the Mentoring Program. The mentor should attach an updated resume and provide 3 references. Mentee applications will require similar information and can be submitted by individuals or engineering supervisors interested in pairing mentors with groups of young professionals.

The Safety and Health Division Mentoring Committee will review all Mentor applications and approve or reject based on experience, reputation in industry, and referral comments. Mentors and Mentees will be paired by the Safety & Health Division Mentoring Committee. Decisions will be based on location, industry, and desired commitment level.

At the time of writing this paper, the pilot mentoring program is in its infancy. Those interested in being either a Mentor or Mentee should contact a member of AIChEs Safety and Health Division Executive Committee.

When it is fully established, we anticipate that this Safety and Health Division Mentoring Program will provide an excellent opportunity for the early career engineer to hit the ground running.

## 6. Conclusion

The early career Chemical Engineer quickly learns that education and competency are an integral part of job duties and in fact, will be an ongoing part of the job during the course of a career. A large portion of this continuous learning is gaining expertise in occupational and process safety; skills that often are not covered in depth at many universities.

It is in the early career engineers interest to gain competency in process safety as the elements of process safety are weaved throughout all aspects of an engineer's job. The CCPS has published many resources that provide excellent background in process safety. Most companies have an onboarding process that details several years of process safety training. Engineers will also gain "hand-on" experience through the early assignments given in common process safety elements such as MOC and PHA. And if successful with the pilot Mentoring Program, the AIChE Safety and Health Division may have an additional tool to help the engineer hit the ground running (safely).

## 7. References


[1] U.S. Chemical Safety and Hazard Investigation Board. *Investigation Report T2 Laboratories, Inc. Runaway Reaction (Four Killed, 32 Injured)*, CSB Report No. 2008-3-I-FL, 2009

[2] U.S. Chemical Safety Board, http://www.csb.gov/csb-applauds-aiche-s-response-to-including-reactive-hazard-awareness-in-college-chemical-engineering-curriculum/ (accessed January 29, 2017)

[3] Crowl, Daniel, A. and Louvar, J. F., *Chemical Process Safety: Fundamentals with Applications*, 3rd ed., Prentice Hall, Englewood Cliffs, NJ (2011)

[4] Center for Chemical Process Safety, http://www.aiche.org/ccps/community/undergraduate-process-safety-learning-initiative (accessed January 29, 2017)

[5] Center for Chemical Process Safety, Introduction to Process Safety for Undergraduates and Engineers, 1st ed., John Wiley & Sons, Inc., 2016

# 478550 Thermal Decomposition of Mono-Nitrated Toluene (MNT) with Additives

**Abstract:**

**Abstract:** On September 21st 1992, an explosion happened in Castleford, UK, at the distillation unit, owned by Hickson & Welch Ltd. The incident killed 5 people and caused nearly 200 injuries. The main reason is due to the decomposition of mono-nitrated toluene (MNT). Nitrated aromatic compounds present a potential run-away reaction threat to the chemical process industry, due to its unstable structure and extensive heat release during upset situations. This study employs the Advanced Reactive System Screening Tool (ARSST) to study the effects of various common contaminants on MNT's thermal decomposition, including sodium nitrate, sodium hydroxide, and sodium carbonate. Experimental parameters and results such as 'onset' temperature, pressure increase rate, temperature increase rate and maximum temperature are present. Sodium hydroxide and sodium nitrate can lower the decomposition onset temperature of MNT and make the consequence more severe. Thermokinetics of the decomposition reaction is shown and SADT prediction of UN25 kg package storage is calculated.

# 479379 Addressing Challenges of Controlling Change By Implementing Best MOC Practices

**Abstract:**

Effectively controlling change in a plant environment results in a cumbersome process with an inefficient paper-based Management of Change process or use of a traditional and complex electronic Management of Change (MOC) system. Recent research indicates that the majority of chemical incidents are the outcome of poor implementation of Process Safety Management (PSM) and the lack of efficient MOC processes. As a result, MOC workflow steps are often skipped or prolonged resulting in an ineffective process that does not capture the change fully or safely; approvals are delayed, Process Safety Information (PSI) is not being updated, Pre-Startup Safety Review (PSSR's) are not being performed, and MOCs are not being closed in a timely manner.

The goal of this paper is to address industry challenges with controlling change, and how Siemens implements best practices for an effective MOC process.

# 479243 Process Change Impact on Relief System Design

**Abstract:**

**Process change impact on relief system design**

**(Understanding equipment impact and the evaluation warranted)**

Various scenarios can cause the pressure within equipment to exceed design pressure. Appropriately sized relief devices are one of the most common methods of protecting equipment and personnel operating this equipment. A key challenge to being in compliance is the dynamic nature of processing plants. Numerous changes can be expected to happen over the lifecycle of a processing plant. Management of Change, one of the Process Safety Management (PSM) elements, requires equipment impact evaluation when these changes occur to ensure adequate pressure relief. This presentation talks about numerous types of changes which will impact equipment and warrant overpressure protection re-evaluation.

# 479152 Identification and Local Impact Analysis of Projectile Hazard in the LNG Industry

**Abstract:**

With a strong growth in natural gas production in the United States, a number of liquefied natural gas (LNG) facilities have been proposed and authorized. These LNG facilities have sufficient safety, security, risk assessment and mitigation plans developed in close coordination with local, state, and federal authorities. However, to continue to ensure safe operation, it is vital to continue to improve the scope of risk assessment and management strategies. Recent incidents have highlighted the need to consider the potential hazard and risk from projectiles. This paper studies the sequence of phenomena that occur as part of the local impact of a projectile. The main objectives of this study were to: 1) list projectile (missile/fragment) parameters such as source, type, and characteristics (mass, diameter, velocity) and 2) conduct a local impact analysis for identified potential projectile hazard scenarios at an LNG facility. The paper concludes that the five different calculated thicknesses for concrete penetration, scabbing, and perforation, and steel penetration and perforation, depend directly on the kinetic energy and diameter of the projectile. Furthermore, the authors recommend that operators conduct a projectile hazard analyses for LNG facility projects, which would be useful in making recommendations for the thickness and type of storage tanks and other equipment and for other appropriate mitigation measures.

# 478563 Thermal Hazard Analysis and Chemical Incompatibility Tests with Custom High-Pressure Crucibles Made from Commonly Used Metals and Alloys

**Abstract:**

The assessment of reactive hazards and chemical incompatibilities is necessary for safe operations in the chemical and pharmaceutical industries. Commercial ARC (Accelerating Rate Calorimeter), RSST (Reactive System Screening Tools) and high pressure crucibles for Differential Scanning Calorimetry (DSC) can be used for this purpose [1-2]. However, ARC and RSST require significant capital investment and require relatively large amounts of sample (at least 1 g).The alternative, commercial high pressure crucibles for DSC, can also be expensive (e.g., several hundred U.S. dollars for each pair of gold-plated high-pressure crucibles) when multiple samples need to be investigated. These requirements make thermal hazard analysis prohibitively expensive for research laboratories that do not focus on chemical process safety, and where materials under development may be available in limited quantities (e.g., milligrams).

To address these short-comings of current hazard screening methods, inexpensive custom DSC high-pressure crucibles made from different commonly used metals and alloys (stainless steel 304, stainless steel 316, titanium, brass and tantalum) have been constructed in-house using a micro-welding system containing a jewelry spot welder and a microscope. Chemical compounds are tested in different crucibles made from different metal/alloys, which simulates different real-world storage, transportation, and process operations in the chemical and pharmaceutical industry. The cost for each pair of crucibles is less than two U.S. dollars. In a typical experiment, only 1-5 mg of sample is required. Kinetic parameters can be extracted from only a few DSC runs with these crucibles, allowing for prediction of the chemical's thermal behavior at larger scales. This method provides an inexpensive and efficient way to perform thermal hazard analysis and incompatibility tests. The application of these crucibles is demonstrated with case studies of ammonium nitrate (AN) and ionic liquids. Predictions of thermal behavior at larger scales have been verified with ARC tests.

References

1. Oxley, Jimmie C. 2003. "Handbook of Thermal Analysis and Calorimetry. Vol. 2: Applications to Inorganic and Miscellaneous Materials M.E. Brown and P.K. Gallagher, Editors 9 2003 Elsevier B.V. All Right Reserved. 349" 2: 349–69.

2. Han, Zhe, Sonny Sachdeva, Maria I. Papadaki, and M. Sam Mannan. 2014. "Ammonium Nitrate Thermal Decomposition with Additives." Journal of Loss Prevention in the Process Industries. Elsevier Ltd, 1–9. doi:10.1016/j.jlp.2014.10.011.

13TH GLOBAL CONGRESS ON PROCESS SAFETY

# Do Dust Hazards Analysis (DHA) Really Differ from a Process Hazards Analysis (PHA) or a Process Analysis (PA)?

**Michael E. Robertson**

**4094 Roslindale Ln**
**Buford, GA 30519**
**processesmatter@gmail.com**

Prepared for Presentation at
American Institute of Chemical Engineers
2017 Spring Meeting and 13th Global Congress on Process Safety
San Antonio, TX
March 26 – 29, 2017

# Do Dust Hazards Analysis (DHA) Really Differ from a Process Hazards Analysis (PHA) or a Process Analysis (PA)?

**Michael E. Robertson**

**4094 Roslindale Ln**
**Buford, GA 30519**
**processesmatter@gmail.com**

**Keywords:** NFPA 652, Dust Hazards Analysis, DHA

## Abstract

Do Dust Hazards Analysis (DHA) Really Differ From a Process Hazards Analysis (PHA) or a Process Analysis (PA)?

This author reviews the latest NFPA 652 requirement on Dust Hazards Analyses (DHA) and compares it to OSHA PSM's Process Hazards Analyses (PHA) requirement and other process analyses required by industry specific NFPA dust regulations. Though process analyses have been a part of NFPA regulations for several years and in the OSHA PSM regulation for 25 years, the 2016 version of NFPA 652 has clearly defined that a process analysis called Dust Hazards Analysis must be done where there is a combustible dust risk. The subject of this paper is the Dust Hazards Analysis.

The aim of the paper is to find commonality in dust hazards analysis, process hazards analysis and process analysis for tips that may limit the number of analyses a company may have to perform. These tips provided will assist a PHA leader in doing their DHA's. This will not be an exhaustive review of standards or regulations, but an overview and guide for those starting DHA's. It has become obvious that NFPA regulations are modeling a PSM framework and some discussion will be made on the subject in the area of risk assessment and management systems. Some comments will be made concerning the software that can be used, but no software will be sold or marketed. The author is not sharing examples of DHA's from his company due to confidentiality and timing of his DHA's.

Some of the Dust Hazards Analysis differences detailed are the emphasis on dust inside and outside of equipment, the use of NFPA dust thickness calculations, building reviews, and frequency of re-validating the analyses. How does a Dust Hazards Analyses affect the process safety information required, team make-up or report after the review? What are the real requirements for a DHA leader?

# 1 Introduction

There have been several dust related incidents that resulted in loss of life and severe damage to industrial properties. Imperial Sugar in Savannah, GA (2008) and West Pharmaceutical in North Carolina (2003) are two that top the list. Since the early 2000's, the Chemical Safety Board (CSB) has been recommending the Occupational Safety and Health Administration (OSHA) develop standards surrounding combustible dust. The National Fire Protection Association (NFPA) has had guidelines in several of its standards concerning dust for years.

**Table 1. Industry Specific NFPA Guidelines Pertaining to Dust**

| |
|---|
| 61 – Agriculture |
| 484 - Metal |
| 654 - General |
| 664 - Wood |

NFPA 654, 664, 61, 484, 86 and 499 have dealt with combustible dust for years.

It is 2017 and while OSHA still has not set forth comprehensive dust standards, NFPA has published a comprehensive guideline on the management of combustible dust as 2016 NFPA 652 Standards on the Fundamentals of Combustible Dust. This guideline requires dust hazards analysis (DHA) to be complete by September 7, 2018. long with renewed regulations, risk assessments are now becoming part of every safety performance system. NFPA has required process analysis and hazards analysis in the industry specific guidelines..

A DHA is similar to a Process Hazards Analysis (PHA), but there are some key differences that need to be understood by study leaders. PHA methodology is well- defined especially from the OSHA 1910 PSM regulation, the CCPS book *Guidelines for Hazards Evaluations* **[2]**,and other information from new and vintage PHA courses like Primatech's Process Hazards Analysis Leaders' Training and Process Improvement Institute's PHA Leader's Training.

The similarities between PHA's and DHA's are in the methodology. The methodologies used to conduct a PHA, such as HAZOP, Checklist, What-If, are applicable for identifying and assessing hazards in dust handling processes. Expertise with PHA's can assist with the completion of DHA's. The differences begin when the building has to be evaluated for combustible dust hazards. Process Analysis is defined in NFPA 664-12 – Standard for the Prevention of Fires and Explosions in Wood Processing and Woodworking Facilities. [3]. However, NFPA did not offer much instruction on how to do a Process Analysis (PA).

There are other tools that have to be employed for the building evaluation. Pictures, audits, root cause analysis and cause and effect diagrams can be used for the out of the process review.

Detailed building structure evaluations may require a code inspection or some other engineering tools.

This paper evaluates the elements of dust hazards analysis. Comparisons are made between dust hazards analysis, a process hazards analysis methodology called HAZOP (HAZard and OPerability Study), and a process analysis as described in older NFPA guidelines. These comparisons are made to assist PSM companies in aligning PHA's and DHA's and to assist non-PSM companies on resources to assist them in conducting DHA's.

## 2 What Is a Dust Hazards Analysis (DHA)

NFPA 652 requires that all facilities handling or creating combustible dust must do a Dust Hazards Analysis. A dust hazards analysis is a systematic review of processes and buildings where dust is created or handled and could create a deflagration.

Per the CCPS process safety glossary: A combustion that propagates by heat and mass transfer through the un-reacted medium at a velocity less than the speed of sound.

This requirement has many businesses scrambling in a similar way to the release of the PSM standard in 1992, which left many oil and gas and chemical companies reeling. The dust hazards analysis is not a PSM requirement nor is it a type of process hazards analysis. While dust hazards analyses are new, analyzing your process for dust hazards has been a part of industrial specific NFPA guidelines for a few years now. Now with NPFA 652, OSHA through Recognized and Generally Accepted Good Engineering Practice(RAGAGEP) has the standards it needs for inspection and enforcement to address the potential for a dust deflagration in a facility.

The dust hazards analysis must be complete by September 2018. The DHA requirement is set forth in Element 7.0 of the 2016 version of NFPA 652. There is a good example of the DHA in Appendix B of NFPA 652. Operating areas are reviewed for the probability of deflagration. Classifications are 'Not A Hazard', 'Maybe a Hazard' and 'Deflagration Hazard'. If the probability exists, then the owner must document how they manage or mitigate that risk. If the owner does not have current controls to prevent the deflagration, steps must be taken to document a hazards management plan with steps to reduce the risk.

Within a process system or building, when there is a sufficient quantity of combustible dust, the following conditions must be considered and documented:

1) Oxidizing atmosphere
2) Credible ignition source
3) Credible suspension mechanism

These are items from the fire triangle used in most fire prevention training. These are key in the dust hazards analysis and must be treated with high importance. Housekeeping also must be properly reviewed and documented to prevent the third item mentioned above – credible suspension mechanism.

# 3 How Do Dust Hazards Analysis (DHA) Compared to Process Hazards Analysis (PHA)

A major difference between the DHA and a PHA is the consideration of dust risk in a building. PHA's do evaluate the risk to occupied buildings in relation to the potential for releases of highly hazardous chemicals covered by the PSM regulation. This risk is evaluated through facility siting studies and more detailed reviews like a Quantitative Risk Analysis (QRA) or Occupied Building Risk Assessment (OBRA).

**Table 2. PHA and DHA Methodology Comparison**

| Process Hazards Analysis (HAZOP) | Dust Hazards Analysis | Process Analysis |
|---|---|---|
| Source: OSHA 1910.119 (e) | Source: NFPA 652 7.0 and Appendix B – Dust Hazards Analysis | Source: NFPA 664 4.0 and Explanatory Material A4.3 |
| Covers the process system | Covers the process system and the building construction | A methodical review of the facility, each operation housed within, and the identification of where a hazard exists. |
| Gather Data – Drawings and Process Information | Gather Data – Lab Data, Drawings and Process Information | |
| Have a Qualified PHA Leader a Team with an Operator and a Process Expert, along with other Subject Matter Experts as needed | Have a Qualified DHA Review Leader and at least one other for a team. | |
| Set Objectives and Boundaries | | |
| Determine the Study Nodes by marking up the Piping and Instrument Diagrams | Determine the nodes on the drawings and sketches | |
| Review HAZOP Guidewords | Review DHA questions from Appendix B of NFPA 652 | |
| Describe Process Node | Describe Process Node | |
| Use all Guidewords in all Nodes | Use all questions in all Nodes | |
| Review Global Node, Human Factors, Management of Change and Incidents | | |

With the PSM regulation, PHA's are required in facilities with highly hazardous chemicals. These chemicals are listed in the PSM requirement. PHA's are a group of tools that are used to review processes handling highly hazardous chemicals. PHA's include HAZOP, Failure Mode and Effect Analysis (FMEA), What If Studies, and Checklists.

The DHA requires buildings to be reviewed for the possibility of a fire from dust build-up. Building construction must minimize the effects of a deflagration or explosion from a combustible dust.

From the table above, you can see some differences in PHA's and DHA's. DHA's do not require a review of management of change and incident review history. The team composition is not set in a DHA. P&ID's are the basis, but DHA's do not have a required standard drawing type although P&ID's are applicable for some aspects of dust handling.

# 4 DHA Toolbox

Similar to PHA's, multiple tools are needed to complete a DHA.A PHA method could be used to evaluate the risk associated with combustible dust in a process. However, beyond typical PHA methods, field audits, pictures and video recordings of the areas may be need of areas where dust can build up.Alternatively, a root cause analysis method like 5 Why method might also be applied in areas where a dust build-up to determine sources of dust leakage.. The key is to locate where dust gets out of a process or accumulates beyond its minimum explosive concentration (MEC) in a process, why the dust gets outside the confines of the process and what can be done to prevent the leakage.

Whereas chemical analysis and laboratory data are helpful in a PHA, parameters like Maximum Explosion Concentration (MEC), Minimum Ignition Temperature (MIT) and Kst are very key in determining dust combustibility and risk for a process and building. NFPA 652 has information in the use of these parameters. There is also good information in NFPA's Guide to Combustible Dust and the AIChE Book on Solids Handling Equipment. This paper will not cover in detail any laboratory or chemical analysis of dust. [4], [5]

# 5 Documentation for Review

Piping and instrument diagrams (P&ID's) have long been the basis of the PHA techniques, including HAZOP, Checklist, and What-If. Supporting documentation may include SDS's, process descriptions, management of change documents, incident history and standard operating procedures. Plant layouts and general arrangement drawings will help orient the team and provide information for facility siting and items in the global node.

P&ID's will help for the process analysis in the DHA. You may also need general arrangement drawings and assembly drawings where machines are involved. Plant layouts and 3D renderings may help with the building analysis portion of the DHA. Pictures and video of areas where dust buildup can occur will also be helpful.

Lastly, plant conveying, processing and finishing OEM information such as screen sizes, motor power, pneumatic flow rates, and duct sizing are key in understanding out the equipment affects the creating or lack of creation of combustible dust. Information on seals and design pressures must be available for comparison with operating parameters.

The bottom line is DHA documentation is as important to the process as Process Safety Information is for PHA. However, it might not be as standard or available as PSI is for a PHA. Some of the information may also be very old since plant layouts and renderings might come from the original start of the plant.

## 6 The Dust Hazards Analysis Process

The DHA process can start like a PHA with an overview, process description and agenda. The DHA must cover the fugitive combustible dust that gets out into the building and the building's ability to withstand deflagrations and explosions.

Documentation must be available to follow the flow of the materials that contribute to creation of combustible dust. Equipment like fans and screens are especially sensitive since those components can create dust suspensions and impact the Minimum Explosion Concentration of the system. Critical laboratory must be available for discussion since it will provide the data to determine combustibility.
As with a PHA, a tour of the area would be very critical to see the areas were combustible dust accumulates the most. A start to the process is chosen and each part is reviewed by asking a series of question. Below are the questions from Appendix B of NFPA 652 used in a DHA.

**Table 3. DHA Process Questions ([1] – B.4.5.1.1 – B.4.5.1.5)**

| | |
|---|---|
| I. | Is the particulate deflagrable (explosible)? |
| II. | Is the particulate suspended in air? |
| III. | Is there sufficient concentration to propagate a flame front? |
| IV. | Are there competent igniters available? |
| V. | What hazard management is in place? |
| VI. | Is there metal detection, spark detection, bonding and grounding, or other hazard management means in place? |

These questions are asked for each piece of equipment, piping and ducting. The answers will should lead the team or competent person to the hazard management system measures in place and the gaps that will need attention from site management. The process must continue until the entire process is analyzed and all the outcomes have been reached.

As mentioned previously, Appendix B of NFPA 652 (2016) provides a good example to do a DHA. For those in the wood industry, the example is actually of a wood processing plant.

## 7 Outcomes

NFPA 652 requires that results from the Dust Hazards Analysis be documented. This includes necessary action items requiring change to the process materials, process itself, operations or facilities. NFPA 652 has requirements for managing the hazards under the heading of Hazards Management Program.

The results from a Process Hazards Analysis are the hazard scenarios, a list of safeguards currently in place, and any action items required to ensure that risk is effectively managed within the process handling highly hazardous materials. This usually involves a risk ranking system as well.

The Dust Hazards Analysis does not require a risk ranking system. Where a hazard exists, identification and evaluation of the fire scenario shall include safe operating ranges, safeguard identification, recommendations for additional safeguards as needed to properly manage the fire event, and lastly a plan for implementation.

## 8 Leaders

A meeting has a leader and teams have leaders. DHA's and PHA's are no different. Each review must have a leader. Her qualifications are listed in Table 2. There are no certifications or extensive training requirements. Many believe that certification is coming for PHA leaders because the competency of those who lead PSM is being discussed in government and industry especially as we enter the next 25 years of PSM.

Many experts agree that an experienced PHA leader can also make an effective DHA leader. This is a good start, but the DHA leader must also be familiar with the hazards of combustible dust. Additionally, it would be prudent to have a leader who is experienced in the process being reviewed or ensure that a process expert is present as a team member. I would review the qualifications listed in CCPS's book on Hazards Evaluation and information.

**Table 4. Comparison of Leader Qualifications**

| **Dust Hazards Analysis Leader** | **Process Hazards Analysis Leader** |
|---|---|
| Must be able to lead DHA's<br><br>Must be familiar with the hazards of combustible dust | Must be trained in PHA techniques<br><br>Must be re-qualified every 5 years |

## 9 Frequency

PHA's are done initially to establish a base evaluation and then every five years after the initial PHA to address changes or revisions to covered processes. . DHA's are done once for the original process and for any major upgrades. There is no renewal period, but the initial reviews have to be done by September 7, 2018.

## 10 Management Systems

What is a dust hazards management system? How does a dust hazards management system compare to the standard elements of PSM? Chapter 9 of NFPA 652 mentions its elements as listed in Table 6. Both PSM and NFPA 652 have the requirement of risk assessment that has been the subject of this paper.

PSM is embedded in many of the NFPA guidelines. Though the thought of PSM to many firms is frightening, PSM offers a good framework for managing risk whether it is combustible dust or other system or process risks.

**Table 5. Management Systems Comparison**

| Dust Hazards Analysis Management System (Ch. 9) | Elements of Process Safety Management |
|---|---|
| 1. Operating procedures and practices | 1. Standard Operating Procedures |
| 2. Inspection testing and maintenance | 2. Mechanical Integrity |
| 3. Training and Hazard Awareness | 3. Training |
| 4. Emergency Planning and Response | 4. Emergency Planning and Response |
| 5. Incident investigations | 5. Incident investigations |
| 6. Management of change | 6. Management of change |
| 7. Management system review, | 7. Compliance Auditing |
| 8. Contractor training | 8. Contractor Safety |
| 9. Employee participation | 9. Employee Participation |
| 10. Document retention | 10. Process Safety Information |
| 11. Ignition Control (Covered in Chapter 8) | 11. Hot Work Permits |
| 12. DHA | 12. Trade Secrets |
| | 13. Pre-Start Up Safety Review |
| | 14. PHA |

## 11 Conclusions

DHA's must be completed. It is similar but different than a PHA. Expertise with PHA's can assist with the completion of DHA's. The similarity between PHA's and DHA's is in the methodology of reviewing processes for dust. The differences start when the building has to be evaluated for combustible dust. Here other tools have to be employed. These tools as previously discussed may vary from audits and inspections to root cause analysis tools. Detailed building structure evaluations may require a code inspection or some other engineering tools.

## 12 References

**[1] NFPA 652-15 – Standards on the Fundamentals of Combustible Dust, National Fire Protection Association, September 7, 2015.**

**[2] CCPS. *Guidelines for Hazards Evaluations.* Center for Chemical Process Safety, American Institute of Chemical Engineers, New York, NY, 2007.**

**[3] NFPA 664-12 – Standards on the Fundamentals of Combustible Dust, National Fire Protection Association, 2011.**

**[4] NFPA– Guide to Combustible Dusts, National Fire Protection Association, 2012.**

**[5] CCPS. *Guidelines for Safe Handling of Powders and Bulk Solids.* Center for Chemical Process Safety, American Institute of Chemical Engineers, New York, NY, 2005.**

# 478758 A Leading Indicators Based Decision Support Tool to Predict Blowout Events

**Abstract:**

Predicting a potential blowout scenario timely and efficiently is often a challenging task due to the complexity of drilling or other well intervention activities. Blowouts are usually preceded by gas kicks and early prediction of gas kick events can provide some precious time window to take control of the well in concern. A leading risk indicators based approach for predicting gas kick or blowout events has been introduced in this work. A causal factors based step-by-step approach for identifying comprehensive sets of leading indicators are presented and discussed. Leading indicators were divided into two broad sections - real-time indicators and long-term organizational safety performance indicators. With the real-time indicators, different decision support algorithms are developed which can be used as training tools. A real-time indicator may not always successfully lead to a well control event but at the same time the situation needs to be assessed carefully considering other factors and indicators. Such uncertainties are taken into account while developing the algorithms. For evaluating the relative importance of different leading indicators Bayesian Network models are developed. The key causal factors for well control barriers failure are identified by conducting fault tree analysis and appropriate indicators are linked with relevant causal factors in a Bayesian model. At the end, a list of efficient leading risk indicators are presented for predicting primary well control barrier failure which can initiate gas kicks.

# Consequence Modeling at the Eastman Kingsport Site

**David Skelton**
**Principal Engineer**
**Eastman Chemical Company**
**P.O. Box 511, Building 18**
**Kingsport, TN 37662**
**davidskelton@eastman.com**

**Michael James**
**Senior Chemical Engineer**
**Eastman Chemical Company**
**P.O. Box 511, Building 18**
**Kingsport, TN 37662**
**mjames@eastman.com**



Prepared for Presentation at
13th Global Congress on Process Safety
San Antonio, TX
March 26-29, 2017

UNPUBLISHED

# Consequence Modeling at the Eastman Kingsport Site

**Michael James**
**Senior Chemical Engineer**
**Eastman Chemical Company**
**P.O. Box 511, Building18**
**Kingsport, TN 37662**
**mjames@eastman.com**

**David Skelton**
**Principal Engineer**
**Eastman Chemical Company**
**P.O. Box 511, Building 18**
**Kingsport, TN 37662**
**davidskelton@eastman.com**

**Keywords**: Dispersion modeling, consequence analysis, risk assessment, dosage, shelter-in-place, temporary haven, evacuation, toxicity, toxic chemical release, toxic load, downwind concentration, blast overpressure, vapor cloud explosion, pressure vessel burst

## Abstract

The intent of this paper will be to review the methods utilized at the Eastman Kingsport site in determining magnitude of consequences of toxic, thermal and blast overpressure scenarios. This will include discussion on dispersion modeling, dose-response methods for toxic and thermal probits as well as assessing occupant vulnerability through the use of BakerRisk's Explosion Risk Cooperative tools.

## 1. Potential Consequences of Loss of Containment

Figure 1 is a typical event tree for potential losses of containment [1].

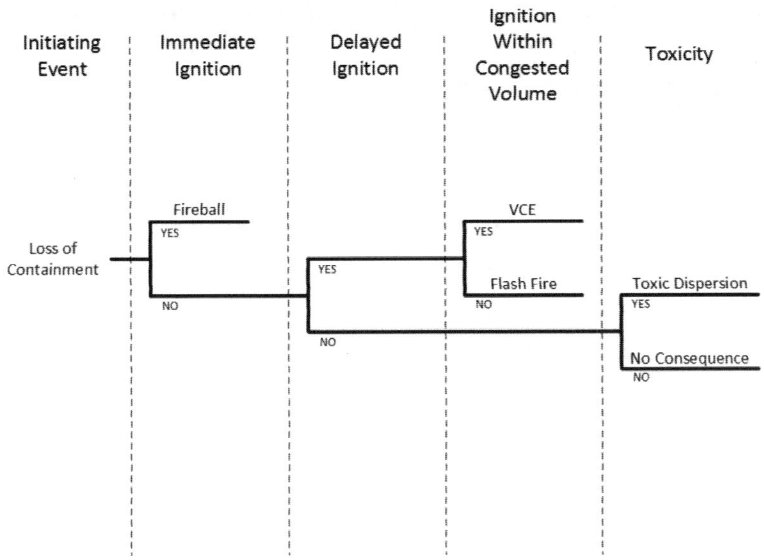


Figure 1. Potential Losses of Containment

The consequences of loss of containment in Figure 1 detail the potential release scenarios that are chiefly of concern. The method for consequence modeling for each type of scenario is described in this paper.

## 2. Source Terms

Figure 2 details the potential source terms that will need to be developed for each type of scenario.

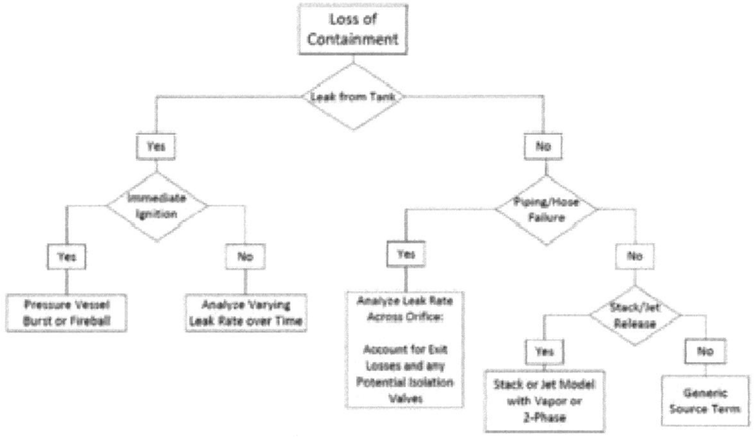


Figure 2. Potential Source Terms.

The source term may have any of the following characteristics:

- Initial momentum
    - Based on the linear velocity across the orifice and other physical properties of the release, the source term may need to account for significant initial momentum of the cloud.
    - In addition, if the linear velocity is sufficiently high, the source of the release may ingest air, resulting in a jetting condition. As a rule of thumb, 10 m/s is a lower limit for potentially jetting releases [2]. There are specific dispersion models to account for jetting conditions during a release.

- Pool Formation
    - Most dispersion models have an integrated pool evaporation model to determine the dynamic evolution of vapor from a liquid spill.
    - If the preferred evaporation model is outside of the dispersion model, the evaporation rate results will need to be analyzed for integration into the dispersion modeling tool.
- Flashing
    - Leaks across an orifice may partially or completely flash from liquid to vapor based on the pressure differential.
    - Modeling tools may report choke conditions at the outlet.
    - Further flash analysis may be required to determine the temperature and liquid/vapor composition from choke to atmospheric conditions.

- Piping Layout between Pressure Source and Leak Point
  - Ensure that all pressure losses are accounted for. This includes entrances, exits, elevation changes, elbows, pipe diameter changes, and other fittings.

- Indoor Release
  - If the release occurs indoors, analyze the concentration changes over time inside the structure based on the source term calculations of the release.
  - Concentration gradients throughout the structure may need to be ignored and perfect mixing assumed.
  - If the vapor portion of the leak will cause overpressuring of the indoor area, determine the amount of vapor required to be vented in order to balance pressure in the room.
  - A flash block can be employed to assess the amount of vapor generated from a leak that will need to be relieved from an indoor area.
  - Flash blocks can also be used to account for vapor suppression systems inside of buildings.
  - This level of modeling is beyond the scope of this paper. However, methods for assessing indoor releases can be referenced from "Consequence Modeling of Dynamic Source Terms" [3].

Presence of the above factors may result in models that require expert judgment and the development of conservative estimates. Most dispersion modeling tools do not have the levels of sophistication required to accurately model these types of source terms. However, the results of a dispersion model employed with conservative estimates may be suitable for site risk assessment.

The following section will discuss the potential leak scenarios and how source terms can be developed, accounting for each of the above factors.

## 3. Blast Modeling

Simple blast models can be used to predict potential consequences of explosions in lieu of more accurate but complicated methods such as computational fluid dynamic (CFD) models. The TNT equivalency model developed by the U. S. Navy and Air Force was largely used for years, even in industrial settings. Over time it became clear that the TNT model did not predict the behavior of all explosions accurately enough, particularly in regard to vapor cloud explosions more commonly occurring in refining and chemical producing industries. Today several simplified are available to estimate the potential of consequences of explosions, each method with its own strengths and weaknesses.

### *3.1 Vapor Cloud Explosion (VCE) Modeling*

Baker Strehlow Tang (BST) is the simplified method predominantly used at Eastman Kingsport site to model vapor cloud explosions [4][5][7][8]. The BST model is based on the premise that only the portion of a flammable vapor cloud that is confined or contained within a congested area will significantly contribute energy to an explosion. Unconfined or uncongested gas clouds are more likely to result in flash fires or deflagrations. The first step to perform a BST analysis is to identify the confined/congested volume of the plant that has the potential to harbor a flammable cloud of vapor. Examples might include piping and equipment, closely spaced railcars or cars, solid walls, ceilings and floors. A flammable cloud is most often generated from a release of material as a result of an equipment failure, a leak, or a spill. Flammable clouds can drift a considerable distance under the right conditions, so it's important to account for all the equipment within a reasonable radius of the congestion. Once the confined/congested volume is calculated, two parameters contribute the greatest influence on explosion strength using BST methodology; the energy contained within the flammable vapor cloud and flame speed as it burns through the available fuel in the congested region. The first parameter is calculated using the spreadsheet shown in Figure 3. It is a simple tool developed by Eastman to calculate volume based on dimensions of the confined cloud, and then estimate the energy contained with the calculated volume.

**ESTIMATION OF VAPOR CLOUD EXPLOSION ENERGY**

The user enters dimensions (in ft.) of vapor cloud subject to confinement & obstructions, ambient temperature (F), and the chemical involved in the designated boxes. Thermodynamic data listed in the lookup table to the right is used, or if not available, an approximation is made based on typical hydrocarbons.

The spreadsheet will output the participating energy to be used in V-CLOUD for prediction of pressure and impulse by the Baker-Strehlow method. The methodology is based on the paper "Vapor Cloud Explosion Analysis" by Baker, Tang, and Scheier; Prepared for the AIChE Loss Prevention Symposium, April 1994.

| | | |
|---|---|---|
| Width (ft) | 200 | |
| Height (ft) | 50 | |
| Depth (ft) | 100 | |
| Volume (ft3) | 1000000 | |
| Volume (m3) | 2.83E+04 | |
| Ambient Temp (F) | 90 | |
| Ambient Temp (C) | 32 | |
| Chemical | Acetic Acid | |
| MW | 60.05 | |
| Hc(kJ/mole) | 919.73 | (heat of combustion in KJ/mole) |
| Hc(J/m3) | 3.67E+07 | (heat of combustion in J/m3) |
| Hc (BTU/lb) | 6.59E+03 | (heat of combustion BTU/lb for ref.) |
| Hcs(J/M3) | 3.48E+06 | (heat of combustion of a stoichiometric mixture with air) |
| Energy (J) | 9.86E+10 | |
| Energy (in-lb.) | 8.72E+11 | (total available energy in confined / obstructed volume) |
| % Participation | 100 | |
| Equiv Energy in-lb. | 8.72E+11 | (net energy of explosion in-lb.) |

Figure 3. Spreadsheet used to determine energy content of vapor cloud

For a single congested area or multiple areas with little separation distance so they can be effectively considered a single area of congestion, the results from the spreadsheet are used as an input in VCloud®. VCloud® is a simple software tool developed by the BakerRisk Engineering Consultant to produce Pressure/Impulse (P-I) diagrams (this is not to be confused with PI Diagrams representing the response curve for structures and might better be referred to as the side-on blast pressure and impulse as a function of standoff distance). Notice the check box denoted "Plot Using Co-op Energy Factor." Membership in the BakerRisk Engineering Consultant Explosion Research Cooperative (ERC) includes access to experimentally verified energy factors that can reduce conservatism and provide a more accurate prediction of explosion effects. The first page of inputs for VCloud® is illustrated in Figure 4.

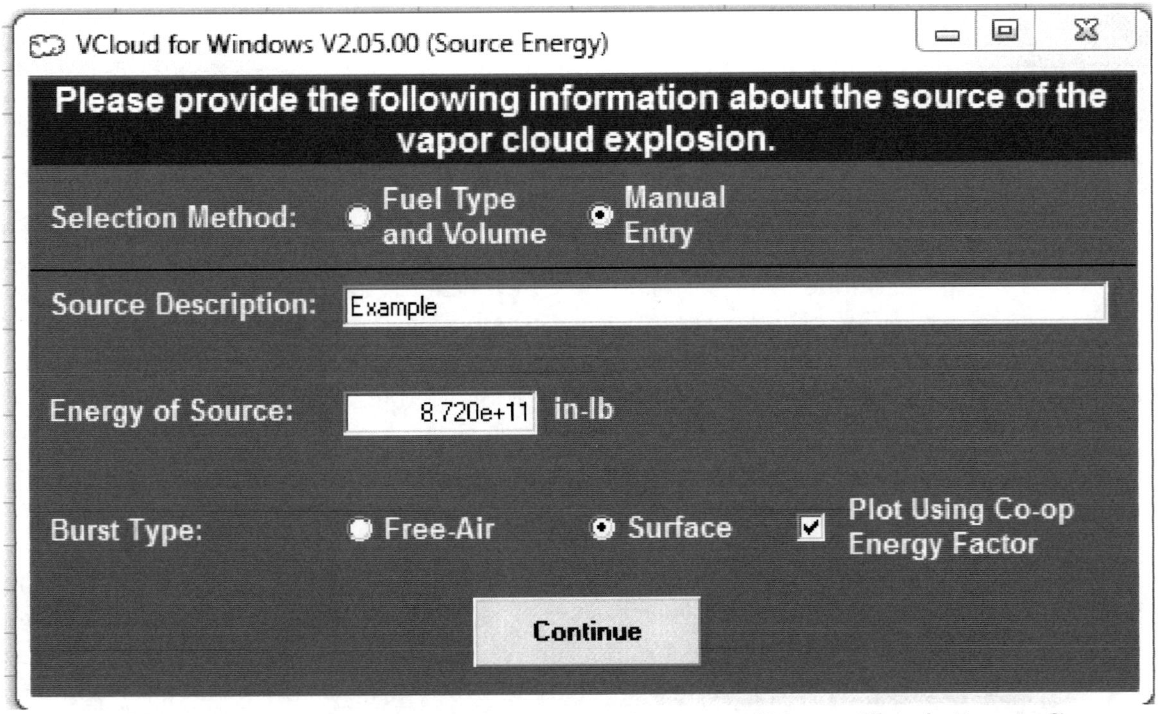


Figure 4. VCloud® Input to Determine Energy Content of Vapor Cloud. Image Courtesy of BakerRisk Engineering Consultant Explosion Research Cooperative.

The second parameter greatly influencing explosive strength when using BST methodology is flame speed, measured as a ratio to the speed of sound or Mach number. Four criterion most influence how fast a flame front burns through fuel: congestion, confinement, fuel reactivity, and the distance to free venting.

- Congestion
  - Burning rate is greatly dependent on the surface area of the flame as it travels through the fuel source. Turbulence causes the flame front to furrow and can greatly increase the surface area which contributes to increasing the speed of the advancing flame. In an industrial setting, turbulence can be generated by piping, equipment, gravel surfaces, vehicles, or any obstacle(s) that inhibits simple linear progression of the burning flame front.
  - The greater the induced turbulence, the faster a flame front will progress increasing the likelihood of detonation. BST method calculations qualify congestion into three categories; high, medium or low congestion. Congestion categorization is the most subjective parameter in BST methodology. Experience is probably the most important tool used to determine degree of congestion.
  - A good rule of thumb is if process equipment occupies an area but can be easily walked through, it is low congestion. If you can manage to slip through an area with difficulty, it's medium congestion. If it is impossible to physically pass through an area, it's high congestion.

- Confinement
  - Any solid surface that prevents the propagation of the flame in a given direction is considered confinement, with the exception of the ground. The BST method models a spherical, free-air vapor cloud explosion, but multiplies the energy of the explosion by a factor of 2 to account for ground reflection of the blast wave from its origin surface.
  - BST calculations categorize confinement in four ways; 3D, 2D, and 2½ D, and 1D.
    - 3D Confinement: A flame front burning through fuel is capable of expanding in all three dimensions with the exception of the ground. It applies to all free-field processing areas not bound by solid walls or ceilings.
    - *2D Confinement:* The propagation of the flame front is free to expand in only two axes because one dimension is blocked by a solid obstacle. A common example is a processing unit located between a solid concrete ceiling with open walls. It also applies if a processing unit is located between two solid walls with no roof. It is important to consider the aspect ratio of 2D confinement, which is the ratio of total width to height of the confined volume. Higher aspect ratios constrict venting capacity and allow additional pressure accumulation over those with a shorter distance to free venting. Typical values are available in a drop down menu in VCloud®, ranging as high as a 25:1 aspect ratio. Anything higher than 25:1 will not provide significantly more pressure accumulation, nor are aspect ratios greater than this commonly found in the field. Any aspect ratio greater than 4:1 should be represented in VCloud as 25:1 to maintain a conservative estimate.
    - 2½ D Confinement: This is a hybrid condition where neither 3D nor 2D accurately describe confinement. For instance, processing equipment contained within a pre-engineered building with lightweight metal walls and roof that will confine the flame front for a time but release at low pressures, before the impending explosion is fully realized. The roof and walls provide some containment, but in short order the blast wave is capable of 3D expansion. Tightly packed equipment on the boundaries of the congested area generating an explosion can also be considered 2½ D confinement, since it provides a small amount of venting, but not enough to consider it 3D expansion.
    - 1 D Confinement: Flame burn in a linear direction, such as inside a pipe.

- Fuel Reactivity
  - The BST method categorizes fuel reactivity in only three ways; low, medium, and high reactivity. Fuel reactivity is based on the laminar burning velocity at a specific concentration, determined by a standard testing procedure in stagnant conditions.
  - Fuels with a laminar burning velocity less than 45 cm/s are considered to have a low reactivity. Fuels with laminar burning velocities above 75 cm/s are considered highly reactive. Everything between is considered to be of medium reactivity.
  - Table 1 illustrates reactivity of a few common chemicals for reference here. NFPA 68 is a good reference for finding laminar burning velocities of many more chemicals so the reactivity category can be determined.

Table 1. Relative Reactivity

| Reactivity | Fuel | Group | Relative Reactivity | Formula |
|---|---|---|---|---|
| Low | Methane | Alkanes | 1 | CH4 |
| Low | Formic Acid | Acids | 1 | HCOOH |
| Low | Acetic Acid | Acids | 1 | CH3COOH |
| Medium | Ethane | Alkanes | 3 | C2H6 |
| Medium | Propane | Alkanes | 3 | C3H8 |
| Medium | Butane | Alkanes | 3 | C4H10 |
| High | Ethylene | Alkenes | 6 | C2H4 |
| High | Acetylene | Alkynes | 7 | C2H2 |
| High | Hydrogen | Inorganics | 7 | H2 |

(Courtesy Eastman Chemical Company)

- Distance to Free Venting
  - Distance to free venting is simply the shortest distance a flame must travel to exit the area of congestion/confinement to find free space.
  - After free venting, turbulence is less likely to accelerate the flame front and under most circumstances it will begin to slow. This reduces the possibility of a transition from deflagration to detonation (DDT).

These four criterion are now input into VCloud® in the format shown in Figure 5. From the provided data, a Mach number ($M_f$) is assigned to the flame front.

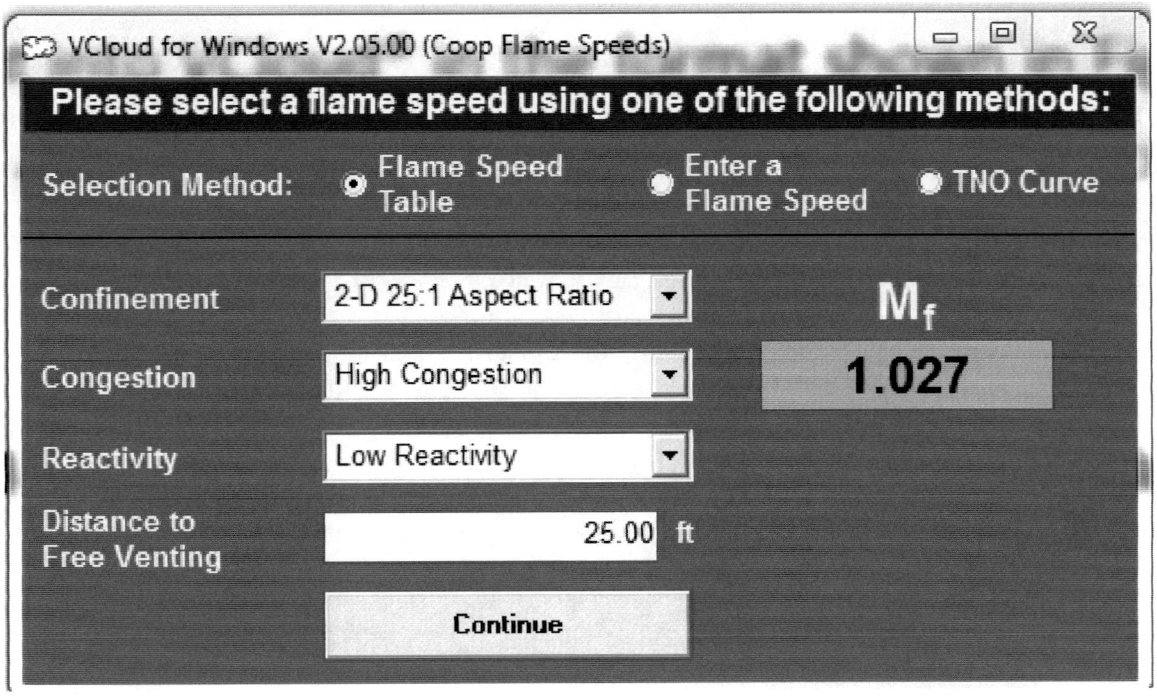

Figure 5. Vcloud® Input to Determine Mach Number. Image Courtesy of BakerRisk Engineering Consultant Explosion Research Cooperative.

A Mach number greater than one indicates the flame front accelerated faster than the speed of sound and beyond, meaning a detonation is more likely to occur. Some typical flame speeds in Mach numbers are provided in Table 2 for reference, with respect to fuel reactivity and degree of congestion.

**Table 2.** Flame Speeds for Soft Ignition Sources [9]. Provided as a reference but not relevant to the example calculation.

Flame Speed in Mach Numbers ($M_w$) for Soft Ignition Sources

| Flame Expansion | Fuel Reactivity | Obstacle Density | | |
|---|---|---|---|---|
| | | High | Medium | Low |
| ID | High | 5.2 | 5.2 | 5.2 |
| | Medium | 2.27 | 1.77 | 1.03 |
| | Low | 2.27 | 1.03 | 0.294 |
| 2D | High | 1.77 | 1.03 | 0.588 |
| | Medium | 1.24 | 0.662 | 0.118 |
| | Low | 0.662 | 0.471 | 0.079 |
| 3D | High | 0.588 | 0.153 | 0.071 |
| | Medium | 0.206 | 0.100 | 0.037 |
| | Low | 0.147 | 0.100 | 0.037 |

VCloud® uses the provided inputs to produce a Pressure/Impulse (PI) diagram. The PI diagram is used to determine the overpressure and impulse experienced by a target at a given distance. Figure 6 is an example of a PI diagram produced by VCloud®. The distance between the target and the explosion is manually input to determine the blast effects at that specified distance. The target at 150 feet from the explosion will experience a 10.12 psig side-on overpressure with a 316.05 psi-ms impulse. This is more than enough to completely demolish an unreinforced masonry building. A key attribute of vapor cloud explosions that differs from high explosives is a much higher impulse. Impulse is more threatening to rigid construction than overpressure, because it tends to distort supporting structural members beyond their plastic limits.

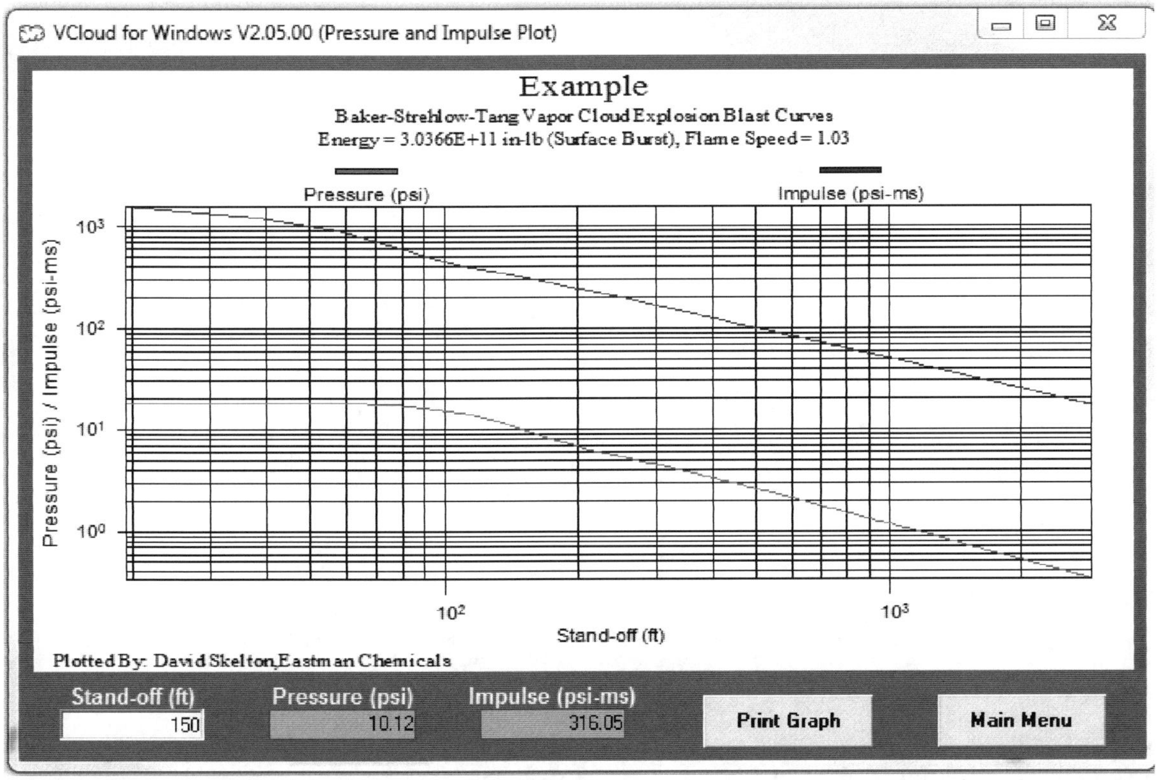


Figure 6. P-I Diagram Output From VCloud®. Image Courtesy of BakerRisk Engineering Consultant Explosion Research Cooperative.

Once a PI diagram is prepared, it can be used to predict the potential damage to conventional industrial buildings from the modeled blast event. Another software tool developed by the ERC can be used for this purpose, the Blast Evaluation and Screening Tool, affectionately known as BEAST®. BEAST® uses empirically based correlations from accidental explosions, shock tube tests, and government sponsored high explosive tests to predict the level of damage imparted by an explosion to a given building or structural component. Originally developed in 1995, BEAST® has been updated over the years to include more building and component damage curves, improve the user interface, and add more building types. The most recent version of BEAST® can evaluate 19 unique building types and many types of individual structural components, such as windows and doors. The first step is to define the building for evaluation. An example using the current BEAST® interface is included in Figures 7a and 7b. Required inputs are type of construction (drop down menu), building dimensions, blast overpressure and impulse (from PI Diagram), and whether wall and roof loading are side-on or reflected.

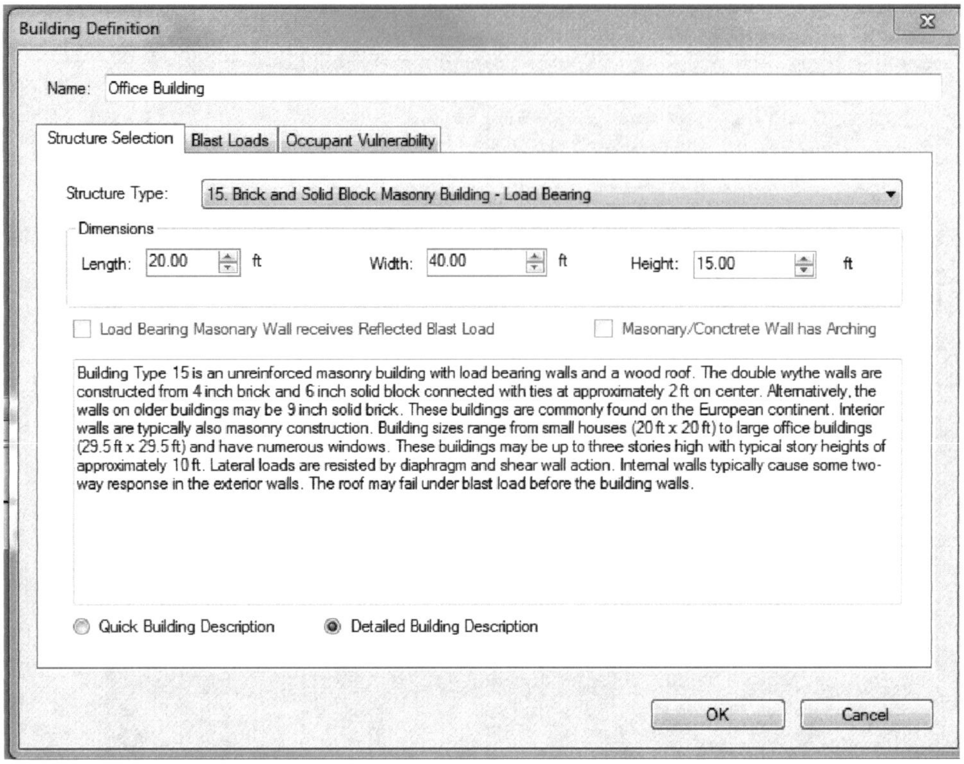

Figure 7a. Beast® Building Selection.

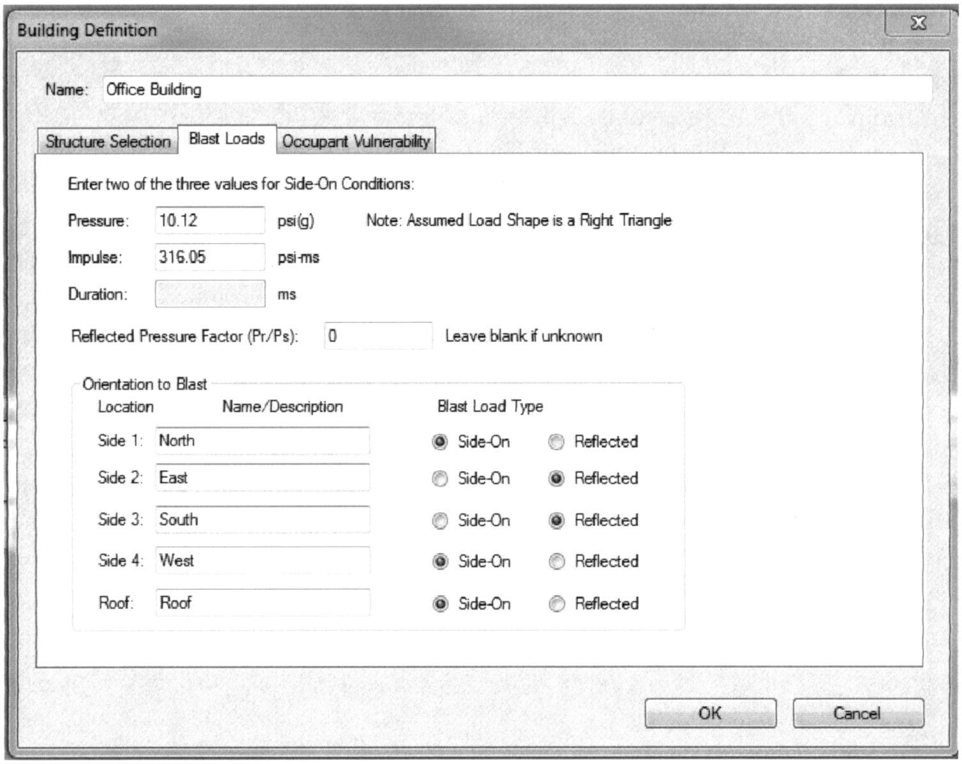

Figure 7b. Beast® Building Selection (cont'd).

Images Courtesy of BakerRisk Engineering Consultant Explosion Research Cooperative.

Given the appropriate inputs, BEAST® predicts a damage level to the building and an occupant vulnerability. Figure 8 illustrates the output provided from BEAST®.

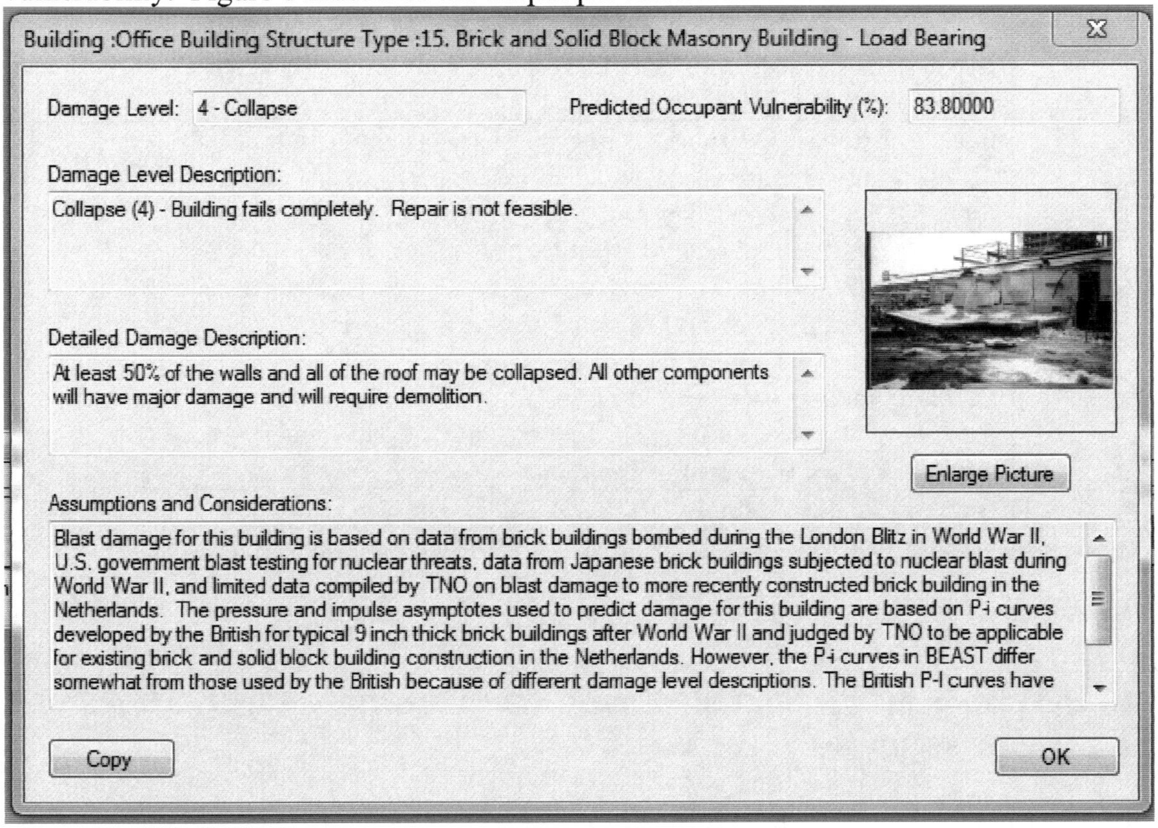


Figure 8. BEAST® Output

Of course, not all explosions result in the complete destruction of a building like the extreme example provided. BEAST® categorizes building damage using the criteria found in Table 3 [6]. A Building Damage Level up to 2B is still survivable by most occupants.

Table 3. BEAST® Building Damage Levels

| Building Damage Level | Damage Description |
|---|---|
| 1 | Onset of visible damage to building walls facing explosive source. Repairs only needed for cosmetic reasons. Building is reusable following an explosion. |
| 2A | Localized building damage, primarily to components on building walls facing the explosive source. Building performs function and can be used; however, major repairs are required to restore integrity of structural envelope. Total cost of repairs is moderate. |
| 2B | Widespread building damage. Building cannot be used until major repairs are completed. Total cost of repairs is significant, approaching replacement cost of building. |
| 3 | Building has lost structural integrity and may collapse due to environmental conditions (i.e. snow or wind). Total repair cost exceeds replacement cost of building. |
| 4 | Building fails completely. Repair is not feasible. |

Courtesy of BakerRisk Engineering Consultant Explosion Research Cooperative

## 4. Development of Source Terms for Potential Vapor Dispersion Release Scenarios

Methods for determining the source term for releases from tanks, liquid pools and piping systems are discussed here. All examples provided here do not reflect actual potential release conditions. For all examples, any relation to true conditions is incidental.

Prior to the modeling methods discussion, it is important to review a tool for determining losses due to piping layout.

IoMosaic SuperChems® v 8.21 (2016) allows for development of piping layouts to detail the piping diameters, elbows, and other potential connections. In simple models, as will be discussed in this paper, this can be used as the outlet piping for a pressure source (e.g., tanks) to a release point.

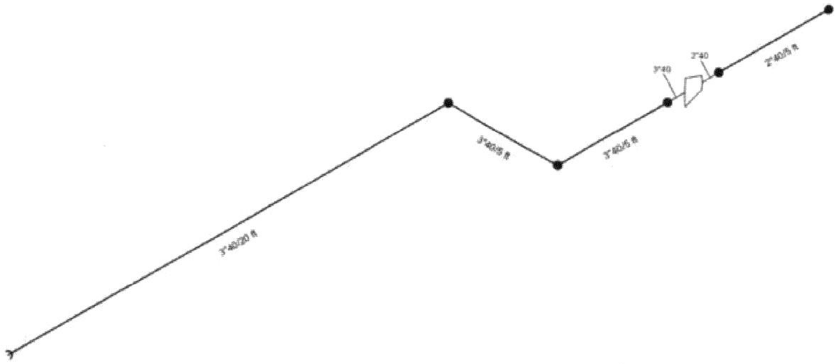


Figure 9. Example Piping Layout.

Figure 3 details a layout from the source at 3 inches with 2 90-degree elbows and a reduction to the 2-inch leak point. The release will be from a guillotine break on the 2-inch line.

### *4.1 Release from Tank*

A reliable model for release from vessels is contained in ioMosaic SuperChems® v8.21 (2016).

This example will model the release of a liquid blanketed by nitrogen. The "Vessels Containing 2-Phases Model" allows for estimation of the vapor and liquid contents of a vessel, accounting for nitrogen blanketing or air in the head space. This same model can be utilized for relief from the vapor phase. There are alternative models for gas-filled vessels.

The overall process for using SuperChems® to model this is as follows:

1. Development of objects for mixture composition, vessel dimensions, piping layouts and other objects. Ensure piping layouts are properly applied.

   The following steps are within the "Vessels Containing 2-Phases" Model.

2. Input the vessel pressure and an initial estimate for overall composition inside of the vessel. The overall composition includes both the vapor and liquid phases. In this example, the modeled vessel is filled to 80% with methyl isocyanate and has a nitrogen blanket.

   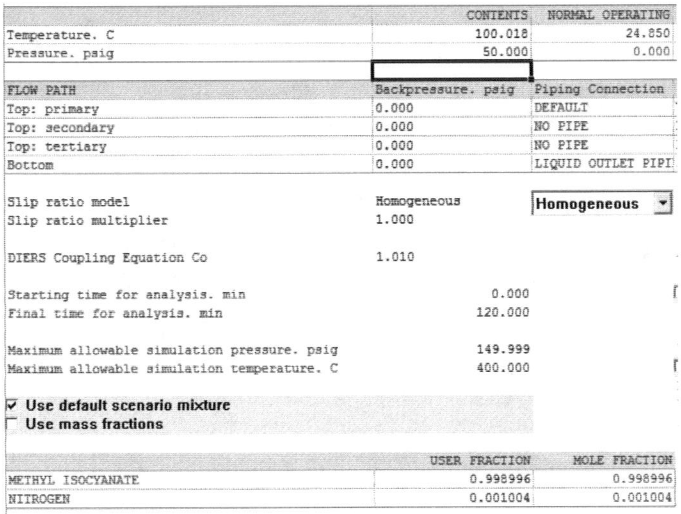

   | | CONTENTS | NORMAL OPERATING |
   |---|---|---|
   | Temperature. C | 100.018 | 24.850 |
   | Pressure. psig | 50.000 | 0.000 |

   | FLOW PATH | Backpressure. psig | Piping Connection |
   |---|---|---|
   | Top: primary | 0.000 | DEFAULT |
   | Top: secondary | 0.000 | NO PIPE |
   | Top: tertiary | 0.000 | NO PIPE |
   | Bottom | 0.000 | LIQUID OUTLET PIPI |

   | | | |
   |---|---|---|
   | Slip ratio model | Homogeneous | Homogeneous |
   | Slip ratio multiplier | 1.000 | |
   | DIERS Coupling Equation Co | 1.010 | |
   | Starting time for analysis. min | 0.000 | |
   | Final time for analysis. min | 120.000 | |
   | Maximum allowable simulation pressure. psig | 149.999 | |
   | Maximum allowable simulation temperature. C | 400.000 | |

   ☑ Use default scenario mixture
   ☐ Use mass fractions

   | | USER FRACTION | MOLE FRACTION |
   |---|---|---|
   | METHYL ISOCYANATE | 0.998996 | 0.998996 |
   | NITROGEN | 0.001004 | 0.001004 |

   Figure 10. Vessels Containing 2-Phase Inputs

   Uncheck the "Use default scenario mixture" as defining this model is an iterative process.

3. Use "Specify Fill Fraction" to input the approximate vessel fill level. This example uses an estimate 80% fill level.

   The model it calculates the vessel temperature at the specified overall composition.

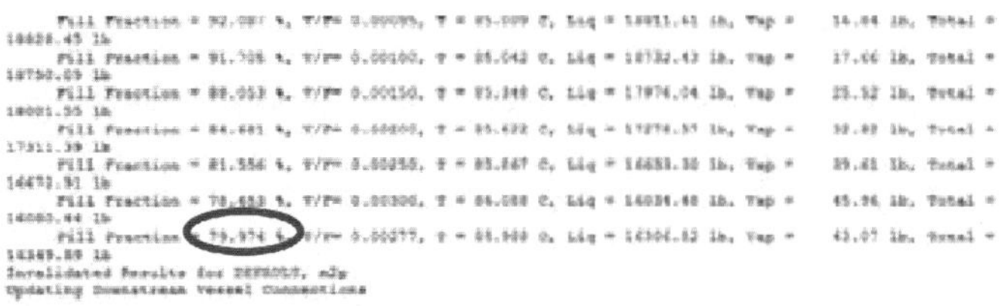

Figure 11. Model Iterations to determine vessel temperature at 80% liquid fill level.

4. Iterate over steps 2 and 3, varying the composition and re-entering the fill fraction until the modeled vessel temperature matches the true conditions. Increasing content of lower boilers will decrease the modeled vessel temperature.

   In the example, the composition was iterated until the storage temperature was approximately 25 deg C.

| | CONTENTS | NORMAL OPERATING |
|---|---|---|
| Temperature. C | 24.918 | 24.850 |
| Pressure. psig | 50.000 | 0.000 |

| FLOW PATH | Backpressure. psig | Piping Connection |
|---|---|---|
| Top: primary | 0.000 | DEFAULT |
| Top: secondary | 0.000 | NO PIPE |
| Top: tertiary | 0.000 | NO PIPE |
| Bottom | 0.000 | LIQUID OUTLET PIPI |

| | | |
|---|---|---|
| Slip ratio model | Fauske | Fauske |
| Slip ratio multiplier | 1.000 | |
| DIERS Coupling Equation Co | 1.010 | |
| Starting time for analysis. min | 0.000 | |
| Final time for analysis. min | 120.000 | |
| Maximum allowable simulation pressure. psig | 149.999 | |
| Maximum allowable simulation temperature. C | 400.000 | |

☐ Use default scenario mixture
☐ Use mass fractions

| | USER FRACTION | MOLE FRACTION |
|---|---|---|
| METHYL ISOCYANATE | 0.991100 | 0.991100 |
| NITROGEN | 0.008900 | 0.008900 |

Figure 12. Final composition.

5. Running the model will determine the flow rate through the piping layout and guillotine break.

6. When the run is complete, a report will appear. The "Profiles-Bottom" tab is the most useful for this source term model.

| TIME. min | VESSEL CONTENTS TEMPERATURE. C | VESSEL CONTENTS PRESSURE. psig | PRESSURE AT VESSEL BOTTOM. psig | BOTTOM PIPING LIQUID MASS FLOW RATE. lb/min | BOTTOM PIPING GAS/VAPOR MASS FLOW RATE. lb/min | Bottom Flow reduction factor (Actual flux with pipe/Ideal Flux) | BOTTOM FLOW NOZZLE PRESSURE. psig | BOTTOM FLOW DISCHARGE PRESSURE. psig | BOTTOM FLOW DISCHARGE TEMPERATURE. C |
|---|---|---|---|---|---|---|---|---|---|
| 0.0000 | 24.92 | 50.000 | 56.394 | 2.005E+03 | 0.000 | 0.405 | 56.392 | -0.001 | 25.094 |
| 0.0246 | 24.88 | 47.142 | 53.517 | 1.966E+03 | 0.000 | 0.394 | 53.516 | 0.000 | 25.051 |
| 0.0506 | 24.85 | 44.305 | 50.662 | 1.871E+03 | 0.000 | 0.383 | 50.660 | -0.001 | 25.006 |
| 0.0782 | 24.81 | 41.472 | 47.809 | 1.775E+03 | 0.000 | 0.372 | 47.808 | -0.001 | 24.960 |
| 0.1077 | 24.77 | 38.642 | 44.961 | 1.679E+03 | 0.000 | 0.360 | 44.959 | -0.001 | 24.913 |
| 0.1394 | 24.73 | 35.817 | 42.116 | 1.583E+03 | 0.000 | 0.347 | 42.115 | -0.001 | 24.864 |
| 0.1735 | 24.69 | 32.996 | 39.275 | 1.485E+03 | 0.000 | 0.334 | 39.274 | -0.001 | 24.814 |
| 0.2104 | 24.64 | 30.181 | 36.440 | 1.387E+03 | 0.000 | 0.321 | 36.439 | 0.000 | 24.761 |
| 0.2506 | 24.60 | 27.371 | 33.610 | 1.289E+03 | 0.000 | 0.306 | 33.609 | 0.000 | 24.706 |
| 0.2946 | 24.55 | 24.568 | 30.787 | 1.190E+03 | 0.000 | 0.291 | 30.786 | 0.000 | 24.649 |
| 0.3433 | 24.50 | 21.773 | 27.971 | 1.090E+03 | 0.000 | 0.276 | 27.970 | 0.000 | 24.588 |

Figure 13. Vessels Containing 2-Phases Output.

The liquid and vapor release rates out of the piping as well as the discharge temperature and pressure are reported. The discharge pressure will be above atmospheric if there is choked flow inside the piping.

If the choke pressure is significantly above atmospheric, running some select points through the "2-Phase Expansion and Aerosol Model" to determine the final vapor/liquid composition and temperature is recommended.

7. As the release is liquid in this case, evaporation from a liquid forming pool will next be modeled. As stated previously, several dispersion models have integrated pool models. The SuperChems® pool evaporation will be employed in this case, and the output from the model will be input to an external dispersion model.

### *4.2 Liquid Pools*

As with tank leaks, SuperChems® will be utilized for determining the source term of a liquid pool formation.

An important consideration is the variety of weather conditions over which a release could take place. As solar incidence is a major factor in pool evaporation, ensure that the most probable conditions for both daytime and nighttime are employed. Also, to account for solar incidence, ensure that the appropriate time of day and site coordinates are input for the scenario in SuperChems®. In addition, diked area dimensions and substrate composition can be specified.

SuperChems® pool evaporation model allows for a static release rate over a set period of time. As most releases will be dynamic in nature, this may involve expert judgment.

Once a characteristic release rate, duration and temperature have been determined, this information can be entered into the pool evaporation model.

| | |
|---|---|
| Liquid discharge temperature. C | 24.209 |
| Liquid discharge rate. lb/min | 583.903 |
| Liquid discharge duration. min | 1.949 |

Use default scenario mixture
Use mass fractions

| COMPONENT | USER FRACTION |
|---|---|
| METHYL ISOCYANATE | 1.0000 |
| NITROGEN | 0.0000 |

| | |
|---|---|
| Average molecular weight | 57.0520 |

Figure 14. Pool Evaporation Model Inputs

The model can perform evaporation on multi-component mixtures. In this case, the evolution of any dissolved nitrogen is not of concern. The liquid rate has been imported from the Vessels Containing 2-Phases model, and the source concentration has been set to pure methyl isocyanate.

Based on the inputs, the model will provide a conservative estimate for the evaporation of methyl isocyanate from the formed pool.

| Time. min | OVERALL POOL EVAPORATION RATE. lb/min | Pool diameter. ft | Pool surface temperature. C |
|---|---|---|---|
| 0.017 | 0.000 | 3.146 | 4.209 |
| 0.017 | 0.258 | 3.146 | -5.182 |
| 0.017 | 0.257 | 3.146 | -5.235 |
| 0.017 | 0.257 | 3.147 | -5.287 |
| 0.017 | 0.256 | 3.147 | -5.339 |
| 0.018 | 0.257 | 3.147 | -5.389 |
| 0.028 | 0.249 | 3.171 | -5.439 |
| 0.038 | 0.242 | 3.241 | -7.525 |
| 0.049 | 0.252 | 3.371 | -8.514 |
| 0.059 | 0.273 | 3.553 | -8.996 |
| 0.070 | 0.302 | 3.765 | -9.216 |
| 0.080 | 0.335 | 3.988 | -9.306 |
| 0.090 | 0.371 | 4.212 | -9.334 |
| 0.101 | 0.409 | 4.431 | -9.335 |
| 0.111 | 0.447 | 4.644 | -9.322 |
| 0.122 | 0.485 | 4.849 | -9.305 |

Figure 15. Pool Evaporation Model Output.

Figure 15 details the most critical data points in the pool evaporation model. Understanding these will allow for straightforward input to a dispersion model.

While SuperChems® does have an integrated dispersion model, the author's preference is SAFER Trace® v10.2 (2012). Implementation of pool models into dispersion models is discussed later in this paper.

### *4.3 Leaks from Piping*

For lengths of piping with isolation, it may be appropriate to treat the isolated volume as a vessel and analyze the varying leak rate over time. This section considers the situation where the pressure differential across the orifice may be assumed constant. In addition to the previous source terms, SuperChems® is utilized in determining the source term.

The "Source Term Wizard" can be employed to specify an upstream pressure and temperature to determine the flow rate across a specified piping layout or idealized nozzle. The example here will determine the release rate of methyl isocyanate pressurized to 50 psig and released through a guillotine break in the 2" segment of piping from Figure 9.

| OPTIONS | |
|---|---|
| Liquid Density | Databank |
| PVT Behavior | Equation of state |
| VLE Search Direction | Dew point |

☐ **Use mass basis data to compute fractions**
☐ **Mixture is soluble in water**
☐ **Use an existing mixture**

| COMPOUND | FORMULA |
|---|---|
| METHYL ISOCYANATE | C2H3NO |

| | |
|---|---|
| Inlet temperature. C | 25.00000 |
| Inlet pressure. psig | 50.00000 |
| Back pressure. psig | 0.00000 |

☑ **Initial state is subcooled liquid**

Figure 16. Liquid Source Term Wizard Inputs

The release is treated as a subcooled liquid in order to allow the model to estimate if any portion of the release will flash across the opening. It is not expected to have any flashing due to the release conditions in comparison to the chemical's vapor pressure at 50 psig. However, this model is employed for consistency.

| ** INLET CONDITIONS | |
|---|---|
| Inlet temperature. C | 25.0 |
| Inlet pressure. psig | 50.0 |
| Surroundings/back pressure. psig | -3.974E-06 |
| Slip ratio exponent eta | 0.001 |
| Vapor quality | 1.E-04 |

| ** EXIT CONDITIONS | |
|---|---|
| Exit temperature. C | 25.160 |
| Exit pressure. psig | -1.7331E-04 |
| Barometric pressure at site elevation. psig | -3.974E-06 |
| Mixture exit velocity. ft/s | 21.213 |
| Exit Mach Number | 0.006 |
| Maximum Mach Number | 0.006 |
| Exit rho*u*u in kg/m/s2 or in Pascals | 38695.90 |
| Maximum rho*u*u in kg/m/s2 or in Pascals | 38695.90 |
| Exit steady state impulse. lbf | 68.146 |
| Maximum steady state impulse difference. lbf | 418.675 |
| Pipe length travelled / total piping length | 0.998 |
| Flow rate. lb/min | 1714.327 |

| COMPONENT | MW | TOTAL FLOW. lb/min | LIQUID FLOW IN. lb/min | LIQUID FLOW OUT. lb/min | VAPOR FLOW IN. lb/min | VAPOR FLOW OUT. lb/min |
|---|---|---|---|---|---|---|
| METHYL ISOCYANATE | 57.052 | 1713.817 | 1713.817 | 1713.817 | 0 | 0 |

Figure 17. Source Term Wizard Outputs.

The results from this model include liquid and vapor release rates as well as exit pressure and temperature. Had flow been choked in the piping, the exit pressure would be above atmospheric. As with previous models, it is recommended to utilize a flash block to determine conditions at atmospheric pressure for releases from systems with choked flow.

For releases from holes or cracks in piping, the piping layout can be developed with an orifice at the end of the layout. The orifice area would be equal to the area of the hole, and the orifice length would be equal to the width of the piping. In addition, due to constraints with the software, a short pipe segment of with the same diameter as the leak should be placed at the exit of the segment.

The results for the above cases can directly be input to a dispersion model, as described in the next section.

## 5. Application of Source Terms to Dispersion Models

This section will describe the inputs to dispersion models based on the results from the two-phase leak from the sour stream.

Section 5 will review how to utilize the data from the dispersion models in order to assess the hazard posed to a hypothetical shelter 500 ft downwind from the release.

### *5.1 Test for Jetting*

In two-phase or vapor release cases, the linear velocity across an opening may be sufficient to cause jetting. As described previously, there are specific dispersion models for jetting conditions. The rule of thumb for jetting regime is 10 m/s [1]. A simple ideal gas calculation can provide an estimate for linear velocity, given a known release rate, hole size and temperature. This can be expanded to a real gas calculation for highly non-ideal cases. The temperature used in this calculation should be the flashed temperature to atmospheric pressure. The rate employed should be that of the vapor phase for two-phase releases as this will be the largest volumetric component once flashed to atmosphere.

In this example, a 200 psig stream of sour hydrocarbons at 150 deg C is released across a 1-inch hole in 6-inch piping. While several VLE models are available for hydrocarbon mixtures, ideality is assumed in this example.

| | |
|---|---|
| Piping Isometric / Layout | 6-INCH LINE WITH 1-INCH CRACK |
| Site Name | DEFAULT |
| Liquid Density Method | Databank |
| VLE/PVT Method | Equation of state |
| ** INLET CONDITIONS | |
| Inlet temperature. C | 150.691 |
| Inlet pressure. psig | 200.0 |
| Surroundings/back pressure. psig | -3.974E-06 |
| Slip ratio exponent eta | 0.001 |
| ** EXIT CONDITIONS | |
| Exit temperature. C | 133.870 |
| Exit pressure. psig | 138.432 |
| Barometric pressure at site elevation. psig | -3.974E-06 |
| Mixture exit velocity. ft/s | 190.056 |
| Exit Mach Number | 1.0 |
| Maximum Mach Number | 1.0 |
| Exit rho*u*u in kg/m/s2 or in Pascals | 393239.785 |
| Maximum rho*u*u in kg/m/s2 or in Pascals | 393239.785 |
| Exit steady state impulse. lbf | 165.061 |
| Maximum steady state impulse difference. lbf | 6037.864 |
| Pipe length travelled / total piping length | 0.998 |
| Flow rate. lb/min | 455.119 |

| COMPONENT | MW | TOTAL FLOW. lb/min | LIQUID FLOW IN. lb/min | LIQUID FLOW OUT. lb/min | VAPOR FLOW IN. lb/min | VAPOR FLOW OUT. lb/min |
|---|---|---|---|---|---|---|
| HYDROGEN SULFIDE | 34.082 | 2.052 | 2.052 | 0.866 | 0.001 | 1.186 |
| n-PENTANE | 72.150 | 347.584 | 347.547 | 268.991 | 0.037 | 78.593 |
| HEXANE | 86.175 | 77.840 | 77.835 | 66.892 | 0.005 | 10.949 |
| n-OCTANE | 114.231 | 27.515 | 27.515 | 26.078 | 0.001 | 1.438 |

Figure 18. Flow of stream across 1-inch crack.

The exit conditions show that the flow has choked. As previously discussed, the "2-Phase Expansion and Aerosol Model" will be employed to determine the composition and temperature at atmospheric pressure. In addition, this model will report the linear velocity and cloud diameter when flashed to atmospheric pressure. The linear velocity can be used to determine if jetting is possible.

**TWO-PHASE EXPANSION / DEFAULT

| | INITIAL | FINAL |
|---|---|---|
| Flow diameter. ft | 8.3333E-02 | 2.3754E-01 |
| Velocity. ft/s | 190.42 | 651.59 |
| Vapor density. lb/gal | 0.24 | 0.02 |
| Liquid density. lb/gal | 4.20 | 5.19 |
| Mixture density. lb/gal | 0.98 | 0.04 |
| Temperature. C | 133.87 | 45.56 |
| Pressure. psig | 138.43 | 0.00 |
| Vapor/Feed (V/F) mole fraction | 0.21 | 0.71 |
| Mass flow rate. lb/min | 455.12 | 455.12 |
| Available energy. BTU/lb | 26.45 | |
| Modified Weber number for thermal breakup | 392.30 | |
| Critical Weber number for mechanical breakup | 12.01 | |
| Viscosity group number | 0.01 | |
| Stable droplet diameter. Microns | 39.50 | |
| Thermal breakup droplet diameter. Microns | 39.50 | |
| Mechanical breakup droplet diameter. Microns | 44.18 | |
| Number of droplets | 4.9421E+10 | |
| Droplet mass. lb | 4.4271E-11 | |

| COMPONENT | TOTAL FLOW | INITIAL LIQ | FINAL LIQ | INITIAL VAP | FINAL VAP |
|---|---|---|---|---|---|
| HYDROGEN SULFIDE. lb/min | 2.053 | 0.866 | 0.038 | 1.186 | 2.015 |
| n-PENTANE. lb/min | 347.681 | 269.076 | 80.158 | 78.605 | 267.523 |
| HEXANE. lb/min | 77.862 | 66.912 | 35.926 | 10.950 | 41.936 |
| n-OCTANE. lb/min | 27.523 | 26.085 | 23.812 | 1.438 | 3.711 |

Figure 19. Sour Stream Contents after Expansion.

Per Figure 19, the temperature of the release cloud will be 46 deg C at atmospheric pressure and will have a linear velocity of 652 ft/s (199 m/s). This is an order of magnitude above the 10 m/s guidance, and should result in significant jetting.

### *5.2 Input to Jet Model*

The dispersion modeling tool utilized for these examples is SAFER Trace® 10.2 (2012).

Figure 20 details the inputs to the jet model based on our example. This is focusing on the vapor portion of the model, ignoring liquid entrainment.

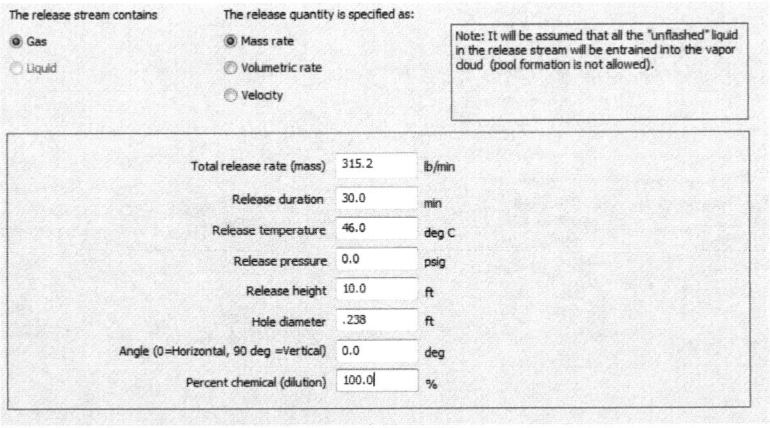


Figure 20. Jet Model Inputs.

The dispersion outputs will report concentrations and plots related to concentration of H2S in air.

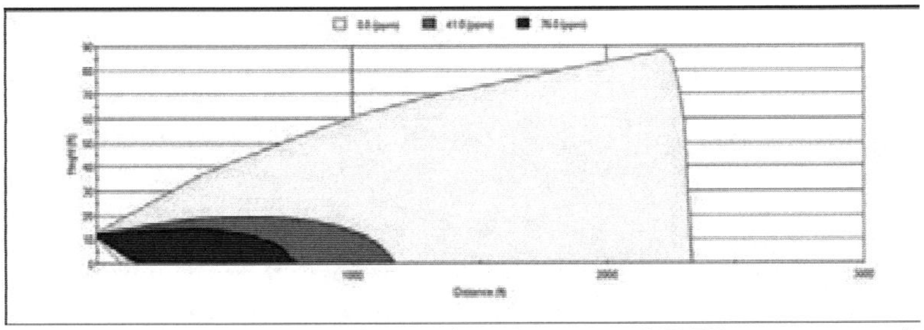

Figure 21. Jet Model Dispersion Output. AEGL concentration limits based on 10-minute exposure to H2S [10].

The H2S concentration profile at the outside of a shelter 500 ft downwind from the release point is shown in Figure 22.

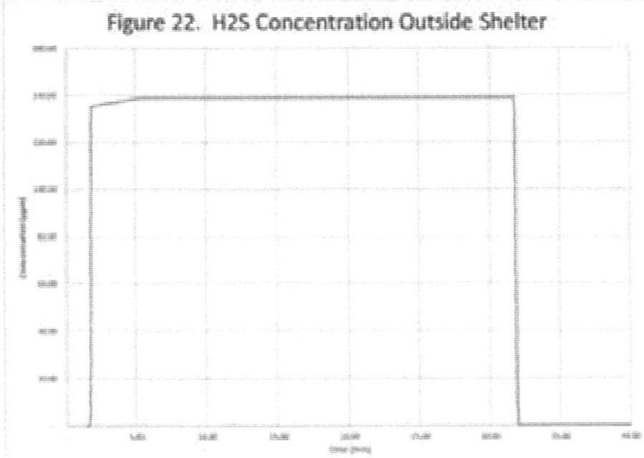


Consequence to personnel in this shelter will be analyzed using a probit function, detailed in Section 5.

### *5.3 Modeling Pool Evaporation*

The liquid portion of the example two-phase release will be treated as the source of a liquid pool.

| | |
|---|---|
| Spill Surface | DEFAULT |
| Liquid discharge temperature. C | 45.560 <<< |
| Liquid discharge rate. lb/min | 139.934 |
| Liquid discharge duration. min | 15.000 |
| ☑ **Use default scenario mixture** | |
| ☐ **Use mass fractions** | |
| COMPONENT | USER FRACTION |
| HYDROGEN SULFIDE | 0.0006 |
| n-PENTANE | 0.6394 |
| HEXANE | 0.2399 |
| n-OCTANE | 0.1200 |
| Average molecular weight | 80.5399 |
| RUN OPTIONS | |

Figure 23. Liquid Pool Model Inputs

The concentration profile of evaporate is highly dynamic over the studied period of the release.

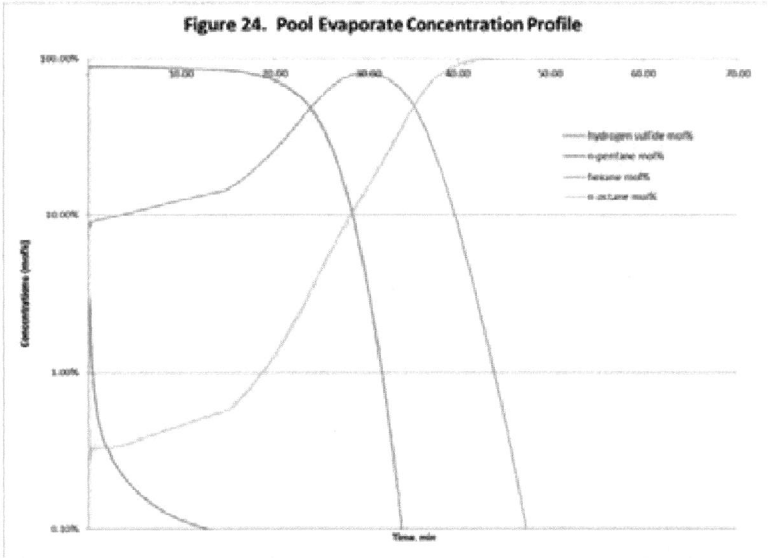
**Figure 24. Pool Evaporate Concentration Profile**

While the concentration is highly dynamic, the main toxicity concern is with H2S. The majority of H2S appears to evaporate in the first 15 minutes. An average rate will be developed based on the total evaporated mass of each component over that time. The effective concentrations and dispersion parameters are shown in Table 4.

**Table 4. Molar Concentrations and Dispersion Parameters**

| Chemical | Mol% |
|---|---|
| Hydrogen Sulfide | 0.1% |
| Pentane | 87.0% |
| Hexane | 12.4% |
| Octane | 0.5% |
| | |
| Average Evaporation Rate | 57.0 lb/min |
| Temperature | 2°C |
| Duration | 15 min |

The data from this model is input to the SAFER Trace® "Single Phase" steady state model, per Figure 25.

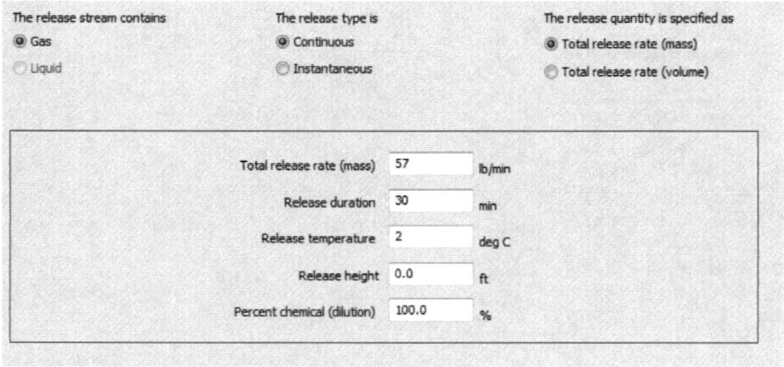


Figure 25. Steady State Model Inputs.

The downwind dispersion of the cloud, focusing on H2S concentrations in air is shown in Figure 26. The concentration is negligible at the 500-ft shelter, so the contribution from evaporation to toxic impact can be ignored.

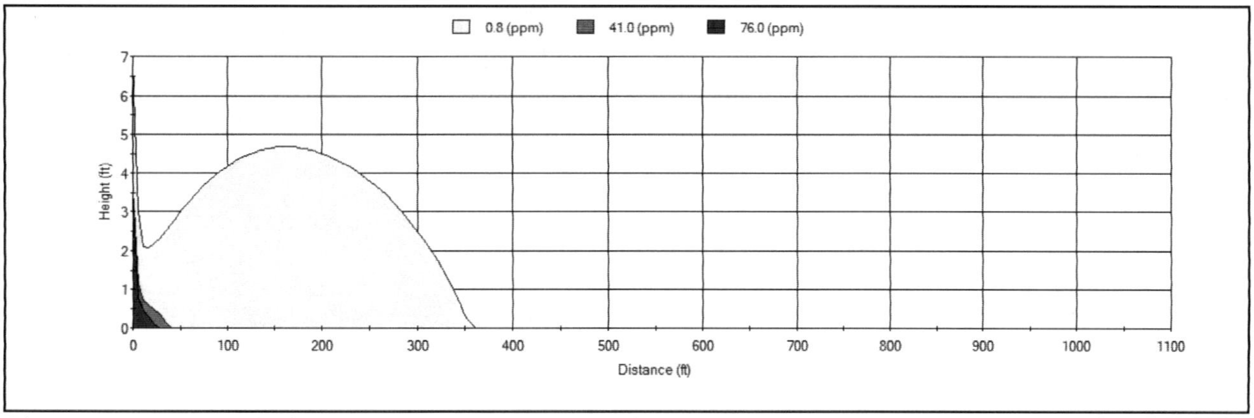


Figure 26. Downwind Impact from Evaporate Dispersion, based on H2S Concentration in Air.

# 6. Magnitude of Consequence Determination Using Probits

The methods utilized here are described in detail in "Simplified Methods of Using Probit Analysis in Consequence Analysis" [1].

To summarize, the impact to individuals may calculated using dose [5].

$$Dose = \int C^n dt \qquad \text{[Eq. 1]}$$

Dose may be transformed to probability of fatality using a probit transformation [5].

$$Y = A + B \log(Dose) \qquad \text{[Eq. 2]}$$

The constants "A", "B", and "n" are specific to the chemical. The Y value, or probit score is referenced on a probit table to determine probability of studied consequence.

**Table 5. Probit table.**

| % | 0 | 1 | 2 | 3 | 4 | 5 | 6 | 7 | 8 | 9 |
|---|---|---|---|---|---|---|---|---|---|---|
| 0 | — | 2.67 | 2.95 | 3.12 | 3.25 | 3.36 | 3.45 | 3.52 | 3.59 | 3.66 |
| 10 | 3.72 | 3.77 | 3.82 | 3.87 | 3.92 | 3.96 | 4.01 | 4.05 | 4.08 | 4.12 |
| 20 | 4.16 | 4.19 | 4.23 | 4.26 | 4.29 | 4.33 | 4.36 | 4.39 | 4.42 | 4.45 |
| 30 | 4.48 | 4.50 | 4.53 | 4.56 | 4.59 | 4.61 | 4.64 | 4.67 | 4.69 | 4.72 |
| 40 | 4.75 | 4.77 | 4.80 | 4.82 | 4.85 | 4.87 | 4.90 | 4.92 | 4.95 | 4.97 |
| 50 | 5.00 | 5.03 | 5.05 | 5.08 | 5.10 | 5.13 | 5.15 | 5.18 | 5.20 | 5.23 |
| 60 | 5.25 | 5.28 | 5.31 | 5.33 | 5.36 | 5.39 | 5.41 | 5.44 | 5.47 | 5.50 |
| 70 | 5.52 | 5.55 | 5.58 | 5.61 | 5.64 | 5.67 | 5.71 | 5.74 | 5.77 | 5.81 |
| 80 | 5.84 | 5.88 | 5.92 | 5.95 | 5.99 | 6.04 | 6.08 | 6.13 | 6.18 | 6.23 |
| 90 | 6.28 | 6.34 | 6.41 | 6.48 | 6.55 | 6.64 | 6.75 | 6.88 | 7.05 | 7.33 |
| — | 0.0 | 0.1 | 0.2 | 0.3 | 0.4 | 0.5 | 0.6 | 0.7 | 0.8 | 0.9 |
| 99 | 7.33 | 7.37 | 7.41 | 7.46 | 7.51 | 7.58 | 7.65 | 7.75 | 7.88 | 8.09 |

Source: Finney, 1952 [11]

The Microsoft Excel (2016) function "NORMSDIST" will convert the probit score to a probability.

The Equation 1 is combined with the "NORMSDIST" function to determine impact to occupants of the shelter, per Table 6. This is assuming that the air handling provides 10 air changes per hour to the shelter.

Table 6. Probability of Consequence to Sheltered Occupants.

| Time (min) | Dosage ($ppm^{1.43}$-min) | % Probability |
|---|---|---|
| 1.45 | | 0% |
| 1.46 | - | 0% |
| 1.47 | - | 0% |
| 1.48 | - | 0% |
| 1.49 | - | 0% |
| 1.50 | - | 0% |
| ... | ... | ... |
| 57.00 | 44,794 | 0% |
| 58.00 | 45,354 | 0% |
| 59.00 | 45,902 | 0% |
| 60.00 | 46,436 | 0% |
| 61.00 | 46,957 | 0% |
| 62.00 | 47,466 | 0% |
| 63.00 | 47,963 | 0% |
| 64.00 | 48,448 | 0% |
| 65.00 | 48,922 | 0% |

In this example, the occupants can shelter for over an hour with 0% probability of consequence. Probit transformations can be utilized to study probability of fatal exposure from toxicity, thermal radiation, and blast overpressure. In addition, they can analyze probability of burn severity.

## 7. Conclusion

This paper has summarized the methods utilized in determining impact to personnel from toxic exposure and vapor cloud explosions. The covered models include pipe breach, tank release, and jetting models. These are highly utilized models for consequence analysis. In addition, the paper summarized methods to utilize dispersion modeling data to fully quantify impact to occupants of shelters downwind from releases.

For blast modeling, this paper provided an overview of one simplified method to predict explosion overpressure, impulse, and damage to buildings within the radius affected by the explosion. The method utilized is Baker-Strehlow-Tang (BST), one of several methods engaged by industry for this purpose. Most of the software tools used to perform the analysis were developed by the BakerRisk Engineering Consultant Explosion Research Cooperative, including VCloud® and BEAST®.

For all consequence models, more rigorous methods such as computational fluid dynamic (CFD) modeling will produce more accurate results. However, the methods discussed in this paper are excellent screening tools whose value lies in their ease of use and are suitable for risk analysis.

## References


[1] Baybutt, Paul, "Using Layers of Protection Analysis to Evaluate Fire and Gas Systems", Process Safety Progress, Volume 31, Issue 3, 2011

[2] "SAFER Trace® Online Help", SAFER Systems, 2003

[3] James, Michael D., "Consequence Modeling of Dynamic Source Terms", Process Safety Progress, Volume 35, Issue 4, 2016

[4] Baker, Q. A., Tang, M. J., E. A. Scheier, and G. J. Silva, "Vapor Cloud Explosion Analysis," Process Safety Progress, Vol. 15, No. 2, 1996, pp. 106-109

[5] A. Sari, "Comparison of TNO Multienergy and Baker-Strehlow-Tang Models," Process Safety Progress 30 (2011), pp. 23-26

[6] BEAST® Help File, 2000 BEAST® Report, 2000

[7] "VCE Evaluation Method Comparison," The Quest Quarterly, Volume 4, Issue 1, Spring 1999

[8] Baker-Strehlow-Tang Methodology, Guidelines for Evaluating the Characteristics of Vapor Cloud Explosions, Flash Fires, and BLEVEs, CCPS, New York, 1994

[9] Pierorazio, A.J., J.K. Thomas, Q.A. Baker and D.E. Ketchum (2005) "An Update to the Baker-Strehlow-Tang Vapor Cloud Explosion Prediction Methodology Flame Speed Table," Process Safety Progress, 24(1): 59-65

[10] "About Acute Exposure Guideline Levels (AEGLs)", 2016, https://www.epa.gov/aegl/about-acute-exposure-guideline-levels-aegls (Accessed Mar 2, 2017)

[11] D. J. Finney, *Probit Analysis*, Cambridge University Press, New York 1952

# Relief Systems Design: The Importance of Accounting for Static Head

**Nicholas N. Cristea**
**Smith & Burgess, LLC**
**7600 W Tidwell Rd, Ste 600, Houston, TX 77040**
**Nicholas.cristea@smithburgess.com**

**Zubin Kumana**
**Smith & Burgess, LLC**
**7600 W Tidwell Rd, Ste 600, Houston, TX 77040**
**Zubin.kumana@smithburgess.com**

**Guomin Wang**
**Smith & Burgess, LLC**
**7600 W Tidwell Rd, Ste 600, Houston, TX 77040**
**Guomin.wang@smithburgess.com**



Prepared for Presentation at
American Institute of Chemical Engineers
2017 Spring Meeting
13th Global Congress on Process Safety
San Antonio, Texas
March 26-29, 2017

# Relief Systems Design: The Importance of Accounting for Static Head

**Nicholas N. Cristea**
**Smith & Burgess, LLC**
**7600 W Tidwell Rd, Ste 600, Houston, TX 77040**
Nicholas.cristea@smithburgess.com

**Zubin Kumana**
**Smith & Burgess, LLC**

**Guomin Wang**
**Smith & Burgess, LLC**

**Keywords:** Relief Systems Design, Static Head, Upstream, Downstream

## Abstract

The process of evaluating an entire facility's relief and flare system design can be time consuming and costly. During the process, there will be systems to be evaluated that have liquid scenarios that may be protecting other assets within the facility. If the static head for the liquid scenarios is not accounted for, then the analysis can become even more costly due to the analysis not being performed correctly the first time. The result could be not meeting code by not adequately protecting system. This paper will look at two different cases where the static head was not accounted for, and how the issue was mitigated in the most cost-effective way. First, when dealing with the liquid heading on an offshore platform, there are many different levels on the platform and ensuring that the static head is accounted for is important to protect the different assets in the liquid handling train. Second, when evaluating a column system, it is important to ensure that the relief device provides protection for every ASME Section VIII Div 1 stamped equipment within the system. These two case studies highlight the necessity of accounting for static head in relief systems design.

## 1 Introduction

In the process safety industry, relief devices are set such that at least one relief device set pressure is equal to the lowest system MAWP within the protected system, to meet the requirement of ASME Section VIII Div 1 to provide adequate protection against overpressure. ASME Section VIII Div 1 UG-125 states "All pressure vessels within the scope of this Division, irrespective of size or pressure, shall be provided with overpressure protection in accordance with the requirements of UG-125 through UG-138, or with overpressure protection by system design in accordance with requirements of UG-140, or a combination of the two". ASME Section VIII Division VIII UG-125(c) and (1) also state that "when a pressure relief device is provided, it shall prevent the pressure from rising more than 10% or 3 psi (20 kpa), whichever is greater, above maximum allowable working pressure" and "when multiple pressure relief devices are provided and set in accordance with UG-134(a), they shall prevent the pressure from rising more than 16% or 4 psi (30 kpa), whichever is greater, above the maximum allowable working pressure". The vessel is normally hydrotested without accounting for the potential impact of inlet piping to the relief devices. Therefore, when the elevation difference between the top of the vessel and the relief device becomes significant [2], the vessel may experience a pressure higher than its allowable accumulation due to the static head, especially during liquid relief scenarios. The equation below is a conversion from the liquid height to the equivalent liquid head (psig).

$$P = 0.433\ x\ SG\ x\ H$$

P: liquid head in psi

SG: Specific Gravity

H: liquid head in feet

***Equation 1: Liquid height to pressure conversion***

This paper will look at two different cases where the static head was not accounted for and how the issue was mitigated for each case. The first case will look at a blocked system liquid outlet scenario on a platform where the vessel of interest has the same MAWP as the protecting vessel on an upper level. The second case will look at the liquid relief case (i.e. overfilling) and the mitigation for a column system that contains several vessels. These two cases will show the importance of accounting for static head and the potential mitigation options for these types of concerns.

# 2 Case 1 – Produced Water Desanders and Accumulators on an Offshore Platform

## *2.1 System overview*

Figure 1 below is an illustration of case study 1. A produced water desander on the upper deck of a platform is fed by a booster pump. The desander then feeds the desander accumulator. The desander and the accumulator both have the same MAWP of 350 psig. The maximum discharge pressure of the booster pump is 415 psig. In the event of a blocked outlet, the relief device PSV-1 (4" P 6") on the desander is intended to provide adequate protection for the desander and the downstream accumulator. The relief device PSV-2 (1" E 2") on the accumulator is intended only for the fire scenario, based on the assumption that the upstream relief device (PSV-1) would limit the maximum deliverable pressure to the accumulator to 350 psig for the blocked outlet scenario. However, the vessels are located on different levels of the platform. The elevation difference was never accounted for in the choice of relief device set pressure. When PSV-1 is relieving at 385 psig, the pressure inside the accumulator would be the relief pressure (385 psig) plus the static head due to the difference in elevation. If the difference in elevation is large, the downstream accumulator may no longer be adequately protected by PSV-1, because the vessel would experience a pressure higher than its allowable accumulation. The elevation difference between the two relief devices illustrated in Figure 1 is roughly 27 feet. The static head from the elevation difference is about 12 psig. This means that when the desander is at the relief pressure, the pressure in the accumulator would be 397 psig, which exceeds the 10% allowable accumulation [2] (385 psig) of the accumulator.

### *2.2 Case 1 mitigation*

The common mitigation option to account for the static head is to lower the relief device set pressure on the desander to account for the static head, so that the downstream accumulator would be adequately protected during the liquid relief event. However, for this particular case study, a more detailed engineering analysis revealed that the installation was acceptable as-is.

However, it was determined that there are technically two relief devices protecting the system in the event a blocked outlet on the accumulator. The allowable accumulation for the accumulator can go up to 16% of the system (406 psig) because of the multiple relief devices. Since the manual valve between the two systems will remain open (the initiating event is the closure of the downstream block valve), the accumulator is still protected. If the manual valve between the two systems were to be closed, then the blocked outlet case on the accumulator would no longer be applicable as the source of pressure would be lost as well.

If the static head is greater than 21 psig (~49 feet), the accumulator may experience overpressure in excess of the allowable (a pressure of 406 psig). The potential mitigation would be to lower the set pressure of the desander relief device to account for the static head such that the maximum pressure did not exceed this limit. For this case study, PSV-1 set pressure is not required to be lowered because the accumulator is still within the 16% allowable accumulation after accounting for the static head. Additionally, if the relief device set pressure were lowered, the potential exists that the relief device may be undersized, and additional mitigations may be required to resolve the concerns.

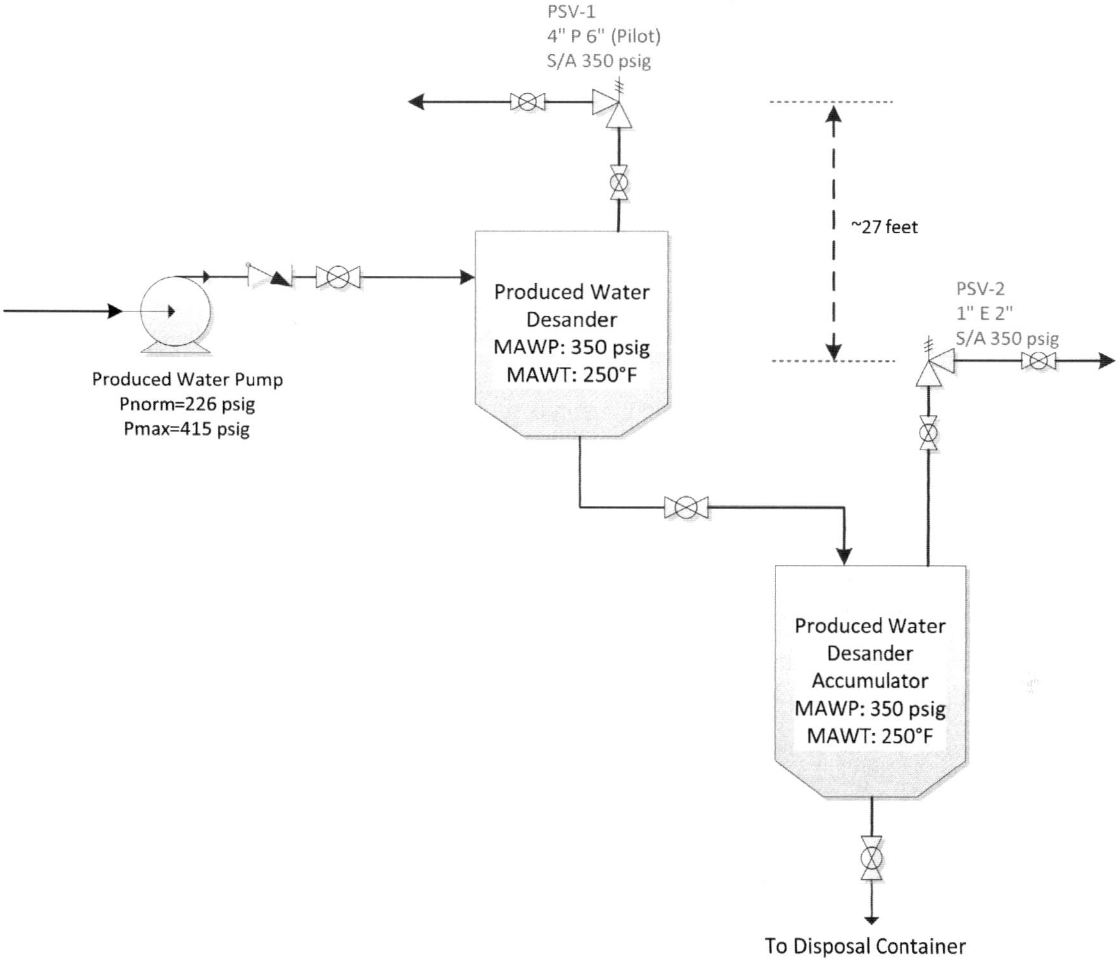


***Figure 1, Case 1 – Produced Water System Overview***

# 3 Case 2 – Common Column System

## *3.1 System Overview*

Figure 2 below is a simplified column system that contains several pieces of equipment. The top of column is 50 feet from grade. The reboiler is located at grade. The relief device (PSV-3) is 5 feet above the top of the column and set at 330 psig. The lowest system MAWP is 330 psig based on the MAWPs of the column and the reboiler (shell side). This case study only looks at the "overfilling" scenario for the column. As the lowest system MAWP and the set pressure are the same, it may appear at first glance that the relief device is set correctly to protect the system. However, during the overfilling scenario, there are 55 feet of static head on top of the reboiler (assuming the reboiler is flooded). Therefore, the pressure that the reboiler (shell side) experiences during the liquid relieving event is higher than the allowable accumulation. For this case study, the specific gravity of the fluid is assumed to be 0.7, which would result in roughly 17 psig of static head. This means that during the relief event, the pressure in the reboiler (shell side) is 380 psig (~15% accumulation), which exceeds the 10% allowable accumulation. Therefore, mitigations may be required as the reboiler is not protected due to the static head.

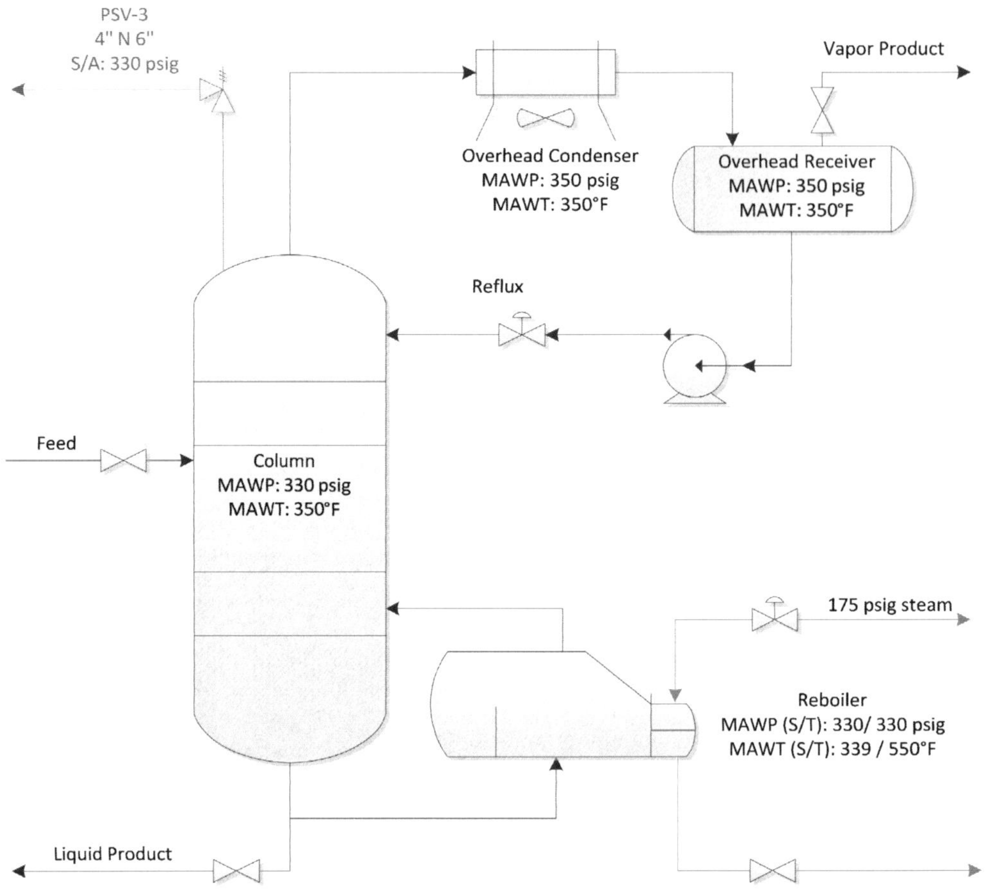


***Figure 2, Case 2 - Common Column System Overview***

### *3.2 Case 2 mitigation*

There are multiple mitigation options for this type of concern. One mitigation option is to re-rate the reboiler to a higher MAWP to account for the static head. Another mitigation is to lower the set pressure of the relief device to account for the static head. Therefore, the new set pressure for this case study can be 314.5 psig or lower if the reboiler MAWP plus allowable accumulation is not exceeded by the relief pressure plus static head. When the relief device set pressure is lowered to account for the static head during the liquid relief events, engineers must confirm the relief device can still provide adequate capacity at the reduced pressure for all applicable overpressure scenarios. Also, the operating pressure of the column needs to be considered to ensure that the lower set pressure would not result in simmering of the relief device during normal operation. In practice, many plants would also have high independent liquid level alarms and fill time criteria to remove the potential liquid relief scenarios by taking credit for operator responses. Another option is to look at risk assessing the concern as the MAWP is not exceeded by 16% accumulation.

There are many ways to mitigate this concern; this case shows one of the most common mitigation option.

# 4 Conclusion

The two case studies discussed in this paper show the need to consider the static head when designing the relief system. The issue of static head should be considered during the design phase to ensure the system is code compliant. The cases studies show that the system may not be protected by code, but may be able to be risk accepted based on what the resulting pressure at the vessel is question becomes. It is important during the design phase to be cognizant on the elevation of the systems and the protecting relief devices to ensure the static head has been considered. This can be especially true when designing an offshore platform with multiple levels. Also, when commissioning columns or other vessels, it is important to consider the MAWP of all the vessels in the system to ensure all aspects of the system are protected by the relief device. The case studies give a brief overview of some of the possible mitigation options for static head concerns. It is the responsibility of the engineer reviewing the information to select the option that best fits the facility's needs.

# 5 References


[1] ASME Section VIII, "Boiler and Pressure Vessel Code," American Society of Mechanical Engineers, 2013.

[2] ANSI/API Standard 520 Part 2, "Sizing, Selection, and Installation of Pressure-relieving Devices," American Petroleum Institute, 6th ed., 2015.

[3] ANSI/API Standard 521, "Pressure-relieving and Depressuring Systems," American Petroleum Institute, 6th ed., 2014.

# 478548 Study of Factors That Affect Dust Explosion Characteristics of Fibrous Dust Materials

**Abstract:**

*Abstract for 13th Global Congress on Process Safety*

*Track: Annual Loss Prevention Symposium (LPS)*

**Study of factors that affect dust explosion characteristics of fibrous dust materials**

Bharatvaaj Ravi, Yan-Ru Lin, Chad Mashuga, M. Sam Mannan

Mary Kay O'Connor Process Safety Center

Artie McFerrin Department of Chemical Engineering

Texas A&M University

College Station, TX-77843-3122, U.S.A

+ 1 (774)-262-5215, vaaj26@tamu.edu

Combustible dust hazard is a serious threat to the process industries. Out of the 281 combustible dust incidents reported by the U.S. Chemical Safety and Health Investigation Board (CSB), 66 incidents have been caused by wood dust. In spite of the National Fire Protection Agency's (NFPA) efforts in specifically catering to the wood working and wood-based products manufacturing industries and Occupational Safety and Health Administration's (OSHA) National Emphasis Program on combustible dust hazards, to create awareness and minimize wood dust hazards, the wood dust incidents continue to occur. The fundamental difference between traditional dusts and wood dusts is that the latter are flocculent and fibrous. They are characterized effectively by an aspect ratio (length to diameter ratio) rather than a single dimension, such as diameter, as in the case of traditional - spherical or near spherical - dusts. Various studies in the literature have revealed that the standard dust explosion testing procedures and the categorization of these non-traditional dust based on particle diameter (mesh sizes) might not be valid. This research is concerned with similar line of thought and exploring the effect of aspect ratio on the time of suspension of these fibrous wood dusts. The likelihood of ignition increases with the time of suspension given other conditions to form a combustible dust hazard. The study also deals with characterizing the participating and non-participating portions when a dust explosion occurs through analysis of samples before and after explosion testing. Does the dust concentration give the entire picture of explosion characteristics such as maximum explosion pressure and explosion pressure rise rate? Or is there a better indicator of these characteristics when it comes to flocculent dust? These are the questions that the current research is trying to answer.

**Key words:** Combustible dust, Fibrous dust, Dust explosion testing, Dust concentration, Dust explosion characteristics, Non-traditional dust.

13TH GLOBAL CONGRESS ON PROCESS SAFETY

# High Integrity Pressure Protection System (HIPPS) and Implementation Challenges

**Janardhanan Kallambettu P.Eng.**
**Principal Engineer-Control Systems**
**Bechtel Corporation**
**3000 Post Oak Blvd, Houston, TX 77459**
**jkallamb@bechtel.com**

Prepared for Presentation at
American Institute of Chemical Engineers
2017 Spring Meeting and 13th Global Congress on Process Safety
San Antonio, TX
March 26 – 29, 2017

# High Integrity Pressure Protection System (HIPPS) and Implementation Challenges

**Janardhanan Kallambettu P.Eng.**
**Principal Engineer-Control Systems**
**Bechtel Corporation**
**3000 Post Oak Blvd, Houston, TX 77459**
**jkallamb@bechtel.com**

**Keywords:** PHA, HIPPS, SIS, SIF, SIL, PLC

## Abstract

Over the years, the process industries have heavily utilized pressure relief devices (PRDs) to protect the mechanical integrity of system components. While PRDs offer a straightforward and practical way of controlling overpressure in process plants, they do have some inherent limitations such as impacts to flare system design, environmental concerns, and some physical limitations in PRD sizing. In these cases, where the use of PRDs is limited, High Integrity Pressure Protection System (HIPPS) can be utilized as a substitute for PRDs.

While HIPPS offers a solution to address the PRD limitations, there are some challenges associated with a successful HIPPS implementation. The HIPPS are classified as Safety Instrumented Systems (SIS) and require compliance with the functional safety standard IEC 61511**[1]**. The aim of this paper is to guide the reader with the necessary tools to successfully perform all the safety lifecycle activities of HIPPS, per functional safety standards and codes. Challenges in the various phases of the safety lifecycle, from process hazard analysis (PHA) and Safety Integrity Level (SIL) selection to operation and maintenance of HIPPS, are discussed in this paper. Additionally, this article describes many types of HIPPS, their pros and cons, to choose the right system for a given application. The paper concludes with an example HIPPS to achieve high-level risk reductions and demonstrates the appropriate use of quantitative tools, such as fault tree analysis (FTA), during the HIPPS design process.

## 1 Introduction

HIPPS is an acceptable solution instead of PRDs as a means of overpressure protection for vessels and pipelines. Codes such as ASME Section VIII, Division 1[**2**] and recommended practices and standards such as API Standard 521[**3**], API 17O [**4**] and HSE, U.K [**5**] provide criteria for the usage of HIPPS. The reader is advised to refer to the cited references to understand in detail the selection criteria for HIPPS.

The HIPPS is a safety instrumented function (SIF) implemented in a SIS and, therefore, the codes and the standards referenced above require compliance with the functional safety standards IEC 61508[**6**] and IEC 61511.

The functional safety standards require adherence to a safety lifecycle model. The various safety lifecycle activities are:

- process hazard analysis;
- SIL selection;
- safety requirements specification (SRS);
- SIS design per SRS;
- design verification meeting the target SIL including systematic capability, and architectural constraints (redundancy);
- installation, commissioning, validation, operation, and maintenance per SRS; and
- change management.

The intent of this paper is to expose particularly process engineers and other non-control systems engineers, about specific aspects of the HIPPS in a straightforward language.

## 2 PHA and SIL Selection

Whenever a HIPPS is indicated instead of PRDs, process and process safety engineers assign an SIL 3 target to the HIPPS as supported by Lawley and Kletz's [7] requirement that the reliability is ten times better than that of a relief device. For clean services, the probability of failure on demand (PFD) of a relief device is typically two orders of magnitude, hence the SIL 3 target for a HIPPS. API Standard 521, while emphasizing the need for an appropriate SIL selection process per established risk tolerance criteria, also suggests SIL 3 target for HIPPS in the absence of a documented risk tolerance criteria.

An assumption, however, that the HIPPS's SIL target is always SIL 3 may not be correct in some circumstances. Applicable codes and standards mandate a requirement for a documented PHA to identify and examine all potential overpressure scenarios. The severity of consequences typically ranges from moderate to catastrophic and, based on the owner's risk tolerance criteria, the HIPPS target can vary from SIL 2 to SIL 4. For example, if the likelihood of an initiating event causing the overpressure is 1 in 10 years and the owner's risk criteria for catastrophic severity is 1in 100,000 years, and in the absence of any other independent protection layers or conditional modifiers, the required risk reduction is 10,000, a SIL 4 target.

Designing a SIL 4 HIPPS is a significant challenge and not recommended. The functional safety standards discourage such high-level risk reduction measures by a single SIF. Every effort should be made to remove or reduce the hazard at the source, introduce other risk reduction measures which are not based on the SIS or minimize the likelihood of the causes of the hazardous event. If the above steps do not help, then use of multiple independent HIPPS, each with a lower integrity, to meet the overall risk reduction requirements should be considered. See section 6 for a hybrid HIPPS design solution meeting an overall risk reduction of 10,000 or more.

# 3 Types of HIPPS

Many in the process industry believe that HIPPS always includes 2oo3-voted pressure transmitters (sensors), Triple Modular Redundant (TMR) safety PLC (logic solver), and 1oo2-voted ball valves (final elements). In reality, there are many types of HIPPS available to choose from based on the specific application.

HIPPS can be broadly classified as **Electronic**, **Mechanical** or **Hybrid** based on the technology used for the logic solver as shown in **Figure 1**.

A brief description, pros and cons, and applications for each of the above types are given in the following subsections.

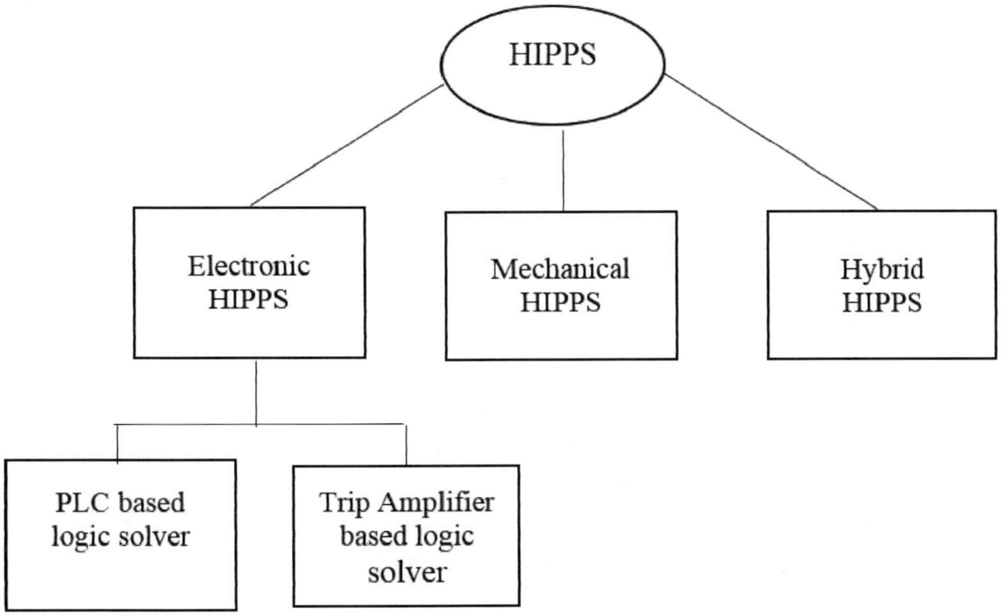


**Figure1. Types of HIPPS**

## *3.1 Electronic-HIPPS*

An electronic-HIPPS uses either a programmable logic controller (PLC) or a trip amplifier system as a logic solver.

### 3.1.1 PLC based HIPPS

There are multiple varieties of PLCs equipped with microprocessor or solid state technology in a multitude of architectures such 1oo2, 1oo2D, 2oo3, 2oo4, and 2oo4D. Typically, all these PLCs will meet SIL 3 reliability. These PLCs are further categorized as complex devices and use application programming languages such as ladder logic or functional block diagrams. Refer to **Figure 2**.

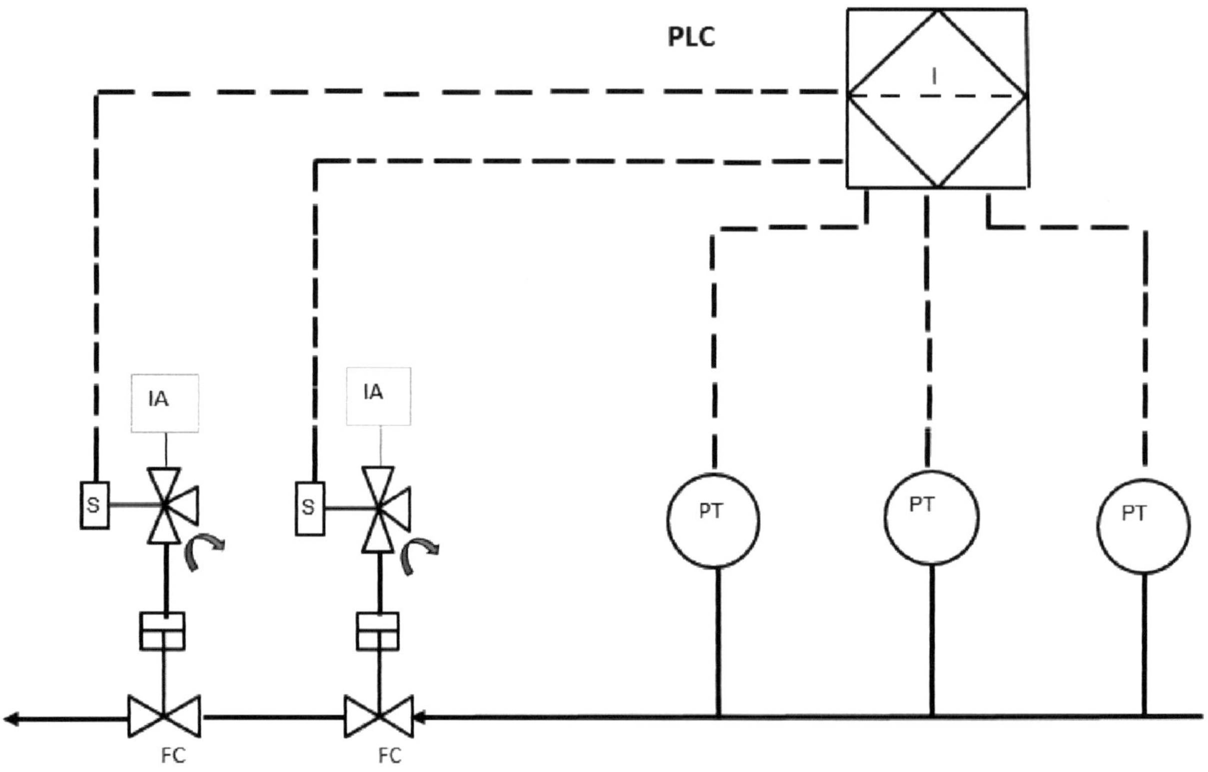


**Figure 2. PLC based HIPPS**

Pros:

- Flexible in developing the logic program and accommodates higher number of field inputs and outputs (I/Os);
- Equipped with special features for application programming such as variable valve closure speeds while closing the valve to mitigate consequences associated with slamming shut.
- Accommodate multiple HIPPS in a single logic solver within a plant facility;
- Facilitates partial stroke testing to meet target SILs or extend the testing interval, when practicable;
- Easily interfaces with the plant operator stations (human machine interfaces, HMIs).

Cons:

- Often needs specialized application programming expertise, in addition to vendor-specific training for selection of hardware and complex control strategies;
- Validation of the vendor claimed SIL against the safety manual is not a trivial exercise;
- Requires protection against cyber security threats;
- Potential for introducing systematic errors in application program;
- Application program using a low variability language (LVL) calls for special considerations per IEC 61508/IEC 61511 standards including application program safety requirements specification;
- Requires rigorous Factory Acceptance Test (FAT);
- Large footprint is needed -typically 800mmX600mmX1000mm panel;

- Extensive installation requirements such as controlled environment and special grounding;
- Requires complex, and redundant large capacity power supply;
- Voluminous documentation is involved;
- Requires a high level of competence in commissioning, validation, maintenance and operation; and
- Expensive for simple HIPPS applications ($50,000 for a logic solver alone).

Applications:

In a large facility, where many HIPPS are to be implemented for different hazardous event scenarios requiring complex functions or features, a PLC based HIPPS is recommended. For example, over pressure protection of multiple vessels, various piping segments, and process equipment within a facility to avoid loading of the flare system.

3.1.2 Trip-Amplifier based HIPPS

Trip-amplifiers are discrete logic solvers of low complexity. A trip amplifier accepts analog 4-20mA input from a field device such as a pressure transmitter and de-energizes contact output(s) when the process value exceeds the trip set point. Each sensor needs a dedicated trip amplifier. Refer to **Figure 3**.

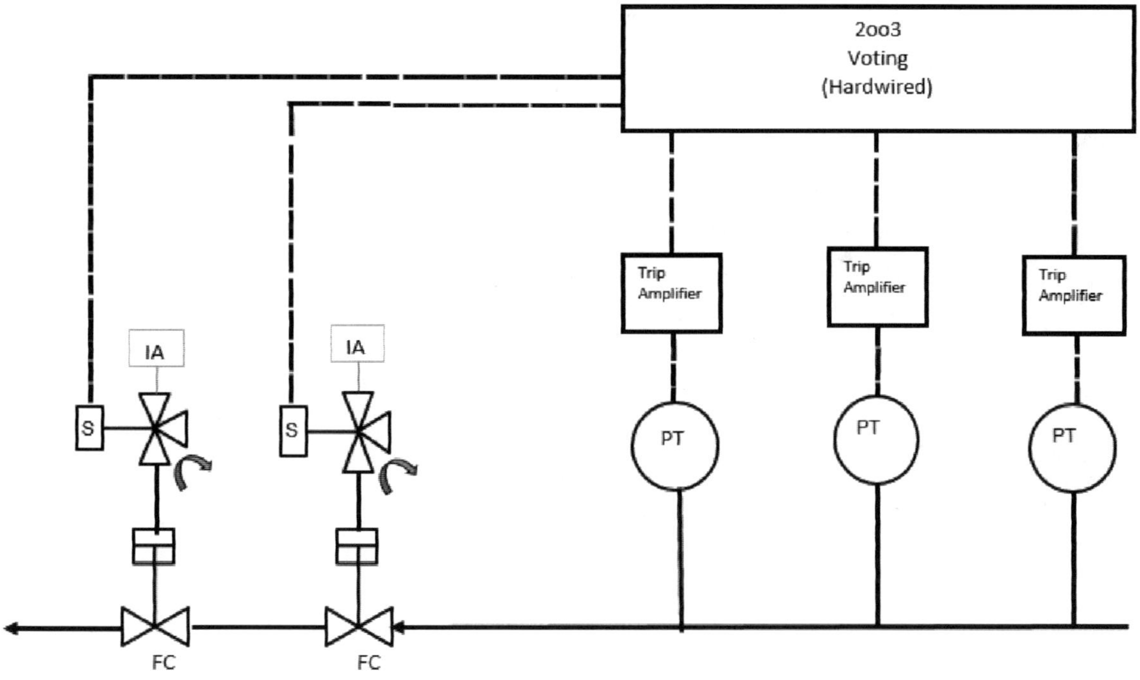


**Figure 3. Trip Amplifier based HIPPS**

Pros:

- Simple, reliable, with low design complexity;
- Meets SIL 2 reliability without redundancy for Trip Amplifiers and SIL 3 reliability with redundant architecture;
- Does not require application programming; classified as a device with fixed programming language (FPL) per IEC 61508/IEC 61511and, therefore, eliminates the need for application program safety requirements specification;
- No cybersecurity related issues;
- Less potential for systematic errors;
- Well suited for simple logic such as HIPPS logic;
- Low space requirement-typically DIN rail mounted;
- Low power device with no special requirements;
- No need for a controlled environment;
- Being a simple device FAT is not required for the logic solver;
- Easy to install and maintain; and
- Low cost; (logic solver range less than $10000)

Cons:

- Requires external devices for complex logic such as variable speed closing times for valves to avoid slam shut scenarios;
- Interfacing with operator stations requires additional I/Os in the plant DCS;
- Difficult to interface and facilitate partial stroke testing;
- Voting among the trip-amplifiers are external to the trip-amplifiers using the trip contacts and physical wiring which is less straightforward; and
- Implementation of maintenance bypasses is not practical;

Applications:

Well suited for applications involving straightforward, simple logic and for facilities which do not use PLCs. Also, appropriate for applications requiring a small footprint, low lifecycle cost and facilities not having multiple HIPPS. For example, remote pipeline compressor stations or storage tanks.

### *3.2 Mechanical-HIPPS*

A mechanical HIPPS uses mechanical initiators. The mechanical initiator not only senses the over pressure but also serves as a logic solver to detect high pressure and actuate the valve. Mechanical HIPPS do not use any electrical or electronic components and, therefore, are inherently safer. Standalone versions of mechanical HIPPS are available which do not require any energy sources, such as instrument air supplies. Refer to **Figure 4**.

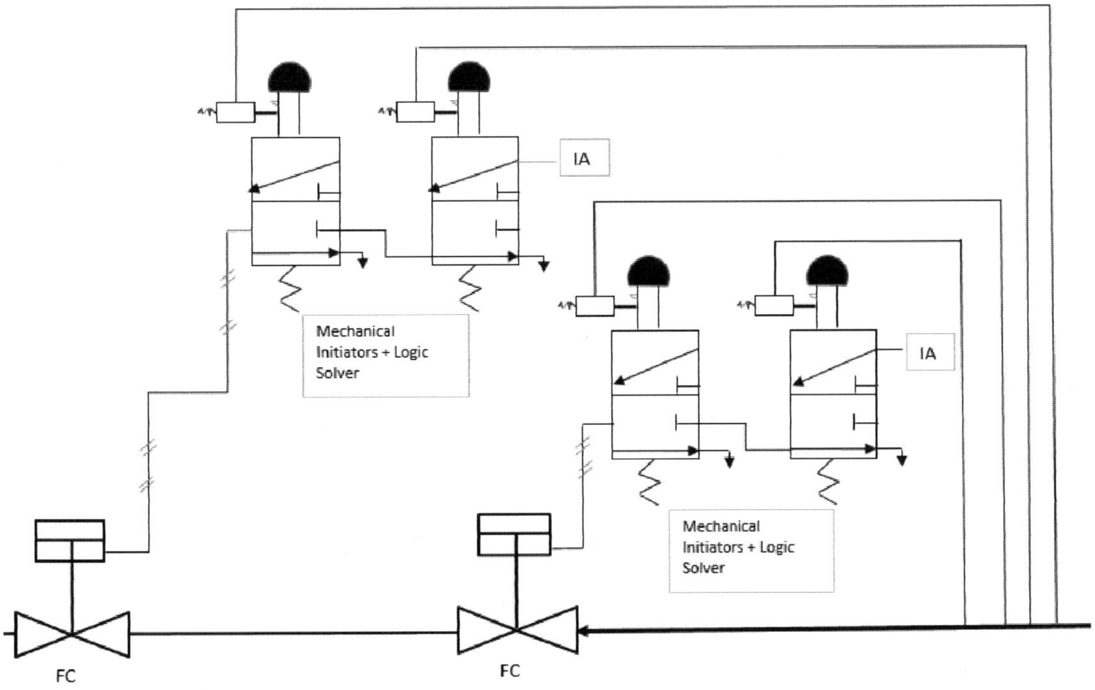


**Figure 4. Mechanical HIPPS**

Pro:

- Simple, reliable and minimally complex;
- Does not involve wiring and programming;
- Less susceptible to systematic errors;
- Cables, cable trays, conduits, etc. are not required;
- Can be made standalone without external energy sources;
- Single valve system can meet SIL 2 targets, redundant valve configuration can even meet SIL 3;
- Third party agencies certify an entire system in contrast to individual elements/subsystems of electronic HIPPS;
- Fewer components to fail;
- Easy to install, commission, validate, maintain and operate;
- Minimal documentation: no wiring diagrams or, installation drawings for logic solvers and initiators, application program SRS required;
- Competence in programming or maintaining a complex electronic system is not required;
- There are no EMC interference issues;
- There are no impacts due to cyber security threats;
- Suitable for harsh environments-no environmental climate control is needed;
- Easy to comprehend by operators and maintenance mechanics;
- Lowest life cycle cost among all the types of HIPPS;

Cons:

- Limited distance between the pressure sensing points, mechanical initiators and the valve- all components must be in the close proximities to avoid transportation time lags impacting the response times;
- Mechanical initiators need to be carefully selected as they are sensitive to setpoints and their accuracies;
- Interfacing with operator stations requires additional transmitters in the field and Inputs/outputs (I/Os) in the plant DCS;
- Requires external devices to facilitate partial stroke testing;
- Implementing maintenance bypasses consistent with the overall facility philosophy is not practical;

Applications:

Mechanical HIPPS are attractive for standalone applications in remote plant locations, for areas where electrical cabling is impractical and in facilities, like pipeline pump/compressor stations, where PLCs are not commonly used.

### *3.3 Hybrid-HIPPS*

Hybrid-HIPPS are a combination of electronic and mechanical HIPPS.

Pros:

- High level of risk reduction- 10,000 (SIL 4) or higher can be obtained;
- Significant reduction of common cause and common mode failures due to diversity (mechanical and electronic);
- Low-level of systematic errors;

Cons:

- Complex system;
- Requires high level of competence in the installation, commissioning, validation, maintenance and operation;
- Requires a sizeable spare parts inventory;
- Many of the Cons of electronic and mechanical HIPPSs described above are applicable for the hybrid HIPPS as well.

Application:

High-pressure feed gas to a facility typically requires a high-risk reduction in the absence of non-instrumented IPLs. When a high level of risk reduction (greater than 10,000) is required, hybrid systems are well-suited as they are diverse systems with minimum common cause and common mode failures.

# 4 Safety requirements specification (SRS)

Functional safety standard IEC 61511 requires the development of a SIS safety requirements specification (SRS), including application program safety requirements, when applicable. HIPPS as a functional SIS, requires an SRS. There are a minimum of 29 SRS requirements specified by the standard. While every element is important, a few present unique challenges that can significantly impact the selection, design, and quantification of random hardware failures of the HIPPS.

## *4.1 Process safety time (PST)*

Process safety time, per clause 3.2.52.1, IEC 61511, is the period between a failure occurring in the process or the basic process control system and the occurrence of a hazardous event, if the SIF is not performed. The SIF has to detect the failure and complete its action soon enough to prevent the hazardous event taking into account any lag in the process.

HIPPS that implements the overpressure protection function must complete the action to bring the process to a safe state within the process safety time. Quite often, the process safety times for HIPPS are less than few seconds. The valve subsystems may need quick exhaust valves, larger size air supply tubing, or solenoid valves to meet the response times. Therefore, the PST determination is an important specification which in turn determines various components within the HIPPS to achieve the response time.

## *4.2 HIPPS output action success criteria*

Output actions of a SIF will prevent a particular consequence of concern only when the action is completed with certain performance criteria. For example, stating an action as to close a valve is inadequate. The criteria for the successful operation should be specified. Many HIPPS applications do not require tight shut-off requirements for valves. However, there are situations where larger leakage rates may impact the process safe states. In such scenarios, the process engineer must specify an acceptable leakage rate for the valves, and ensure the design shall have means to achieve such low leakage rates. The design may require unique valve types and trims. Also, additional pressure transmitters and isolation/bypass valves are required for measuring the extent of leakages.

Low leakage rate specification may impact the proof test coverages for the final elements in the quantification of random hardware failures ($PFD_{avg}$). Achieving a SIL 3 target with tight shut-off requirements is a significant challenge.

## *4.3 Proof test interval and overhaul interval (mission time)*

IEC 61511 defines the proof test as a periodic test performed to detect dangerous (hidden) failures in a SIS so that, if necessary, a repair can restore the system to an "as new" condition or as close as practical to this condition.

The frequency of testing plays a significant role in determining the HIPPS's availability to perform intended safety function on demand. However, frequent testing may affect the production, introduce spurious shutdowns due to human error or affect the life of a device. If only a full stroke test is performed for a valve, then the test is imperfect since it will not detect all the covert failures. The undetected, hidden failures are detected and repaired during overhaul periods (mission time). The longer the mission time, the higher the PFDavg; therefore, it will be difficult to achieve SIL 3 targets with longer mission times.

Bypass valves and isolation valves are required for proof testing intervals shorter than the overhaul periods to avoid process interruption. Provision of bypass and isolation valves increases the complexity of the design, footprint, capital and the life cycle costs of the facilities.

### *4.4 Survivability*

As per IEC 61511, the SRS shall specify any requirements for a SIF to survive a major accident event, e.g. the time required for the HIPPS valve to remain in operation in the event of fire shall be specified. Survivability specification impacts the selection of valve trims and additional fireproofing requirements.

### *4.5 Common cause failures*

The SRS must specify the measures to identify and minimize common cause failures. Common cause failures will significantly impact $PFD_{avg}$ for high SILs, such as SIL 3 and SIL 4. Common cause failures can be due to process conditions or similar redundant equipment configurations or systematic failures.

Physical and electrical separation and use of diverse devices which respond differently to the common stresses will minimize common cause failures. An effective functional safety management plan and verification processes will reduce systematic failures.

## 5 Quantification of random failure of HIPPS

The HIPPS design must be evaluated to ensure that the calculated failure measure (PFDavg) is less than the target related to the SIL as specified in the SRS.

The functional safety standard mandates assessment of the following as part of the SIL verification:

- Systematic capability of all the elements associated with a HIPPS meets the specified SIL target
- Calculated $PFD_{avg}$ is less than the target $PFD_{avg}$ of the selected SIL
- Hardware fault tolerance (HFT) complies with the required SIL target

### *5.1 Systematic capability*

IEC 61511 defines the “systematic capability” as a measure (expressed on a scale of SC 1 to SC 4) of the confidence that the systematic safety integrity of a device meets the requirements of the specified SIL, in respect of the specified safety function, when the device is applied in accordance with the instructions specified in the device safety manual.

Systematic capability assessment is for addressing systematic failures which are not quantified and hence not included in the SIL verification calculations. IEC 61508 and IEC 61511explicitly identify a systematic capability parameter (SC). A systematic capability of "SC N" for a device means that the systematic safety integrity of "SC N" has been met when the device is applied following the instructions specified in the instrument safety manual. If a device is to be used in an SIL 3 application, then it must be assessed for the systematic capability of SC 3.

The systematic capability of a device depends on the nature of the device. If a device has no software, then only hardware systematic failure mechanisms are considered. If a device has both hardware and software, then both hardware and software failure mechanisms shall be considered.

There are agencies, such as TUV, Exida, and SIRA that perform systematic capability assessment of a product as per IEC 61508 and provide a certificate with a report specifying an SC parameter. Following a well-established functional safety management plan during detailed engineering, installation, operation and maintenance phases ensures that the systematic safety integrity of a device meets the requirements of the specified SIL.

## *5.2 $PFD_{avg}$*

Many factors influence the $PFD_{avg}$ of HIPPS function. A few of the factors that pose unique challenges are considered in this section.

### 5.2.1 Failure rates

There is always an uncertainty in the quantification of random failures. The failure rate is, probably, the least precise parameter used in performing SIL verifications [**8**]. Therefore, one must acknowledge that the results of the quantification are not real values, but rather a basis for comparing different design options and for monitoring the reliability performance during the operational phase of the safety lifecycle. The analysis is only a prediction. Historical performance is not the same as future performance.

If the failure rates are conservative, then achieving target SIL will be demanding. However, optimistic failure rates will result in inadequate protection integrity. Therefore, the failure rates of components used in the analysis should be sufficiently realistic.

There are many sources of failure rate data, such as application-specific, site-specific, company-specific, manufacturer-specific, industry-specific, and generic data. These failure rates can vary significantly by several orders of magnitude. While application specific or site specific data is the best option, but it can often be difficult to obtain. A detailed discussion on the selection of failure rates to use in the analysis is beyond the scope of this paper, and the reader is advised to refer to other sources [**9**].

5.2.2 Proof test interval, test coverage and mission time

As explained in section 4.3, a proof test may be imperfect based on the following:

- The extent of tests that could be performed to reveal the hidden failures. For example, if the valve is only full stroke tested without examining the internals then the proof test is imperfect;
- The conditions under which the proof test is carried out may differ from a real demand;
- Lack of competence of testing crew;
- Inadequate procedures.

Perfect testing is not a realistic assumption. IEC 61511/IEC 61508 requires consideration of the impact of imperfect testing in the analysis. In the SIL verification calculations, a proof test coverage (PTC) is considered for incomplete testing. The incomplete testing will increase the PFDavg. The lower the proof test coverage, the higher the $PFD_{avg}$. Undetected dangerous failures due to imperfect proof testing will be detected either during the complete overhaul period (when repaired or refurbished) or when the component is replaced or during an actual demand. Mission time is the terminology used by many reliability specialists for the above overhaul period.

For example, if an overhaul period or demand interval for HIPPS is ten years and the imperfect proof test interval for a valve is one year with a proof test coverage of 90%, then achieving an SIL 3 target for HIPPS is an arduous task. If tight shut-off requirement is specified, then the proof test coverage could be even less than 30%.

From the above, it is evident, that the significance of proof test interval, the effectiveness of proof test, and the mission time cannot be ignored and must be taken into account in the analysis.

5.2.3 Partial stroke testing

Partial stroke test will help to improve the $PFD_{avg}$, extend the proof test interval, or increase the safe failure fraction. A single valve with partial stroke test may achieve an SIL 2 integrity. However, for many HIPPS with short process safety times, partial stroke test may not be useful. Moreover, partial stroke tests may result in spurious trips and cause potential damage/wear to the valve seat due to the frequent testing.

A perusal of literature reveals that the partial stroke testing coverage is in the range of 40-80% [10]. However, the above-stated coverage is widely debated.

If the partial stroke testing is not automated, then it is not considered as the diagnostic testing per functional safety standards. Also, the analyst shall take into consideration the partial stroke testing interval with the repair time in the PFD calculations.

5.2.4 Common cause failures

Common cause failures impact the redundant components such as valves and sensors. The common cause failures occur simultaneously or, at least, within a short interval as to be perceived as simultaneous.

The following are the defenses against the common cause failures:

- Diversity (different types of valves and systems);
- Separation (physically and electrically);
- Staggered testing;
- Use of low complexity devices- simpler equipment less prone to common cause failures;
- Well administered and complied procedures;
- Environmental controls;

When diverse components are used, the common cause factor (Beta factor) is minimal. Using IEC 61508, Part 6 [5], Annex D, Tables, D.1, D.2 and D.3, the Beta factor can be estimated to be as low as 2% with diverse components. However, the diverse components must be suitable for the application with comparable performance without any degradation.

Also, staggered testing reduces the overall $PFD_{avg}$ by a factor of 30 to 50% depending upon the voting configurations [**10**].

5.2.5 Modeling of components

Modeling of HIPPS elements in the analysis must include all the components that are involved in bringing the process to a safe state within the process safety time.

Many HIPPS valves are equipped with quick exhaust valves (QEVs) to reduce the valve closing time to meet the process safety time. Sometimes, redundant QEVs are provided in various voting configurations such 1oo2 or 3oo4, etc. The contribution of QEV failures to the $PFD_{avg}$ is significant and cannot be ignored. Therefore, the reliability analyst must consider the QEVs with their appropriate voting architecture in the $PFD_{avg}$ calculations. Similarly, the likelihood of plugging of sensing lines must be taken into account in freezing applications.

Any interfacing components such as barriers, isolators, interposing relays, and poppet valves, which are part of the safety instrumented function to achieve or maintain the process to a safe state must be modeled and analyzed.

### *5.3 Hardware fault tolerance (HFT)*

Irrespective of the hardware reliability calculated ($PFD_{avg}$) for the design, the functional safety standard IEC 61511 specifies minimum levels of redundancy for each safety integrity level.

There are two approaches:

- Route $1_H$: Determination of HFT based on the calculated safe failure fraction (SFF);
- Route $2_H$: Determination of HFT based on field experience and field data (proven-in-use)

### *5.3.1 Route1$_H$*

The Route1$_H$ approach uses SFF and the SIL target to determine the hardware fault tolerance. The SFF is defined as the sum of the potentially dangerous failures revealed by auto-test and those failures which result in a safe state, as a fraction of a total number of failures.

$$\text{SFF}=\frac{[dangerous\ detected\ failures(self\ test)+safe\ failures]}{Total\ failures}$$

The above is calculated for each element (sensor, complete valve assembly, logic solver). Two tables, Table 2 and Table 3, are provided in IEC 61508- Part 2. Table 2 is for "Type A" devices which are simple devices such as valves (no microprocessors and programming are involved), and Table 3 for "Type B" devices which are complex devices such as electronic transmitters, trip amplifiers, and PLCs. Based on the SFF, type of the device, and the SIL target, the hardware fault tolerance is determined using Table 2 and Table 3.

### *5.3.2 Route2$_H$*

In Route 2$_H$, the hardware fault tolerance is specified by the functional safety standard for each SIL target if the device is selected based on well documented and verified field based failure rate data with 90% statistical confidence. For low demand application with SIL 1 or SIL 2 target, there is no need for any redundancies (HFT=0), and SIL 3 requires one hardware fault tolerance.

Unless the field failure data with 90% confidence levels are available, the Route 2$_H$ cannot be justified. There are third party agencies who provide Route 2$_H$ certifications and should be evaluated carefully before adopting the Route 2$_H$ approach based on those certifications.

## 6 High level risk reductions (RRF> 10,000)

As discussed in section 2, sometimes, PHA/SIL selection may indicate a required risk reduction greater than 10,000 in the absence of any non-instrumented protection layer or, human intervention is impractical due to process safety time considerations or safety reasons.

If there are no options other than providing HIPPS with a risk reduction greater than 10,000, then the functional safety standard IEC 61511 mandates certain requirements. The following are a few of the clauses and their requirements:

Clause 9.2.6: *If further consideration of application and confirmation that a risk reduction requirement >10,000....is still required, then* consideration *should be given to achieving the safety integrity requirement using a <u>number of protection layers with lower risk reduction requirements</u>. If the risk reduction is allocated to multiple protection layers, then such protection layers shall be independent from each other or the lack of independence shall be assessed and shown to be sufficiently low compared to the risk reduction requirements.*

Clause 9.2.7: *If a risk reduction greater than 10,000....is to be implemented, whether allocated to a single SIS or multiple SIS or SIS in conjunction with a BPCS protection layer, then a further risk assessment shall be carried out using a quantitative methodology to confirm that the safety integrity requirements are achieved.*

Clause 9.2.8: *If the risk reduction required for a hazardous event is allocated to multiple SIFs in a single SIS, then the SIS shall meet the overall risk reduction requirement.*

A hybrid HIPPS using the combination of electronic and mechanical HIPPS is a solution to achieve risk reductions greater than 10,000. Each HIPPS can be designed to be better than SIL 2, and the overall risk reduction will be better than 10,000. The performance of electronic and mechanical HIPPS are comparable.

High level of risk reduction is achieved due to the diversity and very low common cause failures between the electronic and mechanical HIPPS. All the subsystems of the HIPPSs are diverse with different technologies. Also, systematic failures are minimal. The integrity of such multiple HIPPS is best analyzed using a fault tree analysis (FTA).

An example fault tree analysis for a hybrid HIPPS using ISOGRAPH's Reliability Workbench™ is shown in **Figure 5**.

The example hybrid HIPPS achieved a risk reduction of 12,815($PFD_{avg}$ = 7.80E-05).

The FTA shows an approach for modeling of various components including the common cause failures. The failure rates assumed are generic for the purpose of demonstration. The actual failure rates and other factors are application specific and cannot be generalized.

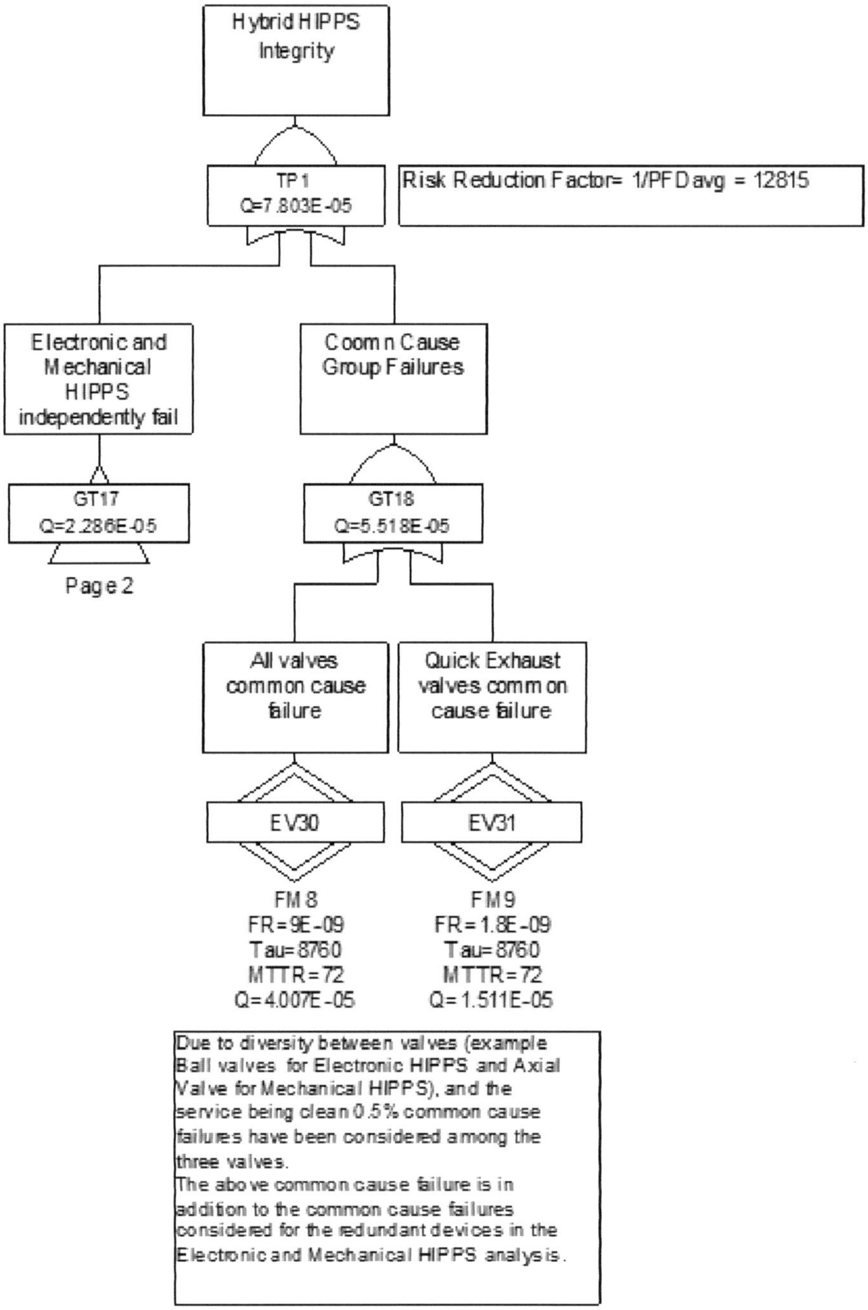


**Figure 5. FTA for multiple HIPPS to achieve high risk reduction**

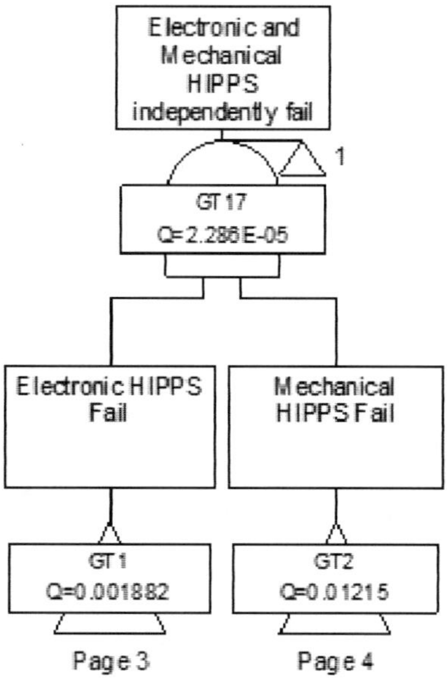


**Figure 5. FTA for multiple HIPPS to achieve high risk reduction (contd..)**

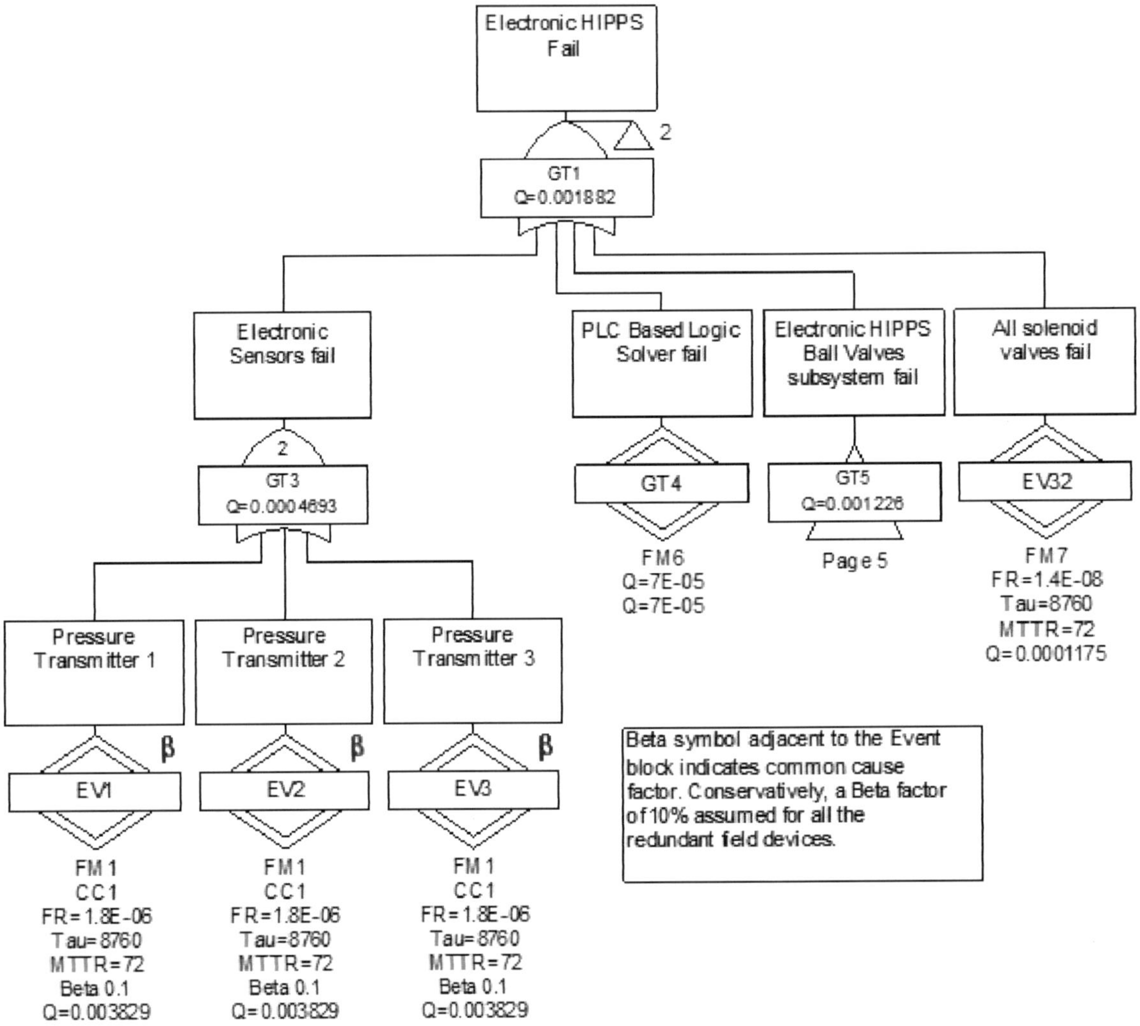


**Figure 5. FTA for multiple HIPPS to achieve high risk reduction (contd..)**

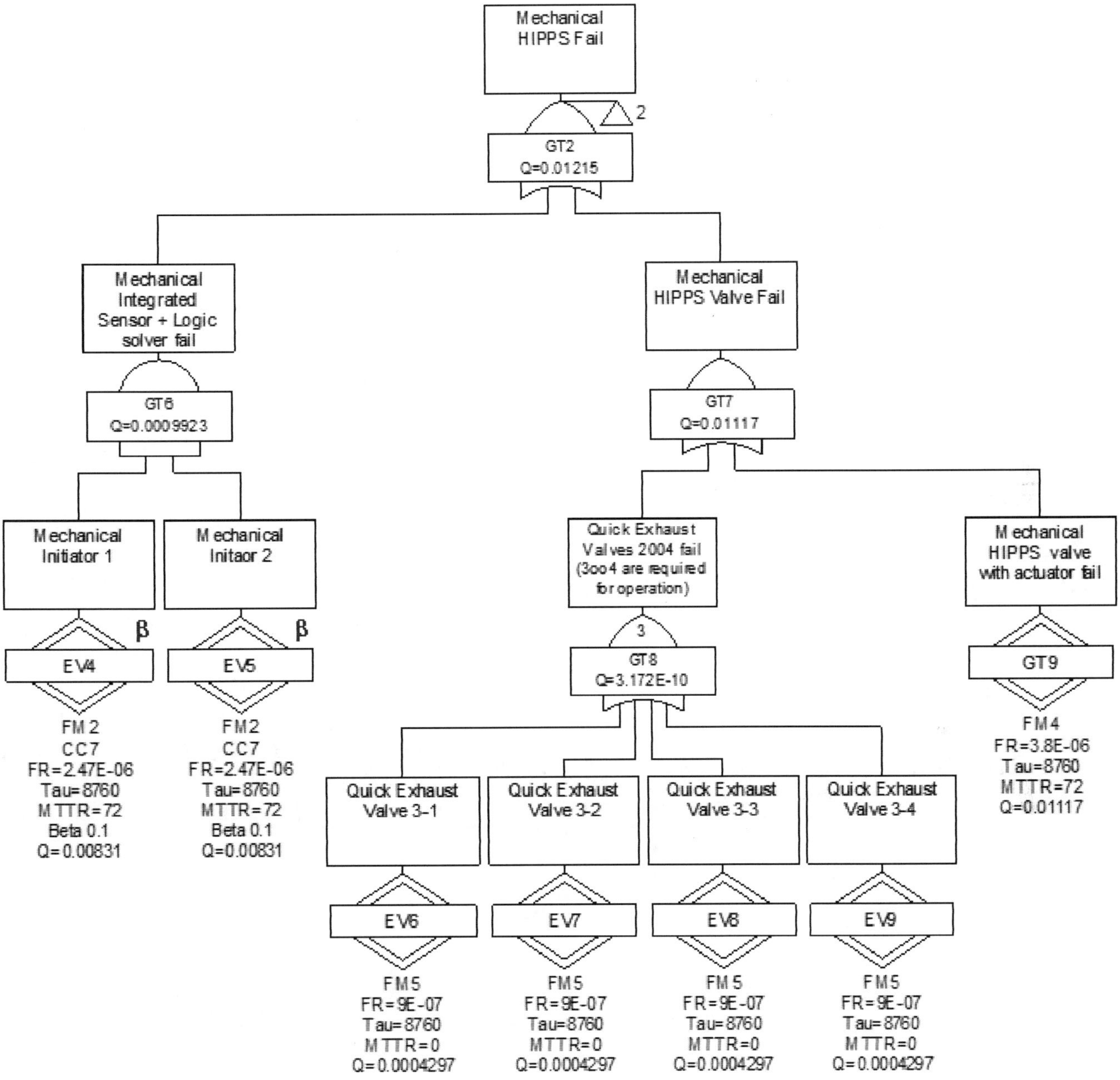


**Figure 5. FTA for multiple HIPPS to achieve high risk reduction (contd..)**

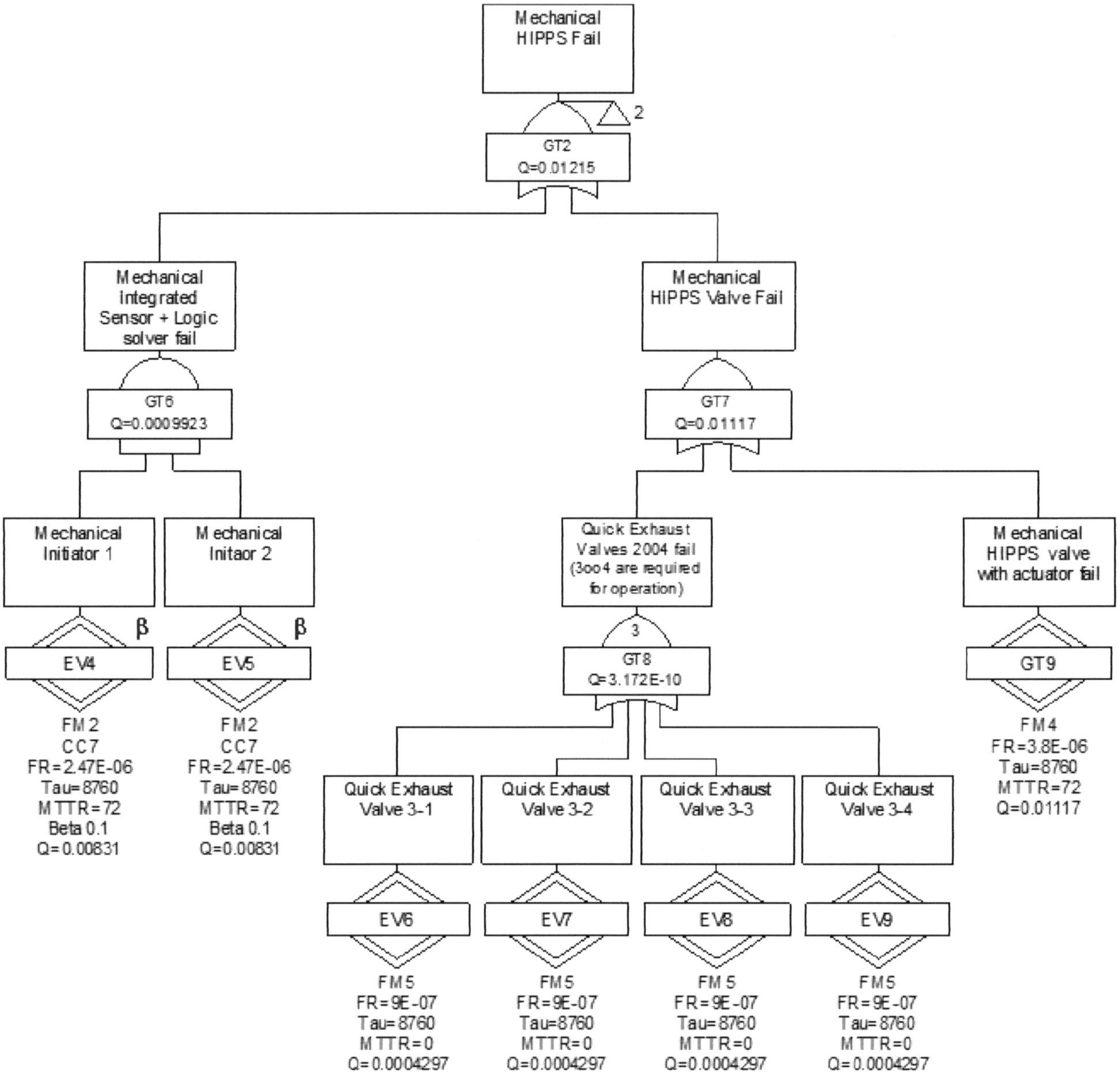


**Figure 5. FTA for multiple HIPPS to achieve high risk reduction (contd..)**

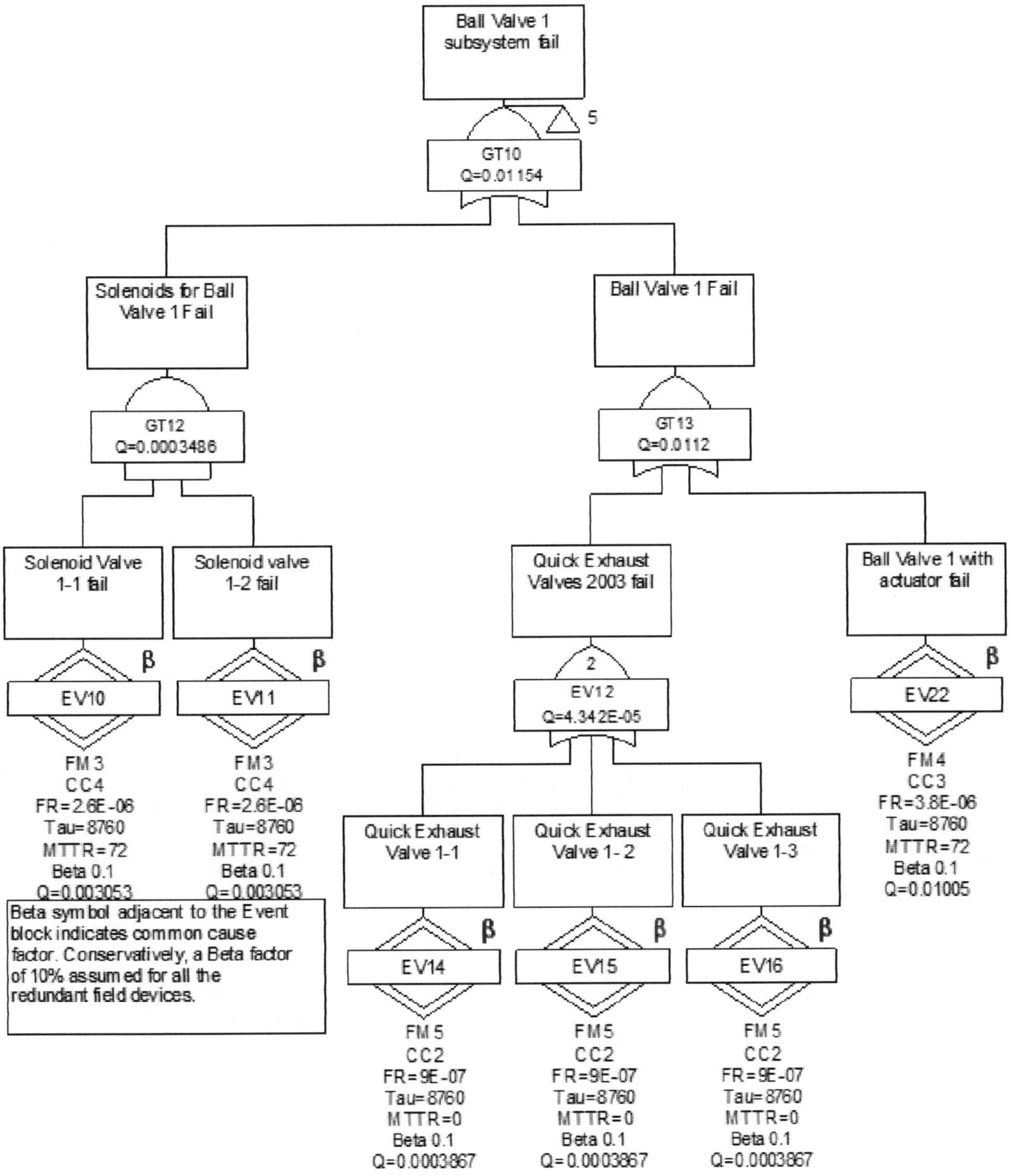


**Figure 5. FTA for multiple HIPPS to achieve high risk reduction (contd..)**

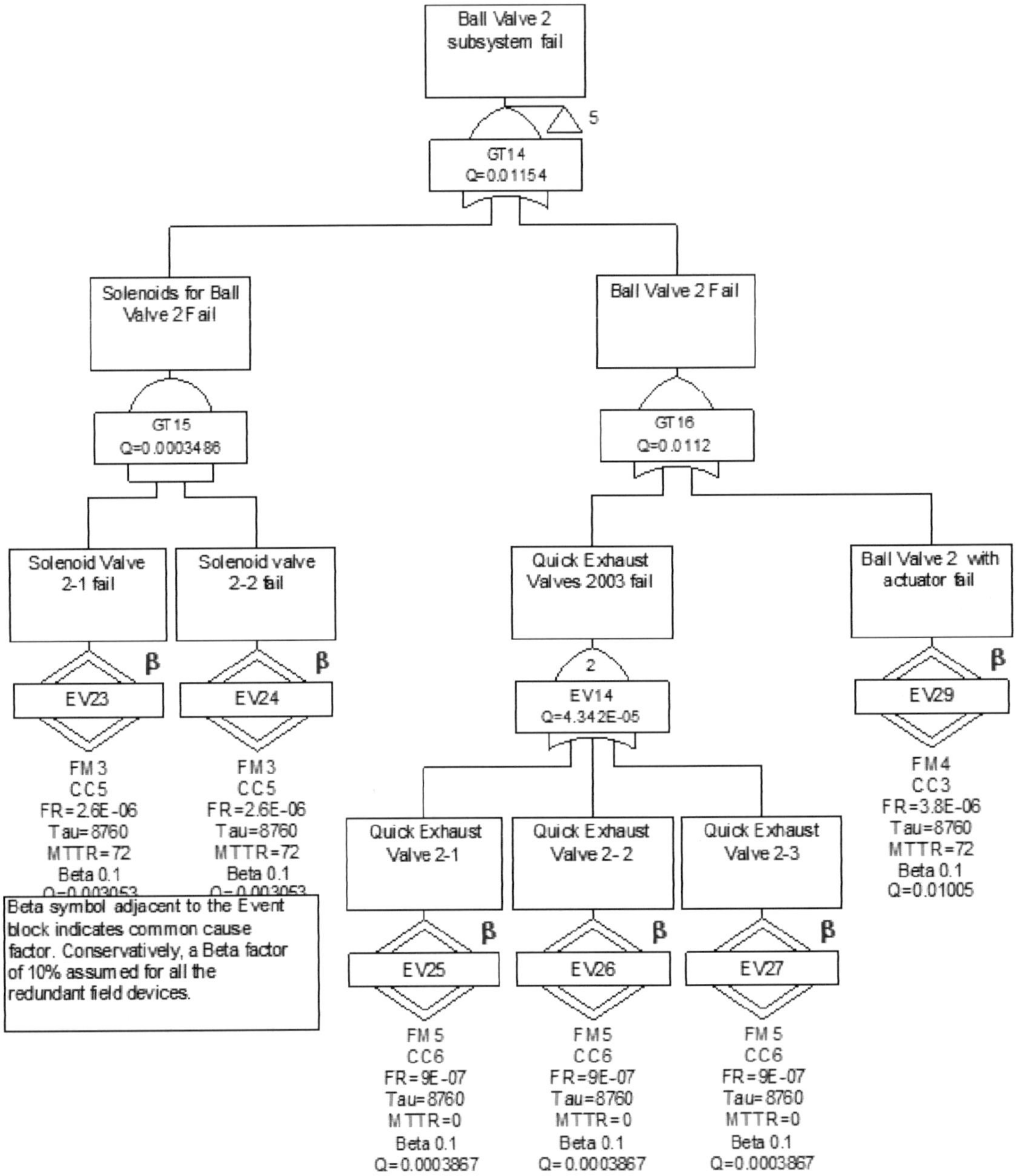


**Figure 5. FTA for multiple HIPPS to achieve high risk reduction (contd..)**

# 7 Installation, commissioning, operation and maintenance

Installation and commissioning of the HIPPS must be completed in accordance with the installation drawings, safety manuals, and the SRS.

Operation and maintenance is the longest period in the safety life cycle. The functional safety standards mandate continuous performance monitoring, including demand rates, failure modes and failure rates of various components. The actual performance must be compared with the predicted performance to identify any gaps so that remedial measures can be provided with appropriate safety review.

A robust change management process must be in place to ensure HIPPS integrity throughout the operation and maintenance phase.

# 8 Conclusion

Assignment of an SIL target and choosing an appropriate technology for a HIPPS is not trivial. In this paper, the significance of PHA and SIL selection for HIPPS has been re-emphasized and a brief account of available options for HIPPS to choose from, their pros and cons, and the selection criteria with examples are provided.

The design, installation, operation, and maintenance of HIPPS shall comply with the requirements of functional safety standards IEC 61508 and IEC 61511. In this article, the challenges in each of the above activities are discussed and appropriate solutions to mitigate the above challenges are provided.

In applications where a high risk reduction greater than 10,000 is required a hybrid HIPPS design approach is proposed. Using FTA demonstrated the modeling of complex hybrid HIPPS to achieve high level risk reduction.

# 9 References

[1] IEC. IEC 61511 Functional Safety: Safety Instrumented Systems for the Process Industry Sector- Part 1-3, Edition 2, Geneva, 2016.

[2] American Society of Mechanical Engineers (ASME), Boiler and Pressure Vessel Code, Section VIII-Pressure Vessels, New York, 2010.

[3] API, Standard 521, Pressure-relieving and Depressuring Systems, $6^{th}$ Edition, Washington, DC, 2014.

[4] API, RP17O, Recommended Practice for Subsea High Integrity Pressure Protection Systems (HIPPS), 1 Edition, Washington, DC 2009.

[5] HSE, UK. SPC/TECH/OSD/31, Safety instrumented systems for the overpressure protection of pipeline risers, Available at

http://www.hse.gov.uk/foi/internalops/hid_circs/technical_osd/spc_tech_osd_31.htm, Accessed on Jan 29, 2017.

[6] IEC. IEC 61508 Functional safety of electrical/electronic/programmable electronic safety related systems- Part 1-7, Edition 2, Geneva, 2010.

[7] H.G. Lawley and T.A. Kletz. High-pressure-trip systems for vessel protection, Chemical Engineering, Vol 82, (1975).

[8] Dr. D. J. Smith, Reliability, Maintainability and Risk, 8th Edition, Butterworth-Heinemann, Waltham, MA, 2011.

[9] S. Hauge, S. Habrekke, M.A. Lundeteigen. Reliability Prediction Method for Safety Instrumented Systems- PDS Example Collection, SINTEF Technology and Society, Safety Research, 2010.

[10] Dr. D.J. Smith, and K.G.L. Simpson, The Safety Critical Systems Handbook, 4th Edition, Butterworth-Heinemann, Cambridge, MA, 2016.

# 475693 Managing Risk Associated w/ Electrical Energy Ignition Sources

**Abstract:**

In facilities that handle flammable hydrocarbons it is extremely important that all forms of potential ignition sources be carefully managed. This presentation will focus on key points associated with process electrical safety, open flame process heaters, and mechanical forms of ignition sources. OSHA 1910.119 has for many years required that hazardous area classification documentation remain in an evergreen state that requires re-certification at least once every five years. For most process facilities this OSHA mandate presents many problems. In most cases documentation has not been maintained in a current state, the facility perhaps has changed owners if not once many times over the years. The presentation attendee will be exposed to a common sense methodology that focuses primarily on minimizing electrical energy from becoming a source of ignition. A process for updating hazardous area classification documentation will be presented with tips on providing the end-user with an audit proof set of documentation. Once the hazardous area map has been developed then methods of properly documenting and conducting a site compliance assessment will be explained. This compliance assessment is a gap analysis that compares the installed electrical systems against the requirements of the National Electrical Code (NFPA 70 Article 500). Methods of presenting this documentation will be explained so the end-user has a clear understanding of the proper steps for remediation.

The presentation attendee will be exposed to the following:

- Managing risk associated with roadways that are impinged by hazardous classified boundaries
- Open Flame Process heaters encroachment by hazardous classified locations
- Methodology for area classification assessment *(it's determined by the process conditions)*
- Hazard's compliance for instrumentation and electrical system installations in hazardous process locations
- How to interpret documentation in support of the area classification assessments which include:
    - Plan Drawings (maps), Elevations, Site Specific Details. Legends, and notes
    - Assessment report documentation that addresses General Considerations, Identification of Point Sources of Emissions, Physical properties of Materials being Handled, Processed, or Stored, and Gas Dispersion Modeling Graphs and plots
- Methods for mathematical determination of hazard radius thru the concept of commercially available computer assisted gas/vapor dispersion modeling software.
- Application of common standards, codes and installation practices i.e. (NFPA, ISA, API and OSHA)

Detailed handouts as well as a complete audio-visual presentation are used to supplement the presentation.

# 479320 Review and Analysis of Recent Cyber Security Attacks

**Abstract:**

Process plants need to ensure that adequate safeguards are in place for a variety of upset conditions, including cyber security attacks, which can result in loss of product, damage to equipment, downtime, and potential safety and environmental disasters. Select cyber security incidents across a variety of industries will be reviewed to determine the modes of cyber security attack and their subsequent impacts. Specific areas of vulnerability will be identified with corresponding mitigation strategies. Possible solutions will be presented for incorporating cyber security into an existing process safety program to ensure a tolerable level of protection for high-consequence events.

# 478316 Resilience-Based Survival Analysis in the Event of a Process Upset: A Case Study of Utility Plant

**Abstract:**

**Abstract:** In a process plant system, safe and reliable operations are highly sensitive to the power or steam failure, loss of cooling water or nitrogen or instrument air. Most of these products and services from utilities plant play an important role as safety barriers. The main focus of this paper is to propose a model for survival evaluation of a process system under an upset condition. This model is based on the holistic resilience-based method that integrates both technical (process parameters variations) and social (policy/regulations, human and organizational) factors. The resilience methodology emphasizes non-linear dynamics, new types of threats, uncertainty, survival through upset conditions and recovery from catastrophic situations [1], [2].

The purpose of this paper is to maximize the performance of safety barriers that will be utilized in case of a process upset condition to prevent the escalation of the event into a catastrophic incident. In this work, a methodology is presented to select the best maintenance strategy [3] while considering the statutory safety regulations for maintenance of equipment like boilers, generators and more. Three different types of maintenance alternatives considered as part of this analysis are: predictive, preventative and condition-based. The key ideas of survival evaluation are presented using a utility plant as the case study.

### References

1. Dinh, L. T., Pasman, H., Gao, X., & Mannan, M. S. (2012). Resilience engineering of industrial processes: principles and contributing factors. Journal of Loss Prevention in the Process Industries, 25(2), 233-241.
2. Jackson, S. (2009). Architecting resilient systems: Accident avoidance and survival and recovery from disruptions (Vol. 66). John Wiley & Sons.
3. Vassiliadis, C. G., & Pistikopoulos, E. N. (2001). Maintenance scheduling and process optimization under uncertainty. *Computers & Chemical Engineering*, *25*(2), 217-236.

# 478816 Model-Based Pressure Relief Design for Reactive Systems

**Abstract:**

Abstract

Reactive systems behave unpredictably under overpressure conditions. Varying composition impacts the physical properties and response of the mixture to temperature and time. Pressure generation in a reactive system is often a function of both vapor pressure and gas generation rate. Standard methodologies applied in non-reactive pressure relief system design cannot be relied upon to give appropriate sizing for reactive systems. Consistent with DIERS methodologies, model-based design provides accurate, reliable vent sizing. A comprehensive approach to modeling is presented which incorporates vapor-liquid equilibrium, reaction stoichiometry, and reaction rate.

13TH GLOBAL CONGRESS ON PROCESS SAFETY

# Dynamic Modeling of Inbreathing Requirements for Low-Pressure Storage Tanks

**Dona Abou-Chakra**
**ioMosaic Corporation**
**93 Stiles Road | Salem, NH 03079**
**chakra.d.nh@iomosaic.com**

**Georges A. Melhem, PhD, FAIChE**
**ioMosaic Corporation**
**93 Stiles Road | Salem, NH 03079**
**melhem@iomosaic.com**

**Ronald J. Willey, PhD**
**Northeastern University**
**360 Huntington Avenue | Boston, MA 02115**
**r.willey@neu.edu**



Prepared for Presentation at
American Institute of Chemical Engineers
2017 Spring Meeting and 13th Global Congress on Process Safety
San Antonio, TX
March 26 – 29, 2017

# Dynamic Modeling of Inbreathing Requirements for Low-Pressure Storage Tanks

**Dona Abou-Chakra**
**ioMosaic Corporation**
**93 Stiles Road | Salem, NH 03079**
**chakra.d.nh@iomosaic.com**

**Georges A. Melhem, PhD, FAIChE**
**ioMosaic Corporation**
**93 Stiles Road | Salem, NH 03079**
**melhem@iomosaic.com**

**Ronald J. Willey, PhD**
**Northeastern University**
**360 Huntington Avenue | Boston, MA 02115**
**r.willey@neu.edu**

**Keywords:** Thermal Inbreathing, API 2000, Storage Tanks, Venting.

## Abstract

A Process Safety Office™ - SuperChems™ component based model is proposed to study the dynamic behavior of the thermal inbreathing of low-pressure storage tanks upon the occurrence of a sudden cooling rainstorm. The objective here is to establish a model that is capable of properly calculating the design relief capacity of the relief device to be installed on such tanks. The impact of phase-change, particularly vapor condensation, on the thermal inbreathing of the tank assessed. The model was validated against two large-scale experiment datasets and some other published theoretical models. The results were then discussed and compared to the relief requirements obtained using the API 2000 7$^{th}$ Edition method, which is the current standard for venting atmospheric tanks.

## 1 Introduction

Low-pressure storage tanks are protected against extreme weather changes following the API 2000 (7$^{th}$ edition) standard. The threat of creating a vacuum occurs when the contents of the tank vapor space are suddenly cooled by an intense rainfall causing the vapor space to contract and condense, hence, rapidly decreasing the internal pressure. This pressure drop will pull outside air into the tank's vapor space and acts as the driving force for thermal inbreathing.

The current limitation with the API standard is that it only addresses tanks storing non-condensable gases and have no liquid residue, where in fact most storage tanks contain condensable vapors such as water vapor which will have a big impact on the thermal inbreathing of the tank. This paper assesses the API 2000 relief requirement calculations for low-pressure storage tanks by comparing them to (1) the results of the dynamic simulations from a proposed SuperChems™ based model and (2) to several methods found in the literature (PTB W-22, Salatino, etc.) for both condensable and non-condensable vapor services.

# 2 API 2000 Methods

The tank contents are affected by the ambient conditions especially when the tank is uninsulated, which is the case for most storage tanks. This method uses air at normal conditions (0°C and 101.325 kPa) as the inbreathed gas. The maximum inbreathing rate, $\dot{V}_{IT}$ , is primarily a function of tank volume, insulation, and a C-factor (found in Table 1 below) as shown in Equation (1) below:

$$\dot{V}_{IT} = C\, V_{tk}^{0.7} R_i \tag{1}$$

where $\dot{V}_{IT}$ is the maximum inbreathing volumetric flow rate of air at normal conditions ($Nm^3/h$), $C$ is the dimensionless factor that depends on vapor pressure, storage temperature and latitude, $V_{tk}$ is the tank volume ($m^3$), and $R_i$ is the reduction factor for insulation.

**Table 1. API 2000 *C*-factor for Thermal Inbreathing**

| Latitude, φ | C-factor | | | |
|---|---|---|---|---|
| | Vapor pressure similar or below hexane | | Vapor pressure higher than hexane or unknown | |
| | Average storage temperature (°C) | | | |
| | < 25 | ≥ 25 | < 25 | ≥ 25 |
| φ < 42° | 4 | 6.5 | 6.5 | 6.5 |
| 42° ≤ φ ≤ 58° | 3 | 5 | 5 | 5 |
| φ > 58° | 2.5 | 4 | 4 | 4 |

The assumptions used for the scenario of "tank cooling by rain" are found in Annex E of API 2000 [1] and are listed below:

- Tank is filled with air only.
- Initial vapor space and shell wall temperature are in thermal equilibrium at 55 °C.
- Rain temperature and intensity are constant at 15°C and 225 $kg/m^2h$ (This rainstorm has a probability of 1/100 year) respectively; rain angle is at 30° from the horizontal.
- Inside free convection heat-transfer coefficient is 5 $W/m^2K$.
- Outside film cooling heat-transfer coefficient is 5,000 $W/m^2K$.

## *2.1 API 2000 Annex A Method*

The method found in Annex A was used prior to 7th Edition (2008) and is quite different than the current sizing method for venting devices. In order to quantify the thermal inbreathing for storage tanks volumes smaller than 3180 $m^3$, Annex A [1] assumes that the tank starts cooling from an initial temperature of 48.9°C and at a constant temperature rate of -56 K/h. The thermal inbreathing rate, $\dot{V}_{IT}$ ($Nm^3/h$) is as follows:

$$\dot{V}_{IT} = 0.169\, V_{tk} \tag{2}$$

where $V_{tk}$ is the tank volume ($m^3$). For tank volumes equal or greater than 3180 $m^3$, the thermal inbreathing is assumed to be limited by a constant heat flux, $h\Delta T$, equal to 63 $W/m^2$. The thermal inbreathing rate in this case becomes:

$$\dot{V}_{IT} = 0.577\, A_{exp} \tag{3}$$

where $A_{exp}$ is the external tank area exposed to rain ($m^2$). Table 2 below summarizes the thermal normal venting requirements as found in Annex A of API 2000. Volumes greater than 30,000 $m^3$ are outside the scope of this method. However, interpolation is allowed for Table 2 as long as the tank volumes lie between the specified volume limits of 10 $m^3$ to 30,000 $m^3$. [1]

**Table 2. API 2000 Annex-A Normal Venting Requirements Due to Thermal Effects [1]**

| Tank Capacity | Inbreathing | Out-breathing | |
|---|---|---|---|
| Column 1 [a] | Column 2 [b] | Column 3 [c] | Column 4 [d] |
| | | Flash Point ≥ 37.8 °C or Normal Boiling Point ≥ 149 °C | Flash Point < 37.8 °C or Normal Boiling Point < 149 °C |
| $m^3$ | $Nm^3/h$ of air | $Nm^3/h$ of air | $Nm^3/h$ of air |
| 10 | 1.69 | 1.01 | 1.69 |
| 20 | 3.38 | 2.02 | 3.38 |
| 100 | 16.9 | 10.1 | 16.9 |
| 200 | 33.8 | 20.3 | 33.8 |
| 300 | 50.4 | 30.4 | 50.4 |
| 500 | 84.5 | 50.7 | 84.5 |
| 700 | 118 | 71.0 | 118 |
| 1,000 | 169 | 101 | 169 |
| 1,500 | 254 | 152 | 254 |
| 2,000 | 338 | 203 | 338 |
| 3,000 | 507 | 304 | 507 |
| 3,180 | 537 | 322 | 537 |
| 4,000 | 647 | 388 | 647 |
| 5,000 | 787 | 472 | 787 |
| 6,000 | 896 | 538 | 896 |
| 7,000 | 1,003 | 602 | 1,003 |
| 8,000 | 1,077 | 646 | 1,077 |
| 9,000 | 1,136 | 682 | 1,136 |
| 10,000 | 1,210 | 726 | 1,210 |
| 12,000 | 1,345 | 807 | 1,345 |
| 14,000 | 1,480 | 888 | 1,480 |
| 16,000 | 1,615 | 969 | 1,615 |
| 18,000 | 1,750 | 1,047 | 1,750 |
| 20,000 | 1,877 | 1,126 | 1,877 |
| 25,000 | 2,179 | 1,307 | 2,179 |
| 30,000 | 2,495 | 1,497 | 2,495 |

# 3 Thermal Inbreathing of Non-Condensable Vapors

One of the major assumptions of API 2000 is that the tank experiencing inbreathing is empty (i.e. only filled with dry air) and the vapor space is considered non-condensable. This section will review recent work on inbreathing dynamics for non-condensable dry air services only.

## *3.1 Salatino's Model*

*Salatino et al.* [2] compare their tank inbreathing model to the older version of API 2000 (1992) that utilized the method of Annex A: a method based on the maximum estimated heat loss rate of 63 W/m$^2$ upon sudden cooling by a rainstorm of 225 kg/m$^2$h rain density. They recognized that all existing thermal inbreathing models assume a uniform vapor-space temperature, thus his team's contribution was accounting for the temperature non-uniformity inside the tank by dividing the tank into two individual volumes: the shell and the roof.

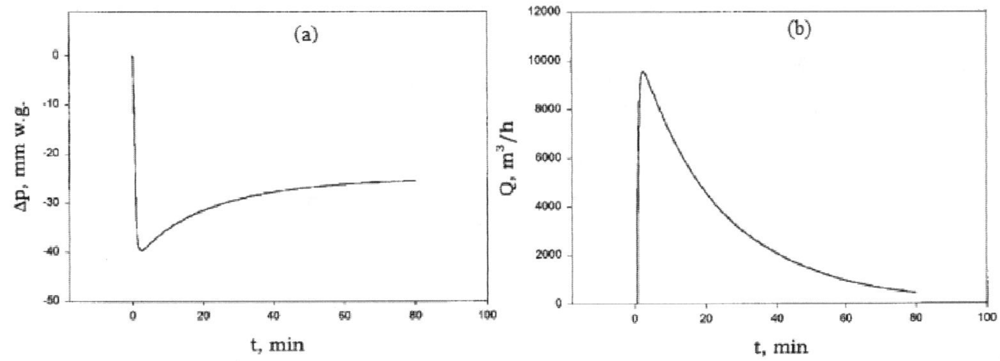


**Figure 1. Pressure drop (a) and thermal inbreathing (b) profiles of Salatino's Model [2]**

Salatino derived an equation for the maximum thermal inbreathing (Equation 4), stating that the maximum occurs at time t=0 (Figure 1) when the sudden weather change causes (1) vanishing of the radiative flux absorbed by the tank, (2) abrupt drop in ambient temperature and finally (3) rainfall.

$$\dot{Q}_{max} = -\frac{V}{T_G}\frac{dT_G}{dt}\bigg|_{t=0} = \frac{V}{T_{G,1}}\frac{1}{\tau}\left(T_{G,1} - T_{G,2}\right) \tag{4}$$

$$\tau = \frac{pV}{RT_G}\frac{c}{\sum A_i h_{iG}} \tag{4a}$$

where $\dot{Q}_{max}$ is the maximum thermal inbreathing rate (m$^3$/h), $V$ is the tank volume (m$^3$), $T_{G,1}$ and $T_{G,2}$ are the gas temperatures (K) before and after the rainstorm, respectively. $\tau$ is the time constant that sums the effects of pressure, volume, heat transfer parameters (Equation 4a).

*Salatino et al.* further simplified the model by assuming an average value for the heat transfer coefficient of 4 W/m$^2$K and a gas temperature drop (after rainfall) of $T_{G1}$-$T_{G2} \cong 25$°C claiming that these values will result in a conservative approximation of the inbreathing rate. Replacing these values in Equation (4) yields the simplified maximum inbreathing rate, $\dot{Q}_{max}$, which is a function of the tank volume, *V*, and the height-to-diameter ratio, *f*:

$$\dot{Q}_{max} = 2.6\, V^{2/3} \left( \frac{1 + 2f}{\sqrt[3]{f}} \right) \tag{5}$$

For a 63,000 $m^3$ cone-roof tank, Salatino predicted a maximum thermal inbreathing value of 9,900 $m^3/h$, using Equation 5. However, using the API 2000 Annex A method, the maximum inbreathing for the same tank was found to be 4,100 $m^3/h$. Salatino proved that his method is more conservative than the API 2000 equations and advised that an oversized thermal relief device is considered a safer design in this case.

### *3.2 Lloyd's Register Energy Americas, Inc. (LR) Model*

Lloyd's Register (LR) Energy Americas, Inc. analytically solved the heat balances of a storage tank that is suddenly cooled by rain [3] and compared their results against ISO 28300, which is an equivalent of API 2000 6$^{th}$ edition (2009). The model results are summarized in Figure 2 below.

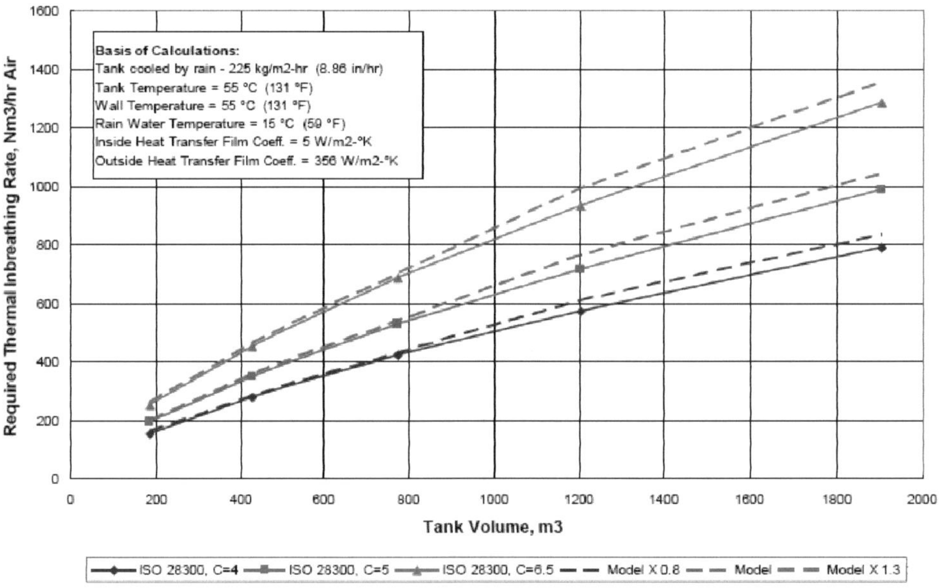


**Figure 2. LR Model versus API 2000 6$^{th}$ ed. (ISO 28300)**

The LR model was slightly more conservative than the API 2000 thermal inbreathing predictions by factors ranging between 0.5 and 1.3. Note that the C value in Figure 2 refers to the C-factor in the API 2000 thermal inbreathing in Equation (1) of this paper.

The model was then used to study the impact of several variables, some of which had minimal impact on the predicted thermal inbreathing (less than 5% error) such as the tank weight and initial gas temperature. Other variables, like the outside heat-transfer coefficient (Table 3) and rain intensity, had substantial impact on the calculated thermal inbreathing rates.

**Table 3. LR Model - Impact of Outside Heat-Transfer Coefficient, $h_o$ [3]**

| Case No. | Tank Volume m3 | Predicted Thermal Inbreathing Rate, Nm3/hr | | |
|---|---|---|---|---|
| | | ho = 200 W/m2- K | ho = 300 W/m2- K | ho = 400 W/m2- K |
| 1 | 186 | 169.6 | 191.2 | 205.1 |
| 2 | 427 | 308.3 | 341.4 | 362.7 |
| 3 | 775 | 475.1 | 520.7 | 548.6 |
| 4 | 1202 | 678.9 | 740.5 | 778.6 |
| 5 | 1903 | 935.2 | 1011.6 | 1058.5 |

The LR model was used to derive a relationship between the rain intensity (mm/h) and the heat transfer coefficient between the outer tank wall and rain (W/m$^2$K) as shown in Figure 3 below. This linear dependence relationship was implemented in SuperChems™ for this study.

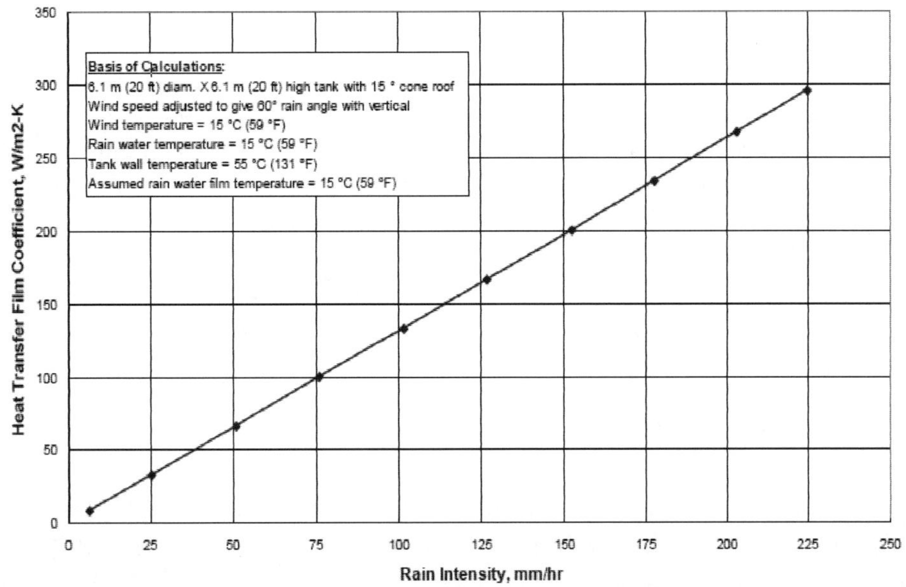


**Figure 3. Rain/wall heat transfer coefficient versus of rain intensity**

### *3.3 PROTEGO© Model*

Scholz and Weber [4] from PROTEGO© derived the API 2000 7$^{th}$ Edition thermal inbreathing formula (Equation 1 of this paper) and determined that the maximum thermal inbreathing was a function of primarily tank volume with a coefficient that is less than or equal to 9.99:

$$\dot{V}_{max} = Cofficient \cdot V^{2/3} \tag{6}$$

The above equation is applicable to both flat-roof and conical-roof storage tanks. The derivation of Equation (6) was obtained by performing energy balances on the stored gas, the tank walls, and on the rainwater film. The differential equation and its solution are not included in this document, but can be found in Scholz and Weber's publication [4].

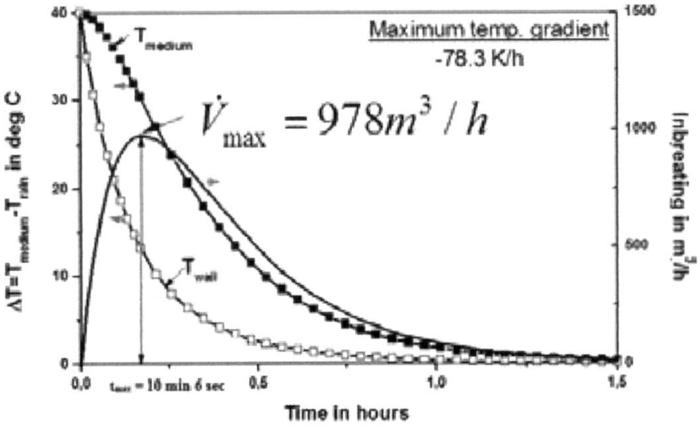


**Figure 4. PROTEGO© Model Results**

PROTEGO© then compared their results with the Foerster's model [5], whose main assumption was a fully immersed tank, which proved to be in good agreement with the published data. Figure 4 contains the results of a simulation using a conical-roof storage tank of 3,978 $m^3$ volume and an aspect ratio f = 0.6. Light rain was assumed (75 $kg/m^2h$) at 15°C. The inside and outside convective heat-transfer coefficients were set to 5 and 37 $W/m^2K$ respectively.

PROTEGO© used their model to study 5 different types of condensable vapors and their effect on the maximum thermal inbreathing rate, however, the results will be discussed in Section 4 of this paper.

## 4 Thermal Inbreathing of Condensable Vapors

API 2000 neglects the possibility of vapor condensation for the thermal inbreathing/out-breathing relief calculation since only dry air is present in the tank i.e. the tank is empty. One reason to consider the effects of having condensable vapors is the steam cleaning that storage tanks undergo. Hot steam will occupy the tank after the cleaning process is complete and if a rainstorm occurs, the water vapor will condense/contract and pull a vacuum inside the tank.

### *4.1 Fullarton Model – Water Vapor*

Fullarton [6] introduced two dimensionless factors in his model: (1) The reduction factor "$\eta$" that accounts for the tank wall thickness by factoring in the metal heat capacity of the walls and (2) the condensation factor "$\chi$" that accounts for the stored-vapor condensation i.e. the additional condensation mass flux and the additional heat of condensation. The formulas of each "$\eta$" and "$\chi$" can be found in Fullarton's paper [6].

It is expected that the tank wall temperature drops much faster than the temperature of the vapor stored inside the storage tank because of the metal's higher heat capacity. However, a thicker metal wall acts as a resistance to heat transfer. This means that the tank walls cool at a slower rate than when the wall thickness is neglected. The maximum inbreathing in this case will be lower, hence the reduction factor "$\eta$" which "reduces" the maximum thermal inbreathing of a storage tank with thicker walls.

**Table 4. Range of $\chi$-factor for Different Condensable Vapors**

| Condensable Vapor | $\chi$-factor |
|---|---|
| Water vapor<br>Heptane<br>Hexane | 1.5-2 |
| Ethanol | 2-2.5 |
| Methanol | 3-4 |

The condensation factor "$\chi$", on the other hand, is primarily a function of the volatility of the condensable vapor being studied. For that reason, Fullarton established a range for the "$\chi$" factor for different vapors as seen in Table 4 above [6].

Therefore, the maximum thermal inbreathing for a condensable vapor space per Fullarton can be obtained by using each of the Naumann approximation, $\dot{V}_N$, the reduction factor, $\eta$, and the condensation factor, $\chi$.

$$\dot{V}_{K,max} = \dot{V}_N \eta \chi \tag{7}$$

One must keep in mind that properties of air at 288K were used to derive this equation in addition to the assumption of the initial temperature gradient ($T_{G,0} - T_A$) equal to 40°C. If other values are to be used, the above equation must be re-derived to account for the appropriate conditions.

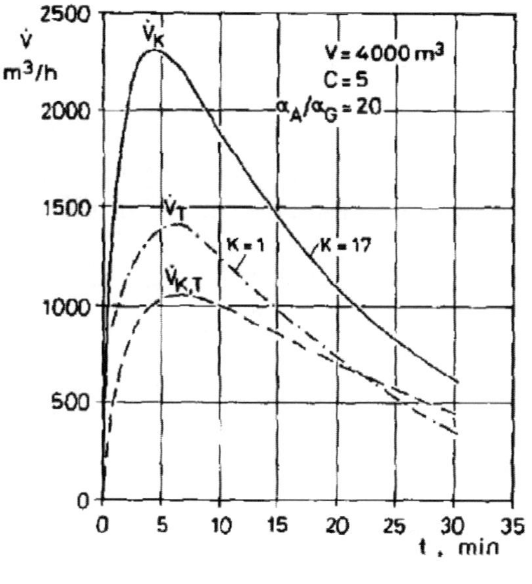


**Figure 5.** $\dot{V}_{inbreathing}$ **profile – Fullarton Model** [6]

Fullarton performed a dynamic simulation on a 4,000 m³ empty tank intended for the storage of ethanol. The initial gas temperature was 55°C and was suddenly cooled by rain at 15°C. The rain fall was assumed to be an extremely heavy downpour (probability of 1 every 100 years) with $\alpha_A/\alpha_G = 20$ and $\alpha_G = 5$ W/m²K. Figure 5 above shows the resultant inbreathing profile for each of dry air, K=1, and saturated water vapor, K=17. The maximum thermal inbreathing occurs at time t = 6 min where the inbreathing of the condensable vapor space ($\dot{V}_{K=17}$) is *1.6 times larger* than that of dry air ($\dot{V}_{K=1}$).

## *4.2 PROTEGO© Model – Condensable Vapors*

PROTEGO© used their model [4] to obtain the inbreathing curves of five different vapors: methanol, ethanol, saturated water vapor, heptane and decane (Figure 6) and comparing them to the inbreathing curve of dry air in the same tank.

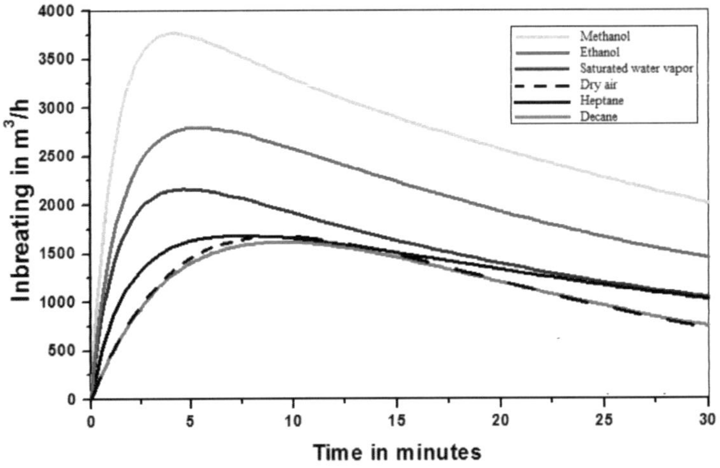


**Figure 6.** $\dot{V}_{inbreathing}$ **profile of different vapors – PROTEGO© Model [4]**

It can be seen that the condensation of volatile vapors has a significant impact on the inbreathing rate causing it to have a twofold increase. On the other hand, non-volatiles such as decane do not affect the inbreathing rate and yields the same inbreathing profile as with dry air.

**Table 5. Condensation Factor "$\chi$" Results**

| Vapor | $\chi_{PROTEGO}$ | $\chi_{Fullarton}$ |
|---|---|---|
| Methanol | 4.76 | 3-4 |
| Ethanol | 2.23 | 2-2.5 |
| Heptane | 1.61 | 1.5-2 |
| Water | 1.29 | 1.5-2 |
| Decane | 1.01 | N/A |

PROTEGO© calculated the condensation factor "$\chi$" for each of the simulated vapors and compared their values to that of Fullarton's (Table 4) and summarized the results in Table 5 above. The condensation factor values from this model are all in agreement with the published values [6] except the $\chi_{methanol}$ which is greater than the range specified by Fullarton (1986). A possible explanation for the latter is that PROTEGO© may have been more conservative (higher safety margin) when dealing with extremely volatile vapors like methanol.

# 5 Large-Scale Experimental Data

A thorough literature review yielded two sets of large scale data from actual experiments to study the thermal inbreathing of low-pressure storage tanks. Large scale datasets are preferred for model validation. The first dataset is for non-condensable gases (air) by *Sigel et al.* [7] and the second dataset is for condensable gases by *Holtkoetter et al.* [8].

## 5.1 Sigel's Experiment

Sigel performed actual data collection for the thermal inbreathing rate from a large tank of 617.5 m$^3$ capacity exposed to high solar radiation. The tank was empty without any trace of product vapor or liquid residue.

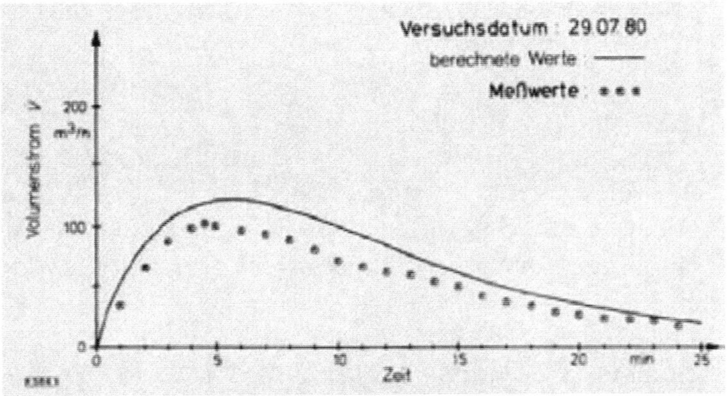


**Figure 7. Sigel Experimental Inbreathing Profile**

Sigel et al. [7] developed their own thermal inbreathing model to compare it with their experimental results. Figure 7 above displays the measured values of inbreathing (markers on the graph) versus the model (solid line). The inbreathing rates predicted by his model proved to be more conservative than the measured rates.

## 5.2 Holkoetter's Experiment

Holtkoetter performed an experiment on a 1.18 m$^3$ tank to measure the actual inbreathing rates and compare the results to his proposed model. This experimental investigation aimed at studying the effects of condensation on thermal inbreathing for three different vapors: water, isopropanol, and methanol.

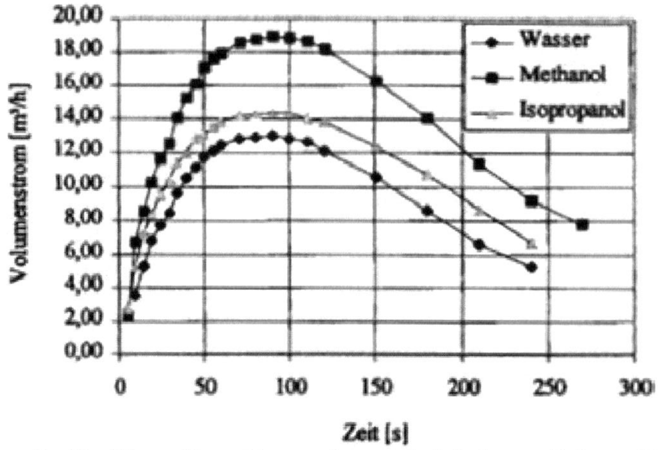


**Figure 8. Holtkoetter Experimental Inbreathing Profile**

The maximum thermal inbreathing is recorded after almost 95 seconds for all three gases. The maximum thermal inbreathing of methanol $\dot{V}_{max} \approx 19$ m$^3$/h is significantly higher than that for both water ($\dot{V}_{max} \approx 13$ m$^3$/h) and isopropanol ($\dot{V}_{max} \approx 14$ m$^3$/h) due to the fact that methanol is the most volatile vapor.

## 6 SuperChems™ proposed Model

The inbreathing profiles were generated using SuperChems™ dynamic simulation of storage tanks experiencing a sudden weather change. Dynamic models continually generate results that are used as the initial conditions for the next time interval calculation which perpetually drives the model to completion.

The proposed model first starts by defining a variety of objects to represent components of the system such as the storage tank vessel, the mixture that is stored, the piping segment and relief device, and the water sprays that are used to simulate a rainstorm. A built-in thermodynamics package is selected (modified Peng-Robinson EOS [9]) and scenarios are then created to collect the objects. Once the scenarios are created, single and multi-phase flow dynamics models are associated with the individual scenarios and the dynamic solutions for the inbreathing profiles are produced.

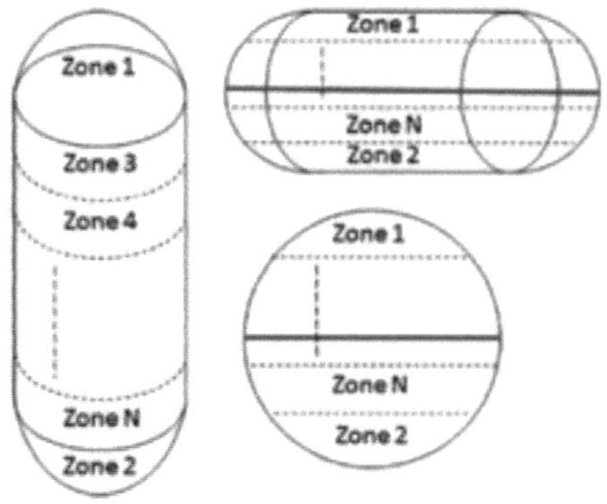


**Figure 9. SuperChems™ Segmentation Scheme**

An important feature of the software is wall dynamics which segments the vessel into multiple zones (Figure 9) allowing the user to account for heat conduction across the wall and between the zones using the effective conduction length method.

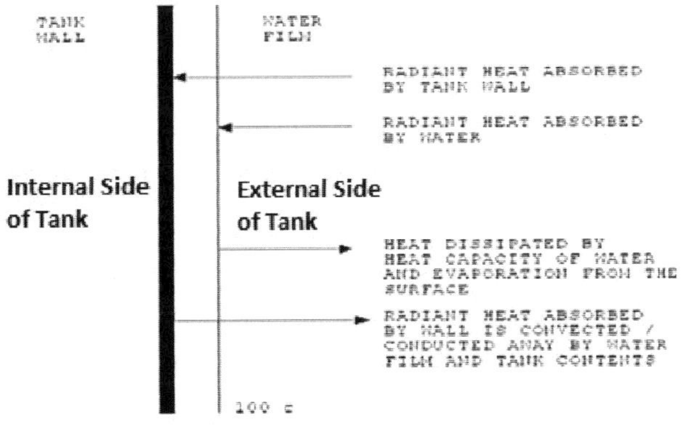


**Figure 10: Heat Transfer Mechanisms for a Tank during a Rainstorm**

Once segmentation is defined, heat transfer analysis is applied per wall segment (Figure 10) where the impact of rain, solar radiation, wind, and insulation is considered. Inside the tank, the wall-to-fluid heat transfer includes radiation and natural convection for the vapor phase; and nucleate, film and pool boiling for the liquid phase.

It should be noted that many parameters from the existing models and large-scale datasets were missing (vessel, dimensions, wall thickness, mixture composition, heat transfer coefficients, rain intensity, etc.) and had to be scientifically estimated in our dynamic simulations.

### *6.1 SuperChems™ Model Prediction for Non-Condensable Vapors (Large-scale Data)*

The dimensions of the 617.5 $m^3$ empty tank used in Sigel's experiment [7] were introduced as input parameters to the SuperChems™ vessel object. The rainstorm used in this simulation had an intensity of 0.38 inch/h and a duration of 30 minutes with a temperature of 15°C. The inside heat transfer coefficient value was set to 5 $W/m^2K$ as specified by API 2000 [1].

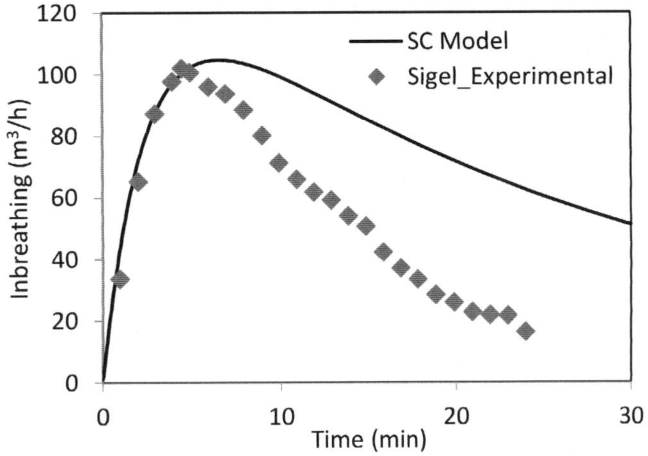


**Figure 11. Simulation of Sigel's Tank using SuperChems™**

The maximum value for inbreathing predicted by SuperChems was 104 $m^3/h$ at time t = 6.6 min, as seen in Figure 11 above. This value is at a 2% deviation from the actual recorded maximum. The model's slope matches the first part of the inbreathing curve up until the maximum. Although the slopes of the decay of the inbreathing are different, they do not contribute to the design of the relief device because the most important value extracted from such studies is the maximum inbreathing that will act as the design relief capacity of the venting device.

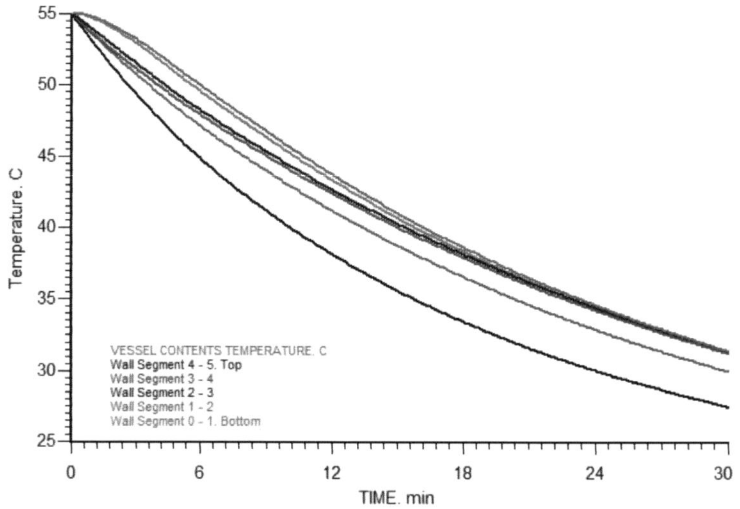


**Figure 12: Temperature Profiles of Sigel's Tank using SuperChems™**

Figure 12 above shows the temperature profiles of the wall segments and air inside the tank from the SuperChems™ simulation. Since Sigel [7] did not disclose the actual temperature profiles, therefore, we could not compare the computed temperature profile from SuperChems™ to the actual readings from Sigel's experiment.

## *6.2 SuperChems™ Model Prediction for Condensable Vapors (Large-scale Data)*

Holtkoetter's experimental setup [8] consisted of a 1.18$m^3$ tank with walls that are 1.2 mm wall thick. The three vapors Holtkoetter studied were: methanol, isopropanol and water. For each type of vapor, the tank was filled with a small amount of liquid which was then vaporized using three infrared radiators (2.5-4.2 kW). After vaporization step was complete, water sprays were activated to cool the tank and the thermal inbreathing was recorded (Figure 8).

Using SuperChems™, the vessel was defined, then the rain-wall heat transfer coefficient, $h_{rain}$, was fixed at 600$W/m^2K$ that corresponds to a rain intensity of 0.1262 $kg/m^2s$ as specified by Holtkoetter. The duration of the rain for this simulation was 300 seconds. The $h_{gas}$ was initially set to the API 2000 value of 5 $W/m^2K$, but was then adjusted to compensate for the vapor condensation inside the tank.

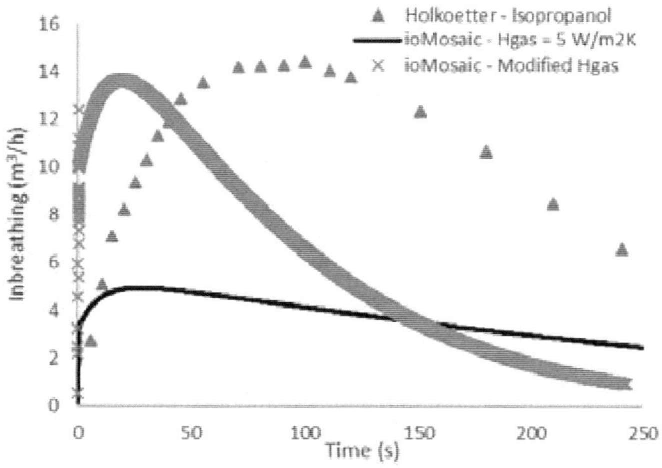


**Figure 13. SuperChems™ Simulation of Holtkoetter's Tank with Isopropanol**

Figure 13 illustrates the inbreathing profile for two different values of the free-convection heat transfer coefficient of the isopropanol vapor, $h_{gas}$, inside the tank. It can be clearly seen that using the API 2000 value $h_{gas} = 5W/m^2K$ underpredicts the thermal inbreathing by almost 62%. After numerous iterations, a modified value for $h_{gas}$ was found to be 30 $W/m^2K$. This higher value makes physical sense since the condensation of isopropanol occurring on the inner walls of the tank enhances the heat transfer between the vapor and the wall.

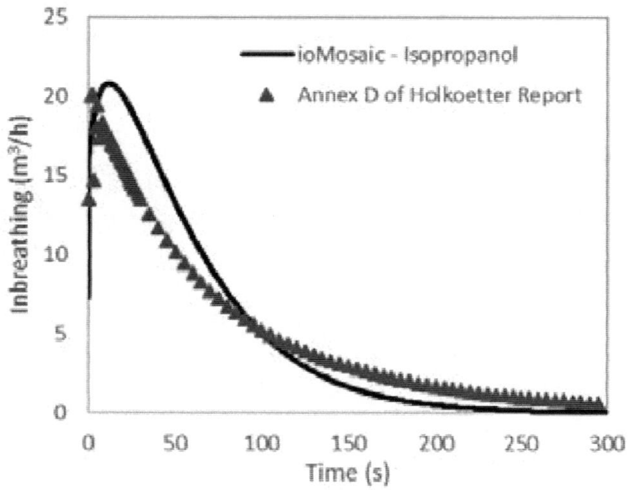


**Figure 14. SuperChems™ Reproduction of Holtkoetter's Inbreathing Model (Isopropanol)**

Although *Holtkoetter et al.* [8] performed experimental work to measure the maximum inbreathing, they did not, however, compare their experimental results to their proposed model. Hence, the SuperChems™ results were compared to Holtkoetter's model found in "Annex D" of his publication [8].

Figure 14 above proves that the SuperChems™ based model shows good agreement with the Holtkoetter model that considers spontaneous condensation. The predicted value for the $\dot{V}_{max}$ using SuperChems™ was 20.7 $m^3/h$ and is at a +3% error from the value reported by Holtkoetter [8].

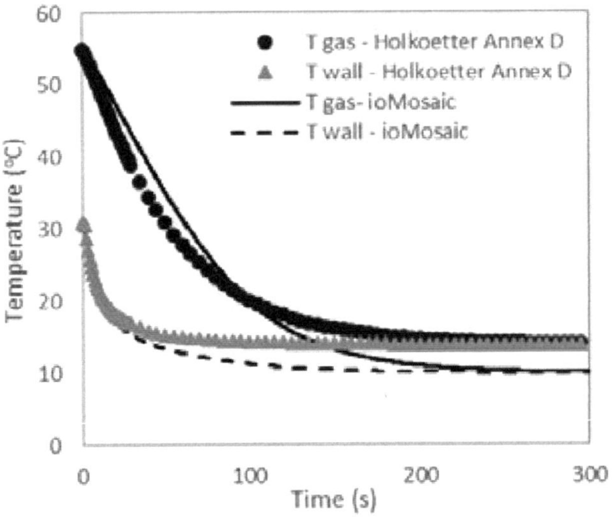


**Figure 15. SuperChems™ Reproduction of Holtkoetter's Temperature Model (Isopropanol)**

The gas and wall temperature profiles were almost accurately reproduced using the SuperChems™ model and are shown in Figure 15. Holtkoetter attributes the discrepancy between his model and his experimental result to the fact that the experiment was done in the month of December where the available water had a temperature of 8°C instead of 15°C. Colder rain will eventually lead to higher inbreathing rate values, therefore, the inbreathing should be inspected as a function of temperature to account for this 7°C difference in the water temperature. For more information about this method, please refer to the actual publication [8].

### *6.3 SuperChems™ Model vs. Published Models*

In addition to validating our proposed model against large-scale experimental data, the SuperChems™ based model was also compared to the published models listed in Section 3 and 4 of this paper [10].

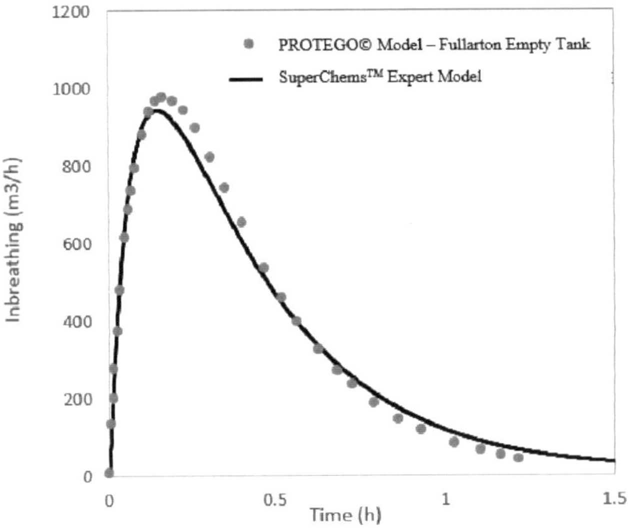


**Figure 16. SuperChems™ Inbreathing Profile for Air (Fullarton's tank)**

Figure 16 shows the predicted inbreathing profile of dry air using the SuperChems™ model and compared against the PROTEGO© model. The maximum recorded inbreathing from the SuperChems™ Expert simulation is around 940 $Nm^3/h$ which has a -4% deviation from $\dot{V}_{max}$ predicted by PROTEGO©. The temperature and temperature-rate profiles were also calculated and can be found in the actual publication along with the fits for the different models and vapors using SuperChems™ Expert [10].

### *6.4 Modified Heat Transfer Coefficient*

The heat transfer coefficient between the stored vapor and the inner-tank walls, $h_{gas}$, was modified for each type of volatile condensable vapor [10]. Table 6 below summarizes the values of $h_{gas}$ used in the two-phase dynamic model for a tank-wall thickness of 5 mm. These results show that condensation "enhances" heat transfer thus providing evidence that use of the API 2000 value heat transfer coefficients $h_{gas}$ will underpredict the maximum thermal inbreathing (Figure 13). Also, note that $h_{gas}$ is directly proportional with the vapor's volatility.

**Table 6. Modified $h_{gas}$ Values for SuperChems™ Two-Phase Model**

| $h_{gas}$ ($W/m^2K$) | Wall Thickness = 5 mm |
|---|---|
| Methanol | 40 |
| Ethanol | 30 |
| Water | 20 |
| Heptane | 5 |
| Decane | 7 |

## 7 API 2000 Conservativeness

The maximum thermal inbreathing rates, $V_{max}$, obtained from the SuperChems™ model were compared to those obtained using the API 2000 thermal inbreathing equation, Equation (1), to assess the performance of the API 2000 inbreathing method.

In order to determine the appropriate value for the C-factor in Equation (1), the vapor pressure and storage temperature of the stored gas are needed.

- Since all the investigated vapors (methanol, ethanol, water vapor, etc.) do not have vapor pressures higher than hexane, the vapor pressure was classified as "Similar to Hexane".
- The storage temperature of the gas is 55°C, which is greater than 25°C, so according to Table 1, the C-factor has a *value of 5*.

The storage tanks in this study were all uninsulated, hence, the insulation reduction factor $R_i$ has a value of one. The API 2000 inbreathing equation used for this case now becomes:

$$\dot{V}_{IT} = 5\, V_{tk}^{0.7} \qquad (8)$$

To determine whether the API 2000 inbreathing relief requirement is adequate, a deviation percentage of the simulated (or experimental) results the API 2000 $\dot{V}_{max}$ method is computed as follows:

$$\%\,Deviation = \frac{\dot{V}_{max,SC} - \dot{V}_{max,\ API\ 2000}}{\dot{V}_{max,\ API\ 2000}} \times 100 \tag{9}$$

where $\dot{V}_{max,\ API\ 2000}$ is the maximum thermal inbreathing obtained from Equation (8) ($m^3/h$) and $\dot{V}_{max,SC}$ is the maximum inbreathing from SuperChems™ model simulation ($m^3/h$).

## *7.1 API 2000 performance – Large-scale Experiments*

*Sigel et al.* [7] used a tank volume of 600 $m^3$, which when replaced in Equation (8), results in a maximum inbreathing value of 440 $m^3/h$. Figure 17 compares $\dot{V}_{max}$ using the API 2000 method to the actual recorded $\dot{V}_{max}$ values during Sigel's experiment. The API 2000 inbreathing prediction is conservative for Sigel's empty tank where the relief device is designed to handle an inbreathing rate 4.2 times larger than the recorded $\dot{V}_{max}$.

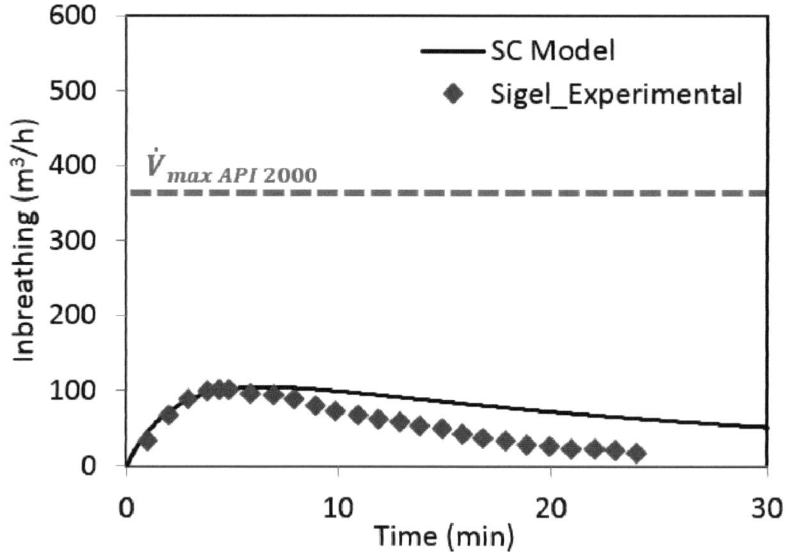


**Figure 17. API 2000 Inbreathing Requirements – Sigel's Tank with Dry Air**

On the other hand, *Holtkoetter et al.* [8] used a smaller tank with a capacity equal to 1.18 $m^3$ which results in a maximum inbreathing rate of 5.6 $m^3/h$ according to the API 2000 method – Equation (8). Figure 18 shows that the API 2000 inbreathing relief requirement is insufficient for water, isopropanol, and methanol vapors in Holtkoetter's tank.

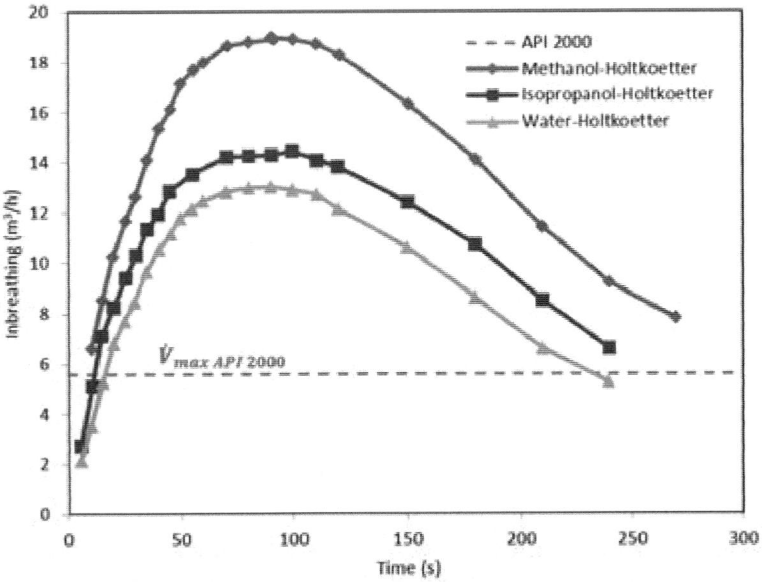


**Figure 18. API 2000 Inbreathing Requirements – Holtkoetter's Tank**

The recorded inbreathing rates are 2.3 to 3.7 times larger than the rate calculated by API 2000; this leads to an undersized relief device that puts the storage tank at a high risk of failure if a sudden change of weather occurs.

### *7.2 API 2000 performance – Computer Model*

The tank size used in Scholz and Weber's model [4] was 6,283 $m^3$ to study the condensable vapors. Replacing the tank volumes in Equation (8), results in a maximum inbreathing, $\dot{V}_{max}$ = 2,278 $m^3$/h. Figure 19 graphically shows the API 2000 $\dot{V}_{max}$ in comparison to the dynamically simulated inbreathing profiles for the different types of vapors, including dry air.

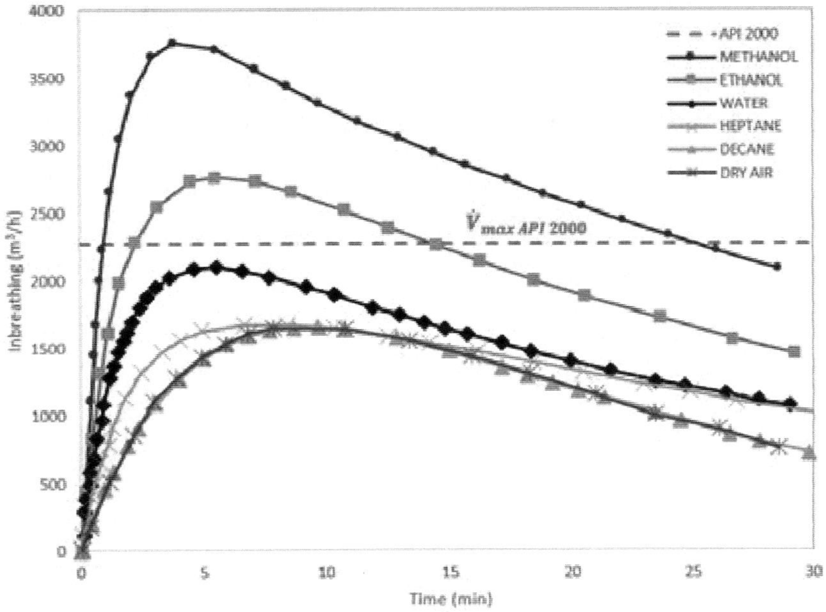


**Figure 19. API 2000 Inbreathing Requirements – Different Types of Vapors**

It can be seen that API 2000 method has a conservative prediction for the maximum thermal inbreathing of: (1) dry air, (2) heptane, (3) decane, and (4) water services. However, the API 2000 method underpredicts the relief requirement for the thermal inbreathing of a tank saturated with *methanol* and *ethanol*. The relieving capacity, for this case, has to be 1.7 times higher for methanol and 1.2 times higher for ethanol.

# 8 Conclusion

A dynamic simulation of the thermal inbreathing of low pressure storage tanks experiencing sudden change of weather conditions was performed using a SuperChems™ model. This model was validated by successfully reproducing the thermal inbreathing and temperature profiles of large scale experimental datasets and other available inbreathing models.

The thermodynamic variables specified in the API 2000 7th Edition standard have been tested for adequacy. It was found that the gas/wall free-convection heat transfer coefficient, $h_{gas} = 5 W/m^2K$, specified by API 2000 was adequate when the tank was filled with non-condensable vapors like air. However, higher values were suggested for condensing vapors with a range of 10 $W/m^2K$ to 45 $W/m^2K$ depending on the type and volatility of the vapor. As for the rain/wall heat transfer coefficient, the API 2000 value of 5,000 $W/m^2K$ proved to be too conservative, and even unrealistic, therefore leading to an oversized relief device. Suggested values for the rain/wall heat transfer coefficient were found to be in the range of 12 to 600 $W/m^2K$ depending on the intensity and angle of the rainfall.

Finally, the API 2000 7th Edition method for calculating thermal inbreathing was compared with several tanks of different vapor services. The method proved to calculate an adequate relief requirement for tanks storing air, heptane, decane, and even water vapor in one case. However, the application of the API 2000 thermal inbreathing method was not conservative enough when the vapor services became more volatile, like in the case of isopropanol, ethanol, and methanol, that resulted in high risk of failure for these types of storage tanks.

# 9 References


[1] API Standard 2000, Venting Atmospheric and Low-Pressure Storage Tanks, 7th Edition, March 2014

[2] P. Salatino, G. Volpicelli and P. Volpe. "On the design of thermal breathing devices for liquid storage tanks", Process Safety and Environmental Protection, 77.6, pp. 354-359, 1999.

[3] D. Brooks, "Thermal Inbreathing Requirements of Low Pressure Storage Tanks at Elevated Temperatures". Lloyd's Register Energy Americas, Inc. Spring DIERS Users Group Meeting, Presentation, May 2012.

[4] R. Scholz and R. Weber, "Cooling Down Industrial Tanks". Prepared for PROTEGO Braunschweiger Flammenfilter at API 2000 Meeting Presentation, Fall 2010.

[5] H. Foerster, K. Schampel, and H. Steen, "Atmungsvorgänge infolge von Witterungseinflüssen an Lagertanks für brennbare Flüssigkeiten", Physikalisch Technische Bundesanstalt (PTB) W-22. Braunschweig Univ. Phys.-Tech. Bundesanst, 1984.

[6] D. Fullarton, J. Evripidis and E.U. Shclünder, "*Influence of Product Vapour Condensation on Venting of Storage Tanks*", Institut für Thermische Verfahrenstechnik, Universität Karlsruhe (TH), 1986.

[7] R. Sigel, B. Kuxdorf, R. Meiss and H. Schwarz, "Praktische und theoretische Untersuchung der Beatmung von Festdachtanks". Chemie Ingenieur Technik, 55.5, pp. 377-379, 1983.

[8] T. Holtkötter, J. Shang and H.G. Scheckerm "Behälteratmung - Experimentelle Untersuchungen und Entwicklung eines Vorhersagemodells" (Master Thesis), 1995.

[9] G.A. Melhem, R. Saini, and B. M. Goodwin, "A Modified Peng-Robinson Equation of State," *Fluid Phase Equilibria,* 47, pp. 189-237, 1991.

[10] D.D. Abou-Chakra, G.A. Melhem, and R.J. Willey, "Dynamic Modeling of Inbreathing Requirements for Low-Pressure Storage Tanks" (Master's thesis), 2016. Retrieved from **http://hdl.handle.net/2047/D20209806**

# 478223 Studying the Relationship Between Inherently Safer Design and Equipment Reliability


**Abstract:**

Risk associated with a process depends on both likelihood and consequence of an incident. Inherently safer design is based on the use of technologies and chemicals with intrinsic properties that reduce or eliminate hazards. This philosophy is mainly focused on reducing the consequences associated with an incident. However, recent research has revealed that applying principles of inherent safety can have an effect on the likelihood element of risk as well. One of the major drawbacks of implementation of an inherently safer design (ISD) is that it can lead to risk migration, Thus, attempting risk reduction through implementation of ISD by mitigating the consequence element of risk can undesirably lead to an increase in risk by affecting the likelihood element of risk. This likelihood element of risk is the mathematical complement of system reliability. This study mainly focuses on assessing the effect of implementation of ISD principles on the reliability of a chemical process as it progresses along different stages of process design and manipulating this relationship to achieve maximum economic benefits while simultaneously minimizing risk associated with the process.

In this study, safety indices are used for quantification of inherent safety of a process. The choice of safety index depends on the stage of chemical project under consideration. Reliability of an equipment involved in a process is determined through the use of offshore and onshore reliability data (OREDA) and center for chemical process safety (CCPS) reliability databases for process equipment. The reliability of process system is then determined from the reliability of the individual equipment involved in the process. Lastly knowing values of safety indices and reliability of systems for competing designs during different stages of a chemical project (such as process selection stage, conceptual stage etc.), the required relationship between ISD and reliability is established for that particular stage.

Applying this methodology to process selection stage reveals that selecting a process that's inherently safer in terms of the involved reactions, chemicals and equipment might lead to risk migration, since the selected process might have lesser system reliability as compared to other competing process. When this methodology is applied to conceptual stage, where-in the competing processes are similar in terms of reactions and chemicals and vary only in terms of the equipment involved, it is observed that a process that has lesser number of equipment might be more reliable and inherently safer as compared to other competing processes. Lastly, when this methodology is applied for a specific equipment such as a reactor where the reliability is modeled as a function of design parameters of the reactor, it has been observed that within a certain range of design parameters, the reactor becomes sufficiently reliable and inherently safer.

Thus, this study provides a framework for designing of process systems and equipment that are reliable, economically profitable and inherently safer during different stages of project.

# 479377 Pressure Relief Analysis - It's about More Than Just Compliance

**Abstract:**

Pressure relief valves and disposal systems are the last layer of defense in petrochemical plants and refineries. Therefore, there are many codes and standards that must be followed to design those systems. Failure to ensure safe plant operations can result in fatalities, injuries, and environmental releases which can result in significant cost and poor public relations.

OSHA's CFR 1910.119 - Process Safety Management of Highly Hazardous Chemicals is designed to manage the hazards associated with facilities which handle highly hazardous chemicals. Many of the requirements in PSM are also valuable for facilities which are not governed by the requirement and application of these principals may improve safety for all facilities.

This paper will present the proposal that Process Hazard Analysis (PHA) be the heart of process for ensuring safe operations to prevent negative impact on companies' image and goodwill with the community.

# 480249 How to Leverage Limted Design Data Along with Historical Incident Data to Assess Inherent Process Safety During Early Stage of Design?

**Abstract:**

Inherent safety level quantification remains as one of the challenges for the inherent safety concept to gain industrial acceptance comprehensively. In current practice, most of the time safety assessment is conducted once the process design has been completed. At this stage of design implemented strategies look to control the hazard level. Any design modification necessary as a result of unfavorable safety performance at this stage of design, would be costly when compared to alteration during conceptual design stage [1]. It is beneficial to design industrial processes that are user friendly, inherently safer and have a level of tolerance for the deviations from ideal performance, which will eventually reduce the number of hazardous accidents. Therefore, both industrial experts and researchers are interested in safety indices that can be applied during the early design stage of a process, where the extent of knowledge regarding the process is very limited. Again, the available indices to quantify inherent safety still leave room for enhancement to represent process stream condition more accurately as a mixture rather than pure component(s) and to account for the interaction of all process parameters in determining the level of inherent safety [2].

In this work, a property based methodology to quantify inherent safety level has been proposed by assessing chemical and process unit risk at earlier stage of design. This will assist process designers in producing safer designs by the application of inherent safety principles in a more efficient and cost effective manner. It would offer process/design engineers an alternative to include safety without waiting until detailed design stage, where safety costs increase. Degrees of freedom at this stage would include alternative route selection, equipment sizing, and selection of operating conditions. The additional benefit of safety consideration in the early design stages is that it decreases the level of dependence on safety systems. The proposed safety assessment method here considers the perspective of five main safety factors associated with chemicals such as flammability, explosiveness, corrosiveness, toxicity & reactivity. These factors were linked to the process operating conditions (temperature and pressure). The process unit risk are calculated by classifying them into six main categories (i.e., pump, fired heat exchanger, pressure vessels, etc.) based on API 14C standard and identifying the undesirable events associated with each category. Major incidents database and the probability of unwanted incidents were used to link the undesirable events to operating conditions; this allows for the calculation of unit safety factors using the limited available information. Lastly, an enhancement technique to reflect the contribution of the individual components in the mixture is introduced based on the mixture properties, where the severity score was calculated for the mixture of chemicals in each unit at the actual operating conditions (whenever applicable) and multiplied by the flow rate of the mixture.

This proposed methodology of safety assessment gives a more intuitive, easy-to-understand way to represent the chemical and equipment safety factors posed by each individual unit. The steps in the methodology calculate, compare and prioritize the level of inherent safety of process streams during simulation work that influences the abovementioned safety factors in order to offer the user an insight into the risk associated with alternative design decisions. By prioritizing the streams based on the potential for chemical and equipment safety factors, the design engineers can easily identify the critical streams/units to be considered for improvement in order to avoid or minimize the risk. This methodology of quantifying inherent safety level will provide users another dimension to consider when selecting optimal design with multi objectives (e.g. techno, economic, and environmental).

The assessment of inherent safety level using the proposed method is demonstrated through a case study. HYSYS simulation is used to design two alternative designs for ammonia production.

1. Thiruvenkataswamy, P., et al., *Safety and techno-economic analysis of ethylene technologies.* Journal of Loss Prevention in the Process Industries, 2016. **39**: p. 74-84.

2. Leong, C.T. and A.M. Shariff, *Process route index (PRI) to assess level of explosiveness for inherent safety quantification.* Journal of Loss Prevention in the Process Industries, 2009. **22**(2): p. 216-221.

---

# Inherently Safer Design from Overpressure Protection Viewpoint

**Chris Ng**
**Sachin Kanade**
**Nancy Faulk**
**Siemens Energy, Inc.**
**4615 Southwest Freeway, Suite 900**
**Houston, TX 77027**
**chris.ng@siemens.com**



Prepared for Presentation at
American Institute of Chemical Engineers
2017 Spring Meeting and 13th Global Congress on Process Safety
San Antonio, TX
March 26 – 29, 2017

# Inherently Safer Design from Overpressure Protection Viewpoint

**Chris Ng, Sachin Kanade, and Nancy Faulk**
**Siemens Energy, Inc.**
**4615 Southwest Freeway, Suite 900**
**Houston, TX 77027**
**chris.ng@siemens.com**

**Keywords:** pressure relief, process safety, inherent safety

## Abstract

Pressure relief and flare system design is mostly evolving or iterative throughout the various engineering process design stages. This iterative nature stems from the need to capture the impact of engineering design changes on this critical layer of protection, which often is designed as the last line of defense for overpressure events.

The systematic review of pressure relief and flare system design basis at various engineering design and project execution phases will not only help to ensure adequate overpressure protection, but can also highlight opportunities for inherently safer processes and project cost savings. Often, evaluating trade-offs between process safety design improvements against project economics may be challenging.

In this paper, several case studies are presented to demonstrate how inherently safer design options and cost-effective solutions could be implemented across the various engineering design and project execution phases to minimize gaps in terms of overpressure protection between early and as-built designs.

## 1 Introduction

Figure 1 depicts the stages of a typical engineering project lifecycle, excluding decommissioning. Each stage has associated activities in terms of selection and refinement of design details which contribute to the overall integrity of the facility and economic feasibility of the project. The general approach involving timely review of pressure relief and flare system design could be applied to both large projects involving major Capital Expenditures (CAPEX) as well as smaller projects with limited scope.

Figure 1: Typical CAPEX Project Life Cycle.

During the early phases of design such as conceptual design and Front End Engineering Design (FEED) phases, the scope, objectives, and requirements are first defined, followed by development of a process basis (i.e., heat and material balance), Process Flow Diagrams (PFDs), and preliminary Piping and Instrumentation Diagrams (P&IDs). Equipment, line and instrument lists are then prepared, and preliminary equipment sizing and specifications for process and utility systems are generated. Design options are selected, and design drawings such as equipment layout plans are completed. A budget is developed and costs are estimated.

During later phases such as Detailed Engineering and Procurement, the remaining design information is added, such as vendor data, drains and vents, and piping details. The actual piping layout is verified and updates are incorporated. In addition, electrical, instrumentation, structural and building requirements to support the project are specified. Often Hazard and Operability (HAZOP) reviews are carried out. Typically procurement of the main equipment and auxiliaries takes place followed by construction, installation, commissioning, and Pre-Startup Safety Reviews (PSSRs), in preparation for start-up, operation and maintenance.

However, what is often overlooked during the early phases is input from pressure relief and effluent handling system design engineers. This paper will describe how early involvement by these engineers could help validate relief system design and reduce overall costs and delays later. The case studies described in Section 3 below offer examples of the potential savings.

Relief systems engineering analysis should be performed during various process design and project implemented phases, which allows identification of potential gaps as well as opportunities for improvement. Typical activities include:

- Overpressure scenario analysis
- Determination of relieving loads
- Selection of relief technology
- Sizing of relief systems
- Evaluation of relief device inlet loss and backpressure effects
- Depressuring analysis
- Hydraulic evaluation of common relief headers
- Flare radiation calculations
- Dispersion analysis, including suitability of atmospheric release
- Acoustic induced vibration (AIV) analysis
- Presenting findings and developing mitigation
- Developing comprehensive documentation
- Reviewing impacts of design changes and updating documentation to capture changes are they are implemented.

It is recommended that the reviewer(s) be well-versed in the appropriate industry standards, e.g., American Petroleum Institute (API) Standards 520 Parts I and II, 521, and 2000 [1][2] [3][4]; American Society of Mechanical Engineers Boiler and Pressure Vessel Code (ASME BPVC) Sections I and VIII [5][6]; as well as corporate guidelines and site-specific requirements. In addition, it is recommended that the reviewer establish and utilize a consistent approach for review of all designs, irrespective of the Engineering, Procurement, and Construction (EPC) companies working on the project.

# 2 Benefits of Early Review of Relief Systems Requirements and Design

## *2.1 Early Involvement Allows Critical Input*

Early involvement of relief system engineers can allow critical input to be incorporated which influences selected design options from an overpressure protection point of view, thus promoting an inherently safer design. By careful consideration of a number of factors, the design team has the potential to reduce overpressure relief and flaring requirements which in turn can reduce sizing of relief devices and associated piping, as well as flare components such as headers, knockout drums, flare stack, and flare tip. Potential reduction of environmental emissions is also realized. The examples below demonstrate some ways inherently safer design options and cost-effective solutions were implemented to minimize gaps in terms of overpressure protection between early and as-built design. In addition, early review can help to reduce or minimize the need for later design changes or mitigating measures, which could add significant costs and delay completion schedule for the project.

## *2.2 Identify Opportunities to Increase Inherent Safety*

### 2.2.1 Centrifugal Pump Design

Consider a centrifugal pump as a pressure source for downstream equipment, as in Figure 2. As it stands, P-1 is capable of overpressuring the downstream vessel, V-1, in the event of a blocked outlet or an inlet control valve failure.

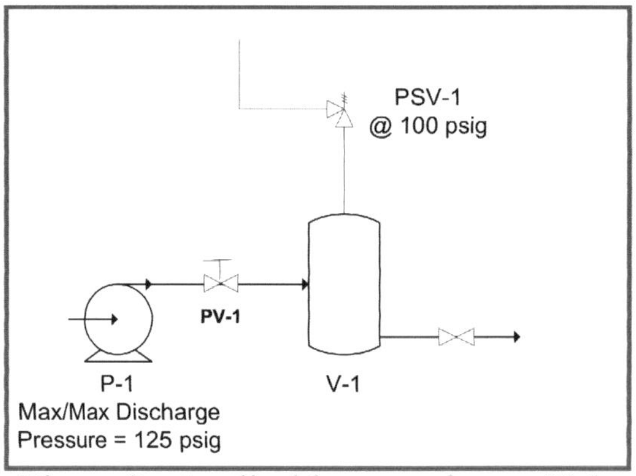


Figure 2: Applicable scenarios on V-1.

By properly sizing this pump during the design phase, however, it may be possible to minimize the relief rates for blocked outlet and control valve failure-open overpressure scenarios of the downstream equipment or even to eliminate the applicability of such scenarios altogether. Working with the process and mechanical engineers as well as rotating equipment designers, the maximum pump impeller size or speed and/ or maximum pump suction pressure could be carefully specified within process operation limits, which in turn limits the maximum pump discharge pressure. By reducing the maximum discharge pressure of the pump to a value less than the downstream equipment Maximum Allowable Working Pressure (MAWP), the above-noted scenarios may not be applicable, as in Figure 3a. An alternative to this, as shown in Figure 3b, is to ensure that the downstream equipment MAWP would be higher than the maximum discharge pressure of the upstream pump. These engineering design choices, which would result in minimizing relief loads or eliminating applicable scenarios, helps make the process inherently safer.

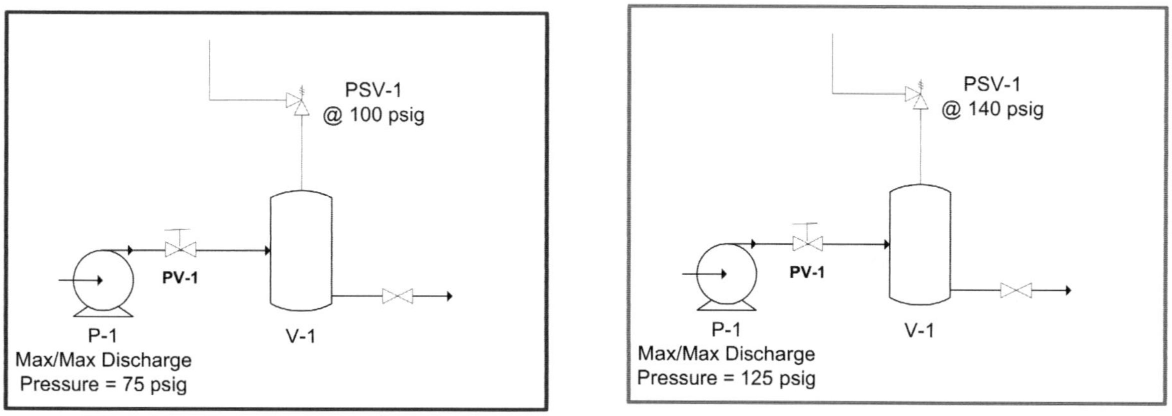


Figure 3: a) Limiting pump discharge pressure. b) Increasing vessel design pressure.

## 2.2.2 Pump and Compressor Discharge Check Valve Design

A related example involving pumps (and compressors, for that matter) is to ensure appropriate discharge check valve(s) are allocated to mitigate potential for reverse flow, impacting upstream equipment. In some instances, such as in Figure 4, feed could continue into an upstream vessel such as a feed surge drum if the liquid outlet pump is stopped with a discharge check valve stuck open. If this relief scenario is not identified by the relief designer, the relief valve on the feed surge drum may not provide adequate relief capacity.

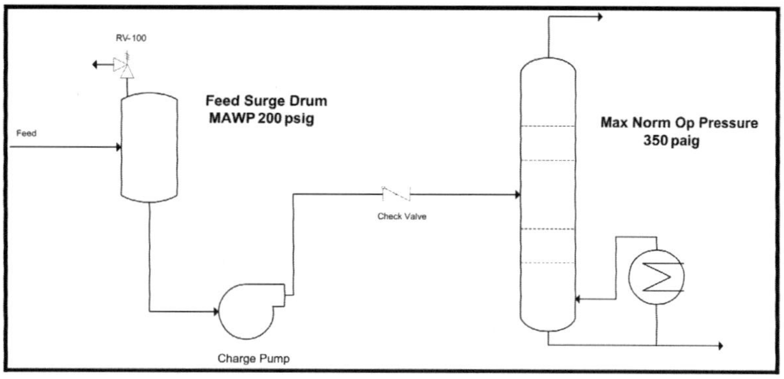


Figure 4: Applicable scenario on Feed Surge Drum

Allocating dual check valves in series of similar (or diverse) design may be required to minimize the backflow rate in the event of a check valve failure; if they meet the requirements set forth in API STD 521, the backflow rate may be limited to check valve leakage instead of check valve failure [7]. Careful specification of the MAWP of the upstream vessel may also help mitigate the overall relief requirement. Consideration of such issues early on is critical to avoid potentially costly design changes later as well as to increase the safety of the process; however, this is commonly missed due to factors such as improper management of change or poor communication between different groups designing upstream and downstream systems.

In a similar fashion, an early design review of several capital projects was performed, focusing on inherently safer design. One recommendation was proposed and accepted to raise the design pressure of the feed surge drum to mitigate the scenario in which a shutdown of a downstream charge pump results in reverse flow to the surge drum, while normal feed to the surge drum continues. By doing so, the feed and reverse flow sources would no longer be capable to overpressure the drum. Early review made the design inherently safer as the scenario was no longer applicable. Without this recommendation, installation of either a safety-instrumented system (SIS) or three 8T10 relief devices would have been necessary.

### 2.2.3 Control and Bypass Valve Design

Early input is critical in the selection and sizing of a control valve and its bypass valve, since the relief designer must account for the scenario in which *both* the control valve and its bypass valve fail open, as discussed in API STD 521 on "Pressure-relieving and Depressuring Systems". Early engagement of a pressure relief systems expert will help optimize the control valve/bypass and pressure relief requirements to provide adequate overpressure protection within the process operation envelope and requirements, while maintaining industry design and safety best practices. In one instance, a 2" gate valve was specified as the bypass for a 1" control valve, as shown in Figure 5. During a later design stage, it was discovered that this installation resulted in a very undersized relief device for the scenario described above, and thus the need to replace the bypass valve with a different and smaller-capacity type of valve, such as a globe valve. As in this instance, if the nominal size of the installed bypass valve is bigger than the control valve size, it necessitates careful review and possible re-design.

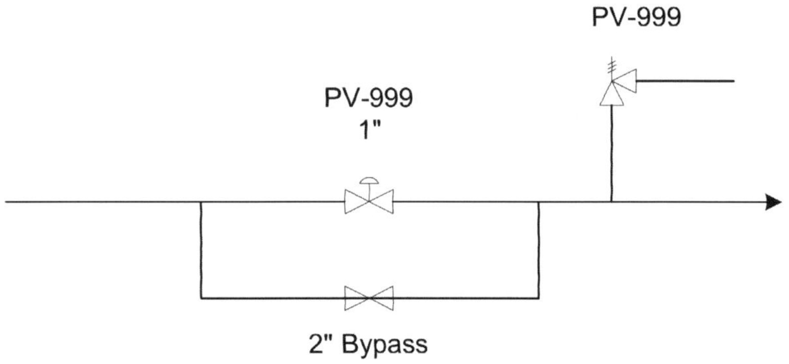


Figure 5: Bypass valve initially specified results in undersized relief device

### 2.2.4 Static Head

Static head is often overlooked in the preliminary design phases as equipment layouts may not yet be available. Relief designers should encourage vessel designers to specify vessels to be designed to be liquid full at the requested MAWP. Alternatively, relief device set pressures can be reduced to compensate for static head effects provided that the vessels can be filled above the top weld seam (into the top head). In another instance, the effects of static head on the set pressure of multiple relief valves had not been accounted for by an EPC company during detailed engineering. The potential for the internal pressure of affected equipment to exceed MAWP plus allowable accumulation per code was identified, allowing the relief valve set pressures to be adjusted prior to field installation.

### 2.2.5 Detailed Design Reviews

An independent pressure relief design audit was conducted for a large operating company. Relief design analysis and calculations had been performed by an EPC company as part of an expansion project. A review was conducted during the detailed engineering phase, during which it was found that most of the relief valve and associated piping upgrades which were being proposed by the EPC company were not required. This review included one application for which the EPC company had recommended upgrades to 6 relief valves (including recommendation to install two additional 8T10 relief valves), based on a recirculation flow rate which had been incorrectly included in a blocked outlet scenario relief rate estimate. Upon review, it was concluded that the existing relief valves were adequate and did not need to be upgraded. This reduced the potential for oversized relief capacity.

### 2.2.6 Design Gaps

In another example of early design review, engineers were able to identify a proposed design gap which would have resulted in costly and unnecessary modifications to a flare system. The proposed design project involved tie-ins of multiple large atmospheric relief valves to the associated flare for a process unit, resulting in inadequate flare capacity and the necessity of making expensive upgrades to this flare system. By performing detailed flare hydraulic modeling, relief engineers were able to show that re-routing the additional relief loads to an alternate adjacent flare having some spare capacity would be more practical.

### 2.2.7 Design of Feed Surge Drum

During a design review of a feed surge drum, additional examples of recommendations were developed to further increase the inherent safety, such as:

- By limiting the maximum discharge pressure of a feed pump to the proposed design pressure of the downstream equipment, a blocked outlet overpressure scenario was no longer credible for that equipment.
- Rerating the feed/bottom exchangers to account for static head in the event of overfilling in the associated distillation column.
- Car-sealing open (CSO) appropriate valves to ensure that various overpressure scenarios would no longer be considered as design contingencies and relief paths would be available for impacted equipment.

#### 2.2.8 Car-Sealing Strategic Valves

In a final example, several recommendations were developed and adopted prior to start-up for a Coker Vapor Recovery Unit (VRU) revamp project. These recommendations include the following:

- Re-rating equipment
- Car-sealing closed (CSC) the control valve bypass valves
- Reducing the control valve flow coefficient ($C_v$) to limit maximum flow through the valve in the event of failure open
- Increasing the vessel design pressure to account for static head
- Adding dual check valves to mitigate reverse flow

These changes helped make the overall design inherently safer, so fewer safeguards were required to reduce the risks associated with overpressure scenarios for multiple systems.

### *2.3 Relief Systems Engineer's Role*

Numerous other examples of inadequate design issues as well as missed opportunities to increase the inherent safety of the design are routinely identified by the relief systems engineers after the design has already progressed to a late stage or even after field implementation. Design review can begin early in the process, when preliminary P&IDs are issued for HAZOP or other types of Process Hazard Analysis (PHA). The relief systems engineer should also validate the design once piping isometrics are issued for approval and construction or after other design changes are proposed later. Taking these steps can help to ensure that design shortfalls can be resolved at the earliest opportunity. Making design changes early on is easier and less expensive than during later design stages and even more so than after commissioning and startup, thereby preventing subsequent shutdown, delay or major re-design work.

## 3 Critical Factors for Success

All pressure relief and effluent handling design documentation and supporting data should be stored in a central repository, for which access can easily be provided to all who have a need for such information. This access can be controlled, so that most viewers will have "read-only" access, and only authorized personnel can make required changes associated with Management of Change (MOC) items, thus keeping overall content up-to-date (evergreen) throughout the life of each facility as MOC activities, major expansion projects, or process modifications are implemented. These activities should trigger design review by relief systems engineers to ensure that relief design continues to be designed to provide adequate relief for all credible overpressure scenarios and that associated documentation is kept up-to-date.

In order for the review process to achieve the desired results, the three parties which are commonly involved (relief systems team, operating company, and EPC company) should work together to understand the goals of each party and how the design inputs from each party should be integrated in the various stages of engineering design and project execution. The following factors have been identified as critical for success:

- Conducting independent relief system design review at the right time in the project scope

- Communication among all parties
- Sharing of critical design data
- Effective management of change
- Discussing and resolving identified issues promptly
- Integrating relief system design review team with project design and execution teams

## 4 Conclusion

As shown in several examples in this paper, the systematic review of pressure relief and flare system calculations at various process design phases will not only help to ensure adequate overpressure protection, but can also highlight opportunities for inherently safer processes and promote overall project lifecycle cost savings. In addition, such a review process can easily be extended to resolve similar design issues commonly addressed during HAZOP reviews and integrated with the MOC workflow.

## 5 References

[1] American Petroleum Institute, *API Standard* 520, *Sizing, Selection, and Installation of Pressure-Relieving Devices, Part I-Sizing and Selection*, 9th Edition, 2014.

[2] American Petroleum Institute, *API Standard* 520, *Sizing, Selection, and Installation of Pressure-Relieving Devices, Part II-Installation*, 6th Edition, 2015.

[3] American Petroleum Institute, *API Standard 521, Pressure-relieving and Depressuring Systems,* 6th Edition, 2014.

[4] American Petroleum Institute, *API Standard 2000*, *Venting Atmospheric and Low-pressure Tanks*, 7th Edition, 2014.

[5] American Society of Mechanical Engineers, *ASME BPVC Sec I, Rules for Construction of Power Boilers*, 2015 Edition.

[6] American Society of Mechanical Engineers, *ASME BPVC Sec VIII-1, Rules for Construction of Pressure Vessels*, 2015 Edition.

[7] API STD 521, §4.4.9.3

**Disclaimer:**

# Three Reasons Major Event Risk Management Fails on Capital Projects

Denise Chastain-Knight

P.E., CFSE

# About the Presenter

- **Denise Chastain-Knight, PE, CFSE, CCPSP**
  - BS ChE, Georgia Institute of Technology
  - Senior Functional Safety Engineer
  - More than 31 years experience in engineering design, process improvements, process risk analysis and safety instrumented system implementation
  - Experienced in the chemicals, resins, coatings, monomers, polymers, pharmaceutical, silicate, bio-fuels, petroleum, power…
- **Affiliations**
  - CCPS Technical Steering Committee
  - Past Chair NCEES PE (ChE) exam development committee
  - Fellow AIChE



3

# Familiar Conversation?

How is the project going?

Great we are commissioning now and startup will begin in a couple weeks

I'll need to review the final SIL Verification and FSA so I can sign off on PSSR

SIL Verification? FSA? What are those?

4

exida

# Familiar Conversation?

Similar conversations occur all to often as capital projects are nearing completion.

Often the topic of the conversation involves an aspect of process safety that is critical for management of high consequence events that pose a significant risk.

Why does this keep happening?

 5

exida

# Missed Opportunity

- Capital projects offer a huge opportunity to make step changes in process safety
- Best practices can be implemented before capital is spent
- Clear guidance will assure that performance standards effecting process safety be implemented in the way the owner/operator desires.

 6

exida

# Operating Company

- Extensive experience managing risk in an operating environment
- Little practice managing process risk within a capital project
- Process Safety model relies heavily on culture, procedures and training
- Assumes the contractor knows how they want to approach complaince

 7

exida

# Engineering Design Firm

- Experienced at project design and execution
- Traditional project execution models are focused on delivering code compliant scope on schedule and budget
- Process safety deliverables are often excluded from project scope
- Knows what performance standards require, but clients approach compliance in different ways

 8

## PS Activities that Usually Happen

- Hazards Identification and Risk Analysis
- Relief system design
- Compliance with code mandated standards (ASME VIII, NFPA 30, NFPA 70)

 9

## PS Activities that Often Fail

- Compliance with RAGAGEP standards (ISA 18.2, ISA 84, ANSI, ASHRAE, Chlorine Institute)
- Inherently Safer Design
- Process Risk Mitigation
- Asset Integrity and Reliability
- Preparation for operation & maintenance and beyond

 10

## Why We Fail

- No or inadequate overall vision
- Insufficient preparation
- Deficient project execution

## Communication

 11

## Lack of Overall Vision

- What is the risk reduction required and what are the tradeoff priorities?
- How will responsibilities be divided?
- What tools and standards to be used?

 12

## Insufficient Preparation

- Philosophy and base procedures not defined by operator owner
- Capital project scope (deliverables) and schedule do not support process safety objectives
- Key role and staff competencies undefined

 13

## Deficient Project Execution

- Underestimate of the workload required and dependencies with other scope objectives
- Process Safety /Risk Management not treated as a discipline
- Reactive not proactive

 14

## How Project work

- Stage gate process
- Scope identified, schedule and budget developed to support scope
- Detail built in layers. Early definition important to minimize rework
- Often involve many contractor teams working on a subset of the scope

 15

## Establish Vision

- Define Process Safety – it is more than a PHA
- Prepare Philosophy and Design basis documents
- Define Competency Requirements
- Define Performance Metrics

 16

exida

## Prepare

- Identify process safety deliverables and expectation for contents of the deliverables
- Set milestones for project stage gates
- Understand the inputs and outputs for PS deliverables and reflect dependencies in the project schedule
- Assess competency and close gaps

 17

exida

## Execute

- Execute PS scope as a project discipline
- Include PS requirements in specifications
- Complete technical review of bid packages

 18

exida

# Pre-Project Documents

- List of Standards for Compliance
- Risk Management Plan & Procedures
- Hazards Assessment Procedures
- Cybersecurity Policy & Procedure
- Functional Safety Management Plan (organization)
- Functional Safety Management Procedures
- Approved vendors list for critical/safety equipment and instrumentation
- Alarm Philosophy
- Inherently Safer Design Philosophy
- Major risk management philosophies (fire, relief, etc)
- Contractor Competency Requirements
- Procedures for required studies (dropped object, etc)
- Maintenance Philosophy (isolation, clearing requirements etc)
- Measurement & Metrics philosophy
- Communications Philosophy
- Spares philosophy

 19

exida

# Preliminary Engineering

- Functional Safety Management Plan (project)
- Chemical Reactivity Matrix
- Hazards identification
- Conceptual SIF list
- Hazards Identification Study
- Define credible accident scenarios
- Major and long lead equipment specifications
- Controls Description
- Cyber Security Risk Assessment & Gap Analysis
- Maintenance Plan

 20

# Basic Engineering

- Project phase HAZOP
- CHAZOP
- HIPPS Studies
- Fire and Explosion Analysis
- Chemical release dispersion modeling
- SIL Determination
- SIS Safety Requirements Specification
- SIF Safety Requirements Specification
- Stage 1 Functional Safety Assessment (FSA1)
- Cyber Security Requirements Specification
- Countermeasure Inspection & Test Procedure
- Penetration Testing
- Cyber Security Assessment (CSA-1)
- Emergency notification, evacuation & response review
- Reliability, Availability, and Maintainability (RAM) analysis




# Detailed Design

- Balance of specifications
- Detail Design documents
- Alarm Rationalization
- SIL Verification
- SIS Factory Acceptance Test (FAT)
- Stage 2 Functional Safety Assessment (FSA2)
- Cybersecurity Factory Acceptance Testing (CFAT)
- Cyber Security Assessment (CSA-1)

## Procure and Construct

- Technical and financial bid reviews
- Operating & Maintenance Procedures
- SIS/SIF Proof Test Procedures
- Stage 3 Functional Safety Assessment (FSA3)
- SIS Site Acceptance and Integration Test (SAT, SIT)
- Pre-startup SIF Validation
- Pre-startup Cybersecurity Review (CSA 3)
- Training
- Operational Readiness Review
- Configure maintenance management system
- Emergency management plan

 23

## Operate & Maintain

- MOC for ALL changes
- SIS/SIF Proof tests & results analysis
- Revalidation of Detailed Risk Assessment
- Cybersecurity Vulnerability Assessment (CVA)
- Cybersecurity Assessment - CSA 4 & 5
- Periodic Revalidation of PHA

 24

25

# A Comparative Analysis of Safety Rating System

Linlin Wang[a]

[a]Mechanical and electrical engineering college, China University of Petroleum (East China)

**Abstract:** With the development of petrochemical enterprises, the scale and petrochemical enterprises frequent accidents, the high risk of petrochemical enterprises has been more and more extensive attention. It's particularly important to establish a sound safety rating system for the timely detection and elimination of risk factors, reduce the probability of accidents appears. However, in China there is less research and application on safety rating system. This paper summarizes the similarities and differences between the various systems and puts forward the development of China's safety rating system.

**Keyword:** Safety Rating System; Petrochemical enterprises

## 1 Introduction

Petrochemical enterprises have a lot of flammable and explosive, toxic substances, and the installation of large-scale and continuous production, can easily cause major accidents. In 2013 Sinopec Huangdao Oil pipeline explosion, 62 people were killed, 136 people were injured. With the occurrence of petrochemical enterprises accidents, the safety production of petrochemical enterprises has become one of the issues that government and national focus on. Evaluation and rating of the status quo of enterprise security can promote the enterprises to find their own problems, develop improvement goals and implement continuous improvement. At the same time, it can also improve the government safety supervision departments to find out the status quo of enterprises, determine the key supervision enterprises, and optimize the allocation of resources, enhance the safe production level of enterprises in China, it is also an important means to improve business efficiency. The Chinese government's security supervision department attaches great importance to the assessment of enterprise safety, since 2006, has issued a "Basic norms of enterprise safety production standards", " General specification for safety standardization of hazardous chemicals", "Guidelines for oil industry safety production standardization "etc., The government promotes the implementation of the safety production standardization assessment throughout the country, take the "enterprise self-evaluation, government review" approach to the enterprise rating, and achieved important work results. But in the implementation a lot of problems were also found. Based on the status quo of safety rating of petrochemical enterprises in China, this paper analyzes and compares the implementation process of enterprise safety rating system such as ISRS (International Safety Rating System) ,NOSA five-star comprehensive evaluation system and VPP project implemented by OSHA. It is expected to provide support for the selection and development of our country's petrochemical enterprise safety rating tool.

## 2 Status Quo of Safety Rating of Petrochemical Enterprises in China

Safety standardization rating is developed, implemented and reviewed by China's relevant laws and regulations, industry standards. It is led by the State Administration of Safety Supervision and implemented on a national scale. It is the most extensive and highest commitment safety rating system in China. In addition, some petrochemical enterprises received the support of government, and independently introduced tools such as ISRS and NOSA. For example, 2014 China Petroleum West Pipeline Company reached the rate 7 of ISRS, and became first enterprise which reached this level in China.

Based on the safety standardization assessment of petrochemical enterprises in China, it is found that there are three main problems in the process of implementing safety standardization assessment in our country:

(1) Safety standardization system and some other implementation systems, such as HSE system, have incompatible problems, resulting in enterprises to repeat engaging in several sets of systems, repeat records, repeat systems. Companies not only feel cumbersome, but also waste human and material resources;

(2) Enterprises have low motivation and a bad attitude to practice safety standardization. Standardization of safety production is a measure imposed by the government to improve the safety level of the enterprise, and enterprises do not really realize its importance. So that the work is easy to become formalize and surface, and did not achieve the purpose of improving the level of safety, but only to let the enterprises to meet the rating and inspection;

(3) China's safety standardization assessment consists of three grades. The grade of the score range is quite large. And the flexibility of the assessment is not sufficient. Besides, in the absence of experimental circumstances, the assessment has a low recognition for some enterprises which take a new process or a new management approach.

# 3 System Introduction

## 3.1 ISRS（International Safety Rating System）

ISRS（International Safety Rating System）is a rating system of international security measure, and improving and showing the level of health, safety, environmental and business performance and other aspects of the organization, designed to help organizations manage risk and promote continuous improvement of the evaluation system. Det Norske Veritas (DNV) international safety rating system (ISRS) eighth edition published in 2009 after decades of practicing and improving.

ISRS [7] international safety rating system has 15 elements, each element is divided into several sub-elements of a total of 119 sub-elements, and list the number of points for each sub-element of the assessment, the total number is 661. 15 elements are: leadership, planning and administration, risk assessment, human resources, compliance assurance, project management, training and skills, communication and marketing, risk control, asset management, procurement and contractor management, emergency preparedness , the event learning, risk monitoring, and review the results, in which risk management (including 3 risk assessment,9 risk control, 14 risk monitoring) system for the trunk. During the assessment process, the customer selects the subroutine according to the grade to be evaluated, as follows::

| Grade | 1 | 2 | 3 | 4 | 5 | 6 | 7 | 8 | 9 | 10 |
|---|---|---|---|---|---|---|---|---|---|---|
| Level range | 1-4 | | | | 5-6 | | 7-8 | | 9-10 | |
| Required number of subroutines | 30 | | | | 60 | | 90 | | 119 | |
| The number of customer chooses | 20 | | | | 15 | | 10 | | 0 | |
| The total number of subroutines | 50 | | | | 75 | | 100 | | 119 | |
| The lowest score of each program% | 10 | 15 | 20 | 25 | 30 | 35 | 40 | 50 | 60 | 70 |
| Average minimum score% | 20 | 30 | 40 | 60 | 60 | 70 | 70 | 80 | 80 | 90 |
| Live tour lowest score % | 65 | 65 | 65 | 65 | 70 | 75 | 80 | 85 | 90 | 95 |

## 3.2 NOSA integrated five-star management system

In 1951, the International Occupational Safety Association (NOSA) created a five-star safety management system by conducting a comprehensive hazardous and harmful factors identification and risk assessment to identify all possible hazard identification and develop appropriate management for reduce hazard and risk. NOSA management system regards hazard identification and risk management as the core, Heinrich's "iceberg theory" as the basis and PDCA manner as its operation mode. System consists of five main elements, 72 elements, more than 300 minimum standards, the implementation of 1200 rules and standard instructions (against NOSA Integrated Five Star system revision and reorganization, the system consists of 75 elements). System evaluation content mainly in the following two aspects:

(1) Performance evaluation

Evaluation content of NOSA five-star rating system mainly consists of five areas, particularly in the following table:

**NOSA integrated five-star rating system**

| project | Unit content | The number of elements | score | Share of the share |
|---|---|---|---|---|
| 1 | Building and housekeeping (factory) management | 11 | 450 | 15 |
| 2 | Mechanical electrical and personal protection | 17 | 600 | 20 |
| 3 | Fire and other emergency risk | 8 | 400 | 13.3 |

| | | | | |
|---|---|---|---|---|
| | management | | | |
| 4 | HSE accident recording and investigation | 5 | 400 | 13.3 |
| 5 | HSE management organization | 31 | 1150 | 38.4 |
| total | | 72 | 3000 | 100 |

Through five major categories of 72 elements of qualitative evaluation, the project almost involved all activities of production and operation. It evaluates each element from the "four levels, three aspects, five grades" to carry out a comprehensive calculation to measure whether the risk of the element has been effectively controlled. "Four levels" means: risk awareness, the file system, compliance, implementation effect. "Three areas" means: safety, health, environment. "Five grades" means: 0, 25, 50, 75, 100 five scoring levels。

（2） Experience assessment

Introducing DIIR (Disabling Injury Incidence Rate) rate of accidents concept of enterprise, and performance assessment combined score, drawn 1 to 5-star level. DIIR shows annual percentage of the total number of employees were injured, according to NOSA claims, statistics should include employees of the enterprise, the tertiary industry and outsourcing construction team personnel. DIIR calculated as follows：

$$\text{DIIR} = \frac{The\ number\ of\ accidents\ caused\ by\ work \times 200000}{Number\ of\ employees\ working\ hours\ per\ year}$$

Where：DIIR——Disabling injury incidence rate；

The number of accidents caused by work - refers to the work caused by or caused by the injury, so that the injured cannot or cannot fully perform the daily designated job work up to one or more working days；

200000-- base by 100 people work for a year (50 weeks), 40 hours a week calculated；

Number of employees working hours per year - Total number of employees × Actual hours per person per year

The final combination of the above two aspects of the results given the final star rating results, the corresponding assessment of the star method as shown in the following table：

**NOSA integrated five - star rating system assessment standard table**

| **Star rating** | **score（%）** | **DIIR（%）** |
|---|---|---|
| NOSCAR | ≥95 | ≤0.8 |
| Five star | ≥91 | ≤1.0 |
| Four star | ≥75 | ≤2.0 |
| Three star | ≥61 | ≤3.0 |
| Two star | ≥51 | ≤4.0 |
| One star | ≥40 | ≤5.0 |

NOTE: DIIR = the number of industrial accidents × 200000 (base) ÷ annual working hours of employees

## 3.3 VPP（The Voluntary Protection Programs）

The Voluntary Protection Programs (VPP) is a safety and health management system to prevent workplace accidents and death, by allowing enterprises to implement additional higher than the US Occupational Safety and Health Administration (OSHA ) Voluntary standards measures to achieve excellent safety and health management goals. OSHA is through a detailed site assessment and comprehensive site accident rate / illness rate to assess whether the corporate site has a good protection system。

The major elements of the program: management leadership and employee involvement; worksite analysis; hazard prevention and control and safety and health training. To assess something like this, there are three main forms of evaluation methods, including the inspection of documents, on-site inspection and interviews。

| | |
|---|---|
| **Leadership and employee participation** | Written safety and health management system, management commitment and leadership, Authority and Line Accountability (authority and responsibility), contractor employees, employees' participation |
| **Workplace analysis** | Baseline Hazard Analysis baseline risk analysis, hazard analysis regular activities, regular inspection, hazard reporting systems, accident / incident investigation, trend analysis |
| **Hazard Prevention and Control** | Whether there are effective measures to combat control and control, as well as written contingency plans safety and health training |

In addition, team members also need to The Total Case Incidence Rate (TCIR)以及The Days Away from Work, Restricted Activity and Job Transfer (DART) Case Incidence Rate evaluated to confirm the company level must be below the national average. After a series of reviews, OSHA decides that the applicant's qualifications are Star, Merit, Demonstration program, if the applicant fails to meet the requirements, return the application, and VPP level is relatively flexible, for those who do not meet the Star designing 1-Year Conditional Star Participants and Star Demonstration Programs.

## 3.4 Safety standardization assessment

Safety production standardization refers to the establishment of safety production responsibility system, the development of safety management systems and operating procedures, troubleshooting and monitoring of major hazards, the establishment of preventive mechanisms to regulate production behavior, so that the production links in line with relevant safety laws and regulations and standards Standard requirements, people, machines, objects, ring in a good state of production, and continuous improvement, and constantly strengthen the standardization of enterprise safety construction.

Safety standardization evaluation mainly taken independent assessment, external review of the way, self-assessed score of 60 points or more companies can apply for formal standardization rating, the final assessment is divided into one, two, three, one is the highest, from the security Production supervision and management departments to review the classification of supervision and management. During the review process, the following aspects are mainly reviewed: objectives, organization and responsibilities, safety production, laws and regulations and safety management system, education and training, production equipment and facilities, operation safety, hidden trouble investigation and management, major hazard monitoring , Occupational health, emergency rescue, accident learning, performance assessment and continuous improvement of the thirteen major aspects of compliance assessment, the final score to score how much. The enterprises that need to be graded need to carry out self-evaluation on a yearly basis to determine the safety standardization status of the enterprises and to re-order them every three years to urge enterprises to continuously improve the standardization work level.

## 4 System comparison results

Through the understanding of the information, compared the above systems, found the following the same and different:

1. Purpose

In essence, the above mentioned rating methods are all aimed at checking and enhancing the enterprise to achieve a certain level of security. However, the basic purpose of the rating system is different because of the different backgrounds of the development period. For example, the NOSA five-star rating system was developed in the 20th century, the rapid development of industrial and mining in South Africa and the quality of personnel and other issues caused the high rate of injury accidents, so its basic purpose is to reduce the high incidence of work accidents. In the final analysis, the rating is only a means and security is the purpose.

2. Applicable scope

ISRS is available in 11 languages and has been implemented on thousands of sites worldwide. The industries involved include marine, oil and gas, energy, nuclear, etc., and ISRS is introduced into the oil industry in China, such as China Petroleum West Pipeline Company reached the international security rating of 7.

NOSA five-star rating also can evaluate several kinds of industries including mining, forestry, railways, health care and natural gas, but its vast application in China is applied to the mining industry.

The VPP assessment program is applicable to all companies that use OSHA standards. VPP's participation companies include construction, agriculture, marine, oil and gas, health care and other industries.

Since the implementation of the Safety Production Standardization, under the guidance of the government, Safety Production Standardization has been applied in most enterprises in China, including industry and trade, power industry, transportation, hazardous industries, non-coal mines, construction industry, fireworks, coal industry, Its application covers most of the production enterprises.

3. Evaluators

Safety Production Standardization rating system take the way of "enterprise independent assessment then external assessment " to carry out the rating, Safety Supervision and Management Departments supervises the classification. The external review is conducted by the relevant experts. Similarly, the NOSA five-star rating system is evaluated by the company's team of experts. In the composition of the assessment team, VPP projects and ISRS systems have made innovations.

The VPP site assessment team is composed of experts from all areas of OSHA and SGE (The Special Government Employee), which has performed on-site evaluation. The implementation of the SGE project is intended for enterprises employees to participate in the process of VPP's on-site assessment. The implementation of this innovative project is more conducive to government and corporate information and ideas sharing.

ISRS needs to form an evaluation team consisting of two or more persons prior to evaluation, and to ensure the depth of the HSEQ field in evaluation, the team members contain relevant experts and do not restrict internal / external members, and All evaluation members will receive DNV training courses.

4. Rating method

Whether it is ISRS system, NOSA, VPP or safety standardization rating system are all required to compare the relevant standards and scoring rules with the actual situation of enterprises, through the document inspection, on-site visits and consultation to decide the company's compliance to standards.

However, NOSA and VPP projects need to combine other indicators before the final results. VPP not only requires companies to meet or exceed OSHA standards, but also requires companies to maintain below the average rate of disease released by National Bureau of Labor Statistics. And NOSA combines the evaluation rate of qualified with DIIR ratio to determine the stars should be given.

5. Follow up later

The "PDCA" model has been adequately applied in several rating systems, "plan, implement, check, action", adopt the infinite loop method of " find problems then solve problems, and then find the problem, and then solve the problem" to promote the continuous improvement of enterprise safety management system and continuously improve the level of enterprise security.

But for the rating implementation unit, the follow-up degree is different from others. Among them, the VPP project's follow-up work is very good. VPP participants need to submit an annual self-assessment report, OSHA's site assessment team will confirm whether the enterprise has made a corresponding improvement, and re-examination time interval varies by the enterprise level, such as star business, every 3-5 years conduct a reassessment. For companies with a one-year correction, if they do not meet the requirements after one year, they will withdraw the VPP application.

**System comparison table**

| | **ISRS** | **NOSA** | **VPP** | **Safety Standardiza tion** |
|---|---|---|---|---|
| **Basic Purpose** | get management | Reduce the | prevent | Promote |

| | | | | |
|---|---|---|---|---|
| | system to work more effectively | incidence of industrial accidents | workplace accidents and death | enterprises to achieve standardization and standardization |
| **Applicable scope** | ISRS is available in 11 languages. The industries involved include marine, oil and gas, energy, nuclear, etc., | Including mining, forestry, railways, health care and natural gas, but its vast application in China is applied to the mining industry | all companies that use OSHA standards | Covers most of the production enterprises, such as industry and trade, power industry, transportation, hazardous industries, non-coal mines, construction industry, fireworks, coal industry |
| **Evaluators** | Two or more people ( DNV training) | Company experts team | OSHA experts in various fields and SGEs who have conducted on-site evaluations | Independent assessment, external review The external review is conducted by the relevant experts |
| **Rating method** | Performance evaluation | Performance evaluation and DIIR | companies to meet or exceed OSHA standards, low disease rate | Performance evaluation |
| **Assessment Frequency** | | Self-evaluati on Twice a year Re-rating depend on clients | Self-evaluation once a year re-examination time interval varies by the enterprise level | Self-evaluation once a year Every three years have re-examination |
| **Grades** | 10 | 6 | Have Merit、Star、Demonstratio n program etc., grade flexibility is high | 3 |
| Voluntary | voluntary | voluntary | voluntary | Forced |

## 5 Conclusion

Combine comparison and thinking above, for the Chinese petrochemical enterprise safety rating system selection, provide the following suggestions:

1. For small private enterprises, Safety Standardization has been fully compatible with security requirements, for other state-owned enterprises, large-scale Sino-foreign joint ventures can choose to introduce foreign security rating system.
2. ISRS system covers the ISO standard and OSHA standard requirements, and also developed a special version of ISRS-Pipe, the rating system also covers the majority of indicators in China oil safety management system, ISRS has been applied in a lot of petrochemical enterprises in China But the ISRS system uses the lowest score of the total score in the first-level elements as the criterion for determining the final grade, so it is not easy to reflect the advantages of the enterprise in the final level judgment.
3. NOSA five-star rating system considers the performance evaluation and DIIR value together in the final evaluation to get the overall level, but when the enterprise finally decided to use the NOSA system, NOSA was originally used for mining assessment should be considered; For the company implementing OSHA standards can apply for the VPP program because VPP determines the enterprise level by checking whether the company is performing a higher-than-OSHA standard.


## Acknowledgement

This work is supported by the National Natural Science Foundation of China (21406271) and the China Scholarship Council “Innovative Talents International Cooperation Program” (Liu Jin Mei [2016] 7988). Thank you very much.


## Reference


1. https://www.dnvgl.com/services/isrs-for-the-health-of-your-business-2458
2. https://www.nosa.co.za/

# 478942 The Difficulty and Prospect of the Development of Quantitative Assessment Technology in the Field of Chemical Safety in China

**Abstract:**

**Abstract:** With the increasingly high demand for safety of chemical industry quantitative assessment technology has been widely recognized in chemical industry safety field because of its more accurate assessment results. Nowadays many existing quantitative assessment techniques,(such as quantitative risk assessment (QRA), reliability assessment, safety integrity level assessment (SIL) .etc)have been gradually introduced into the filed of China chemical safety to provide more assistance for the risk management of chinese chemical enterprises. However, some basic data of the quantitative assessment on chemical industry in China are lost due to the lack of systematic statistical analysis of data( such as equipment failure data, accident event data .etc), which leads to the situation that our country often adopt foreign related data when conducting a quantitative assessment. In result, results of quantitative assessment are out of their authenticity. In addition, because there is no more specific and more operational assessment guidelines for the part of quantitative assessment techniques issued in our country, it also affects the accuracy and scientific of the results. According to analyzing some misunderstandings and difficulties in chinese chemical safety field when carrying out the quantitative assessment technology, the paper puts forward some suggestions on the change and improvement of related industries to make quantitative assessment technology to provide greater help for the development of chemical safety field in China.

# Recipe for a *Sizzling* HAZOP

Poster Presentation to 13th GCPS

Angus Keddie, Process Safety Matters

## Benefits of Running an Effective HAZOP

- Commercial: Meeting schedule; Risk optimisation of facility
- Safety: Pick up all significant potential hazards
- Psychological: Limiting stress on participants
- Chairman: High probability that the chairman will be invited back

PROCESS SAFETY MATTERS

# Recipe Success? – Key Components

- Recipe Disasters......
- Provenance
- Preparation
- Supporting Team
- Kitchen Knives
- Ingredients
- Spice and Flavour
- Oven Conditions
- ....and How to Avoid Them
- Ta Dah!

PROCESS SAFETY MATTERS

# Recipe Disasters

## Recipe Disasters

- Behaviour to avoid if Effective HAZOP to be achieved:
  - Falling asleep
  - Fighting with team members
  - Allowing fighting between team members
  - Silence of participants
  - Permitting interminable discussions on minor issues
  - Scribing yourself
  - Working well in excess of 7 hours per day

## HAZOP Chair Provenance

To paraphrase the Bard "Some or born to HAZOPness, some achieve HAZOPness and some have HAZOPness thrust upon them"

There are around 5000 HAZOP chairmen worldwide all of them being equal, but some are more equal than others.

In order to optimise process safety for existing and new facilities worldwide, excellent PHSERs need to be carried out. A key ingredient of this is the performance of the facilitator. What personal characteristics combine to create this performance?

# HAZOP Chair Provenance

- Undergraduate degree in Chemical Engineering
- Professional exposure to relevant aspects of industry:
  - Process Design
  - Operations/Commissioning
  - Team Leadership/Management
  - Process Safety
- Top up with Masters in Process Safety
- Modal age range: 40 to 60
- CPD of IChemE Training Course
- Observation of Other respected HAZOP chairs to evolve best practice
- But like all great recipes, a little variety doesn't do any harm and may enhance the final product.

# HAZOP Chair Characteristics

Desirable attributes for the leader include:

- wide experience of all the stages of process hazard studies, including quantitative risk assessment;
- extensive experience as a HAZOP study member and, preferably, some as a HAZOP study scribe;
- training in the leadership of HAZOP studies;
- technical competence and the ability to quickly understand the system and its operation;
- meticulous attention to relevant detail;
- good analytical thinking;
- motivational skills including the encouragement of creativity and open speaking;
- independence from the project itself with no other direct responsibility to the project manager other than completing the scope of the HAZOP study.

## Fail to Prepare......

PROCESS SAFETY MATTERS

- Hold a Pre-meeting with key project stakeholders to establish significant draft tenets:
  - Project Background and Scope
  - Stakeholder Drivers
  - Optimal Methodology
  - Key Hazards
  - Putative Assumptions
  - Principal Personalities
- Outcomes and Terms of Reference inform Intro presentation and initial approach

## The Scribe - Sous chef or trusty side kick?

PROCESS SAFETY MATTERS

# LinkedIn HAZOP Discussion:

PROCESS SAFETY MATTERS

- **Does a PHA scribe need to have a technical background? – Ron Brachten, Nov 2014**
- **AK contribution: scribal attributes -**
  - Fast and accurate typing skills
  - Native or excellent English language skills
  - Sufficient *confidence* to capture the consensus and to overlook the inevitable spelling and grammar gremlins
  - *Stamina*
  - Good humour
  - Interest in the process
  - Threshold technical knowledge

  **Technical knowledge is handy but....**

# HAZOP Tools – Sharpen those knives

PROCESS SAFETY MATTERS

- Climate controlled room, good natural light
- A1 PIDs on walls or second screen
- Availability of Flip charts or White boards
- Wifi/LAN access to upload project tech info and wider references
- Good food during breaks and lunch
- Effective HAZOP recording software

PROCESS
SAFETY MATTERS

# Ingredients

- Meat and 2 Veg:
  - Process Description
  - Noded P&IDs
  - Intro Presentatiion
- Side Dishes
  - PFDs
  - Cause and Effect Diagram
  - Report from any previous HAZOP(s), HAZID
  - HAZOP Terms of Reference
  - Facility Design Standards
  - Plot Plan
  - Line List
  - Piping Classification Overview
  - Toxicity/flammability data for materials being handled
  - Relevant National/International Codes/Laws/Standards

PROCESS
SAFETY MATTERS

# The texture and flavour – the local element!

The experience of HAZOP can be vary significantly depending on technical, cultural and stakeholder perspectives:

***Stakeholders***

- Operators focus on operability and consistency of design. **(gooey and sweet)**
- The EPC contractor's focus is on schedule and no changes **(hard and dry)**
- The operator's safety team focus on strict code compliance **(tart and bitter)**

***Technical***

- Upstream – no steady state; little standardisation
- Downstream – Large through puts/inventories; replication

***Cultural***

- "Many nations separated by a common language – HAZOP!"

# Oven Conditions – HAZOP Ambience

The team works better if relaxed and motivated – try to make this happen early on:

- Introduce yourself to team members
- Get everyone to introduce themselves round the table, including what contribution they expect to make
- Directly engage all team members in the first 30 mins
- Allocate responsibilities based on member skills and attributes
- If team is flagging, call a break
- It two team members become adversarial, take a time out
- Smile

# Recipe Disasters – and their remedies

- Behaviour to avoid if Effective HAZOP to be achieved:
  - Falling asleep – stand up and walk around when facilitating
  - Fighting with team members – develop excellent rapport with team from the off;
  - Allowing fighting between team members – call time out; verbal warnings; firing culprit(s)
  - Silence of participants – engage in meetings (and in breaks).
  - Permitting interminable discussions on minor issues – set time limits for discussion; "is there anything we need to record at this point?"
  - Self Scribing – refuse!
  - Working well in excess of 7 hours per day – planning; negotiate schedule; identify key risks up front; prioritise nodes (Shell approach – 6 working hours per day, 4 days per week, 4 weeks out of 5)

# Ta Dah!

**Angus Keddie**
Director
Process Safety Matters
+44 (0) 7515897869
+44 (0) 1344 455800
info@processsafetymatters.com

## 477894 Bridging the Divide - OHS and Process Safety

**Abstract:**

The occupational health and safety field is well established, with a number of professional bodies, undergraduate programs and a comprehensive body of knowledge. Conversely, the process safety field has evolved largely over the past 40 years as an engineering discipline, with a focus on engineering solutions. Both of these fields utilise a risk management basis for decisions and actions. But both fields still largely operate independently of each other. There is much to be learned from sharing knowledge and responsible collaboration to deliver safer plants and operations from an OHS and a process safety perspective. This paper will explore how we can bridge the divide between OHS and process safety to achieve better safety outcomes. This includes better education and information flow between both professions as well as leveraging off existing systems to advance process safety.

# 20 Years Of Personal Process Safety Lessons

**Steve Streblow**
**ACS Engineering**
**10914 Milano Ct.**
**Richmond, TX 77406**
**streblowsr@yahoo.com**



Prepared for Presentation at
American Institute of Chemical Engineers
2017 Spring Meeting and 13th Global Congress on Process Safety
San Antonio, TX
March 26 – 29, 2017

# 20 Years of Personal Process Safety Lessons

**Steve Streblow**
**ACS Engineering**
**10914 Milano Ct.**
**Richmond, TX 77406**
**streblowsr@yahoo.com**

**Keywords:**

Amine tower; carbon dioxide; communication and restriction; competency; compliance; culture; documentation; failure analysis; intimidation; investigation; lessons learned; material of construction; mechanical integrity; pressure; stress; risk assessment; vessel repair.

## Abstract

This paper presents in detail one of the many process safety issues encountered by the author in his 20 years of industrial experience. This issue deals with an amine tower, but touches upon the following topics: carbon dioxide; communication and restriction; competency; compliance; culture; documentation; failure analysis; intimidation; investigation; lessons learned; material of construction; mechanical integrity; pressure; stress, risk assessment and vessel repair.

It describes better ways to address the situation and applicable RAGAGEP. It is hoped that sharing this information will enable others to learn from the past and prevent similar issues in the future.

## 1 Introduction

This paper presents in detail one of the many process safety issues encountered by the author in his 20 years of industrial experience. This issue deals with an amine tower, but touches upon the following topics: carbon dioxide; communication and restriction; competency; compliance; culture; documentation; drawings; failure analysis; intimidation; investigation; lessons learned; material of construction; mechanical integrity; pressure; stress, risk assessment and vessel repair.

It describes better ways to address the situation and applicable RAGAGEP. It is hoped that sharing this information will enable others to learn from the past and prevent similar issues in the future.

## 2 Amine Tower Leak

In one of the operating companies I worked for, I was supposed to be the technical support for many gas processing facilities, which included amine systems for removal of carbon dioxide. I was verbally informed of a small leak in one of the facilities amine regenerator towers which was stopped by placing a screw in the side of the tower, as shown in the photo (taken later) immediately below.

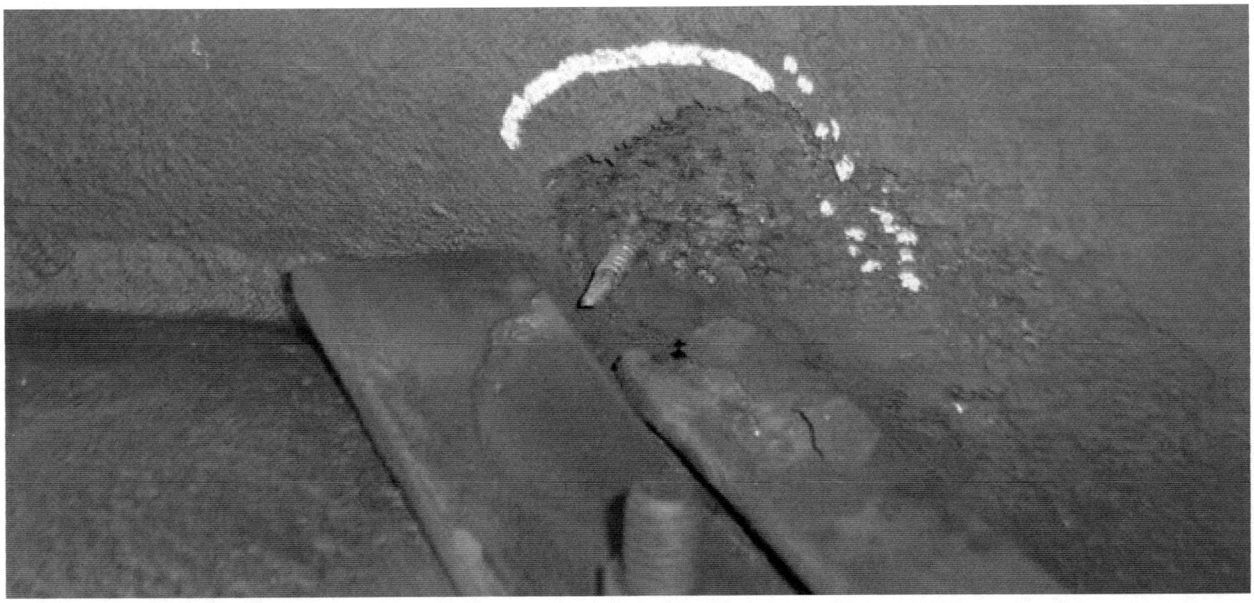

I was shocked for several reasons. First, I was not consulted for my technical opinion on how to proceed. Second, because after I communicated with the plant, they stated they were told by a service provider that they normally use that this type of repair was proper.

So much of this situation seemed wrong to me, so I provided the plant suggestions to restrict access to the area nearby the tower, while I determined how to proceed. The plant did put this area restriction in place.

I started to research what to do. First, I verified the pressure vessel law, as one can do by checking NB-370 at:
**https://www.nationalboard.org/PrintAllSynopsis.aspx?Jurisdiction=Select**
The location was in the state of Texas, which did not have a pressure vessel law. I next tried to find RAGAGEP for this. I called some pressure vessel manufacturers and had some brief conversations, which helped me focus my search. I found out that to keep the ASME code status, the repair should have been performed through a company which has an R stamp and filled out the appropriate paperwork. The R stamp status of a company can be confirmed at:
**http://www.nationalboard.org/ManufacturerDirectory.aspx**
The RAGAGEP to be used to maintain ASME code compliance is NBBI NB 23 (NBIC), National Board Inspection Code, Parts 1-3, reference 1.

If the vessel is a non-code vessel or if ASME code status does not need to be maintained, then one can refer to RAGAGEP API 510, Pressure Vessel Inspection Code: In-service Inspection, Rating, Repair, and Alteration, Tenth Edition, May 2014, reference 2.

After seeing that there was much technical knowledge needed for this activity that I did not have, I checked with management to see if there was someone better qualified than I inside the company, and I was disappointed to learn that there was not. Management did not seem concerned about the lack of expertise or the fact that I was being asked to work outside my area of expertise. It would be good to have a dedicated mechanical integrity expert in the company, since there is a great breadth and depth of technical knowledge needed.

I then tried to get the results of the previous mechanical integrity inspections for the tower, which I was told existed, but I was not provided. This withholding of information without any explanation seemed suspicious to me. This is a weakness in the culture. So, I tried to determine what the inspections should be for this item.

I tried to get the paperwork on the tower screw "repair", but none was provided, again showing a cultural problem.

I informed the plant that the company they were using did not have an R stamp and the screw action was not a proper repair.

I suggested contacting an inspection company to perform some external vessel wall thickness inspections, to see if additional thin spots were present. The external inspection was performed, which showed some additional thin areas.

I was verbally informed of a second small leak in the subject amine regenerator tower which was stopped by placing a screw in the side of the tower, as shown in the photo (taken later) immediately below.

It was disappointing that the facility took the same actions as last time, when they had been advised otherwise. I notified my management, and they did not seem concerned. I again reminded the plant that the company they were using did not have an R stamp and the screw action was not a proper repair.

I was verbally informed of a third small leak in the subject amine regenerator tower which was stopped by placing a screw in the side of the tower, as shown in the photo (taken later) immediately below.

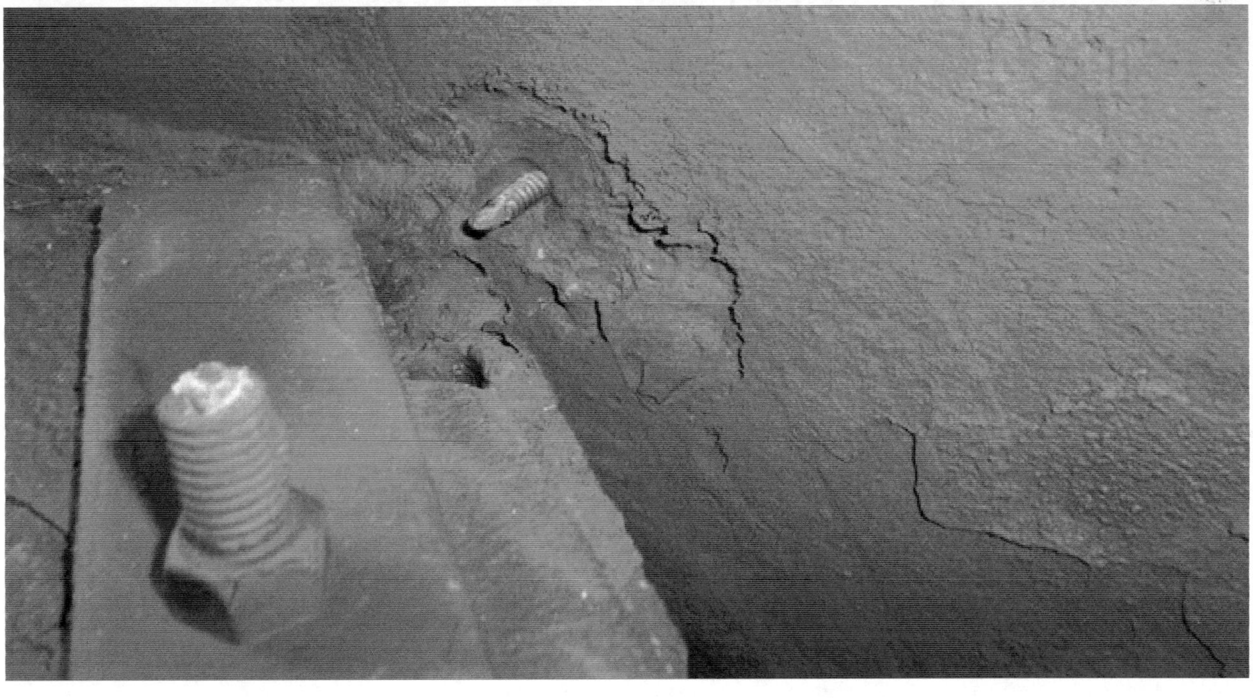

It was disappointing that the facility took the same actions as the last two times, when they had been advised otherwise. I again reminded the plant that the company they were using did not have an R stamp and the screw action was not a proper repair.

This time when I notified my management, they may have over reacted, since they now changed their direction drastically in stating that the right thing to do was shut down the facility, again without trying to get input from me or the plant manager. I was concerned about this drastic change in direction. When I spoke with management about this, they said this was a great success, since the facility was shut down in response to the latest leak. I informed management that I did not view it as a success to take this action in response to the third instance, but would have viewed it more positively if more action was taken on the first issue. A consequence analysis could have been conducted to determine the extent of the consequences. Then the likelihood could have been determined. Combining the consequence and the likelihood results in the risk, which could be used to help make a quantitative risk based determination, rather than a subjective decision.

I then asked management if we could share the learnings from this one site with the other 30+ sites in the company, and they said no, they would not do so until the incident happened at 3 or more facilities. It greatly disturbed me that management would wait for more issues to happen rather than sharing knowledge to prevent future incidents. This shows another cultural weakness.

Since the facility was now shutdown, it presented an opportunity to do some internal inspection, which was done. This effort resulted in many pictures, including the ones in this paper, as well as many wall thickness readings. The wall thickness readings were as thin as 0.110".

The original vessel wall thickness was supposed to be 0.375", with 0.125" corrosion allowance, which shows that the end of life wall thickness should be 0.250", well above the 0.110" measured. The vessel had a 50 psig design pressure, was 78" ID, with an 18 Ft skirt, 64 feet shell height and was made from SA-516-70.

I put together a spreadsheet, based on ASME Boiler and Pressure Vessel Code, Section VIII, Division I, 1980, reference 9, to evaluate what wall thickness was needed to handle different scenarios. Joint efficiencies were taken from reference 9, UW-12. Allowable stresses and overstress for combined loads were from reference 9, UG-23.

Seismic and wind speed were considered by calculating the moment at the bottom head to shell weld, and then converting to metal thickness.

I got the original calculations for the tower, and found that the original seismic calculations were in error. Seismic tension resulted in a required wall thickness of 0.048", which is less than the 0.110" measured wall thickness, so the wall thickness was acceptable. Seismic compression resulted in a required wall thickness of 0.113", which is just above the 0.110" measured, meaning the wall is not thick enough. A useful reference for seismic design is
Guidelines for Seismic Evaluation and Design of Petrochemical Facilities, Second Edition, ASCE, Reston, Virginia, 2011, reference 10.

For the wind loading, the tower was divided into 4 different zones. Wind tension resulted in a required wall thickness of 0.072", which is less than the 0.110" measured wall thickness, so the wall thickness is acceptable. Seismic compression resulted in a required wall thickness of 0.169", which is above the 0.110 measured, meaning the wall is not thick enough. A useful reference for wind loading is ASCE 7-05, Chapter 6, Figure 1A Basic Wind Speed – Western Gulf of Mexico Hurricane Coastline, reference 11. A second useful reference is Wind Loads for Petrochemical Facilities and Other Industrial Facilities, ASCE, Reston, Virginia, 2011, reference 12.

Reference 9, UG-27 was used to determine the minimum thickness for internal pressure and liquid height. The thickness needed for circumferential stress is 0.176", which is above the 0.110 measured, meaning the wall is not thick enough. The thickness needed for longitudinal stress is 0.052", which is less than the 0.110" measured wall thickness, so the wall thickness is acceptable. The thickness needed for compressive stress is 0.085", which is less than the 0.110" measured wall thickness, so the wall thickness is acceptable.

For the combined load temperature, pressure, level, wind and seismic, the thickness needed for longitudinal stress is 0.144", which is above the 0.110 measured, meaning the wall is not thick enough. The thickness needed for compressive stress is 0.306", which is above the 0.110 measured, meaning the wall is not thick enough.

Management tried to direct me to just solve the issue by reducing design pressure until the wall was thick enough. However, this was not a reasonable path forward, for several reasons, such as that if the same corrosion allowance was used, the remaining life (0.110" measurement minus 0.125" corrosion allowance) was 0. Also, there was no mention of the other aspects that would need to be considered, such as increased relief valve sizing at lower pressure. Also, the wall was not thick enough for several other cases listed above, such as seismic and wind, which are independent of the design pressure. Reducing the design pressure from 50 to 15 psig, meant that the wall thickness was still not adequate for the combined loading.

I also did some process simulation to determine if there could be changes made to the process and/or equipment to reduce the corrosion, such as:

- Changing type of amine
- Changing flowrate of amine
- Changing tower diameter
- Changing tower materials of construction
- Changing heat input to the reboiler

I was told to stop on this effort, which meant no better design would be in place when the next tower has problems. Production again won out over process safety when I was told to get a new tower the same as the last one.

A fitness for service assessment could have been conducted, however, production again won out over process safety when I was told to get a new tower the same as the last one.

A risk based mechanical integrity program could have been developed and put into place, but it wasn't. The emphasis was on production over process safety.

A root cause analysis could have been conducted, but the emphasis was on production and appearance over process safety. This is demonstrated by the operations direction to rapidly take the tower offsite to keep the plant appearance attractive. There was not an effort to engage in dialogue to discuss the issue. A better approach is to engage in dialogue with employees and seek their input and comments on suggestions. When management eventually realized the value of the investigation process, the tower was already destroyed. This was unfortunate, because the same type of failures starting appearing at other sites.

I told management that they needed to institute a mechanical integrity group. They eventually did, however, when I was informed of the head of the mechanical integrity group and the recent hire for our area, I could see that they were not qualified. I informed management of this, and I was told that was not my concern. This again shows a poor process safety culture. No attempt was made to resolve my valid concern, just to dismiss it.

I finally put many items in writing to my management, and was provided an individual visit by HR from headquarters. In this closed-door meeting, I was told many unacceptable things, including that I should not write about process safety, but management is glad to discuss orally. This shows a poor process safety culture. I resorted to writing when it was clear that oral was not effective. I did not agree with HR. I was then sent to an individual training session with the head of compliance, which was titled "Write Now, Pay Later." It is clear that the company tried to restrict internal written communication on safety and health, when it is an employee right to write written complaints to external agency, OSHA. This does not show a safety culture.

## 3 Additional RAGAGEP

One good resource is Guidelines for Mechanical Integrity Systems, AIChE CCPS, 2006, Wiley, reference 3. For pressure vessels, a useful item is "Table 9-1 RAGAGEPs for Pressure Vessels".

Another useful item in reference 3 is "Table 9-13 Mechanical Integrity Activities for Pressure Vessels". It has example activities and typical frequencies for:

- New Equipment Design, Fabrication, and Installation
- Inspection and Testing
- Preventive Maintenance
- Repair

Reference 3 also has Technical Basis for Activity and Frequency, Sources of Acceptance Criteria, Typical Failures of Interest, Personnel Qualifications, Procedure Requirements, Documentation Requirements for the following categories:

- New Equipment Design, Fabrication, and Installation
- Inspection and Testing
- Preventive Maintenance
- Repair

Reference 3, Chapter 11 contains an introduction to a risk based approach for mechanical integrity.

Reference 3, Section "12.4 Equipment Failure and Root Cause Analyses" provides useful information such as:

- "Step 1 - Evaluate Conditions at the Site."
- "Step 2 - Perform Preliminary Component Assessment."
- "Step 3 - Preserve "Fragile" Data Sources."
- "Step 4 - Perform Macroscopic Examination."
- "Step 5 - Perform Microscopic Examination (if warranted)."
- "Step 6 - Perform More Detailed Testing (if warranted)."
- "Step 7 - Test Under Simulated Conditions (if warranted)."
- "Step 8 - Make the Failure Mechanism Determination and Develop Corrective Actions."

Listed below are several additional API documents that contain excellent information related to mechanical integrity, with specific applicable sections listed.

API Recommended Practice 571, Damage Mechanisms Affecting Fixed Equipment in the Refining Industry, First Edition, December 2003, is reference 4. See sections:

- "4.2.14 Erosion / Erosion-Corrosion"
- "4.3.6 CO2 Corrosion"
- "5.1.1.1 Amine Corrosion"
- "5.1.2.2 Amine Stress Corrosion Cracking"

API Recommended Practice 572, Inspection of Pressure Vessels (Towers, Drums, Reactors, Heat Exchangers, and Condensers), Second Edition, February 2001, is reference 5. See sections:

- 7 for reasons for inspection
- 8 for causes of deterioration
- 9 for frequency and time of inspection
- 10 for inspection methods and limitations
- 11 for methods of repair

API Recommended Practice 579, Fitness for Service, First Edition, March 2000 is reference 6. Reference 6 states "The ASME and API design codes and standards for pressurized equipment provide rules for the design, fabrication, inspection and testing of new pressure vessels, piping systems, and storage tanks. These codes do not address the fact that equipment degrades while in-service and that deficiencies due to degradation or from original fabrication may be found during subsequent inspections. Fitness-For-Service (FFS) assessments are quantitative engineering evaluations which are performed to demonstrate the structural integrity of an in-service component containing a flaw or damage. This Recommended Practice provides guidance for conducting FFS assessments using methodologies specifically prepared for equipment in the refining and petrochemical industry. The guidelines provided in this recommended practice can be used to make run-repair-replace decisions to help ensure that pressurized equipment containing flaws which have been identified by inspection can continue to operate safely."

API Recommended Practice 580, Risk-based Inspection, First Edition, May 2002 is reference 7. Reference 7 states "1.1 PURPOSE -The purpose of this document is to provide users with the basic elements for developing and implementing a risk-based inspection (RBI) program. The methodology is presented in a step-by-step manner to the maximum extent practicable."

API Recommended Practice 581, Risk Based Inspection Technology, Second Edition, September 2008 is reference 8. In reference 8, "PART 1- INSPECTION PLANNING USING API RBI TECHNOLOGY", see sections:

- 4 for RBI Concepts
- 5 for Pressure Vessels

In reference 8, "PART 2 - DETERMINATION OF PROBABILITY OF FAILURE IN AN API RBI ASSESSMENT", see section 8 for "Damage Factor – Amine Cracking".

Reference 8, "PART 3 - CONSEQUENCE ANALYSIS IN AN API RBI ASSESSMENT" covers hole size, release rate calculation, release inventory, impact of detection systems, flammable and explosive consequences, toxic consequences, non-flammable non-toxic consequences, component damage and personnel injury consequence areas, financial consequences.

## 4 Conclusions

This paper presented in detail one of the many process safety issues encountered by the author in his 20 years of industrial experience. This issue deals with an amine tower, but touches upon the following topics: carbon dioxide; communication and restriction; competency; compliance; culture; documentation; failure analysis; intimidation; investigation; lessons learned; material of construction; mechanical integrity; pressure; stress, risk assessment and vessel repair.

It describes better ways to address the situation and applicable RAGAGEP. It is hoped that sharing this information will enable others to learn from the past and prevent similar issues in the future.

## 5 References

[1] NBBI, NB 23 (NBIC), National Board Inspection Code, Parts 1-3.

[2] API 510, Pressure Vessel Inspection Code: In-service Inspection, Rating, Repair, and Alteration, Tenth Edition, May 2014.

[3] Guidelines for Mechanical Integrity Systems, AIChE CCPS, 2006, Wiley.

[4] API Recommended Practice 571, Damage Mechanisms Affecting Fixed Equipment in the Refining Industry, First Edition, December 2003.

[5] API Recommended Practice 572, Inspection of Pressure Vessels (Towers, Drums, Reactors, Heat Exchangers, and Condensers), Second Edition, February 2001.

[6] API Recommended Practice 579, Fitness for Service, First Edition, March 2000.

[7] API Recommended Practice 580, Risk-based Inspection, First Edition, May 2002.

[8] API Recommended Practice 581, Risk Based Inspection Technology, Second Edition, September 2008.

[9] ASME Boiler and Pressure Vessel Code, Section VIII, Division I, 1980.

[10] Guidelines for Seismic Evaluation and Design of Petrochemical Facilities, Second Edition, ASCE, Reston, Virginia, 2011.

[11] ASCE 7-05, Chapter 6, Figure 1A Basic Wind Speed – Western Gulf of Mexico Hurricane Coastline.

[12] Wind Loads for Petrochemical Facilities and Other Industrial Facilities, ASCE, Reston, Virginia, 2011.

# Identifying Opportunities of Enhancing Safeguard Stewardship through IPL Rationalization

**SreeRaj R Nair**
**Technical Safety, Chevron U.S.A Inc.**
**Houston, USA**
**snair@chevron.com**

**Bernadette Thornton**
**Safety Systems, Chevron**
**ETC, Houston, USA**
**bernadette.thornton@chevron.com**

[Rev2]

Prepared for Presentation at
American Institute of Chemical Engineers
2017 Spring Meeting and 13th Global Congress on Process Safety
San Antonio, TX
March 26 – 29, 2017

# Identifying Opportunities of Enhancing Safeguard Stewardship through IPL Rationalization

**SreeRaj R Nair**
**Technical Safety, Chevron U.S.A Inc.**
**Houston, USA**
**snair@chevron.com**

**Bernadette Thornton**
**Safety Systems, Chevron**
**bernadette.thornton@chevron.com**

**Keywords:** Process Hazard Analysis, Layer of Protection Analysis, Independent Protection Layers, Rationalization

## Abstract

Ensuring effective safeguard stewardship is an integral part of the journey to address the challenge of moving from 'I think" to "I know" that a major accident hazard is managed. One of the steps in achieving effectiveness is to ensure that the process design is robust and that the safeguards provide the necessary protection to reduce the process risk.

Process Hazard Analysis and Layer of Protection Analysis are carried out to identify hazards, determine the potential consequence and evaluate the adequacy of the safeguards. Opportunities for further risk reduction are identified and risks are eliminated or reduced through established risk management programs. Though the study methodologies have been used for decades, there are opportunities for improvement.

This paper specifically addresses the opportunities in the evaluation of safeguards during the design of major capital projects. It explores some of the key concerns including lack of independence and variations in modifier application that could lead to over or under engineering; this could in turn may result in increased risk exposure. Additionally, this paper explains how the use of a defined methodology for the rationalization of Independent Protection Layers can assist in overcoming these challenges. The significance of performing IPL rationalization at appropriate stage of engineering progress is key in ensuring effective safeguards through design. A comparison of the advantages of rationalization exercise performed at various project phases is also discussed in the paper.

# 1 Introduction

A critical starting point in process safety management lies with the identification of facilities with potential for high impact (Safety and Environment) events and an understanding of the high consequence scenarios. For the high impact facilities, a detailed Process Hazard Analysis is conducted to determine the scenarios of concern. The potential impact from realizing those scenarios and the likelihood considering the safeguards (prevention and mitigation) is evaluated. A process safety management system is established with a system of "controls" (including critical safeguards verifications, and validation) to prevent the occurrence of these incidents or reduce their severity.

Layers of Protection Analysis (LOPA) was introduced as a detailed frequency evaluation method compared to Hazard and Operability Studies and as a simpler version compared to Quantitative Risk Assessment. This method is used to assess the adequacy of independent protection layers (IPL) and to ensure that required risk reduction is achievable (i.e., there will be no gaps post implementation of the risk reduction plan). LOPA is a systematic approach for applying IPL which follows guided principles and rulesets as prescribed in the industrial standards (AIChE 2001). CCPS publication (AIChE, 2007) provides guidance on choosing and evaluating initiating event frequency and independent layers of protection.

An independent protection layer (IPL) is a device, system, or action that is capable of preventing a scenario from proceeding to its undesired consequence independent of the initiating event or the action of any other layer of protection associated with the scenario (AIChE, 2015).

An effective IPL should be independent of the initiating causes, survive the initial event, and available to perform at the expected level for the intended duration. In order to determine the effectiveness, each safeguard which has been identified as a potential IPL needs to have the entire core attributes; the attributes are independence, specificity, integrity, reliability, auditability, access security, management of change. IPL can be either 'Active' or 'Passive' with Active IPL being those with the ability to detect (sense a condition or problem), decide (whether an action or intervention is necessary or not) and deflect (action capable of preventing consequence in a timely manner) and Passive IPL being those which primarily deflect to reduce the impact.

The following sections review some of the challenges in determining the effectiveness of active IPL and how to overcome the challenges through a structured approach known as IPL rationalization. It is followed by discussion on the inherently safer approach and its significance in improving the performance of the LOPA results. The potential economic benefits of IPL rationalization (Capital and Operational) are also evaluated with relevant project case studies.

# 2 LOPA Process Concerns

After evaluation of multiple projects, we have determined that the major LOPA concerns tend to fall into three categories: lack of independence, variations in modifier application, and appropriate documentation. The challenges in IPL related documentation can be addressed effectively by implementation of Process Safety Information management plan.

### *2.1 Lack of Independence*

The first issue, lack of independence, relatively an easy one to eliminate, yet it continues to appear in reviews. Most of this is due to a failure to understand packaged equipment controls and a failure to identify the critical actions required by a safety function. Packaged equipment controls typically are not well defined in early Process Hazard Analysis (PHA) / Hazard and Operability (HAZOP) studies. Assumptions concerning the independence of systems and poorly defined functionality tend to lead to IPL credit being given to these systems in error. For example, vibration monitoring systems tend to send requests to other systems, such as the Safety Instrumented System (SIS), to complete critical actions such as the closure of valves or motor stops. If a Safety Instrumented Function (SIF) exists with the same critical actions, then credit cannot be taken for both. The critical actions of each would have to be defined to remove the conflict and the system common cause would need to be reviewed to ensure that the combination will provide the appropriate layers of protection. Failure to validate packaged equipment independence as the project progresses leads to discovery late in the project that the required number of IPL is unavailable.

### *2.2 Modifier Application*

Differences in the team make-up and the facilitation of the LOPA studies leading to differences in the application of enabling events and conditional modifiers and therefore poor application of IPLs. Enabling events and conditional modifiers are applied to change the likelihood of a consequence or severity of a consequence, so these take into account factors such as the likelihood of personnel being in the area and the phase of operation. Failing to apply enabling events and conditional modifiers properly can lead to major operational issues, including alarm floods and excessive instrumentation. For example, if one facilitator applies a conditional modifier to a scenario and another facilitator does not, the required risk reduction is greater for the second facilitator and will require the application of additional IPL. This inconsistency in the application of enabling events and conditional modifiers leads to under-design (with gaps) and over-design (most-often) of instrumented systems and, potentially requiring additional maintenance and testing requirements.

Effective IPL rationalization can reduce the concerns of lack of independence and modifier application by providing an opportunity to look at the application of IPL on a facility basis rather than a scenario by scenario basis.

## 3 IPL Implementation Challenges

Poor application of IPL in the processes could result in facilities operated at elevated risk levels. The key challenges in the implementation of LOPA findings include the following:

- Risk reduction during design primarily relies on engineered systems, leaning to the route of 'Add-on' safety;
- Lack of credits for non-engineered / soft barriers as additional layers of protection;
- Operational inefficiency due to overly engineered protection system (multiple interlocks, trips for dedicated systems designed without considering overall facility reliability requirements);

- Regulatory requirements – prescriptive versus performance based approach;

Kletz and Amyotte [Ref. 3] have said, "Intensification, substitution, attenuation, and limitation of effects produce inherently safer design because they avoid hazards instead of controlling them by adding protective equipment. The term 'inherently safer' implies that the process is safer because of its very nature and not because equipment has been added to make it safer". However, this very point is often overlooked during design and often adding additional layers (mainly instrumented and hardware) are considered as improvement to safety. Industries' traditional solution to increased levels of hazard has been through more controls. For example, an offshore facility designed to be operated as 'normally unmanned' could become 'normally manned' just because of the manning distribution required to ensure the effectiveness (through inspection and maintenance) of the 'added on' engineered safeguards.

Projects have become more complex and personnel involved in projects rarely see the effect of their designs, so design complexity is less of a consideration and the more cost effective solution, typically an instrumented one, is implemented to ensure enough safeguards are present to manage the risk. Unfortunately, this leads to the massive problem of facilities having to manage large numbers of safeguards and IPL.

While industry is moving towards performance based approaches to define risk reduction, prescriptive regulatory environments still exist. Prescriptive environments ensure a minimal level of risk reduction is implemented, but they also can add challenges to the implementation of a performance based approach. Prescribed testing and testing procedures may provide some reduction in the overall risk of a facility, the tests rarely are sufficient to provide assurance of the effectiveness of the prescribed functions as IPL; therefore, additional testing must be performed on top of the regulatory required testing potentially requiring more personnel to complete ensure the functions meet their risk reduction target.

## 4 How to Overcome the Challenge

Eliminating hazards using the inherently safer design route is the most effective way to overcome the challenge of safeguard stewardship simply because there will be only fewer safeguards as IPL (Kletz and Amyotte, 2010). Time and effort can be spent on managing the asset integrity of pipe, vessels, and major equipment rather than managing hundreds of safeguards. In such cases, the projects compare options for managing risk. If elimination is not a viable option considering safety and operability factors, decisions are made to proceed with acceptable levels of risk using adequate safeguards/IPL. The following words of wisdom shall be considered as part of ISD approach (Nair 2009, Nair 2017, Stuart, et. al 2003, Kletz and Anyotte, 2010):

- More is not always safer
- Identify all hazards (you can't eliminate or manage a hazard which have not been identified)
- Design against accidental loads
- Design plants for easy and safe maintenance (*reliable equipment which requires less maintenance*)
- Simplify the process and minimizing the requirement to rely on human intervention

While companies acknowledge that management of IPL is necessary, the number of IPL being identified prioritizes finding ways to reduce the load on maintenance and testing personnel. One way to reduce the load is by using IPL rationalization. IPL rationalization can be defined as a process of defining the most effective set of IPL to provide the appropriate risk reduction while also reducing the number of management activities and operator tasks and the maintenance and testing load a facility must endure. So the question of how to define the most effective set of IPL must be answered.

IPL rationalization uses the results of a LOPA to identify IPL which are unnecessary to provide the facility with the risk reduction required. The process requires a functional safety Subject Matter Expert (SME), a process engineer and an operations representative to review the application of enabling events and conditional modifiers, identify SIF with multiple Safety Integrity Level (SIL) and ensure that SIF, critical alarms and basic process control system (BPCS) loops identified as IPL are, in fact, effective. During a LOPA, many of these items may be addressed, but as projects require more and more facilitators to make large HAZOP more manageable, more and more issues creep into the design. The smaller group allows for a more in-depth review of independence, consistency in identification of the SIF and its actions, and a way to compare the results of one study with another study to ensure IPL chosen between similar units and similar scenarios provide the appropriate risk reduction while applying the same enabling events and conditional modifiers. Most importantly, IPL rationalization provides a way for the project to review the IPL from a bird's eye view as well as scenario by scenario to identify whether IPL are appropriate to mitigate the risk for the overall facility as well as the individual scenarios. The most surprising element of the process is that it requires very little time to complete, especially with the correct pre-work completed.

The process of IPL rationalization is simple. First, the team must first identify the order of preference for implementation of IPL. This preference will take into account the facility's maintenance staff size, the competency of the staff and the operating basis of the facility. For example, some owners may prefer to take more SIFs as they have a good management program for them and no management program for other engineering IPL. Other owner may prefer other instrumented IPL over SIF as they do not expect to have the competency they would prefer to manage the IPL.

Secondly, the team reviews the application of enabling events and conditional modifiers to guarantee their effectiveness or applicability. Next, the IPL are reviewed in order of preference and where different IPL have been provided for scenarios with the same consequence, the use of different IPL is either rationalized or the IPL are revised to provide the most effective set. Unused IPL are either removed or removed from the IPL list.

Finally, the team reviews any recommendations to identify if there are LOPA gaps that may be addressed using existing IPL or whether new IPL must be created.

Experience has demonstrated that the process can be effectively used for brownfield facilities as well. However, it need to be noted that there could be differences in the IPL preferences and that unused IPL are less commonly found as facilities choose to use existing IPL rather than add new ones, even if the final result may be less efficient.

## 5 A Representation of the Benefits

The benefits of completing IPL rationalization depends on three things:

1. The objective of the project
2. The timing of the study
3. The regulatory environment of the facility

The objective of the project may affect IPL rationalization results as looking for ways to eliminate or reduce the numbers of a particular IPL type or may require the addition of IPL in other categories. This activity may not be a reduction of IPL, but rather a reallocation of IPL. Alternatively, the project may be looking to reduce the number of IPL overall by ensuring the application of IPL is streamlined. Illustrations of IPL rationalization benefits in terms of IPL and potential economic saving are given in Figure 1 and Figure 2. A recent project looking to reduce the number of alarms was able to achieve a 90% reduction in alarms with little effect to the number in other IPL categories due to the proper application of conditional modifiers. If one assumes that maintenance of these alarms requires approximately $100 per year and that the number of alarms was reduced from 1000 to 100, then it can be found that the facility saves approximately $90000 per year in alarm maintenance alone (*representative figures only*).

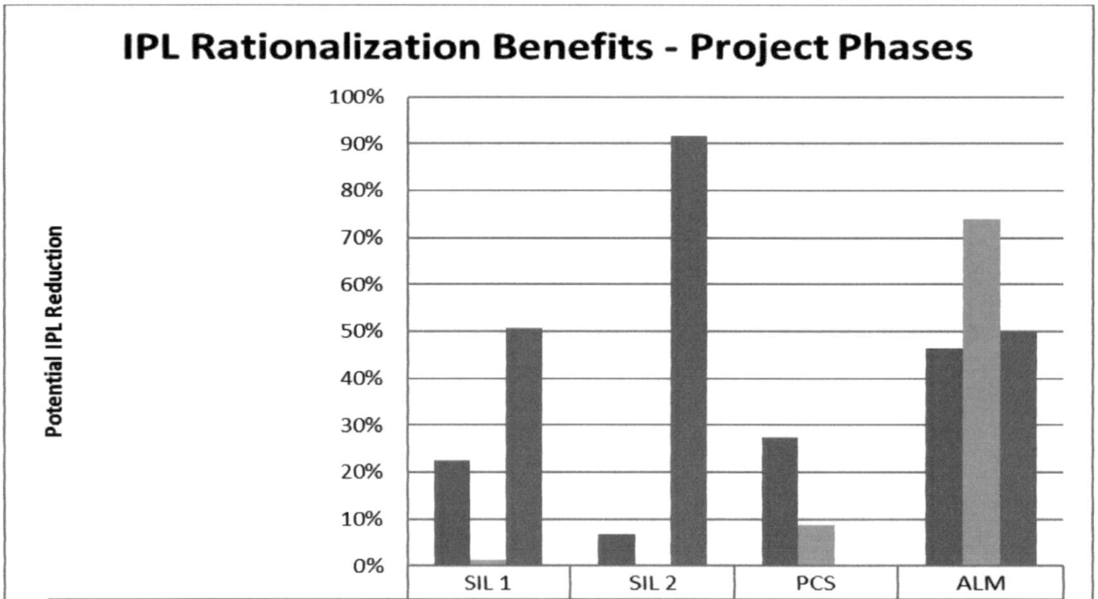


| | SIL 1 | SIL 2 | PCS | ALM |
|---|---|---|---|---|
| Project 1 - Basic engineering | 23% | 7% | 27% | 46% |
| Project 2 - Detailed design | 1% | 0% | 9% | 74% |
| Project 3 - Post commissioning | 51% | 92% | 0% | 50% |

**Figure 1: IPL Rationalization Benefits – At Different Phases of Project**

The timing of the study has a significant effect on the outcome of the IPL rationalization. Performing rationalization before the design is complete (i.e., before construction) provides opportunity to eliminate IPL, such as SIF, that are found to be unnecessary (Figure 1). A project may choose to keep the functions or alarms for operational or reasonable risk reduction justification/demonstration; but take the opportunity to reduce the testing or maintenance associated with these. In addition, the team may be able to identify designs which could be modified to provide a more inherently safer facility. A study performed late in the project will not achieve the same benefits. They may be able to make reductions in some areas, but if the design or installation is almost complete, there is only limited opportunity to remove unnecessary IPL (Figure 1). Alarm priorities may be reduced and maintenance and testing requirements may be relaxed, but the facility will likely be forced to retain most IPL. Performance of IPL rationalization post-commissioning provides even less opportunity for savings. IPL have not only been designed and installed, but they have already been commissioned and, in many cases, changing maintenance and testing requirements will require changes to maintenance managements systems that likely have already been completed, potentially adding cost short-term. A review of projects showed potential IPL reduction of 25-30% for projects in Basic Design / Front End Engineering Design (FEED) and 10% reduction for those in Detailed Design or later.

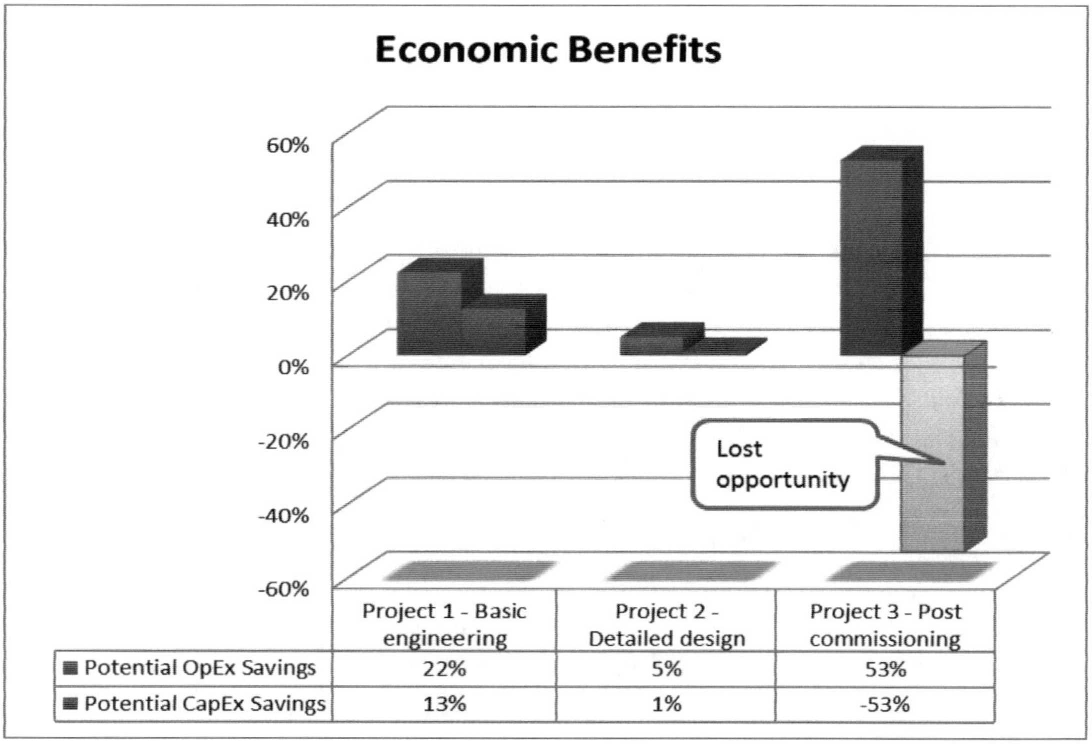


**Figure 2: IPL Rationalization and Potential Economic Savings**

When taking into account both Capital Expenditure (engineering design, construction, installation) and Operational Expenditure (inspection, testing and maintenance), the savings to a facility can be significant if the appropriate evaluation is carried out in early project stages; even in later stages of the project, the suggested approach can reduce IPL requirement and remove some of the maintenance and testing cost of the so-called "gold plating" that occurs on projects (Figure 2).

Finally, the regulatory environment can have an impact on the study. Where regulators follow a prescriptive regimen for safety instrumented systems (e.g., U.S. Gulf of Mexico), then the benefits of IPL rationalization are realized through the removal of extensive testing and maintenance related to IPL, but not through removal of IPL or through the removal of regulatory required testing. The potential to gain savings still exists though, so the exercise provides benefit by removing the "gold plating" of the facility.

## 6 Conclusion

Critical safeguards identified, performance requirements established and an evergreen process to ensure safeguards remain in place are key to keep high consequence scenarios from occurring. Through safety studies and by following industrial codes and standards, multiple layers of protection are factored in to prevent or mitigate potential events. However, it is experienced that often the designs and facility become complicated due to 'add-on' safety. Such facilities could experience operational inefficiency, high cost of maintenance and potentially higher risk exposure to personnel.

IPL rationalization should be part of the design engineering process to overcome the challenges and a systematic approach will help in:

- establishing 'effective' safeguard
- optimizing IPLs
- lifecycle benefits

The paper reviewed some of the challenges in determining the effectiveness of IPL and how to overcome the challenges through a IPL rationalization exercise. The paper discussed the inherently safer approach and its significance in improving the performance of the LOPA results. The timing of the exercise has a significant effect on the outcome of the IPL rationalization. The potential economic benefits (Capital and Operational) are also evaluated with relevant project case studies.

## 7 References


1. CCPS. "Guidelines for Independent Protection Layers and Initiating Events", American Institute of Chemical Engineers, New York, NY, 2007
2. CCPS, "Guidelines for Initiating Events and Independent Protection Layers in Layer of Protection Analysis", American Institute of Chemical Engineers, New York, NY, 2015
3. CCPS, "Layer of Protection Analysis – Simplified Process Risk Assessment", American Institute of Chemical Engineers, New York, NY, 2001
4. Nair SreeRaj, "Determining the criteria for evaluation of toxic hazards, Journal of HSE & Fire Engineering", ASFE, Issue 2 March 2009
5. Nair SreeRaj, "'Inherently Safer' – the Concept and its Application in Oil & Gas Sector," International Conference on Safety and Fire Engineering, SAFE-'17, 2017
6. Stuart Chia, Kieran Walshe and Ed Corpuz, "Application of inherent safety challenge to an offshore platform design for a new gas field development – approaches and experiences," Symposium Series No. 149, IChemE, 2003
7. Trevor Kletz and Paul Amyotte, "Process Plants: A Handbook for Inherently Safer Design," Second Edition, 2010

13TH GLOBAL CONGRESS ON PROCESS SAFETY

# Process Safety Site Assessment Program (PSSAP) – 5 years of Assessments

**Andrew P. Broadbent**
**American Petroleum Institute (API)**
**1800 West Loop South, Suite 475**
**Houston, TX 77027**
**broadbenta@api.org**

Prepared for Presentation at
American Institute of Chemical Engineers
2017 Spring Meeting and 13th Global Congress on Process Safety
San Antonio, TX
March 26 – 29, 2017

# Process Safety Site Assessment Program (PSSAP) – 5 years of Assessments

**Andrew P. Broadbent**
**American Petroleum Institute (API)**
**1800 West Loop South, Suite 475**
**Houston, TX 77027**
**broadbenta@api.org**

**Keywords:** Process safety, assessments, benchmarking, protocols, scoring, rating, weighting

## Abstract

The API Process Safety Site Assessment Program was developed in 2011 as part of the AFPM and API Advancing Process Safety Programs. PSSAP began conducting assessments in 2012. By the end of 2016, the API Site Assessment program will have conducted 52 General (7 protocols) and 27 HF Alkylation/RP 751 assessments. These assessments have been conducted 59 different refineries and petrochemical facilities. The areas assessed are: Process Safety Leadership, Management of Change, Mechanical Integrity, Safe Work Practices, Operating Practices, Facility Siting, Process Hazards Analysis, and HF Alkylation/RP 751.

This presentation will provide a brief overview of the program and present some the learnings, trends and benchmarking developed from the assessments based on the five-year history of the program.

In 2010, the American Petroleum Institute (API), American Fuel & Petrochemicals Manufacturers (AFPM) and industry joined together to address improving process safety performance with the goal of preventing process safety events at refineries and petrochemical plants. API, AFPM and the industry wanted this effort, known as Advancing Process Safety, to be:

- Focused on higher risk process safety activities;
- Refining & petrochemical focused;
- Voluntary; and
- An avenue to communicate and share experiences and knowledge.

Advancing Process Safety (APS) consists of six different programs which are:

- Process Safety Metrics and Analysis (RP 754)
- Process Safety Hazard Identification
- Event Sharing
- Focused Improvement
- Regional Meetings
- Process Safety Site Assessments Program (PSSAP).

The intent of all the programs is to raise the level of process safety performance at the sites to prevent process safety events.

The management of these programs is split between API and AFPM. API manages the Process Safety Site Assessment Program, Process Safety Metrics and Analysis, and the Focused Improvements effort. AFPM manages Process Safety Hazard Identification, Event Sharing and Regional Networks.

<u>Benefits/Opportunities</u>

The Process Safety Site Assessment Program is perhaps the most hands-on of the six programs since it involves a team of assessors being onsite at a refinery or petrochemical plant for an entire week. The program uses credible, third-party assessors who have specific expertise in one or more of the eight areas being assessed. The team utilizes a set of protocols that are applicable to both refineries and chemical plants. These protocols were developed through a collaborative effort of subject matter experts from a wide range of companies.

Through these site assessments, the program provides the sites with feedback regarding what they are doing well and what improvements they can make to their process safety programs. Additionally, the program provides the sites with benchmarking data so the sites can compare how they are doing versus their industry peers.

The assessments are not process safety management compliance audits, and so we refer to them as assessments and not audits. We speak to process safety rather than process safety management so people will understand that these assessments are not the three-year process safety management (i.e., 29 CFR 1910.119) compliance audits, rather they are assessments of how sites are performing relative to 'what good looks like" based on the experience of industry and the expert assessors. As you will see, the assessments only cover seven areas of process safety (compared with the 14 areas in OSHA 1910.119) as well as HF alkylation/API RP 751.

***Protocols***

The eight protocols address:

- Process Safety Leadership
- Operating Practices
  - Procedures
  - Shift Handover
  - Operator Rounds
  - Alarm Management, and
  - Workplace Conditions
- Mechanical Integrity – Fixed Equipment
- Safe Work Practices
  - Procedures
  - Work Permit Procedures
  - Utilities, etc.
- Management Of Change
- Process Hazards Analysis
- Facility Siting
  - Permanent Buildings
  - Portable Buildings, and
  - Tents
- HF Alkylation/API Recommended Practice (RP) 751

The first seven elements are areas that could prevent a process safety event from occurring or reduce the severity of an event (i.e., Facility Siting). The eighth area – HF alkylation – deals with process safety in this specialized operation. Assessments involving the first 7 protocols are referred to as "General" Assessments and the HF Alkylation/RP 751 ones are known as "HF In-Depth RP 751" Assessments.

Note: Since the HF Alkylation protocol specifically follows API RP 751, the HF-In-Depth RP 751 Assessment does meet the RP 751 requirement to conduct an audit once every three years.

Sites may also choose to do just some of the protocols (for example, MOC and Facility Siting) rather than all seven at a time. These are known as "Partial" assessments.

These protocols were developed by industry for industry utilizing 54 subject matter experts from a wide range of companies. The protocols were developed, balloted using the API standards development consensus process, and then piloted at three sites. The protocols were then reviewed and modified based on learnings from the pilot assessments, including input from the assessors.

Keys to the Program

There are three very important components the PSSAP program, which are:

- Benchmarking
- Mentorship
- Expertise of the Assessors

**Benchmarking**

Refineries and petrochemical plants always like to see how they compare to their peers in the industry. In the PSSAP program this is done by providing a score for each protocol and each protocol question.

The 7 General Assessment protocols have a total of 577 questions and the HF In-Depth protocol has 394 questions.

When developed, each protocol question was given a weighting indicating the question's importance to preventing a process safety event from occurring. The question's weighting is a fixed number between 1 and 4 in half increments (i.e., 0, 0.5, 1, 1.5, etc.) with 4 indicating the most importance. During the assessment the assessor will rate each question based on the information and evidence the site provides. This is done on a scale of 0 to 4, also in half increments. The score is a simple calculation of the weighting times the rating.

Keeping the scoring objective is done through ensuring the protocols are well written so that different assessors will understand and interpret the question the same way. Consistency is also addressed through the training provided and calibration sessions with the assessors at the start of the assessments. Additionally, the PSSAP Program Manager provides oversight so that a consistent approach is used by the assessors.

Mentorship

One of the most important pieces of the program is the information exchange the sites get with the assessors. As mentioned, the assessors have considerable experience and expertise based on their years of service in the industry and through performing PSSAP assessments.

We inform the sites (we typically visit the site six months prior to the assessment to set up all the logistics), to get the most value out of the assessment, that their employees should be interviewing the assessors as much as the assessor is interviewing the employee.

As an example, on one assessment the assessor sat down with three site employees for an hour and discussed various ways to manage car seals.

We encourage sites to take advantage of the time with the assessors throughout the week, including the daily debriefs, lunch time discussions, and the closing conference of Friday.

Assessors

As mentioned, the assessors are major piece of the PSSAP program. They are handpicked based on their expertise and industry background/knowledge. Typically they are retired from industry companies, though some work for consulting firms (or have their own firm) but worked in the industry at some point in time.

An assessor candidate with appropriate qualifications and references is selected to attend a 1-1/2 day training course which explains the API Process Safety Site Assessment Program to them. The candidate must undergo an interview with a three-person panel – the PSSAP Manager and two industry representatives before being selected to be an assessor. Not all candidates that attend the training get selected. Also, assessors are picked for specific protocols (i.e., we do not look for Mechanical Integrity assessors to be Operating Practices experts and vice versa). Currently, over 70 people have inquired about being an assessor and we have around 20 in our assessor pool.

To date, a typical assessment team is averaging 39 years of industry experience. Typically the team has 7 assessors.

An in-depth HF alkylation unit assessment follows the API RP 751 Recommended Practice in detail. This assessment can be done at the same time as a "basic" assessment or completely separate and consists of 5 assessors. As noted, an In-Depth RP 751 assessment can be used by the site as their 3-year RP 751 audit.

Assessments

An assessment gets started when the site contacts the PSSAP Program Manager to schedule a week for the assessment. It is important that the site assessment date, be selected well in advance – at least 6 months – so appropriate time is available for planning, including scheduling the assessors.

A pre-assessment visit is then scheduled and conducted at the site by the PSSAP Program Manager. This meeting serves two purposes. First, it allows the site management team to learn more about the program and ask any questions. The second purpose is to meet with the site contact (the person who will work with the PSSAP Program Manager to ensure all arrangements are made prior to the start of the assessment) to go over the logistical needs for the assessment - including conference rooms, scheduling interviews, documents needed prior to the assessment, etc.

Conducting the Assessment

On the week of the assessment the team arrives on Sunday afternoon/evening ready to go. The assessors will have already spent about 8 hours reviewing site documents, provided by the site through an FTP file share site, so they will be well versed on the site's programs relative to the protocols they are assessing.

Monday begins with an opening conference to review the ground rules and usually a quick tour of the site. After the tour, each assessor interviews key people to learn more about the site's management systems, programs and procedures. These interviews have been pre-scheduled by the site using a template provided to them by the PSSAP Program Manager.

The assessors also work to spend as much time in the field as possible. The intent is to not only review whether the site has good management systems, but to provide them with insight on how well their systems are being implemented in the field.

On Tuesday, Wednesday, and Thursday, the assessment team meets with site staff, spends time in the field, and holds a debrief with the site personnel, typically around 4:00 pm. These daily debriefs have been extremely beneficial as they are excellent opportunities for site personnel to continue to interact with the assessors. Often when the debrief has been completed, there are follow-up conversations with the assessors so site staff can gain more insights.

Thursday night and Friday morning is when the assessors compile all the information and the protocol ratings. The assessors enter observations and ratings for each line item in each protocol, thus the facility is provided with a considerable amount of information.

On Friday, an hour to an hour and a half closing conference is held with the site personnel, including the site management team. The site receives a PowerPoint overview, the completed protocol spreadsheets, and some preliminary benchmarking data. The draft report is provided to the site the following week. Distribution of the reports is as designated by the site.

When the assessment team leaves the site on Friday, the assessors return all hard copy documents and delete any electronic documents that were provided to them by the site.

Handling Report & Data

The site will typically be given two to three weeks to comment on the draft final report. This review period is not a discussion of the scores, but whether the wording of the Observations and Good Practices is clear. A final report is then issued by the program manager to the site.

Once the final report is issued, the data and information including the good practices and observations, with the permission of the site, are blinded and entered into the database for trending and benchmarking purposes. API does not ask for corrective action plans or updates.

Program Status, Experience & Keys

As of year-end 2016, we have conducted 53 General site assessments, 27 HF alkylation/API RP 751 in-depth assessments and 5 Partial assessments. The program has been to 28 different companies at 60 different locations (note: several companies have had a site assessment at more than one site.)

| Item | Total | Average per Assessment |
|---|---|---|
| Protocol Questions | > 41,200 | 577 Basic<br>394 HF |
| Interviews | > 8,000 | >100 |
| Observations | > 2,800 | ~35 |
| Good Practices | > 2,400 | ~30 |

Below are a couple of comments received from sites that have been assessed:

- "You cannot fool these Assessors — they have

done *your* job for 30 years"

- *"In one week the assessment team found issues that our internal teams could not find in 3+ weeks, and it was stuff we needed to know and address. Well focused, deep dive process."*

Benchmarking Data

The benchmarking data is provided in several formats. For example, the data is analyzed and presented by:

- Protocol
- Sub-protocol (some protocols have multiple areas of review – Shift Handover, Operator Rounds, etc.)
- Specific protocol sections (groupings of questions for specific areas)
- Management Systems vs. Implementation (having the procedure vs. how is it being done in the field)
- Size of site
- Size of company
- Quartile

Benchmarking data updates are provided generally every six months to sites that have had an assessment.

Data Review – Good Practices and Observations

Below are some examples of the types of Good Practices and Observations that the assessors have been identifying through the site assessments. If you note the last bullet on both slides this is an example of exactly what we are after. At some sites it's an item that needs to be addressed while at other sites we have seen them doing it well.

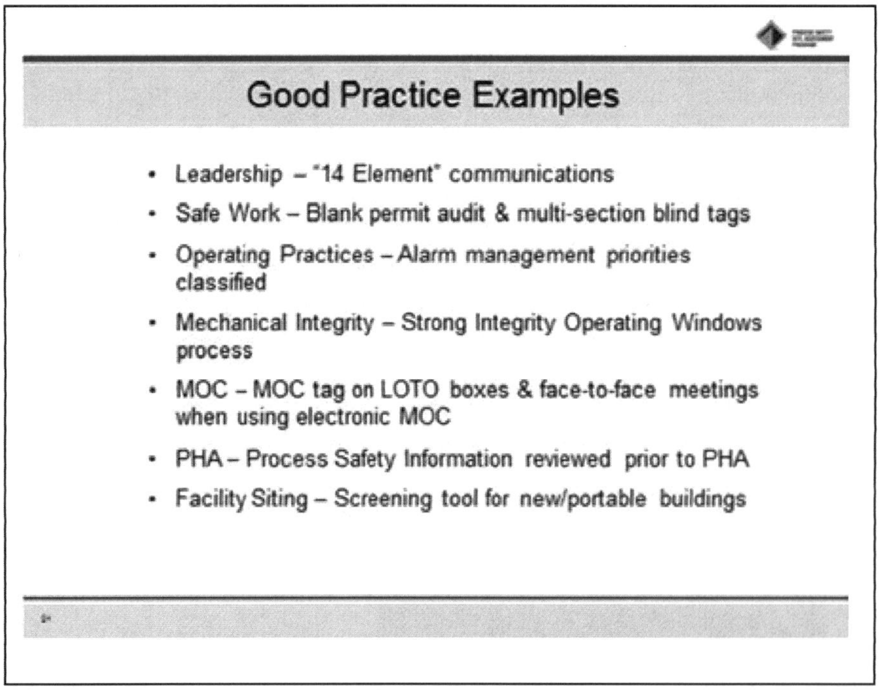

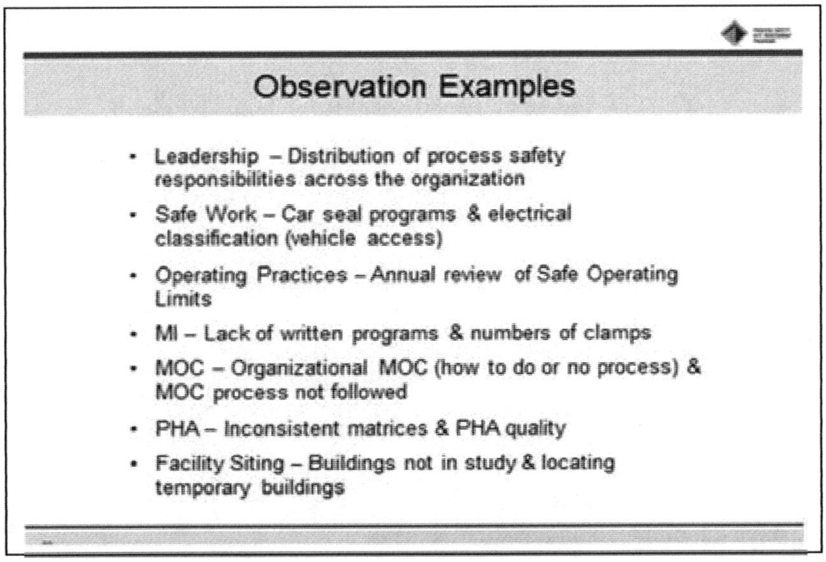


This points to a key element of the site assessment program - sharing the information from sites doing it well with others that are working to put systems/programs/procedures in place at their site. A variety of methods are used for sharing the practices learned through the site assessment program including:

- The Advancing Process Safety Program
  - Regional networks
  - Hazard Identification program
- Industry Conferences and Seminars
- API standards

The data, information, and learnings from the site assessments have been used to focus resources to develop informational and guidance documents regarding areas identified as improvement opportunities. For example, an informational document for mechanical integrity regarding small bore piping was developed based on observations during PSSAP site assessments and data from the 754 metrics and Event Sharing database..

<u>Conclusion</u>

In conclusion, the "boots on the ground" performance of site assessments began April of 2012. Feedback from these assessments has told us that the process of preparing for an assessment (going through the protocols prior to the assessment) and the mentorship/interaction with the assessors has been very beneficial to the sites. And now that we have done enough sites to allow the sharing of benchmarking data the sites are getting even more value from the program.

# Alarm Management Meets SIS

## SIS Alarm Requirements

***How many alarms must be configured per the SIS Standard (ISA-84 Part 1) for an SIF?***
Zero

***How many alarms must be configured per the AM Standard (ISA-18.2) for an SIF?***
Zero

***How many alarms are necessary for an SIF to correctly perform its function?***
Zero

## Pre-alarm – What makes sense?

***Pre-alarm before an SIS impending trip?***
True – This gives the operator an opportunity to prevent the trip, keeping the plant running and reducing the demand frequency on the SIF.

***Pre-alarm should always be on the BPCS control or indicator tag rather than the SIS transmitters?***
False – Choice of where to put the pre-alarm is up to the facility, and it's a good idea to document this in the Alarm Philosophy.

***Each transmitter in a voting arrangement should alarm individually?***
False – Are we trying to keep our plants operating safely or do we just throw multiple alarms at the operator?

***SIS and BPCS transmitter disagreements among redundant SIS transmitters should be alarmed?***
True – With this alarm in place, there is no need for a pre-alarm on more than one transmitter.

## Introduction

Most plants focus on meeting the alarm management requirements laid out in ISA 18.2. Even if the proper approach is taken in configuring the BPCS, there is one area that is sorely neglected, the SIS system. SIS is addressed based on the requirements laid out in ISA 84. Unfortunately, this standard does not address alarming related to SIS events. ProSys contends that a successfully implemented alarm management philosophy is only effective when taken as a holistic view of the BPCS and SIS systems.

## Unified Alarming Scheme

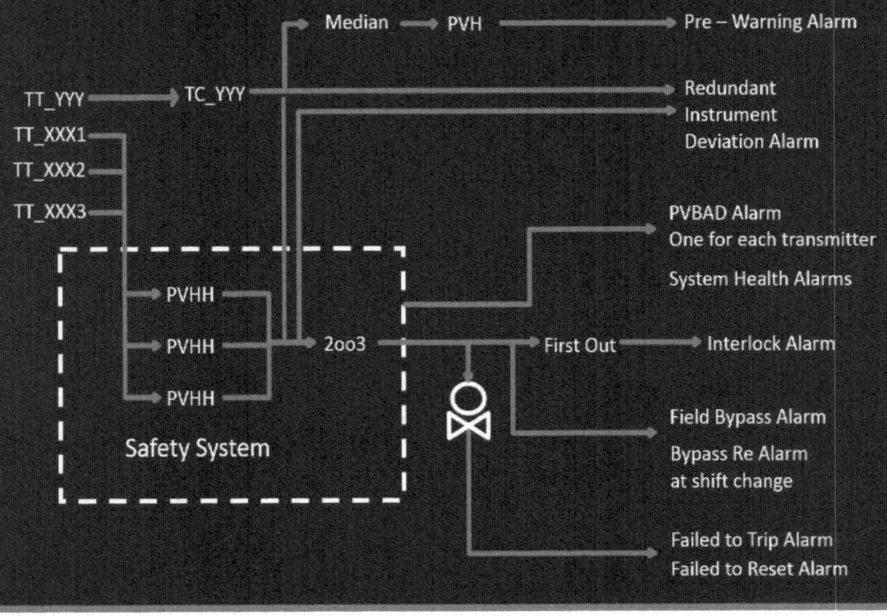


## Trip – What makes sense?

***First out alarm indicating the reason for a trip?***
True – There is no better way to announce to the operator that a trip has been initiated.

***Alarm SIS valve moving to trip position?***
False – Going to the trip position is the expected response to an SIS trip – better, provide a failed to trip alarm, which makes noise if the valve doesn't work as expected.

## Bypass – What makes sense?

***Alarm board operator initiated SIS bypass?***
False – Although a common practice, this alarm adds little or nothing to the integrity of the system: "Hey Joe/Mary/Sam, you just pressed a button!"

***Alarm field operator initiated SIS bypass?***
True – This may come as a surprise to the console operator, so it's not a bad thing to annunciate.

***If a bypass is in place, trip point alarm on the SIS transmitter?***
False – Why give the operator an alarm that isn't reliable or won't be believed? Find an alternative monitoring plan.

***Re-alarm on SIS Bypass?***
Neutral – Bypass and shift changeover procedures exist, but do we follow them 100%? This is a facility decision. If used, re-alarm shortly after shift change.

# ENHANCING PROCESS SAFETY MANAGEMENT INTEGRATION TO IMPROVE PROCESS SAFETY AND PROFITABILITY IN THE FUELS AND PETROCHEMICALS INDUSTRY

**Thomas D. Watson, P.E.**
**TDWatson Consulting LLC**
**Pittsburgh, PA USA**
**twatson@tdwatson.com**

**Ronald D. Fuchs, P.E.**
**RD Fuchs Engineering LLC**
**Lake Jackson, TX USA**
**rdfuchs13@aol.com**

**Harold Phillippi, P.E.**
**Independent Process Safety Consultant**
**Charleston, SC USA**
**butch.phillippi@gmail.com**

**Keywords: PSM, Process Safety Management, Relief Systems, Incident Investigation, Mechanical Integrity.**

## Abstract

Process Safety Management (PSM) in the Fuels and Petrochemicals industry has the paradox of being well established and yet constantly challenged to improve its effectiveness and impact on process safety and profitability. Such a paradox is a natural consequence of an industry exhibiting a positive process safety culture that maintains a sense of vulnerability and sees the imperative to improve process safety. This paper addresses the frequently asked question, "How do we continue to improve our organization's process safety culture?" and presents opportunities and examples for organizations to enhance their existing PSM systems and culture by further integrating PSM elements both within an organization's established PSM systems and its other systems, processes, and procedures. The paper illustrates integration and enhancement opportunities using Relief System Design Basis (RSDB) documentation, a component of the PSM element - Process Safety Information (PSI). Illustrations show RSDB integration with the PSM elements of Mechanical Integrity (MI), Process Hazard Analysis (PHA), and Incident Investigation (II). The paper then illustrates opportunities to increase profitability by highlighting RSDB integration with partial downtime throughput opportunities and improved project definition. The illustrations challenge organizations that truly "own" a process safety culture to look for additional opportunities to further enhance and integrate established PSM systems into their organization to further improve safety and profitability.

## 1. Introduction

The Fuels and Petrochemicals Industry in partnership with governing regulatory agencies have established effective Process Safety Management (PSM) systems to handle hazardous materials and energies over the past several decades. Examples include the U.S. Occupational Health and Safety Administration (OSHA) PSM Standard and the European Commission Seveso II Directive. The Center for Chemical Process Safety (CCPS) offers an extensive reference list of regulatory/industry/system requirements.[1] The catalyst for these industry systems have often been the tragic circumstances around industry incidents and catastrophes. Today, PSM in the Fuels and Petrochemicals Industry has the paradox of being well established and yet constantly challenged to improve its effectiveness and impact on process safety and profitability. Such a paradox is a natural consequence of an industry exhibiting a positive process safety culture that maintains a sense of vulnerability and sees the imperative to improve process safety.[2] The vulnerability and sense of urgency are well founded as the industry continues to experience preventable incidents. The Marsh Report states the 100 largest property damage losses in the hydrocarbon industry from 1974 to 2013 had a combined property damage cost of approximately US$34 billion.[3] Organizations demonstrating a committed process safety culture continually challenge and address the question, "How do we improve our organization's process safety culture?"

Organizations that have successfully managed the PSM implementation and sustainment phases have already experienced and moved past the common misconception that process safety and profitability have conflicting objectives. These organizations further realize they must move beyond a "separate but equal" PSM system established among their current business management systems, and fully integrate PSM into their systems to further improve their organization's impact on process safety and profitability. Recent efforts in the Fuels and Petrochemicals Industry to improve process safety through management system integration include industry efforts such as:

- The American Fuel and Petrochemical Manufacturers (AFPM) "Advancing Process Safety Programs," which promotes sharing PSM learnings across member organizations of AFPM
- The CCPS publication containing guidelines for integrating management systems and metrics to improve process safety performance.[4]

This paper identifies and presents specific examples and opportunities for organizations to continue to enhance their existing PSM systems and process safety culture by further integrating PSM elements both within an organization's established PSM systems and its other business systems, processes, and procedures. The paper illustrates integration and enhancement opportunities using Relief System Design Basis (RSDB) documentation, a component of the PSM Process Safety Information (PSI) element. Specifically, the examples illustrate RSDB integration with the PSM elements of:

- Mechanical Integrity (MI)
- Process Hazard Analysis (PHA)
- Incident Investigation (II)

RSDB integration with other business processes also present opportunities to directly improve profitability. Two examples presented here include RSDB integration with:

- Production planning (partial downtime throughput opportunities)
- Improved project definition

The following sections provide a general introduction to RSDB and illustrate the five integration opportunities in more detail.

## 2. An Introduction to RSDB and its Relationship to PSM

Within the Fuels and Petrochemicals Industry, many processes involve highly hazardous chemicals. To adequately assess and address a process' hazards an organization must adequately define the Process Safety Information of the process. Process Safety Information (PSI), includes information regarding the process' hazards, technology, and equipment.[5] Examples of hazard information include toxicity and permissible exposure limits. Examples of technology information include a process' chemistry and operating limits. Finally, examples of equipment information include materials of construction and the Relief Systems Design Basis (RSDB). There are many references for what constitutes RSDB and how to effectively manage RSDB documentation.[6] The American Petroleum Institute states RSDB documentation should demonstrate that all pressure-containing equipment has been identified and the relief requirements adequately analyzed. Documentation should include specific engineering guidance, methodology, analysis, and reference information that collectively reflects Recognized and Generally Accepted Good Engineering Practices (RAGAGEP). [7]

RSDB is a key component of PSI, but RSDB is also an important component in maintaining other Process Safety Management (PSM) elements such as Mechanical Integrity (MI), Process Hazard Analysis (PHA) and an Incident Investigation (II). Furthermore, RSDB can be a significant aspect of process production constraints and project opportunities associated with processes. The next sections discuss integrating RSDB into the aforementioned PSM elements, production planning, and project development processes.

## 3. RSDB Integration with Mechanical Integrity (MI)

Integrating Relief System Design Basis (RSDB) analysis with Mechanical Integrity's (MI's) Risk Based Inspection (RBI) through a combined risk assessment can identify acute process equipment vulnerabilities. Mechanical Integrity (MI) activities ensure the proper design, installation, operation, and maintenance of process equipment. Risk Based Inspection (RBI) is one component of an effective MI program that links risk assessment and mitigation activities with appropriate inspection and process control activities. RSDB analysis, on the other hand, identifies the applicable relieving scenarios and required relieving rates affecting pressure-containing process equipment to ensure the relieving scenarios do not expose the equipment to operating conditions outside of design limits. The American Petroleum Institute's Recommended Practice (RP) 581 identifies the interdependency of the Pressure Relief Device (PRD) with the fixed equipment the PRD is protecting.[8] RP 581 evaluates PRD criticality to set mechanical integrity activities such as inspection frequency. The RBI methodology combines the PRD's anticipated demand rate with the probability of failure on demand of the PRD to determine an overall risk for the PRD and the process equipment the PRD protects. While the API RP 581 methodology is sound, when the RSDB documentation is not available or the RSDB analysis has identified deficiencies in the PRD design, the RBI methodology develops shortcomings that should be addressed.

As an example, consider a process vessel containing a flammable hydrocarbon liquid. An RBI of the vessel defines a risk associated with the vessel, the vessel's PRD and the anticipated overpressure scenarios. The RBI analysis is predicated in part, however, on the PRD adequately protecting the vessel in the anticipated overpressure scenarios should the PRD open. The RSDB documentation identifies fire as the only applicable overpressure scenario but concludes the PRD is undersized. The RSDB can provide additional insight into the level of overpressure that would occur (and the anticipated consequence). The additional insights would significantly impact the RBI risk assessment and potentially warrant a reassessment of inspection and mitigation actions for the PRD and the protected vessel.

The example illustrates how integrating RSDB with MI's RBI can identify acute equipment vulnerabilities and aid in further prioritizing MI efforts and RSDB deficiency resolution, including risk mitigation efforts. Consequently, integrating both RSDB and MI can further enhance PSM integration and improve process safety and profitability.

## 4. RSDB Integration with Process Hazard Analysis (PHA)

Integrating RSDB with PHA can reduce costs and improve the effectiveness of both the PHA and the RSDB. As most readers are aware, a Process Hazard Analysis (PHA) in the United States and many other countries, is a requirement of a Process Safety Management (PSM) program, (for example, 29 CFR 1910.119 requires PHAs and periodic revalidations). No matter which methodology a PHA employs [Hazard and Operability study (HAZOP), "what-if," Failure Modes and Effects Analysis (FMEA), etc.], capable teams must perform the analysis. The teams typically range from 4 to 8 (or more) persons, consisting of owner's personnel and often outside consultants/specialists. No matter who is present, the participants represent a necessary and significant staff resourcing cost. Furthermore, each PHA finding should be well documented, especially those requiring additional evaluation and stewardship before the findings are declared either a confirmed issue or not. The PHA finding documentation and corresponding follow-up/stewardship also represent a significant staff resourcing cost. Finally, actions required to address the confirmed issues, whether facilities changes, work process changes, or administrative changes represent significant costs.

With the PHA costs sufficiently identified, now consider how integrating RSDB with PHA reduces costs and improves PHA and RSDB effectiveness. PHAs always identify questions and concerns about the adequacy of the over (or under) pressure protection systems as well as items such as Maximum Allowable Working Pressure (MAWP), hydrotest pressure, materials of construction, etc. Furthermore, PHAs require the most recent drawings, heat and material balances, and other Process Safety Information (PSI). Similarly, the most recent RSDB should have many (if not all) of these items assembled and documented, with any changes since the most recent RSDB tracked and provided as a part of the facility's Management of Change (MOC) program. The overlap of PSI and resources between RSDB and PHA is substantial. Consequently, completing RSDB analysis prior to and integrating it with the PHA process and follow-up activities introduces cost savings. Common savings occur relating to over (or under) pressure findings. If properly completed and documented, the RSDB can provide the PHA participants with immediate feedback regarding the adequacy of pressure protection devices, which in most cases are code items and must be addressed. A benefit of having a quality RSDB done prior to the PHA is that the over or under pressure concerns can be quickly resolved. On many occasions, no action item follow-up work is necessary. This reduces costs both during the PHA and the PHA follow-up. Conversely, the PHA serves as an audit check of the RSDB as a capable team reviews the RSDB basis as needed.

As a quick example, a PHA team questioned the adequacy of an exchanger shell's design pressure during a blocked-in condition. The maximum pressure generated from a blocked-in condition from a manual valve closure was greater than the design pressure by approximately 10 percent. The PHA team referenced the RSDB. The RSDB previously identified the same issue and investigated the situation in more detail. The RSDB identified the exchanger's MAWP to be approximately 20 percent more than the noted design pressure, and the hydrotest pressure was based on 1.5 times the MAWP. The U-1 form that was part of the RSDB package clearly indicated this. The RSDB package also noted the exchanger shell was adequate based on ASME VIII, Appendix M, M-5.7, in this situation, because the pressure was below 116% of MAWP and it would be immediately noted by operators, the valve closure was not identified as an issue and the PHA follow-up item was closed. Because the PHA utilized the RSDB, the PHA saved costs on discussion and documentation of an action item and the necessary follow-up and tracking of a code related item. Conversely, the PHA team validated the RSDB, enhancing PSM integration and improving process safety and profitability.

## 5. RSDB Integration with Incident Investigation (II)

Incident Investigation (II) is also a mainstay of any effective personnel or Process Safety Management (PSM) program. II is an opportunity to learn from mistakes, system failures, equipment failures, and/or overlooked vulnerabilities. II is also an opportunity to evaluate the effectiveness of preventive and control measures installed to prevent or mitigate an incident. In the CCPS Vision 20/20 Characteristics of Companies with Great Process Safety Performance the authors write: "To reduce incidents, employers and employees must enthusiastically support a culture that is driven to learn from many sources including benchmarking, near misses and incidents, and jobs done well."[9] For incidents involving release from a relief device and/or a flaring event, II intersects Relief System Design Basis (RSDB) and together they provide a learning opportunity that enhances PSM and improves process safety.

Yogi Berra said; "When you come to a fork in the road, take it." When an incident occurs investigate it. Definitions may vary for what constitutes an incident. Indeed, there are legal requirements for investigation of safety and environmental incidents. These form the baseline for what organizations must do but do not limit organizations from doing more. For instance, some organizations with very effective PSM programs investigate all flaring events. The organization calling the event an incident, a near miss, or flaring event is least important, while the organization's learning opportunity from the event is most important.

By integrating RSDB with II organizations can do real case study evaluation of the RSDB's assumptions and technical predictions. Integration can help answer such questions as:

- Did the scenarios of the RSDB represent the incident that occurred?
- Did the relief valve effectively handle a release rate lower than the controlling case without issues such as chattering?
- Were the release rate, pressure limitation, and system temperature change what the calculations predicted they would be?
- Did the flare handle the relieving event without an environmental release or safety issue? For example: flare smoking, toxic release, or high temperature exposure of personnel and adjacent equipment.

RSDB integration with II may facilitate getting to the root cause of the incident. The RSDB contains valuable information about the relief system response and may help identify the original process design intent. If the relief system response was not as intended, the relief system may be the cause of the incident. Additionally, if there are layers of protection designed to make flaring or release valve opening an infrequent event but flaring is occurring often, the preventing layers of protection may need further evaluation.

The RSDB and II integration may also help identify additional inspections and corrective actions. In this case II also integrates Mechanical Integrity (MI). For example, damage to the relief system may have occurred from too-high or too-low temperature or pressure. Mechanical damage resulting from corrosive chemicals or relief valve chattering may have occurred. Additionally, the RSDB may shed light on damage that the incident caused to other process equipment. Finally, if the RSBD does not accurately predict what will occur during such an incident there should be a corrective action to change the RSDB and possibly modify the relief system. Consequently, the II serves to further validate the RSDB and the RSDB enhances the II. An important point to note: incorporating RSDB into II requires someone knowledgeable in RSDB be on the II team.

In conclusion, a relief device release or flaring event II that does not consider the RSDB leaves a valuable tool unused. Integrating RSDB with II enhances PSM integration and improves process safety and profitability by gaining additional benefit from both RSDB and II.

## 6. RSDB Integration with Production Planning

Opportunities often present themselves in complex facilities to ensure process safety and increase profitability by integrating Relief System Design Basis (RSDB) with production planning. Many facilities integrate processing units in a complex manner. Such facilities often have multiple production constraints preventing individual processing units from running maximum capacities simultaneously. Production constraints may include downstream processing capacity, feed qualities, processing unit downtimes, permitting limits, or even facility RSDB constraints. When a facility RSDB constraint limits production, understanding the relationship between other production constraints and the facility RSDB constraint becomes critical to ensure process safety and optimal profitability. Hence the need and opportunity to integrate RSDB with production planning.

Before presenting RSDB and production planning integration examples, there are a few points to note. First, RSDB analysis for an individual processing unit typically establishes the unit's safe operating window at a maximum capacity. A facility RSDB, however, reflects multiple processing units operating simultaneously. Second, due to the myriad of production constraints a complex facility may have, the facility RSDB often reflects a limited set of facility production constraints (often with a very conservative basis). Third, in some instances, the facility RSDB may constrain various processing units to a fraction of their maximum production capacity. Finally, when production constraints setting the facility RSDB change, production planning needs to evaluate alternative production options that the facility RSDB does not adequately reflect.

In one example, a complex facility relief system constraint of excessive relief system backpressure for low set point relief devices exists. The excessive backpressures result from a potential utility failure scenario impacting multiple processing units. This facility RSDB constraint limits the facility's production. When the processing unit with the low set point relief devices is down for maintenance, however, the facility RSDB constraint relaxes. Thus, opportunities present themselves for other processing units to increase production. In another example, a complex facility schedules a portion of the facility relief system maintenance downtime with processing units having a substantial impact on the facility relief system constraint. During the downtime, however, the startup timing for the relief system is delayed due to discovery work and opportunities arise to increase profitability by minimizing production cuts within a newly defined facility RSDB constraint.

## 7. RSDB Integration with Project Development

Integrating Relief System Design Basis (RSDB) with projects presents opportunities to improve project definition, increase project profitability (through project cost savings), and enhance Process Safety Management (PSM). Having a preliminary well documented RSDB at the on-set of a project — often called the Front End Engineering Design (FEED) stage — will identify the various scenarios that should be a part of the sizing of not only the individual relief devices, but also the relief device collection system functional requirements. The potential costs of the project's impact on the relief devices and systems can then be reviewed and optimized in coordination with the project's functional equipment. The following examples best illustrate these points.

In a larger project a liquid feed drum was required. The feed drum's initial design was 50 psig, and the feed sources ranged from 35 psig to 100 psig. Because the feeds to the drum were greater than the drum's proposed design pressure, the overpressure device and corresponding system had to be designed for liquid overfill. All the sources that were greater than 50 psig represented an overfill potential of around 70,000 barrels a day, and consequently the relief device and the corresponding collection system size and costs were unattractive. Upon further investigation and review of the options, the project team increased the design pressure of the feed drum to 110 psig, which had minimal cost differential when compared to the necessary relief system design. Thus, the 110 psig design was chosen and the relief device was then sized for fire and some potential back-flow on stoppage of the feed pumps. Cost savings were in the millions of dollars.

A second project involved design and installation of equipment that had to be tied into the flare system. A preliminary RSDB was done early in the project. Upon analysis of some typical global scenarios such as loss of power, loss of cooling water, etc., the new project's primary relief device(s) would require increasing the collection system's piping and flare size. Furthermore, the sizing of the relief device and piping alone had a large estimated cost.

The relief device's size was driven by the main tower's selected design pressure and steam reboiler pressure available. The project team, in coordination with the RSDB personnel, were tasked with optimizing the reboiler and the relief system. During the optimization phase, the team identified 2 key items: 1) the reboiler steam had only one pressure available, but it could be let down to a lower pressure, and 2) at a slightly lower pressure the steam could easily provide the available heat necessary for the process.

Additional investigation by the team revealed that the tower material and thickness had not initially taken advantage of the potential MAWP. The material thickness needed was ~0.82 inches, and 0.875 inches was to be selected. This allowed for a MAWP higher than design. However, because the steam pressure was reduced, good engineering practice would require that the design consider failure of the steam pressure control valve.

Failure of the steam pressure control valve would provide sufficient heat to overpressure the tower based on the new design pressure and the relief device to the flare was still needed. To address this failure, the steam system (downstream of the pressure let-down valve) was provided with its own pressure relief valve which could be vented to atmosphere.

With the team's recommended changes noted above, the only scenario for the main tower relief device was fire, which had a lower cost of installation and did not require an update of the flare system. Having the RSDB done early in the project allowed the designers and project team to make some easy (paper) cost effective changes, and did not require any modifications to the existing collection and flare systems.

The examples above demonstrate organizations integrating RSDB during project development to improve project definition, reduce project cost, and even make processes inherently safer. Organizations realize further PSM enhancements by integrating a process unit's existing RSDB documentation with project development activities. In addition to the benefits illustrated above, organizations incorporate project impacts on a process unit's RSDB with minimal duplication of effort (minimal cost), and enhance the process unit's RSDB integrity. From an overall PSM perspective, the project development integration enhances both the PSM and project development system.

## 8. Conclusion

This paper presents opportunities for organizations to enhance their PSM systems and process safety culture by integrating Relief System Design Basis (RSDB) with the following PSM elements and business processes:

- Mechanical Integrity (MI)
- Process Hazard Analysis (PHA)
- Incident Investigation (II)
- Production planning
- Project development.

Organizations that truly "own" a process safety culture should use each successful opportunity implementation as a spring board to further enhance PSM integration and improve process safety and profitability.

## 9. References


1. CCPS, "Guidelines for Integrating Management Systems and Metrics to Improve Process Safety Performance", Appendix A, Center for Chemical Process Safety, American Institute of Chemical Engineers, New York, NY 2016.

2. CCPS, "Vision 20/20 Booklet – Committed Culture", Center for Chemical Process Safety, American Institute of Chemical Engineers, New York, NY 2016.

3. Marsh and McLennan Companies, "The 100 Largest Losses", 23rd edition (2014).

4. CCPS, "Guidelines for Integrating Management Systems and Metrics to Improve Process Safety Performance", Center for Chemical Process Safety, American Institute of Chemical Engineers, New York, NY 2016.

5. Occupational Health and Safety (OSHA) "Process Safety management of Highly Hazards Chemicals" 29CFR1910.119(d).

6. Thomas Watson, "AM-13-24 Considerations for Effectively Managing Emergency Relief Systems Design," American Fuels & Petrochemicals Manufacturers Annual Meeting, 2013.

7. American Petroleum Institute, "API Standard 521 – Pressure-relieving and Depressuring Systems", section 4.7.3, sixth edition (2014).

8. American Petroleum Institute, "API RP 581 – Risk-Based Inspection Technology, Part 1", Section 7, second edition (2008).

9. Center for Chemical Process Safety's Vision 20/20: "Characteristics of Companies with Great Process Safety Performance" (2013, McCavit, Berger, and Nara).

# 479098 Reactive Hazards Management: Lessons Learnt from Major Incidents

**Abstract:**

Chemical process industry has long suffered from major incidents due to lack of proper understanding of reactive hazards. The methyl-isocyanate leak in Bhopal, India, has been by far the worst industrial incident in history which led to the development of many process safety regulations. However, the formation and implementation of regulations has not prevented the occurrence of these incidents. Many incidents such as Concept Sciences Inc. where a hydroxylamine manufacturing unit was destroyed, the explosion at the Sterigenics International Inc., ethylene oxide sterilization facility, Bayer CropScience Pesticide Chemical Runaway Reaction Explosion and West Fertilizer Explosion have highlighted the importance of having a better understanding of the reactive hazards. The focus of this presentation is to identify different types of reactive hazards through examples of chemicals incidents and analyze how a better understanding of these hazards could have avoided these incidents.

# 479245 The Study of Safety Evaluation Technique for High Temperature Process

**Abstract:**

There are many accidents of a firing furnace and kneading machine by overheating and the abnormal injection of the product. When the gas which suddenly expanded by thermal decomposition destroys plant equipment and enters the explosion concentration range, a gas explosion is caused and damage spreads. The material which it is easy to decompose is invested by mistake by a firing furnace and has the trouble that the pressure of the furnace abnormally rises to in our factory. Also, troubles such as an explosion of the plumbing by the thermal decomposition of the polymer by the overheat, vent explosion by the thermal decomposition by the false injection of an additive having low heat stability has broken out with kneading machine

These troubles are caused by the thermal decomposition by a sudden temperature rise. The temperature program of the analytical instrument which is generally used by a thermal analysis is 10-20 /min. The conventional thermal analysis equipment cannot evaluate the temperature rise rate at the time of the trouble exactly to become than 20 /min. It becomes the problem that the slow temperature rise may be different from rapid heating up in a decomposing reaction.

We produced a device capable of quickly raising temperature to solve these problems.

This device is comprised of the vertical plumbing which it sealed and heating furnace, and it is very simple structure. The sample holder is fixed to the upper part in the plumbing by a magnet. And the holder drops in the heating zone by keeping a magnet away from the plumbing and is heated rapidly. The materials of the plumbing are SUS316L, and, as for the use upper limit temperature, 500 , the use upper limit pressure are 1MPa. Because the deterioration of SUS materials by the heat is intense when it is raised the temperature of the device to 750 , the plumbing is disposable.

The inside of the plumbing is replaced with inert gases such as nitrogen, argon. The maximum temperature rising rate is 150 /min (furnace temperature: 750 ).

We introduce an example of the trouble analysis of a firing furnace and kneading machine using this device.

[Case example 1] Study of firing furnace trouble

After burning with a new additive at 720 , abnormal pressure rise happened in a furnace, and a plant stopped. After measuring quantity of outbreak gas using the rapid heating device, it was recognized that gas of 7 times was generated and a gasification rate tripled in comparison with a conventional additive.

[Case example 2] Study of kneading machine trouble by overheating

The additive which had low heat stability was spent mistakenly in large quantities to a kneading machine, and a vent board exploded. After performing a rapid temperature rise examination under a kneading temperature condition, as for the pressure raising rate when an additive disintegrated, 1.8MPa/min, the quantity of gasification became 190L/kg. The additive concentration became (mixing ratio with the resin) more than at least 75%, and it was recognized that a large amount of additives were invested mistakenly.

Fortunately, this example does not become the big accident with resulting in injury or death. In this explosion example, a vent board penetrated a checker plate. It might become the casualty when there was a person near a kneading machine.

In comparison with the plant of the petrochemical and fine chemistry that heat risk analytical technique establishes, the firing furnace and kneading machine is often run without detailed risk assessment. The rapid heating device becomes the powerful tool to do the heat risk evaluation of these high temperature processes.

In future, it is going to be improved the rapid heating device changes the plumbing materials, and to use it to the temperature ranges more than 1,000 .

# Handling uncertainty in LOPA and SIF design

*Akshat Khirwal*
*IRESC*
*Unit 2801, Enterprise Square 3*
*39 Wang Chiu Road*
*Kowloon Bay*
*Hong Kong*
*Akshat.khirwal@irescglobal.com*

*Arjan Abeynaike, Venkatesh Sourirajan*
*IRESC*
*Arjan.abeynaike@irescglobal.com, Venkatesh.s@irescglobal.com*

**Keywords:** Uncertainties, LOPA, SIL, SIF design, reliability

## Abstract

Layer of protection analysis (LOPA) is a widely used semi-quantitative method for determining the reliability requirement for safety instrumented functions (SIFs). An advantage of LOPA is that it specifies a numerical reliability target - the average probability of failure on demand ($PFD_{avg}$). However, uncertainties in the input parameters and simplifications within the LOPA method lead to a margin of error in the target $PFD_{avg}$ that is often not accounted for in the design of the SIF. Further uncertainties are introduced when carrying out a reliability analysis of the SIF design to ensure that the target $PFD_{avg}$ is met. The result of overlooking such uncertainties is that the target $PFD_{avg}$ may be set low and the calculation of the actual $PFD_{avg}$ achieved by the design may be optimistic, leading to a SIF that is under-designed. The need to account for uncertainty is emphasized in the latest version of IEC 61511, the international standard for SIF design.

This paper investigates and highlights the sources of such uncertainties in LOPA and SIF reliability analysis. In particular, the quality of the data used as input parameters to LOPA and reliability models has a large effect on the accuracy of the results. These parameters include failure rates, conditional modifiers, beta factors for common cause failures, and proof test coverage factors. Available data for these parameters can be sparse and may not be representative of the conditions in which the SIF is applied. Where the confidence intervals of the data are available, these can be incorporated into the calculations to reach a more conservative result. Alternatively, a safety factor can be applied to try and capture all the uncertainties. In this approach, the target $PFD_{avg}$ obtained from LOPA is reduced by the safety factor to give a more stringent target $PFD_{avg}$ for the SIF design to meet. The methods for handling uncertainty are reviewed in this paper.

## 1 Introduction

SIL Determination is the process of assigning a Safety Integrity Level (SIL) to Safety Instrumented Functions (SIFs) following any of the methods i.e. Risk Graph, Risk Matrix or Layer of Protection Analysis (LOPA); as laid down in IEC 61511 and IEC 61508 [1] [2]. Of these methods, the semi-quantitative LOPA approach is increasingly being adopted for SIL Determination as compared to other qualitative approaches such as Risk Graph or Risk Matrix. An advantage of LOPA is that it specifies a numerical reliability target for a safety instrumented functions (SIF), which is expressed in terms of the target average probability of failure on demand (target $PFD_{avg}$), or target risk reduction factor (RRF, where RRF = 1/ $PFD_{avg}$). This target $PFD_{avg}$ then forms the basis for design of the SIF wherein a reliability analysis is carried out to ensure that the target $PFD_{avg}$ is met. However, uncertainties in the input parameters and simplifications within the LOPA method lead to a margin of error in the target $PFD_{avg}$ that is often not accounted for in the design of the SIF. Further uncertainties are introduced when carrying out a reliability analysis of the SIF design. Without accounting for these uncertainties in the assessment, the SIF may be under-designed since LOPA may result in under prediction of the target $PFD_{avg}$ or the reliability analysis may show that the target $PFD_{avg}$ has been achieved although this may not be the case. This paper has reviewed the sources of uncertainties in the LOPA method and design of the SIF (SIF reliability analysis) and provides a way forward for handling such uncertainties.

## 2 Uncertainties in LOPA

Initiating event frequency, consequence severity, and likelihood of failure of Independent Protection Layers (IPLs) are the key parameters used in LOPA to approximate the risk of a scenario in case the SIF fails to act when the demand arises. In addition to the above, conditional modifiers [3] may also be used to factor in specific parameters pertaining to the demand on the SIF. The simplicity of LOPA allows it to be easily understood by engineers of all disciplines, leading to its widespread use. The drawback of this simplicity is the greater uncertainty in the results, as described in the following sub-sections. The governing equations and the LOPA worksheet used for assessing the scenarios discussed in this paper are referenced from IEC 61511-3 [1] and CCPS, Simplified Process Risk Assessment [4].

### *2.1 Initiating event frequency*

The demand on any SIF may arise due to one or more events, each with a frequency in the range of once per year to once in plant life (for example, 30 years). There is some approximation involved in the selected initiating event frequency values especially when using generic databases. For instance, the source data in the generic database may not be appropriate for application to the facility under review or there may not be enough confidence in the dataset. Power failure or failure of other utilities such as cooling water or steam is a good example. The frequency of these failures from generic databases is typically 1x $10^{-1}$ per year (once in 10 years) [4]. However, in the author's experience, based on feedback from many operators, these values are likely to be somewhere between once per year to once in ten years. Similarly, the frequency of pump trip based on generic databases is 1x $10^{-1}$ per year, however operational experience suggests that a value in the range of once in 2 years to once in 3 years may be more appropriate. An analysis of the OREDA [5] data also provides a value close to once in 3 years. Additionally, when there are multiple demand scenarios for a particular SIF, failure to consider the cumulative frequency of all the demand scenarios can also lead to under-estimation of the overall initiating frequency. A simple approach is to consider a suitable approximation for the cumulative demand frequency [6] [7]. In addition, common cause events specific to the facility is also a contributor which needs to be identified. Without such considerations, the required target $PFD_{avg}$ may be under-predicted. All of the above introduce uncertainties in the LOPA assessment noting that the initiating event frequency is the starting point of a LOPA assessment. Therefore, care must be taken in selecting an appropriate value for this parameter.

To better understand the dependence on the initiating event frequency, consider a typical distillation column as shown in Figure 1. There are many scenarios such as overhead condenser failure, reflux pump failure, reboiler control failure, etc. which may lead to a demand on the high high pressure SIF provided on the column. For the sake of brevity, only the case of Reflux Pump trip is used further in this example.

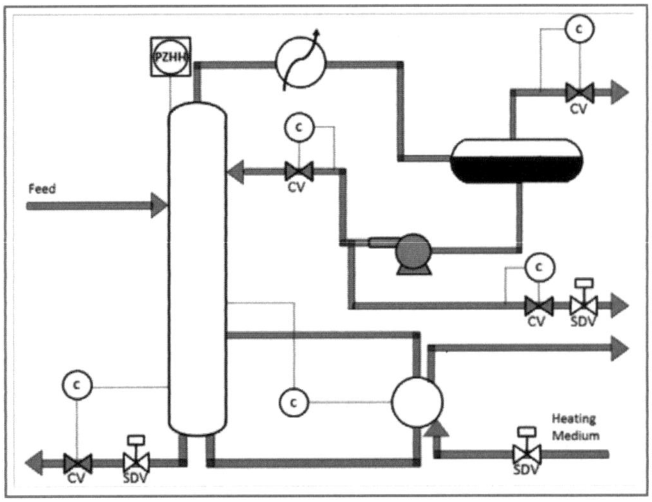


**Figure 1. Typical Distillation Column**

Assuming an initiating event frequency of 1 x $10^{-1}$ per year for pump trip and considering a Target Mitigated Event Likelihood (TMEL) of 1 x $10^{-5}$ per year for multiple fatalities along with an ignition probability of 0.1 and taking credit for pressure safety valve (PSV) on the column ($PFD_{avg}$ of 1 x $10^{-2}$), this would result in a SIL 1 with Risk Reduction Factor (RRF) 10 (target $PFD_{avg}$ of 1 x $10^{-1}$). However, if the initiating event frequency for pump trip is changed to 3.33 x $10^{-1}$ per year (once in 3 years), this translates to SIL 1 with RRF of 33 (target $PFD_{avg}$ of 3 x $10^{-2}$). This example demonstrates the dependence of initiating event frequency selection on the target $PFD_{avg}$.

### *2.2 Consequence Severity and Conditional modifiers*

The TMEL assigned to the SIF is directly associated with the degree or the severity of the consequence assumed (e.g. single fatality: 1 x $10^{-4}$ per year, multiple fatalities: 1 x $10^{-5}$ per year). This in turn has a direct impact on the SIL assigned to the SIF as well as its target $PFD_{avg}$. Therefore, careful consideration must be given to the selection of the consequence severity during a LOPA review. For instance, the impact area or hazard zone due to a loss of containment event can be determined using consequence modelling [7], the result from which can be used to assign the severity (e.g. single fatality vs multiple fatalities). Additionally, this can be used to determine the relevant conditional modifiers such as occupancy factor within the impact zone. This approach is more reliable than making coarse assumptions which may introduce uncertainties in the analysis. Such refinement can also be applied to other modifiers such as ignition probability where using a more specific value representative of the hazardous fluid and/or the facility features may be appropriate as compared to use of generic values.

If we revisit the example of typically distillation column presented in Section 2.1; considering an initiating event frequency of 3.33 x $10^{-1}$ per year and assuming a TMEL of 1 x $10^{-4}$ per year (single fatality), an ignition probability of 0.1 and taking credit for PSV on the column ($PFD_{avg}$

of 1 x $10^{-2}$), this would result in a RRF of 3 (target $PFD_{avg}$ of 3 x $10^{-1}$), which is below SIL 1 requirement. However, if a higher severity (TMEL: 1 x $10^{-5}$ per year corresponding to multiple fatalities) is selected based on consequence modelling results and an ignition probability of 0.3 is selected based on refinement, the assigned SIL will increase significantly to SIL 2 with RRF 100 (target $PFD_{avg}$ of 1 x $10^{-2}$). Although this example shows an increase in the assigned SIL, the application of detailed assessment for determining severity could also show a reduction.

### *2.3 Independent Protection Layers*

Independent Protection Layers (IPLs) are devices, systems or actions that prevent the realization of an undesired consequence (e.g. the consequence the SIF is safeguarding against). Each IPL has an associated $PFD_{avg}$ which is assigned and used in the LOPA methodology for SIL Determination. For example, process deviation alarm with sufficient time for operator intervention has a $PFD_{avg}$ of 1 x $10^{-1}$, an independent BPCS control has a $PFD_{avg}$ of 1 x $10^{-1}$ while a PSV has a $PFD_{avg}$ of 1 x $10^{-2}$.

#### 2.3.1 Effect of Common Cause Failures

One of the fundamental assumptions in the LOPA approach is that these IPLs are completely independent and do not share any common cause failures (CCF). In reality, multiple protection layers may be impacted or their performance may be affected due to similar environmental conditions or deviations from desired process parameters. Corrosion or deposition leading to simultaneous degradation of a pressure safety valve and a shutdown valve is one such example. Other examples of CCF include scenarios where multiple PSVs are required to meet total relief requirements, or where an alarm and control, although independent safeguards, both reside on the BPCS. Additionally, dual check valves for preventing backflow could also be impacted by CCF. Although such scenarios may take a period of time to develop and the likelihood may be low, such CCF can have a direct impact on the $PFD_{avg}$. For example, consider a case where the safety system comprises a SIF and an IPL and the target $PFD_{avg}$ for this system is 1 x $10^{-3}$. The $PFD_{avg}$ of the IPL is 5 x $10^{-2}$. If the SIF and IPL fail independently only, then the target $PFD_{avg}$ required of the SIF is (1 x $10^{-3}$) / (5 x $10^{-2}$) = 2 x $10^{-2}$. However on inclusion of CCF, even if a low probability of CCF is assumed (1% of the $PFD_{avg}$ of the IPL), the target $PFD_{avg}$ for the SIF then becomes 1 x $10^{-2}$.

The authors do acknowledge that for the purpose of time-effective LOPA review, it may not be practical to dwell deeper into CCF for IPLs as it may require an in-depth analysis (for which sufficient details may also not be available at the time of LOPA review). However, it is important to be aware of the inherent uncertainties introduced in the analysis by not accounting for the CCF associated with IPLs.

#### 2.3.2 Effect of Multiple Protection Layers

There is another inherent uncertainty associated with IPLs apart from CCF. The $PFD_{avg}$ for any protection layer (whether mechanical protection, alarm with operator response, BPCS or SIF) is not static but increases over time until a required test or maintenance activity is conducted to check for faults [4]. This relationship between PFD and time is shown graphically in Figure 2. In LOPA, the $PFD_{avg}$ values of the IPLs are multiplied together to estimate the $PFD_{avg}$ of the overall safety system [4]. This calculation, while simple to perform, may tend to underestimate the overall $PFD_{avg}$. In a more quantitative analysis of reliability, the time-dependent PFDs of each protection layer are multiplied together to yield a result similar to Figure 3, which shows the PFD of the safety system. The $PFD_{avg}$ for the system can then be calculated.

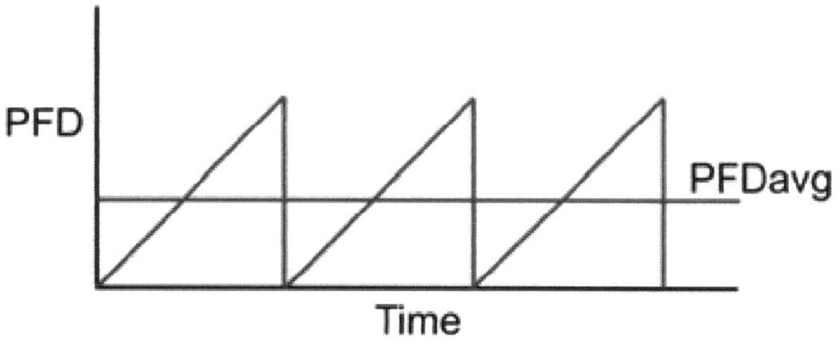


**Figure 2. PFD as a function of time for a single protection layer. Periodic testing / maintenance was assumed to detect all faults in this example.**

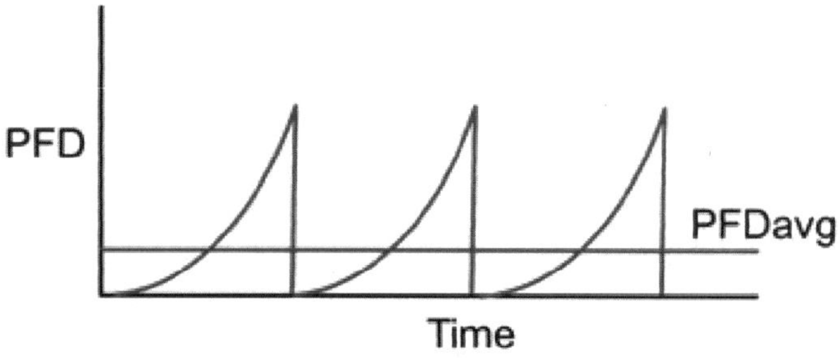


**Figure 3. PFD as a function of time for a safety system with two protection layers (one SIF and one other IPL). Proof testing was assumed to detect all faults in this example.**

To illustrate the difference in the results provided by LOPA and a more rigorous analysis, the same example from Section 2.3.1 is used. LOPA estimates the $PFD_{avg}$ of the SIF to be $2 \times 10^{-2}$ (excluding CCF), considering one IPL with a $PFD_{avg}$ of $5 \times 10^{-2}$ per year and a target overall PFD for the system of $1 \times 10^{-3}$ (see calculation in Section 2.3.1). Using the PDS method of reliability analysis [8] gives a SIF $PFD_{avg}$ of $1.5 \times 10^{-2}$, i.e. LOPA overestimates the target $PFD_{avg}$ by 33%. As the number of protection layers increases, this error becomes more significant, as shown in Table 1. For two or more IPLs with a SIF, the error in the target $PFD_{avg}$ for the SIF is at least 100% using the PDS method; i.e. the LOPA value for target $PFD_{avg}$ is double that given by the PDS method. This gross overestimation in the target $PFD_{avg}$ for the SIF suggests that the LOPA method may not be sufficiently accurate where two or more IPLs are considered. In such cases a more rigorous analysis may be conducted outside of the LOPA session (refer Section 3). Such rigorous approach, if adopted, may also help reduce the error for cases where only one IPL is applicable.

**Table 1: Effect of the number of IPLs on the error in $PFD_{avg}$ for a safety system as calculated by LOPA**

| Number of IPLs | Error in target $PFD_{avg}$ for SIF |
|---|---|
| 0 | 0% |
| 1 | 33% |
| 2 | 100% |
| 3 | 220% |
| $n$ | $\frac{2^{n+1}}{n+2} - 1$ |

## *2.4 Assessing the cumulative uncertainty for multiple parameters in LOPA*

The above sections have discussed how the uncertainty of each individual parameter affects the target $PFD_{avg}$ calculation. In practice, however, the cumulative effect of the uncertainties for multiple parameters also needs to be considered. As a simple illustration and keeping in line with the discussion above, the same example of the distillation column high high pressure SIF is used. As shown in the table below, the changes in various parameters results in a significant difference in the target $PFD_{avg}$ requirements. One additional parameter presented in the table below is the use of $PFD_{avg}$ of 3 x $10^{-2}$ for IPLs considering multiple PSVs are required to meet relief requirements. As expected, the required RRF is much more stringent when the uncertainties are modelled. However, as highlighted in earlier sections, depending on the SIF under review, in some cases, the requirements may also be less stringent upon further refinement.

**Table 2: Effect of cumulative uncertainty of multiple parameters in LOPA**

| Description* | Initiating Event Frequency | IPL | Ign. Prob. | Total Mitigated Event Frequency | TMEL | Target $PFD_{avg}$ | RRF |
|---|---|---|---|---|---|---|---|
| Case 1 | 1.00E-01 | 1.00E-02 | 1.00E-01 | 1.00E-04 | 1.00E-04 | 1.00E+00 | 1 |
| Case 2 | 3.33E-01 | 1.00E-02 | 3.00E-01 | 1.00E-03 | 1.00E-04 | 1.00E-01 | 10 |
| Case 3 | 1.00E-01 | 1.00E-02 | 1.00E-01 | 1.00E-04 | 1.00E-05 | 1.00E-01 | 10 |
| Case 4 | 3.33E-01 | 1.00E-02 | 3.00E-01 | 1.00E-03 | 1.00E-05 | 1.00E-02 | 100 |
| Case 5 | 1.00E-01 | 3.00E-02 | 1.00E-01 | 3.00E-04 | 1.00E-04 | 3.33E-01 | 3 |
| Case 6 | 3.33E-01 | 3.00E-02 | 3.00E-01 | 9.00E-04 | 1.00E-04 | 1.11E-01 | 9 |

# 3 Approaches for handling uncertainty in LOPA

Based on the various modes of introduction of uncertainties in the LOPA analysis as discussed in the preceding sections, the question that arises is whether the direct use of target $PFD_{avg}$ from LOPA as input to SIF design is optimal or is it desirable to investigate possible ways to refine the target $PFD_{avg}$ from LOPA to ensure the SIF is not under-designed. Some ways to refine the target $PFD_{avg}$ are discussed in the following paragraphs.

One proposed approach for dealing with uncertainty without compromising the simplicity that makes LOPA an attractive method is to conduct a more rigorous risk assessment outside of the LOPA session for high consequence / high risk scenarios to confirm that the SIF reliability target $PFD_{avg}$ obtained from LOPA is sufficient. Such an assessment may include a quantitative reliability analysis such as a Fault Tree Analysis or Markov Model to obtain a more accurate estimate of the required target $PFD_{avg}$. These methods can also take into account common cause failures and the time dependency of PFD. In fact the most recent version of IEC 61511 [1] requires that a quantitative methodology be used for cases where the required risk reduction is greater than 10,000 (target $PFD_{avg}$ from LOPA is equal to or lower than $1\times10^{-4}$ per year). By using LOPA as a screening tool, this strategy can ensure that high consequence / high risk scenarios receive the more rigorous assessment they deserve, while for lower risk scenarios the more time-efficient LOPA can be considered sufficient.

Alternatively, a number of different cases can be considered for each LOPA scenario, such as in Table 2. Each case considers different values for the various parameters used in the LOPA calculation, resulting in a range of RRFs, from the more conservative to the more liberal. Taking Table 2 as an example, the results range from an RRF of 1 to 100. A final RRF at some point in between these two extreme cases may be selected for the SIF and carried forward into the SIF design. In this way, a level of uncertainty can be accounted for without being overly conservative.

Another approach is to introduce a "Safety Factor" to the reliability target set during LOPA. In this approach, it is recommended that the target Risk Reduction Factor obtained from LOPA should be increased by a safety factor to give a more stringent target for the SIF design to meet. For instance, if the target based on results from LOPA is RRF 100, the SIF design can conservatively consider a target $PFD_{avg}$ associated with RRF 300. Similarly, if the target is RRF 500, then SIF design may consider RRF 800. However, when the RRF is at or close to borderline values, the SIF design may consider the next interval, e.g. for RRF 800 and above, the SIF design may consider RRF 1300 or alternately the RRF used can be justified by conducting a more rigorous quantitative analysis as mentioned above. Some engineering judgement should be applied when selecting this safety factor depending on the level of conservatism adopted in the initial LOPA assessment. As can be inferred from Table 2 above, when uncertainties are accounted for, the target $PFD_{avg}$ can actually vary by one order of magnitude. Applying a safety factor to the results will therefore account for the uncertainties.

# 4 What $PFD_{avg}$ to assume when Risk Graph Method is adopted for SIL Determination?

Risk Graph Method [1] is the alternate method commonly used for SIL Determination. Unlike the semi-quantitative LOPA approach, which specifies a numerical reliability target ($PFD_{avg}$) for the SIF in addition to the assigned SIL, the Risk Graph approach relies on the qualitative definitions for the parameters involved to deduce the assigned SIL without setting an actual numerical target ($PFD_{avg}$). Furthermore, this target $PFD_{avg}$ value from Risk Graph is provided as a range. Hence, where Risk Graph method is used, it is not an uncommon practice to use the upper bound of the $PFD_{avg}$ associated with the assigned SIL. For example, the LOPA method may provide an RRF of 40 ($PFD_{avg}$ of $2.5 \times 10^{-2}$ per year) for a SIL 1 function, while the same SIL 1 function following the Risk Graph method may adopt the upper bound RRF of 10 ($PFD_{avg}$ of $1 \times 10^{-1}$ per year).

Consider the same example of a distillation column as discussed in Section 2. If the column high high pressure SIF is assessed using the Risk Graph approach assuming that a trip of Reflux Pump could be expected once a year, taking credit for the pressure safety valve on the column as an IPL (credit of 100) and taking no credit for occupancy and avoidance; the resulting analysis show that Risk Graph method also determines a SIL 2 which is similar to the results from LOPA are accounting for uncertainties. The key difference being the overall $PFD_{avg}$ of $3.33 \times 10^{-3}$ per year (RRF 300) obtained from the LOPA which will be more stringent than the $PFD_{avg}$ of $1 \times 10^{-2}$ per year (RRF 100) in case the upper bound value of SIL 2 is used.

The above example shows that using the output from the Risk Graph directly may lead to under-prediction of the target $PFD_{avg}$ and thereby leading to under design of the SIF. The concept of "Safety Factor" as discussed in Section 3 may therefore also be applied where Risk Graph is used. Based on this, it is recommended that the SIF design can use the mid-point of the upper bound and lower bound values based on $PFD_{avg}$ range applicable to the SIF corresponding to the assigned SIL. The safety factor may also be selected on a case by case basis depending on the severity of the associated consequences (for example, using a value above the mid-point of upper and lower bound may be justified for the most severe consequences).

# 5 Uncertainties in SIL reliability analysis

The uncertainties in terms of obtaining an appropriate target $PFD_{avg}$ for a SIF using LOPA methodology are discussed in Section 2. It is however, worth noting that potential for introduction of uncertainties into the analysis is not limited to this step alone. Moving on in the safety life cycle, the next step after LOPA involves a reliability analysis to ensure that the target $PFD_{avg}$ is met for each SIF. This activity is commonly termed as "SIL Verification" in the industry, but strictly with reference to IEC 61511, it is considered within the domain of "SIS Design and Engineering". The target $PFD_{avg}$ that is considered for the reliability analysis should be the same as the target $PFD_{avg}$ that was calculated in LOPA (a safety factor may be applied as described in Section 3). Where Risk Graph methodology is applied, the target $PFD_{avg}$ can be a mid-point value in the assigned SIL band as suggested in Section 4.

This reliability analysis also involves a number of input parameters, some of which may introduce uncertainties into the analysis. It is important to understand and evaluate the nature of these uncertainties associated with input parameters such as failure rate data for SIF components, beta factor for common causes and proof test coverage factors in SIF reliability analysis as discussed in the subsequent sub-sections. In this section, $PFD_{avg}$ results that were calculated using exSILentia software are presented to illustrate the effects of the input parameters on the $PFD_{avg}$. The numerical $PFD_{avg}$ results are only applicable for the specific systems and assumptions used in the calculations.

## *5.1 Failure Rate Data*

Failure rate data for the individual SIF components is the basis of the reliability analysis. It is therefore of utmost importance that the data used is based on sufficient sample size and is representative of the conditions in which the SIF is applied. The quality of the data used as input parameters to reliability models has a large effect on the accuracy of the results.

Ideally, an appropriate method for determining the failure rate is to collect field data from a large, randomized sample of units. The operational time until failure should be recorded for each unit. Such data mining however; is a long term project, requiring the co-operation of multiple operators, ideally from a wide range of different facilities and operating environments. For this reason, such studies are rare and normally "generic data", i.e. data applicable for a broad type of component (such as pressure transmitter or shut-down valve) is used rather than data for a specific model or a specific environment. The OREDA database [5] is one example, which provides such generic data. Alternately, some vendors may commission a field study to collect failure data specific to their components or build their own failure data based on reported failures. However, given the smaller sample size as well as the gaps in ensuring that each failure is reported appropriately by the operating facilities, there will be uncertainties in the reported data and these values may tend to be on the optimistic side. For example, while the failure rate for a ball valve from a generic data source is $9.8 \times 10^{-7}$ per hour [9], several vendors of ball valves report failure rates of less than $1.0 \times 10^{-7}$ per hour. Another issue is that such data may only be sourced from a limited number of facilities operating in a specific environment and hence, this data may not be applied to a range of facility types or operating environments.

One approach to iron out these uncertainties is to conduct a Failure Mode and Effect Analysis which is a workshop activity bringing together a group of experienced personnel (operations, maintenance, instrumentation, design, etc.) for brainstorming the possible failure modes associated with each component and its associated failure rate using operational experience and field data. The quality of such an exercise is highly dependent on the experience of the participants and their engineering judgement. Hence, even though conducting such an exercise will be fruitful, some margin of error is still expected.

### *5.2 Beta Factor*

Typically, the overall reliability of a SIF is estimated based on both independent and dependent failures. For functions with redundant components, multiple components may fail simultaneously due to a common failure mode (e.g. stress, maintenance errors, etc.). These dependent failures are the percentage of the failures attributed to the common causes and are denoted as Beta Factor in the reliability analysis.

Taking note of the uncertainties in failure data for components as highlighted in earlier sections, it is worth highlighting that the chances of availability of detailed field data for beta factors covering a large sample size is even lower, thereby leading to a greater degree of uncertainty. IEC 61508 [2] advocates a method for estimation of the beta factor using a questionnaire and a scoring system. Based on the scoring system, typically the beta factor value ranges from 1 - 10%. This is a somewhat subjective approach. In reality, beta factors higher than 10% have also been reported in the industry [10]. This may however represent an extreme operating environment pertaining to a particular facility. Therefore, some engineering judgement is required.

It is to be noted that beta factor has a direct impact on the $PFD_{avg}$ achieved for systems with redundant components. Since common cause failures are often the dominating contributor to $PFD_{avg}$, an increase in beta factor leads to a significant increase in $PFD_{avg}$. This relationship can be seen in the $PFD_{avg}$ results given in Figure 4, for SIFs with various configurations (1oo2, 2oo3, etc.). Hence, selection of a low beta factor may lead to underestimating the $PFD_{avg}$ and potentially also under design of a SIF. For configurations without redundancy (e.g. 2oo2), the beta factor has little impact on the $PFD_{avg}$.

In the author's experience, some references and guidelines from operating companies provide recommended beta factor values. However, taking note of the findings presented in this section, it should be ensured that such recommended values have a sound basis and are as accurate as possible. Over-estimating beta factor value may lead to over-design of the SIF which in turn also has an associated cost impact. Conversely, under-estimating the beta factor may lead to SIF being under-designed.

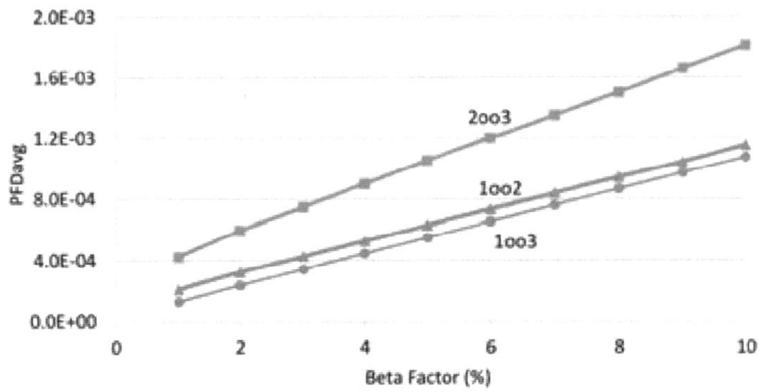


Note: Calculation performed using exSILentia software with the following parameters: λ = 2675 FIT; MTTR = 8 hrs; PTI = 1 yr; PTC = 100%; Mission Time = 10 yrs.

**Figure 4. Impact of Beta Factor on $PFD_{avg}$ calculation**

### *5.3 Proof Test Interval*

A proof test is performed periodically to detect failures and restore the system / component to an "as good as new" condition. The actual proof test interval may be governed by factors associated with SIF design and operating philosophy specific to a facility. However, the selected interval forms the basis of SIF design since the reliability of the SIF can be compromised with regard to the desired target $PFD_{avg}$ in case the proof test interval used for SIF design is not followed. The uncertainty in this regard lies in whether the recommended proof test interval is rigidly followed or is only conducted on as-need basis in the actual operational phase. Operators may tend to postpone a shutdown for testing purposes if it may result in a loss of production, thereby compromising the SIF reliability. The emphasis to rigidly adopt the recommended proof test intervals therefore rests with the Operating Company and the kinds of safety culture they advocate or follow.

### *5.4 Proof Test Coverage*

One of the important parameters associated with proof tests is the coverage factor which represents the fraction of the potential failures that will be identified during a full proof test. A perfect proof test (i.e. coverage factor of 100%) is unlikely. As far as the authors are aware, there is also no field data available to determine the effectiveness of proof tests. Vendors may recommend a coverage factor based on a detailed FMEA exercise for the associated components. Such an FMEA can identify which failure modes will be successfully captured by a proof test, and thus estimate the coverage factor. Proof test coverage factors can sometimes be lower for components with good diagnostic coverage because the remaining failure modes that are undetected by diagnostics are more difficult for a proof test to detect. In the author's experience, typical coverage factors range from 80 - 99% for sensors and logic solvers and 70 - 80% for final elements (e.g. shutdown valves). While this is currently the best approach for determining the coverage factor, the uncertainty in the result is usually not estimated.

The proof test coverage factor has a significant impact on the calculated $PFD_{avg}$ of a SIF element, as can be seen in Figure 5. For example, for an element with a 1oo1 configuration, decreasing the coverage factor from 90% to 80% resulted in a 46% increase in the $PFD_{avg}$. Given this effect, caution should be exercised when selecting coverage factors for use in SIF design since they will impact the risk reduction that can be achieved. The actual test performance should ideally be verified to match the assumptions (this becomes especially critical when higher factors e.g. 90% or above are used).

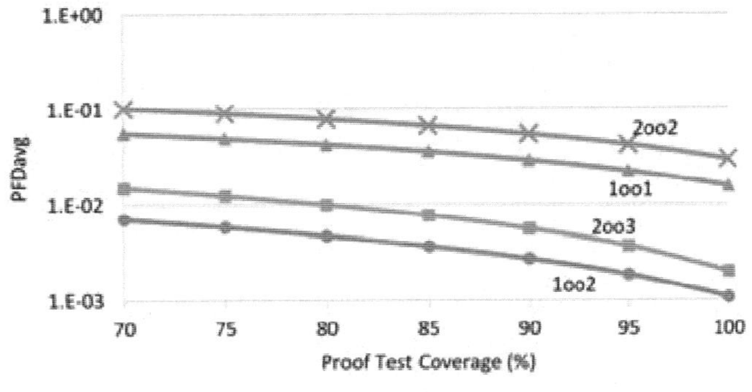


Note: Calculation performed using exSILentia software with the following parameters: λ = 1810 FIT; MTTR = 8 hrs; PTI = 2 yrs; Mission Time = 20 yrs; beta factor = 5%

**Figure 5. Effect of Proof Test Coverage Factor on $PFD_{avg}$ for various voting configurations**

An alternative method for considering failures that are not detected by a proof test is to include a "probability of test independent failures", or $P_{TIF}$. This approach, which is advocated in the

PDS method [8], differs from the proof test coverage approach in that $P_{TIF}$ is a constant contributor to the $PFD_{avg}$, whereas the effect of proof test coverage on the $PFD_{avg}$ increases over time. Even when using $P_{TIF}$, however, there remains uncertainty in the value used for this parameter, with implications on the accuracy of the calculated $PFD_{avg}$.

### *5.5 SIF Operational Life*

Operational life of a SIF refers to the life time of the various components of the SIF, sometimes also denoted as mission time. This also serves as a key input in SIF design. In the author's experience, typically a mission time in the range of 20 to 30 years is assumed. Lower mission times such as 5 or 10 years may be impractical in terms of actual implementation on site since this implies that the corresponding components will be replaced or completely overhauled after a 5 or 10 year period, which also has a significant cost impact. This assumption on operational life of the SIF has a direct impact on the risk reduction achieved by a SIF, provided that all other parameters are kept constant. Figures 6 and 7 show that a higher mission time is likely to result in a higher $PFD_{avg}$ (or in other terms, a lower risk reduction). Figure 7 shows that the relationship between mission time and $PFD_{avg}$ is almost linear, with a steeper gradient at a lower proof test coverage. This shows that the mission time has a greater impact on $PFD_{avg}$ when the proof test coverage is lower. If the proof test coverage is taken as 100%, the mission time has no effect on $PFD_{avg}$. However, as discussed in Section 5.4, 100% proof test coverage is not likely to be realistic.

While operators may typically consider the cost of replacing equipment when deciding upon mission time, consideration should also be given to the effect that mission time will have on the reliability of SIFs. A shorter mission time can help SIFs to meet the target $PFD_{avg}$ and may even allow a lower level of redundancy to be used for the SIF components. This is particularly relevant for SIL 3 rated SIFs, which may require a high level of redundancy as well as high reliability of components, resulting in a high capital cost. Reducing the mission time may help optimize the capital cost of the SIF. A cost benefit analysis may be conducted to justify optimization of the mission time. However, some engineering judgement must be exercised in order to keep the mission time within practical limits. In this regard, 15 years may be considered as a practical lower limit for mission time but this may vary on a case by case basis also depending on the owner's operating philosophy. In such cases, it is important to be aware that in case the capital replacement is not performed after 15 years, this will potentially leave the system exposed to higher probability of failures.

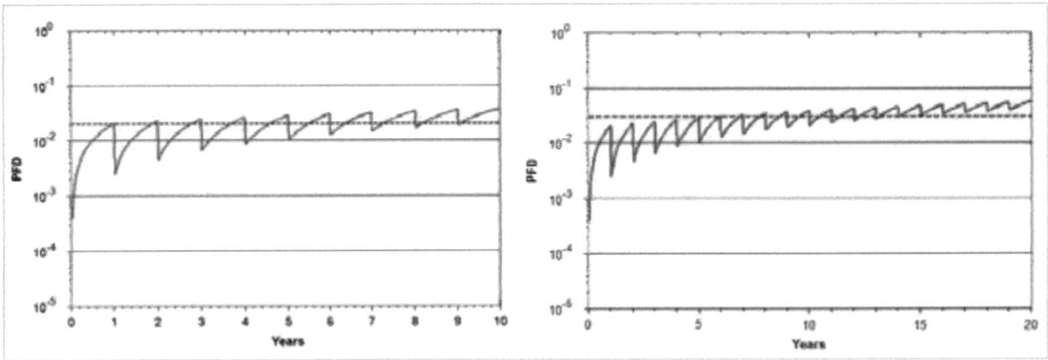


**Figure 6. $PFD_{avg}$ vs SIF Operational Life / Mission Time (Years)**

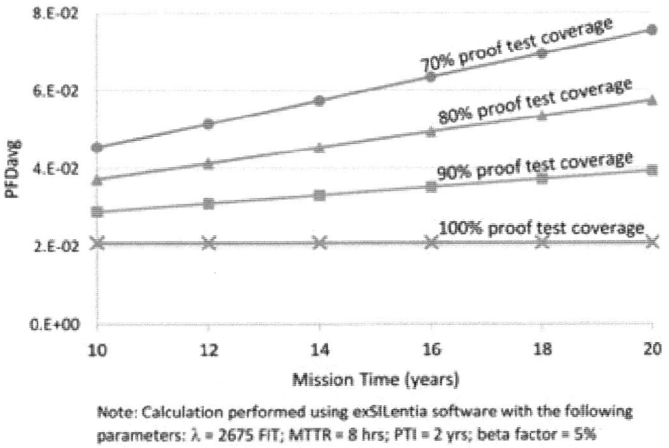


**Figure 7. Effect of Mission Time on $PFD_{avg}$ for a SIF with 1oo1 configuration**

### *5.6 Assessing the cumulative uncertainty for multiple parameters in SIF Reliability Analysis*

The above sections discuss how the uncertainty of each individual parameter affects the $PFD_{avg}$ calculation. In practice, however, the cumulative effect of the uncertainties for multiple parameters also needs to be considered. As a simple illustration, Table 3 presents $PFD_{avg}$ values calculated for a high pressure SIF with 1oo2 component configuration, using three different combinations of beta factor and proof test coverage, representing a best case, base case and worst case scenario. The beta factor was varied between 1% and 10%, while the proof test coverage was varied between 70% and 90%, which are ranges that can be reasonably expected for these parameters. The $PFD_{avg}$ for the worst case scenario was more than double that of the base case and more than five times that of the best case. In these scenarios, beta factor was found to be the main contributor to the variation in $PFD_{avg}$, as the uncertainty in the beta factor may be +100% / -60%, whereas the proof test coverage may typically have an uncertainty of only ±13%.

**Table 3: $PFD_{avg}$ for a high pressure SIF in three scenarios**

| Scenario | Beta Factor | Proof Test Coverage | $PFD_{avg}$ | Variation from base case |
|---|---|---|---|---|
| Best Case | 1% | 90% | 2.93E-04 | -70% |
| Base Case | 5% | 80% | 9.88E-04 | 0 |
| Worst Case | 10% | 70% | 2.10E-03 | +113% |

Note: Calculation performed using exSILentia software with the following parameters: λ = 600 FIT; MTTR = 8 hrs; PTI =3 yrs; Mission Time = 15 yrs.

## 6 Conclusion

In this paper, the authors have highlighted multiple modes through which uncertainties are introduced into the LOPA review and subsequently in the reliability analysis for SIF design. Through working examples, this paper demonstrated that without accounting for such uncertainties, the SIF can easily be under-designed or over-engineered. The emphasis on accounting for such uncertainties is also highlighted in the most recent edition of IEC 61511. This paper investigated further into the nature of uncertainties associated with various input parameters associated with LOPA and SIF reliability analysis including approaches using which such uncertainties can be accounted appropriately.

## 7 Acknowledgement

The authors wish to acknowledge the support received from IRESC including some of the information included in this paper, as well as the contribution from Ms. Nicole Yeung.

## 8 References

[1] The International Electrotechnical Commission (IEC), Functional Safety – Safety Instrumented Systems for the Process Industry Sector, IEC-61511, Second Edition

[2] The International Electrotechnical Commission (IEC), Functional Safety of Electrical/ Electronic/ Programmable Electronic Safety Related Systems, IEC-61508, First Edition

[3] Guidelines for Enabling Conditions and Conditional Modifiers in Layer of Protection Analysis, CCPS

[4] Layer of Protection Analysis, Simplified Process Risk Assessment, CCPS

[5] Offshore and Onshore Reliability Data (OREDA), Volume 1 – Topside Equipment, 6th Edition, 2015

[6] Akshat Khirwal, Venkatesh S., Consideration of Cumulative Frequency of Demand Scenarios in SIL/LOPA Review, Hazards Asia Pacific, Kuala Lumpur, Malaysia, 2015

[7] Akshat Khirwal, Venkatesh S., Inherently Safer Design & Robust Protection Layers for High Consequence Hazards, Mary Kay O'Connor Process Safety Center, 19th Annual International Symposium, College Station, Texas, USA, 2016

[8] Reliability Prediction Method for Safety Instrumented Systems, PDS Method Handbook, SINTEF, 2010 edition

[9] Safety Equipment Reliability Handbook, Exida, Third Edition, 2007

[10] S. Hauge et al., Common cause failures in safety-instrumented systems: Using field experience from the petroleum industry. Reliab. Eng. Sys. Saf. 151 (2016), 34-45.

# 479264 Fundamental Research at Purdue Process Safety Research Center

**Abstract:**

Fundamental Research at Purdue Process Safety Research Center

Significant industrial incidents continue to occur such as fires, explosions and gas releases resulting in serious injuries, loss of life, and damage to the environment and facilities. Such incidents are not solely germane to the oil & gas industry, but occur in a wide variety of industries including agriculture, pharmaceuticals, and manufacturing. Two years ago the 'Purdue Process Safety and Assurance Center' (P2SAC) was established within the School of Chemical Engineering to conduct fundamental research to prevent such incidents. The presentation will address the four research projects the Center is currently pursuing with funding from its initial Sponsors: BP, Chevron, ExxonMobil, Honeywell, Eli Lilly and Shell.

The initial projects include:

- **optimal placement of gas detectors to prevent fires and explosions**: the research uses CFD-based dispersion simulations and rigorous mathematics to perform optimal placement over hundreds of thousands of leak scenarios and potential detector locations.

- **design & synthesis of catalysts to prevent undesired hazardous byproducts**: zeolite catalysts are widely used in the refining industry and certain structures can promote undesired byproducts leading to potential over-pressures and explosions. Zeolite synthesis and characterization methods are being developed.

- **real time process optimization to reduce risk**: Novel continuous manufacturing processes can reduce the risk related to powder handling. Continuous manufacturing, as with pharmaceuticals, requires efficient process monitoring and control.

- **application of inherently safer separation technology** to significantly reduce processing volume, solvent use, and extent of unit operations. Advanced mixed solvent extraction & simulated moving bed techniques are used to recover polycarbonates and flame retardants from plastic waste.

Indirect benefits from the Center are the depth of the core Chemical Process Safety course required of all undergraduates, as well as undergraduate process safety research initiatives in such as areas as dusts, reactive chemicals and nano-materials.

Current focus is on growing the Center and encouraging participation by a wide variety of industries such as pharmaceuticals, agriculture and manufacturing, in addition to oil & gas, and chemicals.

10-13-16

# 479277 Bringing New Life to Process Safety


**Abstract:**

<>Abstract:

Whilst risk cannot be eliminated, it can be better managed. The world of Process Safety Management continues to evolve as organizations find better ways of managing risks. In the simplest form, managing the risk is to reducing inherent process risk to the tolerable level.

First step in managing risks is to identify source of Hazards and quantifying risk associated to Hazards. Traditionally organizations heavily rely on expert judgment and brainstorming methodologies like HAZOP for finding operational and design risks. In addition, quality of such analysis are directly related to the precision of assumptions and information from operation and maintenance of the equipment under control.

In this session we will explore how organizations are using new techniques such as dynamic risk visualization, cumulative risk techniques, operational risk assessments and cyber Hazop's to advance the boundaries of process safety management.

This paper will explore how existing systems and data sources, risk visualization and intelligence techniques are being used to bring static process safety data to life. Flat "1 dimensional" information such as lists, forms and databases are now being transformed into "2 dimensional" dynamic information such as plot plans, barrier models, risk matrix, bow ties etc. that provide greater context of relationships between information.

This information will resonate with engineers and technician who rather working with meaningful information of operation and maintenance systems instead of some abstract numbers and concepts that need extensive expertise in the reliability engineering.

We also will look at how advances in Operational Risk Assessments and Cumulative Risk allow operators to determine in real time if the overall cumulative risk is tolerable before actions are taken, devices are removed from service, maintenance activities are deferred etc. In other words, leading indicators and metrics will be in disposal of operators to proactively facilitate decision making for managing risk. This could avoid incidents that developing in the operation due to degradation of control and safety systems or deviations from initial operation and maintenance procedures and may lead to severe incidents.

And finally, since new technologies and architectures for instrumentation, control and safety systems introduced new risks of cyber intrusions and attacks that need to be considered in the risk management and the world of security standards (ISA 62443-3-2 Security Risk Assessment and System Design) and safety standards (ISA-TR84.00.09-2013 Security countermeasures related to Safety Instrumented Systems and Associated IACS) advanced in the last few years, we will look at how Cyber Process Hazard Analysis (PHA) techniques are being applied to provide the same systematic approach to security as traditionally applied to safety.

"Good process safety management does not happen by chance and require constant active engagement", in this paper we suggested novel approaches that can revive and improve the process safety.

# 479367 Developing an Effective System for Action Item Resolution

**Abstract:**

PSM action items are are regularly identified through many different means in a typical chemical processing facility. Knowing what to do with these action items and how best to proceed is no easy task. Not all action items are created equally, or even with the same goal in mind. An action item identified in a PHA may present a direct hazard to an employee, while an action item generated from a pressure relief systems study may present a documentation concern, a deviation from current practices, a direct hazard to an employee, or any combination thereof. Many facilities have a system in place for shepherding action items to completion, but these systems typically assume all action items are created equally. In this paper, we will demonstrate the importance of having a system that handles each action item in the most effective way; from evaluating the nature of the concern, considering the risks associated with the action item, developing the action plan, to completing the action plan. We will also cover whose input should be considered at each stage in the process to help ensure the best solutions possible are developed.

# Where Does Management of Change (MOC) Begin?

**Juliana Schmitz**
**Praxair, Inc.**
**Av. Pastor Martin Luther King Jr., 126, bloco 10/ala A,**
**Corporate Shopping Nova América, Rio de Janeiro, RJ, Brazil**
**juliana_schmitz@praxair.com**

**Patti Jones**
**Praxair, Inc.**
**175 E Park Dr, Tonawanda, NY, 14150, USA**
**patti_jones@praxair.com**

**Gabriel Drummond**
**Praxair, Inc.**
**Av. Pastor Martin Luther King Jr., 126, bloco 10/ala A,**
**Corporate Shopping Nova América, Rio de Janeiro, RJ, Brazil**
**gabriel_drummond@praxair.com**

Prepared for Presentation at
American Institute of Chemical Engineers
2017 Spring Meeting and 13th Global Congress on Process Safety
San Antonio, TX
March 26 – 29, 2017

# Where Does Management of Change (MOC) Begin?

**Juliana Schmitz**
**Praxair, Inc.**
**Av. Pastor Martin Luther King Jr., 126, bloco 10/ala A, Corporate Shopping Nova América, Rio de Janeiro, RJ, Brazil**
**juliana_schmitz@praxair.com**

**Patti Jones**
**Praxair, Inc.**
**175 E Park Dr, Tonawanda, NY, 14150, USA**
**patti_jones@praxair.com**

**Gabriel Drummond**
**Praxair, Inc.**
**Av. Pastor Martin Luther King Jr., 126, bloco 10/ala A, Corporate Shopping Nova América, Rio de Janeiro, RJ, Brazil**
**gabriel_drummond@praxair.com**

**Keywords:** MOC, engineering, process safety

## Abstract

Management of Change is often affiliated with an operating facility. However, several projects have the common challenge of being developed within the constraints of an aggressive schedule, considering the time from project kick-off until unit start-up. In order to comply with the deadlines, early releases of engineering design documentation become subject to changes. During project development, these changes may come as scope changes, supplier updates, etc. that may affect more than one engineering discipline and any on-going construction/assembly activities. Another change source may be non-conformances identified during the construction/assembly phase. In this case, considering project milestones and costs, an option taken into account is to leave the system or portion the way it is instead of re-working to match the original design. For all of the given cases, a robust Engineering Management of Change (EMOC) process must be in place throughout the project life cycle to guarantee modifications are properly identified, communicated, and evaluated by technical experts for possible effects on process and design safety prior to implementation, or at a minimum, prior to startup of the system/subsystem. In addition, an EMOC process gives the opportunity to analyze the reasons for the modifications in order to avoid increased costs and design errors on future projects, leading to improved quality control. This paper presents some previous accidents related to management of changes, and addresses EMOC implementation, work process and elements of a successful EMOC program.

# 1 Introduction

Preliminary design activities begin early in the process to define scope, cost and product delivery requirements. Changes are a natural part of these efforts.

As the project progresses through the engineering detail design phase, a process hazard analysis (PHA) is conducted. At this moment, a multidisciplinary team evaluates if the process is safe and if any modifications are required. The project team is informed of the revised documentation and proceeds with each skill activity.

Subsequent changes may occur as, for example: updated supplier documentation, the team that will effectively operate the plant in the future may raise a new need for the project, some may be identified during detail design requiring changes and change requests during the construction phase. Any modification after the PHA may not be properly communicated to the adequate professionals and analyzed, resulting in a possible unsafe condition.

This paper focuses on project execution phases until plant startup and turnover to operations where the Management of Change (MOC) work process begins. It also explores the challenges identified when developing an Engineering Management of Change (EMOC) work process and shares an approach taken within Praxair, Inc.

# 2 Why is the Engineering Management of Change Important?

Process Safety is a relatively new engineering area, being developed from the early 60's and mostly based on catastrophic incidents, which negatively affected the industry. Kerin [5] confirms this point of view and lists a series of landmark events, their key learnings and influences in process safety. These events had different causes, mostly related to poor management of various safety areas, such as hazard identification, training, maintenance, management of change (MOC) and others.

The numerous and notable MOC related accidents that happened in the industry over the last century showed the world the need of a well-developed and maintained MOC procedure.

## *2.1 Titanic, 1912*

The sinking of the Titanic, in 1912, is a well-known catastrophic incident. "The design of the ship included 44 water tight compartments giving rise to the perception that the ship was unsinkable and in the event of an accident would be its own lifeboat." [2] An accelerated construction schedule was the catalyst in making changes that proved to be fatal. A robust management of change process and a cultivated safety culture may have prevented the incident.

## *2.2 Flixborough, 1974*

Venart [3] describes the Flixborough accident, which occurred on June 1$^{st}$, 1974, causing 28 fatalities and several property damages near the Nypro plant in North Lincolnshire, UK. In this plant, there was a process area responsible to produce caprolactam, an intermediary for the Nylon production. Caprolactam production was conducted through the oxidation of the cyclohexane, a highly flammable chemical. There were six reactors in series where this oxidation reaction was performed.

Two months before the explosion, a crack was observed in the 5th reactor of the chain. With the objective of reducing the impact in production, the responsible plant personnel decided to connect directly the 4th and 6th reactors while the 5th was removed for maintenance. To proceed with the change, a “dog-leg” pipe was used to connect the reactors 4 and 6, while the original installation considered a straight pipe between these reactors. This is a modification that clearly changed the characteristics of fluid flow between reactors 4 and 6. The dog-leg line support was a temporary installation, made of scaffolding bars. These changes were never evaluated in detail and during start-up. On June 1st, the bellows connected to the reactors flanges failed due to an overstress and tons of cyclohexane was released into the atmosphere. With the vapor dispersion, an ignition source was found in the plant, and a vapor cloud explosion occurred resulting in fatalities, as well as significant property damage. [4]

### *2.3 CONDEA Vista Company, 1998*

U.S. Chemical Safety and Hazard Investigation Board [5] promotes a discussion about the reaction vessel explosion in CONDEA Vista Company, on October 13th, 1998. In this unit, linear alkyl benzene was produced by mixing aluminum chloride and hydrocarbons. Around 3 months before the incident the process was modified, involving a mixing of powdered aluminum with hydrogen chloride to form the aluminum chloride. After this change was implemented, the bottom nozzle of the reactor got plugged due to the formation of a sludge residue, which the operators were not able to reach and remove during maintenance. Considering this scenario, the chemist of the site and the process engineer decided to use water to dissolve the sludge and to absorb the heat generated due to an exothermic reaction between the water and the sludge. This process was started and once it was apparently under control, the process engineer and chemist left the plant.

The shift supervisor decided to make a new change and inject steam at the nozzle. The process supervisor who was not in the plant agreed with the new procedure. At the shift turnover, the day shift supervisor talked with the night shift supervisor about the steam injection. The night shift supervisor took responsibility to inject steam and oriented an operator to execute this procedure.

After a few minutes of continuous steam injection, the reactor exploded, leaving three workers injured and property damage estimated in $13 million USD. Among other shortcomings related to the catalyst change that resulted in sludge formation, U.S. Chemical Safety and Hazard Investigation Board **[5]** affirms that this accident highlights the value of having an authorization or approval step as part of an MOC system for abnormal situations. If such a procedure had been in place, a technical manager would have reviewed the proposed procedure involving steam and may have detected its deficiencies.

### *2.4 Prevention*

Among others, these accidents demonstrate the importance of Management of Change in process safety. This was emphasized when MOC was included as one of the main pillars of the Process Safety Management (PSM) standard, issued by the U.S. Occupational Safety and Health

Administration and of the Risk Management Plan of the Environmental Protection Agency (EPA).

However, not only changes performed after the plant startup will result in safety accidents. Engineering Management of Change (EMOC) is also a key area to prevent accidents and should be addressed in a proper manner by industrial companies.

One example of a poor EMOC related event can be an updated drawing sent by an equipment supplier during the project execution phase indicating a decrease of a vessel Maximum Allowable Working Pressure (MAWP). If this modification is not communicated to the proper professionals, the pressure relief device set point designed for the equipment may not be adjusted as required, which can lead to lack of vessel overpressure protection and possible vessel rupture.

Another example can be a change of a pipe routing during project phase, without revision of the flexibility study. This could result in pipe over stress, leading to a line rupture and energy release.

Several other examples of poor EMOC related accidents could be discussed showing proper management of changes during engineering is indispensable for the safety of the plant, workers and environment. Figure 1 indicates some sources of modifications that could impact the safe startup/operation of the system.

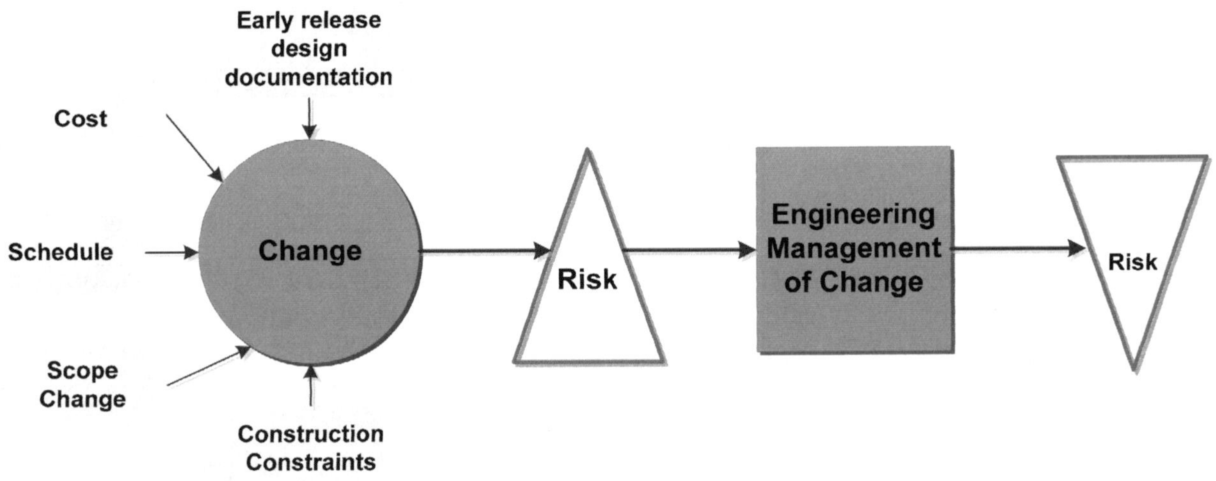


**Figure 1 Integrated management of change system**

Unfortunately, EMOC use is not as widespread as MOC, which still gives space for industrial accidents to happen. The main purpose of defining a work process for managing changes, even during the project development phases, is to guarantee all changes are adequately communicated, reviewed, new hazards are identified and analyzed and a mitigation plan is defined to ensure that a safe installation and operation is delivered to the customer.

# 3 EMOC Implementation

A change is known as a deviation from the original design. Modifications may come as scope changes, supplier updates, or other sources, and they may affect more than one engineering discipline and any on-going construction/assembly activities.

"Changes in construction projects are very common and likely to occur from different sources, by various causes, at any stage of a project, and may have considerable negative impacts on item such as costs and schedule delays." [6] If there is no formalized change management process, impacts not only on cost and schedule may be observed, but also process safety events, as mentioned on the previous section of this paper.

According to CCPS [7], the level of detail covered by the management of change system should be based on

- The hazards or risks of the process
- The existing process safety culture at the location

The company should develop a system for the control of modifications implemented on the project under design, construction, commissioning or startup phases. This work process should include approval of the change by a qualified professional or team, depending on the complexity of the hazards associated.

Based on the hazards involved in the specific process, the type of changes that need evaluation by an engineering expert from the appropriate discipline, as well as changes that do not, shall be defined when designing the specific EMOC work process for a certain company. For example; changes that can affect safety systems, critical operating parameters, shutdown set points, relief valve settings, materials of construction, etc. should undergo EMOC approval. Minor corrections to documentation or clarifications may not.

The management of change work process may not be the same in every industry and this does not mean one is better than another. The culture of the company and the location may require different applications of the MOC activities. [7]

## *3.1 Critical Phase*

Changes during construction inherently introduce unexpected risk, especially under schedule and cost pressure. Some changes that can affect process safety may not be obvious to construction personnel. A simple yet important example is relocating a support on the discharge of a relief valve due to interference. Another may be substituting flange bolt material. Regardless of the origin of the change, evaluation and approval should take place prior to starting up the system/subsystem. Identification of all sources which can generate an engineering change must be considered as a natural part of the overall work process.

# 4 Proposed EMOC Work Process

Typically, during the engineering phase, several design safety meetings are conducted, such as P&ID Reviews, Layout Reviews, PHAs and FSSs. These reviews shall be conducted with the participation of all engineering disciplines related to the plant design. After these project landmark meetings, several changes may occur and their evaluation is a critical part of risk management throughout the project life cycle. At this moment, after the main safety reviews are performed, the EMOC work process is expected to begin.

As shown on Figure 2, all changes related to the project design are identified and communicated by the professional who initiated the modification, to a single point of contact within the project team.

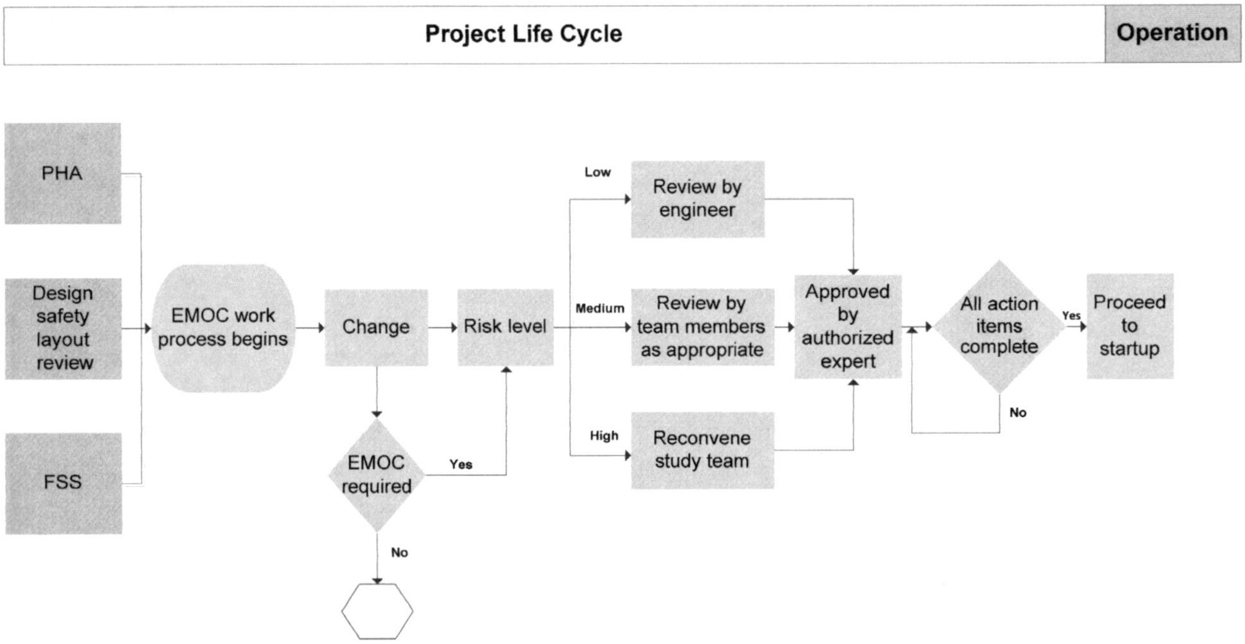


**Figure 2 High-level EMOC Workflow Diagram**

In cases where a modification is only a correction, clarification or replacement in a kind, there is no need to go through the full EMOC work process. The change can be recorded in a simple spreadsheet containing the type of change, a brief explanation and a mitigation plan to make sure nothing has been missed.

The modification should then be classified as low, medium or high risk per the EMOC owner and the level of review is defined based on the associated risk. The risk ranking criteria shall be described on the company's EMOC program; e.g.:

- Low: addition of a local pressure indicator on the design for maintenance purposes only.
- Medium: manual isolation valve removal, while valve not indicated at PHA as part of a hazard scenario.

- High: customer addition of a tie in point near the pipeline battery limit, which could lead to misdirected or reverse flow, with potential for process safety impact.

After reviewing the change, an approval should be required by a Subject Matter Expert (SME) before proceeding to change implementation and mitigation plan. The plant will be allowed to start-up only after all EMOC items were reviewed and approved.

It will be responsibility of the company implementing EMOC to determine which kind of changes will require a complete review, the risk categories that will be available to be assigned, both qualitatively and/or quantitatively and the required level of review according to these risks. In the example presented in Figure 2, a low risk results in a review by an engineer, a medium risk results in a review by the team members and a high-risk change will require the engineering team involved in the HazOp, Design Safety Layout review or the FSS evaluations to reconvene.

Several different approaches are possible, as long as they are simple, easy to understand by those responsible for implementation and provide a robust EMOC work process.

## 5 Elements of a Successful Engineering Management of Change Program

Successful EMOC programs define the key elements applicable to their industry. Some will be common such as commitment from leadership, roles and responsibilities, etc. Included below are some elements to take into consideration.

- Commitment from Leadership;
- Roles and responsibilities as they apply to specific skill sets (i.e. process engineering, controls system engineering, manufacturing, construction management, etc.);
- Clear definition of changes that must undergo engineering evaluation, including examples;
- Definition of changes that are not considered EMOC related;
- Risk level associated with the change;
- Approvers limited to subject matter experts;
- A streamlined work process focused on function over form, aligned with the various phases of the project life cycle;
- Identification of change sources;
- In process verification at predetermined milestones;
- Robust training that includes internal and external examples accompanied by a test with feasible scenarios;
- Associated PSI updated accordingly;

- Auditable;
- Evergreen
  - Periodic calibration of the work process;
  - Input from those responsible to implement the process;
  - Captures internal data to enhance understanding of what constitutes a change that requires EMOC review.
- Lessons learned captured and implemented on future projects;
- Leverage key learnings from external events in where MOC was identified as the cause or a contributing factor.

# 6 Conclusion

An aggressive schedule may lead to early release of design documentation that is still subject to change, which could result in the execution of PHA studies earlier than the optimal period. Although the early study may provide an opportunity to identify key risks earlier in the process, a significant number of action items and changes will probably occur, which may affect the integrity of the PHA.

A well-defined and implemented EMOC program is an essential part of risk management. The methodology applied allows systematic control of project changes which helps to ensure process, personnel and environmental safety. An additional benefit is facilitating a proactive means to analyze the underlying cause for unplanned changes. These lessons learned can assist on future projects to manage cost and improved quality control.

The main purpose of the management of change is to avoid loss of primary containment (LOPC), especially of hazardous chemicals, which could lead to catastrophic consequence. Once the EMOC process is successfully implemented, the reduction of quality issues will become a natural part of the process, resulting in a safer and higher quality facility/assembly/operation. An effective Engineering Management of Change work process is critical to the safe commissioning and startup of a chemical plant.

# 7 References

[1] T. Kerin. The evolution of process safety standards and legislation following landmark events – What have we Learnt?, Process Safety Progress 35 (2016), 165-170.

[2] CCPS. *Incidents that Define Process Safety*, Center for Chemical Process Safety, American Institute of Chemical Engineers, New Jersey, 2008.

[3] J.E.S Venart, Flixborough: The explosion and its aftermath. Process Safety and Environmental Protection 82 (B2): 105 – 127 (2004).

[4] C. Sadee, D.E. Samuels and T.P. O'Brien, The Characteristics of the explosion of cyclohexane at the Nypro (UK) Flixborough plant on June 1974, Journal of Occupational Accidents, 1 (1976/1977) 203 – 205.

[5] U.S. Chemical Safety and Hazard Investigation Board, Management of Change (No. 2001-04-SB) (2001).

[6] Q. Hao et all, Change Management in Construction Process, International Conference on Information Technology in Construction, Santiago, Chile, 2008

[7] CCPS. *Guidelines for Management of Change*, Center for Chemical Process Safety, American Institute of Chemical Engineers, New Jersey, 2008.

# 479400 Effectively Utilizing Calarp/RMP/PSM Contractors to Assist in Optimizing Process Safety Culture, Compliance with Standards, and Workforce Involvement

**Abstract:**

Since the inception of Occupational Safety and Health Administration (OSHA) Process Safety Management (PSM) in 1992, facilities have frequently contracted process safety consultants to satisfy regulatory needs, as well as assist in complying with industry best practices.

The primary objective of this paper is to provide tangible suggestions for effectively utilizing process safety contractors for assisting in optimizing process safety culture, compliance with regulatory requirements and industry best practices, as well as tips to improve workforce involvement and coordination. The core of this paper will detail various management styles of the prevention program and an evaluation of influences on process safety culture based on varying degrees of involvement from the facility and usage of contractors. This paper aims to provide an in-depth analysis of the pros and cons of contractor involvement and its efficacy relative to the scale of the facility and intricacy of corporate structure based on anecdotal evidence and industry experience.

Key topics include:

- Overview of different types of process safety culture in relation to prevention program management and contractor usage,
- The Challenge - Striking an optimal balance between contractor use and facility involvement whilst maintaining economic viability,
- The Challenge for Multi-Facility Corporations - Ensuring changes in corporate standards are successfully communicated to all children facilities ranging from documentation updates and organizational changes,
- Suggestions and Improvements - Implementation of effective prevention program management systems to ensure workforce involvement and accountability, as well as practical allocation of contractor resources to accomplish critical tasks whilst balancing facility resource commitment.

# A Streamlined Approach for Full Compliance with SIF Implementation Standards

**William G. Bridges, President**
**PROCESS IMPROVEMENT INSTITUTE, INC. (PII)**
1321 Waterside Lane, Knoxville, TN 37922
Phone: (865) 675-3458
Fax: (865) 622-6800
**e-mail: wbridges@piii.com**

**Art Dowell III, Principal Engineer**
**PROCESS IMPROVEMENT INSTITUTE, INC. (PII)**
2437 Bay Area Blvd PMB 260
Houston, TX 77058
Phone: 713-865-6135
**e-mail: adowell@piii.com**


Prepared for Presentation at
13th Global Congress on Process Safety
San Antonio, TX
March 27-29, 2017

UNPUBLISHED



13TH GLOBAL CONGRESS ON PROCESS SAFETY

# A Streamlined Approach for Full Compliance with SIL Implementation Standards

**William G. Bridges, President**
**PROCESS IMPROVEMENT INSTITUTE, INC. (PII)**
**e-mail: wbridges@piii.com**

**Art Dowell III, Principal Engineer**
**PROCESS IMPROVEMENT INSTITUTE, INC. (PII)**
**e-mail: adowell@piii.com**

**Keywords: Safety Instrumented System, SIS, Safety Integrity Level, SIL, LOPA, Layer of Protection Analysis, Independent Protection Layers, IPLs, human factors**

## Abstract

Many companies put FAR too much redundant effort into determining what SIL (safety integrity level) is needed and then verifying the SIF (safety instrumented function) design will give the SIL targeted. This paper shows how to apply the qualitative definition of independent protection layers (IPLs) within the setting of a process hazard analysis (PHA) to get most of the gain from Layer of Protection Analysis (*LOPA)* without doing a LOPA (without using numerical values). We show how we use a PHA team to identify when a SIF is needed and to select the proper target SIL. This portion of the SIL evaluation and the identification and labeling of the IPLs during the PHA/HAZOP does not take any longer than a normal PHA/HAZOP, once the right habits are established. Note that this approach eliminates the need for a separate SIL Evaluation Study to identify the SIFs and select the target SIL. Then, this paper describes how to perform the SIL Verification and Safety Requirements Specification (SRS) remotely, again without the need for a redundant team meeting. This approach has been used at many sites and for thousands of SIFs.

## Background

Typically, hazardous consequences, their causes, and safeguards are identified in a PHA/HAZOP (process hazard analysis / hazard and operability analysis)[1]. The PHA may determine if additional protection layers are required or if additional analysis is required.[2]

In many organizations, a separate LOPA (layer of protection analysis)[3] meeting is held to determine which HAZOP safeguards are independent protection layers (IPLs) and to determine if additional IPLs are needed to meet the risk tolerance criteria. If SIFs are needed, the required risk reduction (PFD, probability of failure on demand) is determined in LOPA, which then sets the SIL.

Some organizations may even have a separate SIL evaluation study, which is certainly redundant to a well-run PHA and is wasteful if the PHA is properly staffed and managed and/or if any subsequent LOPA are performed well.

## Fundamentals of SIL Assessment

The lifecycle requirements for safety instrumented functions are specified in industry standards IEC 61511 (ANSI/ISA 84.00.01-2004, Part 1)[4,5]. The required steps to specify the SIS (safety instrumented system) are:

1. Safety Instrumented Function (SIF) identification.
2. Determining the Safety Integrity Level (SIL) for each SIF.
3. Designing the SIF to meet the required SIL.

   SIL Verification Calculation (actually, this calculation is iterative with step 3, but the end calculation is a deliverable that proves 3 is correct). The SIL standards allow the following methods to use to determine the required SIL (ANSI/ISA 84.00.01-2004, Part 3, Section 3.8)[4]:

   - Fault Tree Analysis (FTA), Full Quantitative Risk Analysis (QRA)
   - LOPA – simplified quantitative analysis
   - Qualitatively within PHA (HAZOP, etc.)

   > "A qualitative method may be used as a first pass to determine the required SIL of all SIFs. Those which are assigned a SIL 3 or 4 by this method should then be considered in greater detail using a quantitative method to gain a more rigorous understanding of their required safety integrity." – *ANSI/ISA 84.00.01-2004, Part 3, Section 3.8*[4]

Table 1 compares risk analysis approaches for Steps 1 and 2.

FTA and QRA are very detailed and labor-intensive and would be required for complicated systems, particularly where there is not complete independence between safeguards.

LOPA is a simplified analysis tool that requires fewer resources, but is heavily dependent on the requirement that all the safeguards be completely independent.

However, for many processes, there are a number of scenarios (cause leading to a consequence) where the PHA team can readily determine sufficient independent protection layers are already in place.

**Table 1: Comparison of Risk Analysis Approaches**

| Approach | Assessment Methods | Risk Judgment Method | Risk Judgment Method | Estimated Range of the Results |
|---|---|---|---|---|
| Qualitative Only | HAZOP, FEMA | Expert Voting, focusing on site data | Capable >95% of time | Plus or minus 1/2 order of magnitude |
| Simplified Quantitative | LOPA, Risk Graph | Multiplication of statistical averages of general failure rate data; with broad assumptions on management systems | Needed on about 5% of the scenarios | Plus or minus 1 order of magnitude |
| Full Quantitative | FTA, ETA, QRA, HRA | | Needed for less than 0.01% of the scenarios | Plus or minus 1 order of magnitude |

Figure 1 contrasts the typical approach to SIL assessments with the optimized streamlined approach.

## Typical SIL Determination versus Streamlined Approach

In the typical approach,

- **Step 1**: The PHA/HAZOP is performed using a multidisciplinary team of engineers, operators, instrumentation staff, maintenance staff, and perhaps vendors. Then in a separate SIL study, a multidisciplinary team of engineers, instrumentation staff, maintenance staff, and perhaps vendors brainstorm the hazard scenarios (again) and decide which are candidates for SIFs. A semi quantitative approach, such as RiskGraph, or layer of protection analysis, determines if an SIF is required.
- **Step 2:** The RiskGraph dictates the required SIL, or the LOPA compares the mitigated risk to the risk tolerance criteria and determines the SIL from the order-of-magnitude risk reduction needed.

In contrast, in the streamlined approach,

- **Step 1:** The PHA/HAZOP is performed using a multidisciplinary team of engineers, operators, instrumentation staff, maintenance staff, and perhaps vendors (to this point the same as the typical approach). Within the PHA/HAZOP, the team estimates the risk of each scenario using qualitative judgment (based on experience, knowledge, and memory of site-specific data). The PHA/HAZOP team then qualitatively determines if the best risk reduction method is an SIF.
- **Step 2:** The PHA/HAZOP team qualitatively determines the SIL needed (typically by judging the level of redundancy needed based on the value they place on each IPL and the new or modified SIF needed)

**Figure 1: Comparison of the Typical and the Optimized Streamlined Approach from PII for SIL Assessments (Steps 1 and 2)**

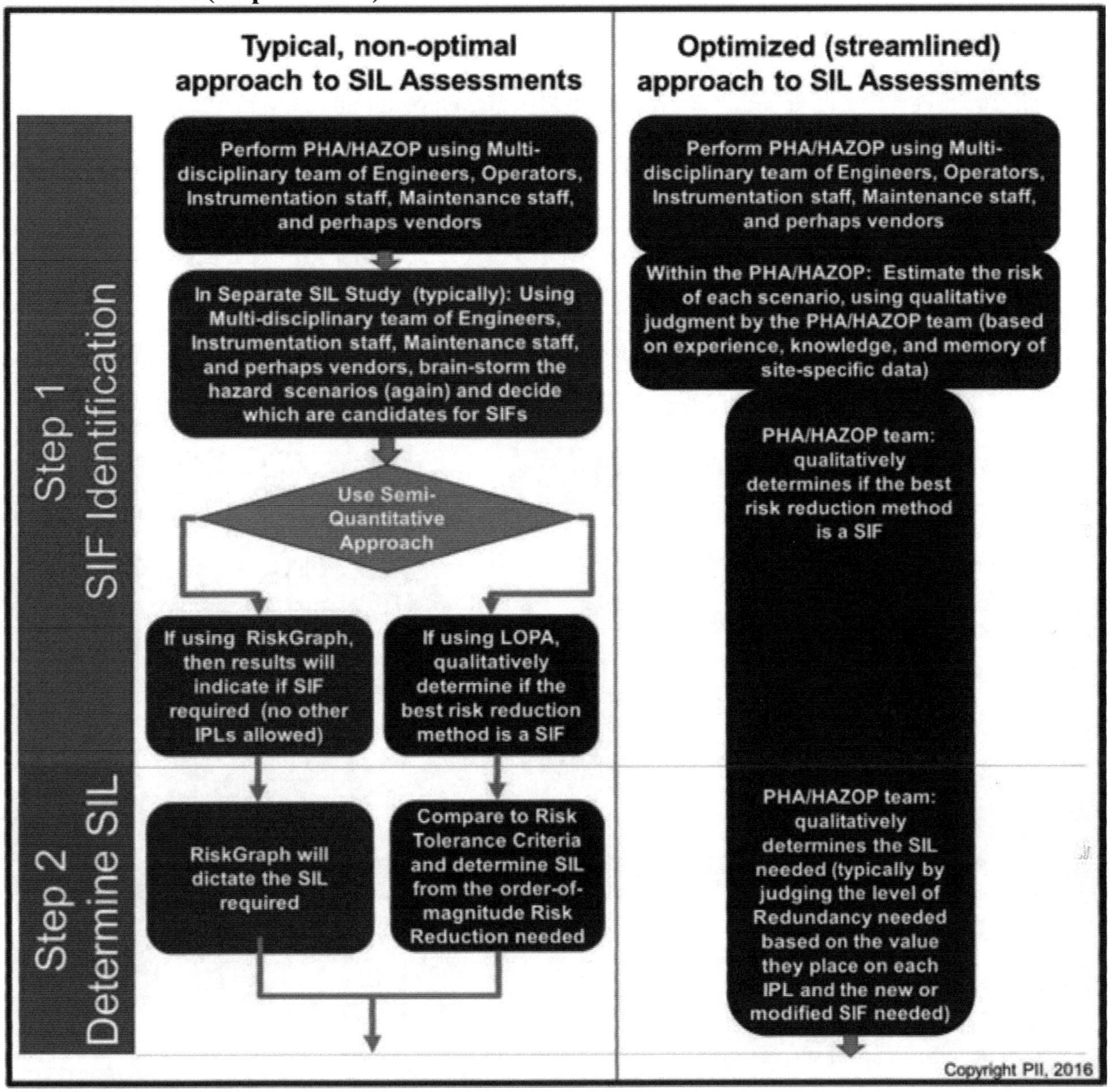


Figure 2 also contrasts the typical approach to SIF design and verification with the streamlined approach. Steps 3 and 4 are iterative.

In the typical approach,

- **Steps 3 and 4**:
    - Decide on basic architecture of SIF, from experience with other SIFs.
    - Decide on vendors of components
    - Provide full design of SIFs from experience

- Perform trial SIL verification calculations using software such as exSILentia™ using actual component data
- If the target SIL is achieved, issue the SIL verification calculation report and the safety requirement specification (SRS)
- If the target SIL is not achieved, revise the design and recalculate.

**Figure 2: Comparison of the Typical and the Optimized Streamlined Approach from PII for SIL Assessments (Steps 3 and 4)**

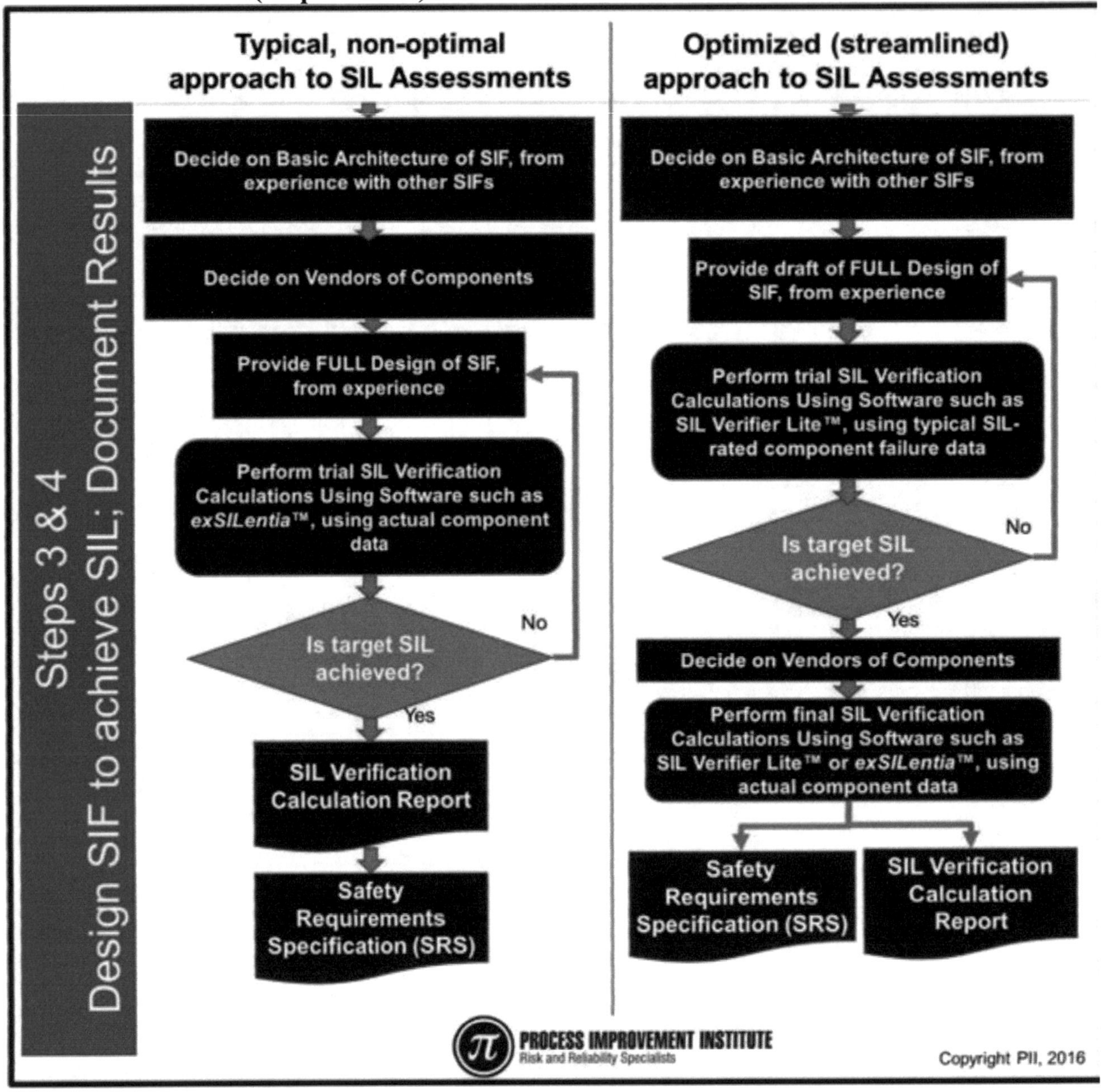


In the streamlined approach,

- **Steps 3 and 4**:
  - Decide on basic architecture of SIF, from experience with other SIFs.

- Provide draft of full design of SIF, from experience.
- Perform trial SIL verification calculations using software such as SIL Verifier Lite™ using typical SIL-rated component failure data.
- If the target SIL is not achieved, revise the design and recalculate.
- If the target SIL is achieved,
  - Issue the SIL verification calculation report and the safety requirement specification (SRS).
  - Decide on vendors of components.
  - Perform final SIL Verification Calculations Using Software such as *SIL Verifier Lite*™ (from PII) or *exSILentia*™ (from exida) using actual component data.
  - Issue the SIL verification calculation report and the safety requirement specification (SRS).

The advantage of the streamlined approach is determination of the need for an SIF and of its SIL quicker and with fewer resources than the typical approach.

## Requirements for the Qualitative Streamlined Approach for SIL Determination

For the qualitative streamlined approach to be used effectively, there are important requirements for the PHA/HAZOP team.

- PHA/HAZOP team leader is knowledgeable and proficient in LOPA.
- PHA/HAZOP team members have the following information in memory:
  - Consequence severity.
  - Initiating cause frequency.
  - Independent protection layer (IPL capability).
  - Risk tolerance criteria.

An experienced PHA team, who is knowledgeable in the process technology, can effectively and efficiently analyze a scenario and mentally perform a layer of protection analysis. The team determines the consequence severity and the initiating cause frequency. A highly competent team understands which safeguards are truly independent protection layers and can judge the risk reduction provided by each. Then the team can determine if additional IPLs are required to meet the risk tolerance criteria. For some scenarios, an SIF may be required and the team can determine the required probability of failure on demand for the SIF and can thus determine the required SIL.

A best practice is that the PHA team recommends the functional description and the SIL for the proposed SIF. The functional description does not specify the details of the SIF design, rather it specifies what process variables should be considered and what action should be taken to detect the upset and to prevent the consequence. For example, the functional description may say "monitor the level of a vessel and stop the flow into the vessel to prevent overflow".

If the PHA team is uncertain, or if the scenario is complex, then the PHA can recommend a detailed LOPA be done for the scenario.

These conclusions are based on more than 8000 PHA/HAZOPs, 3000 LOPAs, and about 100 QRAs and HRAs. We have observed PHA participants becoming knowledgeable about LOPA and IPL requirements. The participants then began to apply the principles of LOPA spontaneously in the PHA. With good leadership, the PHA team could then easily move to the full streamlined approach.

The experience and competency of the PHA leader and the PHA team is extremely important. If the PHA team is not comfortable with the principles of LOPA, and especially the principles of IPLs, the streamlined approach is not appropriate. Likewise, the PHA team must have the required information and understanding in its memory. Otherwise, the qualitative approach can degenerate into the "arm wrestling" arguments that PHA teams used qualitatively to determine SIL requirements before the introduction of LOPA.

## Suggestions for the Streamlined Approach for SIF Design

Likewise, there is a streamlined approach for the SIF design. Organizations have developed a recipe book of typical SIF designs for SIL 1, SIL 2, and SIL 3. Depending on the required SIL, the designer can choose a design from the recipe book (as described in Figure 2, "from experience"). The recipe book provides a starting point for the detailed SIF design. Figure 3 shows a small sample of SIF designs that could be included in a recipe book.

**Figure 3: Typical SIF Architectures for Information**

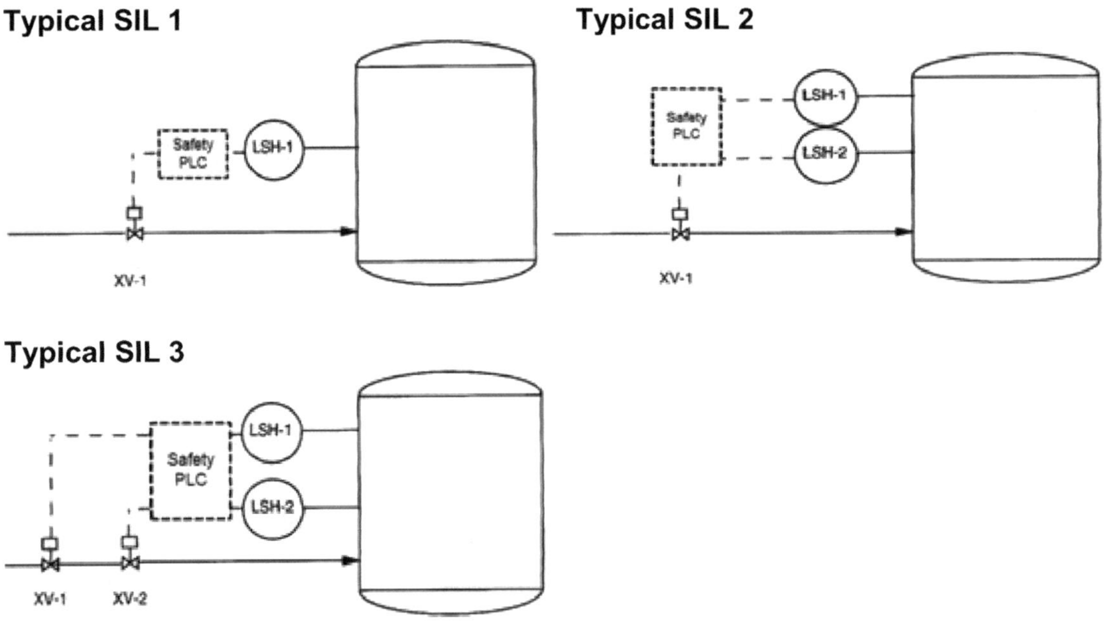

## Case Study 1

An example of the streamlined approach is summarized in Table 2 for a facility with storage spheres (in China; the PHA and other analyses were performed in 2014). The table shows two deviations from the PHA. IPLs are identified and existing SIFs are assigned an SIL, based on the architecture. (Note that the achieved SIL for each of the SIFs (existing or recommended) must be confirmed by a SIL verification calculation that includes the failure rate of the components, the voting architecture, and the proof test interval.)

For the high-level deviation, the high-level SIF was assigned SIL 1. The overflow through the pressure equalization line to other spheres was determined to be an IPL.

For the low level deviation, the low level indication and low level alarm were determined not to be an IPL because they are part of the initiating cause. The PHA team concluded that feeding from two spheres at all times is an IPL for this facility because of the unlikelihood for both spheres to have low level at the same time. The PHA team also thought that an additional SIF level alarm with operator response to switch tanks within 60 minutes could be a possible IPL, if the action of the operator is quick enough. The team wrote a recommendation to ensure that the operator response could be an IPL.

In this case study, the PHA team was able to determine in the PHA meeting which safeguards were IPLs. They also determined the nominal SIL for existing SIFs (to be confirmed in a later SIL verification). And the team determined that operator response could be an IPL if there was a documented procedure and an annual drill. For these two deviations, the PHA team did the work that would be done in a later LOPA meeting in the typical approach.

Next, the SIL Verification was performed for the SIL 1 SIF (and all other SIFs in the unit). This SIL Verification was accomplished using exSILentia™ software and the analysis and documentation took less than 2 hours.

Thus, the entire requirements for a SIL identification, SIL target level determination, and SIL Verification were accomplished very quickly compared to typical approaches.

**Table 2: Storage Sphere PHA**

| **No.:** 2 XXXX storage spheres xxx-T-XX A/B/C/D/E/F/G/H/I/J/K/L (1 of 12) | | | | | |
|---|---|---|---|---|---|
| **#** | **Dev.** | **Causes** | **Consequences** | **Safeguards** | **Recommendations** |
| 2.1 | High level | Too much flow to one sphere from XX Plant (through their pump; about 40 bar MDH) | High pressure (see 2.5) | High level SIF with level sensors voted 2oo2, to close inlet valve - **SIL 1**<br>Overflow thru pressure equalization line to other spheres (through normally open [NO] valve) - **IPL** | |
| 2.2 | Low level | Failing to switch from the sphere with low level in time (based on level indication) | Low/no flow - Liquid from spheres through high pressure product pumps to the vaporizer (see 4.2)<br>Low/no flow - Unqualified liquid from spheres back to Plant (see 6.2) | Level indication and low level alarm, inspected each year, per government regulation **(not IPL; part of the cause)**<br>9 other spheres with possibly enough level to switch to<br>Feeding from two spheres at all times, so unlikely for BOTH spheres to have low level at the same time - **IPL**<br>Two level indication from SIS level transmitter, with low level alarm, with more than 60 min available to switch tanks (SIF driven alarm and response) - **possible IPL, if action of the operator is quick enough** | Rec 4. Make sure the Human IPL of response to low level in all spheres and tanks is described in a trouble-shooting guide (like an SOP) and practiced once per year per unit operator. This will make this response a valid IPL. |

## Case Study 2

In 2014 and 2015, an identical approach to that of Case Study 1 (the streamlined approach of this paper) was performed for a large gas producer in South America. In that work, the PHAs for more than 500 nodes plus all procedures of all operating modes was completed in about 10 weeks of PHA meetings. The PHA team identified more than 800 SIFs, along with their target SILs, using the approach in Steps 1 and 2.

Next, staff within PII but separate from the PHA team performed about 60 LOPA to clarify certain issues, some related to SIFs. But less than 5% of the SIFs were established with LOPA; 95% were established with the PHA alone.

Finally, staff separate from the PHA performed SIL Verification of the 800+ SIFs, using exSILentia™. The verification was for a combination of SIL 1 and SIL 2 SIFs; a couple of SIL 3 SIFs were also analyzed and debated. The relatively few LOPAs took about 2-3 hours each (due to the complexity of the 5% of scenarios that went to LOPA) and each SIL Verification took about 2-3 hours to perform and document.

## *Caution – the issue of human factors for SIL 2 and SIL 3 SIFs*[6]

There is an important aspect of important aspect of SIFs and the possible achieved SIL that is not addressed in this paper. Relying on a high integrity SIF, without accounting for human error during interventions, is also a waste of resources. You may believe that you have reduced the risk by a factor of 100 for a SIL 2 SIF, or a factor of 1000 for a SIL 3 SIF, when in fact the human factors during commissioning and proof testing may degrade the overall risk reduction factor to about 10. If the human factor is accounted for in the SIF design phase and the SIL verification phase, then the human error can also be prevented to allow the higher reduction in risk that is targeted.

## Conclusions

Based on experience in a multitude of PHAs, LOPAs, QRAs, and HRAs, we have seen that experienced, competent PHA teams with an experienced, competent leader can effectively apply the principles of LOPA to many scenarios during the PHA meeting. The team can determine the approximate SIL of existing SIFs, the team can determine which safeguards are truly IPLs, and the team can recommend additional IPLs, including SIFs with specific SILs. If the scenario is too complex, or the PHA team is uncertain, the team can recommend a detailed LOPA. The streamlined qualitative makes better use of resources and time than the typical approach with separate meetings for PHA and LOPA.

It is critical that the PHA leader is experienced and competent in LOPA. It is critical that the PHA team has in its memory the data for consequence severity, initiating cause frequency, IPL requirements, and the risk tolerance criteria.

## Acronyms Used

**AIChE**– American Institute of Chemical Engineers
**CCPS** – Center for Chemical Process Safety (an AIChE technology alliance)
**HAZOP** – Hazard and Operability Analysis
**HRA** – Human Reliability Analysis
**IEC** – International Electrotechnical Commission
**IEF** – Initiating Event Frequency
**ETA** – Event Tree Analysis
**FMEA** – Failure Mode Effect Analysis
**FTA** – Fault Tree Analysis
**HRA** – Human Reliability Analysis
**IPL** - Independent Protection Layer
**LOPA** – Layer of Protection Analysis
**PHA** – Process Hazard Analysis
**PII** – Process Improvement Institute, Inc.
**QRA** – Quantitative Risk Analysis
**SIF** – Safety Instrumented Function
**SIL** – Safety Integrity Level
**SIS** – Safety Instrumented System
**SRS** – Safety Requirements Specification

## References


1. *Guidelines for Hazard Evaluation Procedures*, 2008, CCPS/AIChE.
2. Bridges, W., & Dowell, A., "Identify SIF and Specify Necessary SIL, and other IPLs, as part of PHA/HAZOP," *12th Global Congress on Process Safety*, AIChE, 2015.
3. *Layer of Protection Analysis: Simplified Process Risk Assessment,* 2001, CCPS/AIChE.
4. *Functional Safety: Safety Instrumented Systems for the Process Industry Sector - Part 3: Guidance for the Determination of the Required Safety Integrity Levels*; ANSI/ISA 84.00.01 Part 3, 2004.
5. IEC 61511, *Functional Safety: Safety Instrumented Systems for the Process Industry Sector - Part 1: Framework, Definitions, System, Hardware and Software Requirements*, International Electrotechnical Commission
6. Bridges, W., and Thomas, H., "Accounting for Human Error Probability in SIL Verification Calculations," *8th Global Congress on Process Safety,* Houston, AIChE, April 2012.

# 479509 The Art of the Audit: Being Audit Ready, Everyday


**Abstract:**

Have you ever made an appointment for an annual physical and suddenly join a gym? We all know we should be exercising every day, but it becomes more important before going in for that checkup. In reality, not exercising has consequences to your health and not your doctors, so why do we do this? To impress them, instead of doing it to take care of ourselves? This phenomena of doing things you do not normally do in preparation for an event is unfortunately how too many of us treat audits. We hear an auditor is coming and end up running around trying to organize paperwork, convince the operators to clean up the plant, and everyone in management becomes much more attentive. Does this behavior really set the proper tone to develop a safety culture? Auditors are there to check up on our compliance of rules that benefit only us to follow, particularly in the field of safety.

A successful operation will be 'Audit Ready' every day. 'Audit ready' means that operations are in compliance with rules, personnel are safe and exhibit safe behaviors, and all of the team members understand why they perform the functions of their job. Compliance tools, personnel awareness and routine self-audits are three key topics in becoming ready for an audit every day.

Personnel awareness is among the most vital areas to ensuring safe and effective operations, as well as successful audits. Audits involve people from across almost all departments, and as such mirror the requirements of multifunctional teams in everyday operations.

Compliance tools are important to track work that is being completed, and create convenient locations to store data which proves compliance. This can include tracking the progress of MOCs through weekly notifications, using databases to track when procedures need to be reviewed, and more. Visibility of what needs to be accomplished is important.

Routine self-audits help us to find gaps in compliance, and reduce complacency. These can include a simple spot check of one area of compliance, or be more comprehensive, and there are pros and cons to both. A spot check keeps everyone on their toes and takes a short enough time such that they can be done more frequently.

Ensuring sustainable records and checking everyday habits will go a long way to creating a safe operating environment, and with that a workforce that is audit ready, every day.

# 479546 Economic Justification for Better Maintenance in a Safety System

**Abstract:**

It is not difficult to think of numerous ways in which human factors might impact functional safety. For example, imperfect repair, improper calibration, faulty installation, etc. can all contribute to decreases in functional safety performance. However, if actions are to be taken to mitigate these impacts then it is essential to have some quantitative measure of how this impacts the PFDavg, so that the actions' effectiveness can be assessed. We will show how a simple measure system called the Site Safety Index$^{TM}$ (SSI) is used to adjust safety metrics, including the PFDavg and the MTTFS, on a site specific basis. The lifecycle economic impact of these metrics is then used to show how improved operations and maintenance can be justified. Economic justification for better maintenance can be easily verified through this approach being further mindful of the safety culture.

# 479547 How Important Is Realistic Failure Rates for Safety Instrumented Functions?

**Abstract:**

A given Safety Instrumented Function design must be verified to meet the required SIL level by calculating probability of failure on demand. However, this does require realistic failure rate data for each device. Fortunately, many device manufacturers are getting their products certified per IEC 61508 and many of the certificates now have failure rate data. Other manufacturers supply analysis reports with failure rate data. But how good is this data supplied by the manufacturers? All such data must be checked to make certain it is applicable for the process industries as some data may have been derived based on machine safety or automotive applications.

This paper will show some examples of manufacturer provided failure rate data and a methodology created to validate device type failure rate data for the process industries.

# 479917 Enabling Sustainable Relief System Management Programs through Independent Reviews

**Abstract:**

Overpressure protection and flare systems usually represent the last line of protection for personnel and the environment in industrial facilities operating under high temperatures and pressures. During the design phase of either Greenfield or Brownfield projects, assessing the basis for major relief systems and making clear recommendations to reduce the risks associated with each system has helped companies realize substantial safety and economic benefits. Independent and safety focused reviews, with an emphasis on value added solutions has enabled organizations to set the right foundation for a sustainable relief system management program. This, in turn, has helped enable the facilities in meeting the ever-changing regulatory needs in a timely manner.

This paper will give a typical breakdown of the workflow process for each of the entities involved in the project - the client, the contractor, and the independent reviewer. Four distinct work activities are typically identified for such a review - development of preliminary overpressure protection, engineering design of relief systems, validation of the design, and validation of as-built documentation. After each stage, the output from the independent analysis is compared with the contractor's work and updates are performed before continuing to the next stage. This paper will also give an overview of these stages and discuss the various steps and value added for each stage of the project.

# Jack Rabbit II: Experimental Large-Scale Chlorine Release Trials

Mark Whitmire[1], Shannon Fox[2], Thomas Spicer[3], Leo Stockham[4], Steven Hanna[5], Joseph Chang[6], Michael D. Sohn[7], Thomas Mazzola[8], Ronald Meris[9], Damon Nicholson[10], Andy Byrnes[11]

[1]Noblis, Falls Church, VA, USA; [2]DHS S&T Chemical Security and Analysis Center, Aberdeen Proving Ground, MD, USA; [3]University of Arkansas, Fayetteville, AR, USA; [4] Leidos, Albuquerque, NM, USA; [5]Hanna Consultants, Kennebunkport, ME, USA; [6]ANSER, Falls Church VA, USA; [7]Lawrence Berkeley National Laboratory, Berkeley CA, USA; [8] Engility, Lorton, VA, USA; [9]DTRA, Fort Belvoir, VA, USA; [10]Dugway Proving Ground, UT, USA; [11]Utah Valley University, Provo, UT USA

## INTRODUCTION

Chlorine is a potent toxic inhalation hazard (TIH) chemical, and each year millions of tons of it are transported in bulk via road, water, and rail. Consequently, an accidental or deliberate chlorine release near a populated area may cause a considerable number of fatalities, and would create other health, safety, and environmental problems. Hazard prediction models are inconsistent with the evidence and observations from chlorine disasters, and the understanding of such releases is otherwise insufficient to guide and inform an effective emergency response. So beginning in 2010, DHS and its Health, Safety and Environment (HSE) partners and stakeholders planned the Jack Rabbit (JR) Program. The program was led by the DHS Chemical Security Analysis Center (CSAC) located at Aberdeen Proving Grounds-Edgewood Area MD.

## OVERVIEW

The Jack Rabbit (JR) program is comprised of a series of projects that includes JRI (2010), JRII Phase 1 (2015), and JRII Phase 2 (2016). The JRI (Fox and Storwold, 2011) project was completed to develop, test, and evaluate a mechanism for the controlled, rapid release of liquefied, pressurized gases (specifically, ammonia and chlorine), to observe the transport and dispersion of a disseminated vapor and aerosol cloud, and to investigate chemical deposition and reactions with water and soil. The project also evaluated instruments, test methods, and strategies that may be used to study more industrial scale tests. Using either chlorine or ammonia, two 'pilot' tests and eight 'record' tests of either 1 or 2 tons were completed. For chlorine, among other findings, the results demonstrated that downwind transport and turbulent mixing are initially resisted by a dense persistent gas and aerosol cloud, that source phenomena were non-linear with increasing release volumes, and that the reactivity with soil containing water and organic matter is an important removal mechanism.

## JACK RABBIT II

JRII leveraged and extended the work of JRI with controlled chlorine field experiments on a scale corresponding to railcars and tank-trucks; that is, 5 to 20 tons each. The goals were to collect data on the release source, measure chemical reactions with the environment, observe and evaluate exposure effects on equipment and infrastructure, and assess urban impacts using a Mock Urban Environment (MUE) or an unobstructed dissemination and downwind dispersion environment, both located on the Urban Test Grid (UTG) at Dugway Proving Grounds UT. Data was collected in real time via ground based and unmanned aerial vehicle systems, video cameras, and spectroscopic instruments, in situ point detectors, path integrated detectors, and concentration and dosage determination instruments. The data was thoroughly and qualitatively reviewed and shared with all project participants in prescribed and common formats. The generated data and findings will improve the understanding of and operational hazard prediction modelling for TIH releases, and lead to more effective emergency response and training, national preparedness, and mitigation strategies.

URBAN TEST GRID (UTG)

The UTG (Nicholson, Hedrick, and Lian, N; 2017) was the designation for the JRII chlorine release site surveyed and built specifically for the JRII tests; it was at Dugway Proving Grounds west of Granite Mountain on Goodyear Road. It included the playa and adjacent support areas, concrete pad, gravel pad, and surveyed grid extending 11 km downrange and aligned roughly south to north.

The UTG included a circular, 0.5 ft. deep, 82 ft. diameter, rebar reinforced concrete disseminator vessel pad with 3 embedded temperature sensor stations, and a 1 inch steel lip to contain chlorine liquid. The center of the circular pad was the center of the UTG site, Northing 4445633.945, Easting 288109.182 (Latitude, 40.134347; Longitude, -113.487076). The concrete pad rested on a 24 inch deep, 400 ft. × 600 ft. gravel pad level with the concrete pad and symmetric about it.

*Figure 1. Chlorine release from the disseminator on the concrete pad, with CONEX container array on the gravel pad of MUE, looking upwind (south) from atop the 2 x 3 CONEX container stack. The dense cool cloud remains low as it moves downwind and laterally around and through the Mock Urban Environment made from CONEX containers.*

Winds predominate from approximately 165° from June to September, so the UTG was oriented about a centerline from that direction. The UTG included the playa within a 90° angle bisected by the UTG centerline with vertex at the disseminator vessel location, and extending to the 11 km arc. The "Near Field" was the UTG region from the disseminator up to and including the 2 km arc and inside the 90°

angle; the "Extended Field" was the UTG region past the 2 km arc to the 11 km arc and inside the 90° angle.

The Mock Urban Environment (MUE) was constructed on the gravel pad and surrounding the concrete disseminator pad using a staggered row array of 86 CONEX containers (CONtainer EXpress shipping containers, either 18 or 40 ft. long) to observe and study the transport and dispersion of dense gas around and above urban structures. The array included a two wide by three high (2 x 3) stack of six 20 ft.-CONEX containers located directly downwind of the source, and two 2-axle trailers nearby with insulation, dry wall, and trailer skirting for infiltration studies. Aside from the CONEX containers and trailers, up to seven vehicles were included in the array. Three fire trucks, one ambulance, and three cars were located on the north edge of the gravel pad during Trials 1-3. During Trial 4, four vehicles were placed on the playa past the 250 m arc; during trial 5, three vehicles were moved between the rows of CONEX containers but adjacent to the concrete disseminator pad, and three vehicles remained on the playa past the 250 m arc.

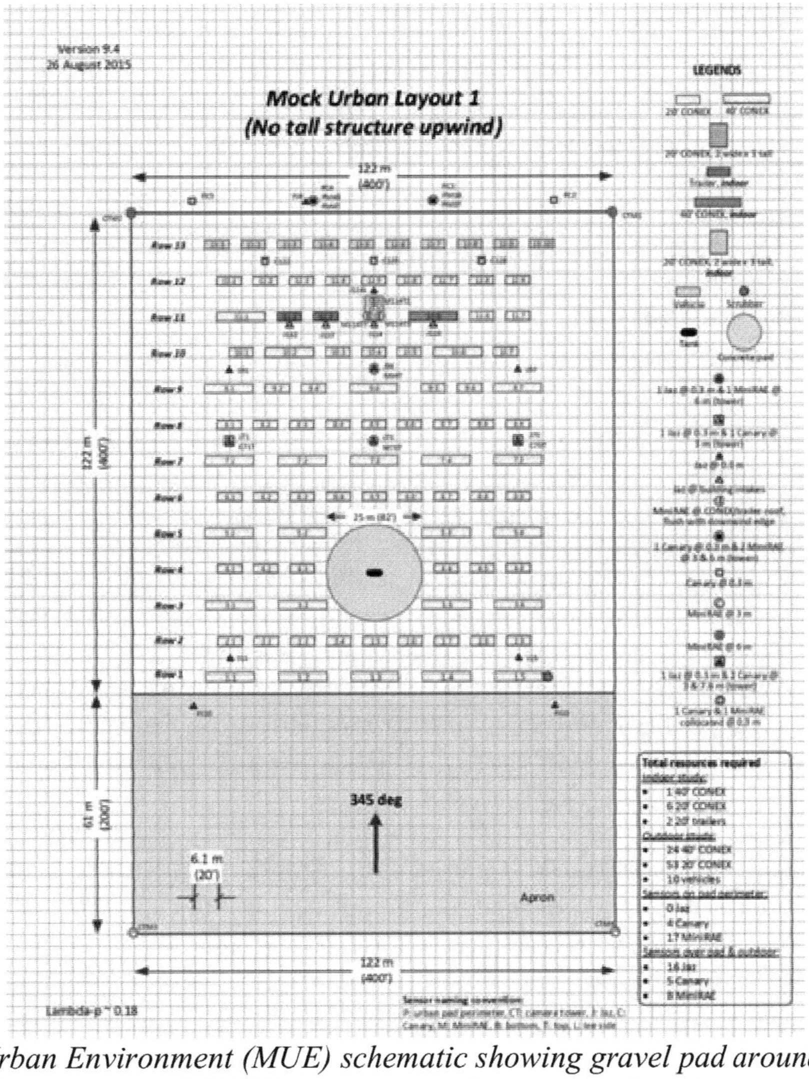


*Figure 2 Mock Urban Environment (MUE) schematic showing gravel pad around circular pad for Chlorine disseminator; CONEX containers and instrument locations are also shown.*

Instrument locations for the UTG "Near Field" and "Extended Field" were surveyed, recorded, and reported. Point sensors included about 170 MiniRAE 3000s equipped with volatile organic compound

(VOC) photoionization detectors; about 60 ToxiRAE Pro units; a Gasmet™ DX-4000 (FTIR) gas analyzer; 28 Ultraviolet (UV) Canary units for UTG monitoring and indoor studies; and 16 Ocean Optics Jaz instruments for UTG monitoring and indoor studies. Standoff Detectors included two Cerex Monitoring Solutions UV Sentry chlorine-specific integrated path detector, three Ultraviolet (UV) Light Detection and Ranging (LiDAR) systems for monitoring the moving chlorine gas cloud, and elastic backscatter LiDAR systems for monitoring the moving aerosol cloud. A GRIMM Aerosol portable laser aerosol spectrometer and dust monitor was also used for particle sizing model using laser scattering techniques. Other equipment included remotely operated high-definition (HD), high-speed (HS), and FLIR® IR Camera SC660 micro-bolometer cameras positioned throughout the test grid to visualize and record the chlorine release and resulting plume.

Meteorology equipment included 34 Portable Weather Information Display Systems (PWIDS) collecting data at 2 m; three 32 m towers each equipped with sonic anemometers collecting data at heights between 2 and 32 m; and 31 3D ultrasonic anemometer to gather wind and turbulence statistics, and fluxes of heat and momentum. Energy Balance Station (EBS) provided in situ measurements of the surface vertical fluxes of sensible heat, latent heat, net radiation, and soil surface heat flux. The MiniSODAR™ high-frequency Doppler acoustic sounding system designed to measure the atmospheric vertical wind profile from 15 to 200 m. Numerical gridded forecast meteorology using embedded observations and real-time meteorological data from the test grid were used for HPAC modelling leading up to each trial and to determine "GO/NO GO" test status.

DOWNWIND DISPERSION AND TRANSPORT ENVIRONMENT

For tests in 2016, and other than two vehicles, a CONEX container, and a modular office trailer used for follow-up infiltration experiments, the UTG was cleared of the simulated buildings and obstructions used in 2015. The project goals were to monitor and collect the cloud concentration data from the initial phase of a very dense, two-phase cloud near the source to dispersion of the vapor cloud farther downwind and up to 11 km away using point detectors and stand-off spectroscopic instruments.

DISSEMINATION

The disseminator vessel for use as the chlorine release source during the tests was located at the center of the concrete pad with its major axis perpendicular to the UTG centerline. The 2000 gal. tank was mounted in a 3-dimensional dynamic mass measurement load cell system. Other vessel mounted instrumentation included a redundant system for determining the remaining mass of chlorine in the vessel, differential pressure sensors on the vertical axis at various depths in and below the vapor space, guided wave radar liquid level measurement, and instrumentation for absolute pressure and direct contact temperature profiles inside the vessel. At 80-90% operating capacity, the vessel could contain from 8.5 to 10 tons chlorine.

Each release was accomplished using a blind flange with exploding bolts rated to safely hold the pressure of a vessel loaded with liquid chlorine at ambient temperature. To blow the flange and release the chlorine, the bolts were remotely detonated; when the bolts broke, the blind flange was blown off with the aid of the chlorine weight and head pressure, and chlorine flowed through the vessel's circular opening. A quick release valve was not used to simulate the release because of the likelihood of internal flashing flow in the piping and valves; that is, the mass release rate would be less than what would be observed from a breach of containment. In 2015, every release was directly down (180°) through a six inch hole. In 2016, each release was either directly down (Test 6); at 135° and downwind (Test 7), or directly up (Test 8). After Tests 7 and 8 any liquid chlorine remaining in the vessel was discharged via a remotely operated valve in the bottom of the tank. Before inert gas purging and chlorine refill operations, the ejected blind flange was replaced using standard bolts. The last release was from a tanker truck that had been removed from service, and containing up 19 t chlorine. The tanker truck's skin was breached with a remotely activated explosive shape charge.

*Figure 3. Test Release 6; 10 Tons chlorine on August 31, 2016. A moment after the release the liquid chlorine may be seen splashing onto the concrete pad. As the chlorine aerosolizes and vaporizes it cools the surroundings and condenses water vapor.*

## OBSERVATIONS AND FINDINGS

From the JRII trials the program observed that:

- Upon breaching the tank, liquid chlorine exits the tank and rapidly flashes to a two phase aerosol and vapor cloud with a liquid layer deposited on the pad, ultimately all vaporizing and transporting downwind.
- The escaping chlorine was found to auto-refrigerate to its -34 degree Celsius boiling point and lower due to further evaporative cooling outside the tank.
- The force of the escaping chlorine jet generated strong initial turbulence and momentum to drive the expanding cloud over nearby Conex structures.
- Evacuation of the tank during 5-ton releases took approximately 35 seconds.
- Evacuation of the tank during 9-ton releases took approximately 50 seconds.
- Structures were observed to channel the escaping and expanding cloud laterally.
- Significant upwind movement of the chlorine clouds was observed due to the momentum of the escaping gas, expansion of the gas, and gravity-slumping.
- Visible clouds were observed to persist in UTG for approximately 20-30 minutes depending upon release amount and winds.
- Long visible clouds observed to stretch from UTG to approximately 2 km as downwind dispersion progressed.

- Data sets from Jack Rabbit II have been quality-assured, finalized, and recently released, enabling in-depth studies to begin on cloud concentrations, movement, and behavior to be quantified.

EMERGENCY RESPONSE EXPERIMENTS

For each test, researchers from Utah Valley University placed two "Witness Boards" on surveyed arcs adjacent to the disseminator in the urban environment to observe the effect of liquid chlorine and high concentrations of chlorine gas (approaching 25%) on common building components, urban surfaces, and fire equipment. Go-Pro® and HD color surveillance cameras captured the exposure on video. The selection of the test materials emphasized high organic content (creosote treated telephone pole and railroad tie, road asphalt, shingles, rubber, and bare wood), but copper, aluminum, galvanized steel, and carbon steel coupons were included. The copper and carbon steel surfaces began to be discolored 1-2 hours after the exposure, and the chlorine sometimes dissolved the petroleum based materials.

INDOOR INFILTRATION STUDIES

The Defense Threat Reduction Agency (DTRA) is evaluating various hazardous airborne contaminant applications for transition to DOD CBRN consequence models. Their current indoor application models presume that any contaminant is inert and neutrally buoyant, thus the models may not be useful for predicting the indoor effects of a dense, reactive gas like chlorine. Lawrence Berkeley National Laboratories (LBNL) proposed a series of experiments to DTRA and the JR program to assess the applicability of the models for predicting the indoor concentration of chlorine in the event of an outdoor chlorine release.

LBNL successfully completed a series of indoor infiltration experiments in 2015 in support of the DTRA effort. The LBNL experiments were completed to understand and characterize the dominant infiltration pathways for reactive dense gases into buildings, the uptake of such gases into structures, and the re-emission of them from the structures after the peak of the plume had passed. Three single-story CONEX structures, a single three-story CONEX structure, and 2 modular office trailers placed on the gravel pad were used to represent indoor office, school, or home environments. Follow-on studies were completed in 2016 using a single-story CONEX structure at the same, but unobstructed site. The CONEX structures were configured for controlled experiments of airflow and transport in single- or multi-story structures, and modified to include insulation and temperature-control. Chlorine penetration into and through the structure, and transfer of it across the building envelope was monitored via controlled artificial "leaks" and exhaust fans. The model offices were furnished with realistic insulation, surface materials, windows, and HVAC system characteristics, thus the simulated modular offices helped to relate the data from the controlled CONEX structures to real buildings. The data may be used for calibration, verification and/or validation of the indoor model, and to support DOD concepts of operation for sheltering in the event of an outdoor CBRN release.

PROGRAM STATUS

The program is entering a new phase after completing field testing. In 2017, the program will be processing and reporting data from 2015 and 2016, and transitioning it for permanent data storage, and to stakeholders. CSAC will also be presenting a JR overview, narrative, and plans in presentations to interested groups. The program will be proposing, designing, managing, and performing small and large scale experiments during chamber testing, and reducing and analyzing data to demonstrate or elucidate the reaction of chlorine with various substrates to suggest large-scale toxic inhalation hazard (TIH) chemical release mitigation strategies, inform emergency response strategies that may reduce the impact to affected populations and infrastructure, calculate the value of modeling parameters, and otherwise support stakeholders interests.

ACKNOWLEDGEMENTS
This research has been supported by the US Department of Homeland Security (DHS) and the US Defence Threat Reduction Agency (DTRA). Additional support (funding, personnel and instruments) is coming from numerous partners, with the list growing with time. We appreciate guidance from many scientists who are members of the planning groups and/or have provided comments on the scientific approach.

REFERENCES

1. Nicholson, D., Hedrick, A., and Lian, N., Final Data Report for Jack Rabbit (JRII), 2017

2. Fox, S. and D. Storwold, 2011: Project Jack Rabbit: Field Tests. Chemical Security Analysis Center, Science and Technology Directorate, U.S. Department of Homeland Security. CSAC 11-006, 162 pp.

# 479394 Functional Safety - Pitfalls and Good Practices

**Abstract:**

The implementation of effective functional safety is valuable in ensuring that safety critical control and protective systems are appropriately specified, designed, installed and maintained. Through the International Electro-technical Commission (IEC) 61508 and 61511 suite of functional safety standards, the concepts of the Safety Lifecycle and Safety Integrity Levels (SIL) are now well known. However, DNV GL's experience from numerous projects across different industry sectors suggests that there remain challenges to be overcome.

DNV GL has conducted a number of Functional Safety Assessments (FSAs) on various electrical/ electronic and programmable electronic safety systems in the process sector. The FSAs are a mandatory element of IEC 61511 which provide an independent judgment on the integrity of the safety system and should be performed by competent independent Functional Safety Practitioners. At least one FSA must be performed on a Safety Instrumented System and should be completed prior to the introduction of hazards. Through DNV GL's experiences in functional safety assessments this paper will discuss;

- How to avoid common mistakes in the management and implementation of functional safety;
- Examples of good practice in the application of functional safety;
- How good practice can help optimize test regimes, increase plant availability and reduce maintenance workload whilst ensuring the system continues to provide the required risk reduction.

# General Method for Uncertainty Evaluation of Safety Integrity Level (SIL) Calculations – Part 2 Analytical Methods

**Raymond "Randy" Freeman**
S&PP Consulting

For Presentation at the American Institute of Chemical Engineers
2017 Spring Meeting in San Antonio, Texas -- March 26-30, 2017

**Abstract**

The IEC 61511 standard requires a verification calculation that a proposed design for a safety instrumented function (SIF) achieves the desired safety integrity level (SIL). The evaluation of the safety integrity level of a new or existing safety instrumented system requires detailed calculations based on the failure rates of the device and the planned maintenance/testing cycle for the system. The failure rates of the devices are often taken from standard failure rate tabulations of equipment. The maintenance and testing plans are developed based on plant experience. All of the data used in the SIL calculations are uncertain. This paper develops a general method for uncertainty analysis of the SIL calculations. The general method is based on the application of probability theory - variance contribution analysis (VCA) – to the equations presented in ISA TR 84.00.02-2115. An example is worked to demonstrate the methodology.

## Background

The calculation of the probability of failure on demand (PFD) is a common engineering task when designing an interlock or safety system that is to be in compliance with IEC 61511 [Ref. 1]. The calculation of the PFD is often done using approximate equations defined in the ISA TR84.00.02 technical report [Ref. 2]. The simplified equation method in the ISA report is commonly used and is based on the use of reliability block diagrams where the field sensors, Safety Instrumented System (SIS) logic solver and final control elements are considered independent of each other in the sense of not sharing common devices or systems. The PFD is then calculated using the failure rates of the devices, planned test intervals, vendor supplied estimates on diagnostic coverage of the devices and an allowance for the potential for common cause failures. Almost all of these parameters are uncertain. The failure rate data is often taken from generic data sources which show wide ranges in the observed values.

Because of the uncertainty in the parameters, the design engineer makes allowances in the design by the use of safety factors or rules of thumb to improve the chances that the final interlock installation will work as intended. Since each engineer has a different set of safety factors and rules of thumb, two designs may differ significantly in the way a hazard is controlled.

A more formal method for handling the underlying uncertainty in the calculation of the PFD of an interlock is needed. Previously, Freeman and Summers [Ref. 3] published an uncertainty analysis of the PFD of an interlock. Two different methods were used in this analysis:

- Monte Carlo Simulation
- Variance Contribution Analysis (VCA)

Monte Carlo Simulation requires that the engineer build a model of the interlock using specialized computer software. The use of VCA requires that the sensitivity of the interlock model be determined either by numerical methods or by direct analytical calculations. The Freeman paper demonstrates that VCA can be used for the uncertainty analysis. However, the paper does not present a complete analysis method that can be applied to any system defined in the ISA technical report TR84.00.02. The goal of this paper is to develop a general set of analytical equations that will allow VCA to be used in uncertainty analysis of any interlock developed per the IEC 61511 standard.

**Review of Interlock Design**

The IEC standard looks at an interlock as a series of three major components (see Figure 1). First is the sensor set which sends an indication of an abnormal event to a logic solver. The logic solver determines is the signal from the sensor meets the conditions required to activate (trip) the interlock. If the interlock is to be activated, a signal is sent to the final control elements to take action (stop flow, close valve, etc) to prevent a process safety event from occurring.

Based on the devices selected the design engineer then evaluates whether the proposed interlock design meets the design criterion for probability of failure on demand. Typically, the design engineer is given the target PFD as a statement such as "provide a SIL-2 interlock to stop flow on high level in a vessel." A SIL-2 interlock requires a PFD of at most 0.01 or a risk reduction factor of at least 100. The needed risk reduction is often determined in a Layer of Protection Analysis (LOPA) [ Ref. 4] and represents the managements decision on how risk is to be managed in a system. The resulting SIL-2 interlock design must have a high likelihood of achieving the target risk reduction as the LOPA team may be relying on it to be part of an overall risk reduction plan to prevent a process safety incident.

The ISA technical report (TR 84.00.02) [Ref. 2] provides shortcut methods for the evaluation of the PFD of an interlock design. Tables 1 and 2 are taken from the ISA technical report [Ref. 2]. Typical parameters need to compute the PFD include:

- DC is the diagnostic coverage;
- DI is the diagnostic interval;
- TI is the proof test interval,
- $\lambda^D$ is the dangerous failure rate;
- MTTR is the mean time to restore the system to operation
- $\beta$ is the common cause failure parameter that is always is between 0 and 1

In practice most if not all of these parameters are uncertain. Using parameter values obtained from plant records, vendor data or generic data bases the design engineer proceeds to develop the model for the interlock.

What is needed is a means to rapidly evaluate the impact of parameter uncertainty on the interlock PFD. The remainder of this paper applies the methods of variance contribution analysis (VCA) to determine the mean (expected value) and the standard deviation of the interlock PFD for a given design.

| Table 1. Simplified $PFD_{avg}$ formulas for Non-Repairable System without considering CCF, Diagnostics or MTTR. | | | |
|---|---|---|---|
| **Configuration** | **Function** | **PFD based on "Average before" failure rate** | **Appendix A Equation Number** |
| 1oo1 | $F_1$ = | $\left[\lambda^D \times \frac{TI}{2}\right]$ | A-1 |
| 1oo2 | $F_2$ = | $\left[(\lambda^D)^2 \times \frac{TI^2}{4}\right]$ | A- 2 |
| 1oo3 | $F_3$ = | $\left[(\lambda^D)^3 \times \frac{TI^3}{8}\right]$ | A-3 |
| 2oo2 | $F_4$ = | $\left[\lambda^D \times TI\right]$ | A-4 |
| 2oo3 | F5 = | $\frac{3}{4} \times \left[(\lambda^D)^2 \times TI^2\right]$ | A-5 |
| 3oo3 | F6 = | $3 \times \left[\lambda^D \times \frac{TI}{2}\right]$ | A-6 |

Where TI is the proof test interval, $\lambda^D$ is the dangerous failure rate.

| Table 2. Simplified $PFD_{avg}$ Formulas for Repairable System considering CCF, Diagnostics and MTTR | | | |
|---|---|---|---|
| **Configuration** | **Function** | **PFD based on "Average before" failure rate** | **Appendix A Equation Number** |
| 1oo1 | $F_7$ = | $\left[\frac{(1-DC)\times\lambda^D\times TI}{2}+\frac{DC\times\lambda^D\times DI}{2}+\lambda^D\times MTTR\right]$ | A-24 |
| 1oo2 | $F_8$ = | $\left[\frac{(1-DC)\times(1-\beta)\times\lambda^D\times TI}{2}+\frac{DC\times(1-\beta)\times\lambda^D\times DI}{2}+(1-\beta)\times\lambda^D\times MTTR\right]^2+\left[\frac{(1-DC)\times\beta\times\lambda^D\times TI}{2}+\frac{DC\times\beta\times\lambda^D\times DI}{2}+\beta\times\lambda^D\times MTTR\right]$ | A- 25 |
| 1oo3 | $F_9$ = | $\left[\frac{(1-DC)\times(1-\beta)\times\lambda^D\times TI}{2}+\frac{DC\times(1-\beta)\times\lambda^D\times DI}{2}+(1-\beta)\times\lambda^D\times MTTR\right]^3+\left[\frac{(1-DC)\times\beta\times\lambda^D\times TI}{2}+\frac{DC\times\beta\times\lambda^D\times DI}{2}+\beta\times\lambda^D\times MTTR\right]$ | A-26 |
| 2oo2 | $F_{10}$ = | $2\times\left[\frac{(1-DC)\times\lambda^D\times TI}{2}+\frac{DC\times\lambda^D\times DI}{2}+\lambda^D\times MTTR\right]$ | A-27 |

| Table 2. Simplified $PFD_{avg}$ Formulas for Repairable System considering CCF, Diagnostics and MTTR | | | |
|---|---|---|---|
| **Configuration** | **Function** | **PFD based on "Average before" failure rate** | **Appendix A Equation Number** |
| 2oo3 | $F_{11} =$ | $$3\times\left[\frac{(1-DC)\times(1-\beta)\times\lambda^D\times TI}{2}+\frac{DC\times(1-\beta)\times\lambda^D\times DI}{2}+(1-\beta)\times\lambda^D\times MTTR\right]^2+\left[\frac{(1-DC)\times\beta\times\lambda^D\times TI}{2}+\frac{DC\times\beta\times\lambda^D\times DI}{2}+\beta\times\lambda^D\times MTTR\right]$$ | A-28 |
| 3oo3 | $F_{12} =$ | $$3\times\left[\frac{(1-DC)\times\lambda^D\times TI}{2}+\frac{DC\times\lambda^D\times DI}{2}+\lambda^D\times MTTR\right]$$ | A-29 |

**Review of Variance Contribution Analysis (VCA) Methodology**

The mean and variance of a function of random variables can be approximated using the method described by Haugen [Ref. 5] and applied by Freeman [Ref. 3, 6, 7]. Define an arbitrary function of a set of random variables, $x_i$, as:

Let

$$Y = F(x_i) \quad \text{(Eq 1)}$$

The mean of Y can be estimated using the following approximation:

$$E(Y) = F[\,E(x_i)] \quad \text{(Eq 2)}$$

Where:
$E(Y)$ = expected value of random variable Y = mean of Y
$E(x_i)$ = expected value of random variable $x_i$ = mean of $x_i$

The variance of Y can likewise be estimated as:

$$V(Y) = \sum_{i=1}^{n} \left[\frac{\partial Y}{\partial x_i}\right]^2 V(x_i) \quad \text{(Eq 3)}$$

Where:

$V(Y)$ = variance of random variable Y as defined above in Equation 1

$V(x_i)$ = variance of random variable $x_i$ as defined above in Equation 1

Note that the variance is simply the square of the standard deviation. Using the variance will simplify the mathematics that is described below. The contribution of each independent random variable to the overall variance in the function is:

$$V(Y \text{ from } x_i) = \left[\frac{\partial Y}{\partial x_i}\right]^2 V(x_i) \quad \text{(Eq 4)}$$

The relative contribution of each term to the overall variance V(Y) is a measure of the importance in the uncertainty in the particular random variable, $x_i$. In effect, this is a sensitivity analysis combined with a uncertainty evaluation. The variance contribution combines the sensitivity in the answer to changes in the uncertain random variable, $x_i$, with a measure of the uncertainty in the random variable, $x_i$. The overall variance in Y is found by summing the sensitivity weighted variances from each random variable.

## Recommended Interlock PFD Uncertainty Analysis Method

Previously, Freeman and Summers [Ref. 3] suggested a framework for the inclusion of uncertainty analysis in the calculations completed for a new interlock per the ISA TR [Ref. 2]. The following uncertainty analysis method has been expanded to incorporate critical decisions that the process management must make in the design process.

1. Complete the interlock design using the methods outlined in IEC 61511 [Ref. 1].

2. Review with the process system management and determine if the proposed interlock should be considered repairable or non-repairable. A simple flow chart for this decision making is presented in Figure 2. The basic question is: "can the interlock be repaired safely while the process operates." This is a management question and management should be the one that decides the answer to this important question.

3. Create interlock performance equation as the mathematical model for the combination of sensor, logic solver and final control elements using the methods outlined in ISA technical report (TR84.00.02) [Ref. 2]. Equations needed to evaluate the uncertainty of the models specified in the ISS report are developed in Appendix A of this paper. For non-repairable systems, Table 3 can be used. For more complex systems such as non-repairable redundant systems with the potential for common cause failures, use the repairable equations of Table 4 and set DI, DC, and MTTR all equal to zero. For systems where common cause failures (CCF), Diagnostics and repair are to be considered, use the recommendations of Table 4. There are design situation where the equations developed in Appendix A will not exactly apply. Specifically, designs where the system contains redundancy subject to common cause failures (CCF) that should be considered non-repairable during operation. The failure models and the equations needed to evaluate the uncertainty in these systems are developed in Appendix B to this paper.

| Table 3. Roadmap for Mean and Variance of Non-Repairable System Without Considering CCF, Diagnostics and MTTR | | | |
|---|---|---|---|
| **Configuration** | **Function** | **Mean PFD Using Appendix A Equation** | **PFD Variance Using Appendix A Equation** |
| 1oo1 | $F_1$ = | A-9 | A-18 |
| 1oo2 | $F_2$ = | A- 10 | A- 19 |
| 1oo3 | $F_3$ = | A-11 | A-20 |
| 2oo2 | $F_4$ = | A-12 | A-21 |
| 2oo3 | $F_5$ = | A-13 | A-22 |
| 3oo3 | $F_6$ = | A-14 | A-23 |

| Table 4. Roadmap for Mean and Variance of Repairable System Considering CCF, Diagnostics and MTTR | | | |
|---|---|---|---|
| **Configuration** | **Function** | **Mean PFD Using Appendix A Equation** | **PFD Variance Using Appendix A Equation** |
| 1oo1 | $F_7$ = | A-30 | A-44 |
| 1oo2 | $F_8$ = | A- 31 | A- 86 |
| 1oo3 | $F_9$ = | A-32 | A-94 |
| 2oo2 | $F_{10}$ = | A-33 | A-48 |
| 2oo3 | $F_{11}$ = | A-34 | A-102 |
| 3oo3 | $F_{12}$ = | A-35 | A-50 |

Note that in all of the calculations indicated by the equations referenced in Table 4, the expected value of any the random variables is used to complete the calculations.

4. Define the uncertainty in the parameters and variables of the interlock model specified in step 2. The uncertainty can be given as the upper and lower range of the possible values (uniform probability distribution), as the upper, lower and recommended values (triangular distribution), or as a mean and standard deviation (normal distribution). See the example above for guidance in the evaluation of safety instrumented system interlocks.

5. Compute the expected value of each variable in the interlock performance equation. The equations for the mean and variance for the uniform, triangular and normal probability distributions are presented for various probability distributions in Vose [Ref. 8]. The example interlock calculations used a triangular distribution to represent the uncertainty in the parameters.

6. Compute the expected value or mean of the interlock PFD using the mean value of each of the variables in Step 5. See Tables 4 and 5 for recommended equations.

7. Compute the sensitivity of the result from the interlock performance equation by use of the partial derivative of the basic interlock performance equation with respect to each of the variables as presented in Appendix A.

8. Compute the variance of the interlock performance equation PFD by use of the variance contribution using equations from Tables 3 or 4. This entails multiplying the variance of each of the uncertain variables in the basic interlock performance equation by the square of its sensitivity (obtained in step 7), as evaluated at the variable mean. Sum the resulting terms to obtain the overall variance of the PFD in the interlock performance equation. See Tables 4 and 5 for recommended equations.

9. Determine the level of risk that the owner/operator wishes to take that the final interlock will not work. Note that this is a management decision not an instrument engineer decision! In this paper, the 95% level of risk reduction has been used:
   - 5% chance of failure or a 95% chance of the interlock achieving the desired risk reduction

10. Assuming that the interlock owner operator wishes to take a low risk (5%) of the interlock failing to achieve its design target PFD, compute the 95% upper confidence limit on the computed PFD by use of the standard normal factor, Z, [Ref. 9] as:

$$Z = \left[\frac{xi - E(x)}{\sigma}\right] \qquad \text{(Eq 5)}$$

Where:

$\sigma$ = standard deviation of the PFD of the interlock of interest from the interlock performance equation obtained from step 7. Note that the variance of a random variable is the square of the standard deviation of the random variable.

E(x) = the expected value of the PFD of the interlock of interest from the interlock performance equation obtained from step 5

For the 95% upper limit, Z = 1.645. Rearranging Eq. 5 allows for the direct calculation of the corresponding value of the 95% upper confidence limit on the PFD as:

$$X95\% = 1.645\,\sigma + E(x) \qquad \text{(Eq 6)}$$

Where:

X95% = the upper 95% limit on the computed PFD of the interlock of interest from the interlock performance equation.

Compare the 95% upper confidence limit on the PFD of the interlock of concern with that established as the desired PFD for risk reduction. If the 95% confidence of the RRF is greater than the desired RRF, the design is complete. If not, revise the design or change inspection test intervals to achieve the desired RRF. If it is not possible to achieve the desired target RRF economically, revisit the LOPA study accordingly to incorporate better information obtained in the uncertainty analysis. Improve the integrity of the LOPA IPLs or identify additional IPLs to drive the risk to a tolerable level. Continue this process until the computed RRFs are greater than the desired RRFs for risk reduction and risk management.

**Example Interlock Problem**

Previously, Freeman and Summers [Ref. 3] published an example in the use of VCA to determine the likelihood that the interlock would provide the risk reduction desired by the management and the LOPA team. The process system is shown in Figure 3. The process uses a compressor to increase the pressure of a process stream prior to additional processing. The gas being compressed is toxic and flammable. Of concern is a slug of liquid being sent to the compressor. If a liquid slug is sent to the compressor, significant damage to the compressor is expected with probable seal damage and a subsequent release of flammable and toxic material into the work area. A large fire and/or explosion could result if the release were to be ignited. If the release is not ignited, the nearby workers could be exposed to the toxic gas resulting in death or injury. A Layer of Protection Analysis (LOPA) review of this system has been completed. Among several recommendations, the LOPA team recommended the installation of a SIL-2 high level interlock in the Compressor Knock Out Drum to stop the compressor prior to liquids entering the system.

Figure 1 shows a simplified diagram for the interlock. Three level sensors are provided in the Compressor Knock Out Drum. The SIS logic solver will use 2oo3 voting to detect high level in the Compressor Knock Out Drum. The SIS logic solver will also monitor the difference in level signal from each of the three level sensors and will activate an alarm if the deviation is excessive. Two independent methods of stopping the compressor are provided. The SIS will directly signal the motor controller on the compressor to stop. In addition, the SIS logic solver will also signal two additional relays to open the power supply to the compressor motor. These two different shutdown will both be activated in the event of high level in the Compressor Knock Out Drum. Either one of the two shutdowns is capable of stopping the compressor by turning off the electric power supply to the motor.

Models

The first step in the calculation of the "goodness" of an interlock is to establish the model to be used in the calculations. Note that the sensors are 2oo3 voting and the final control elements are each 1oo1. Appendix A presents the equations for various models that can be used for describing this system.

The overall probability of failure on demand (PFD) of the interlock is given as:

PFD = PFDs + PFDsis + PFDfce (Eq 7)

Where:

PFD = Probability of failure on demand of the interlock as a whole

PFDs = Probability of failure on demand of the sensors (voting as 2oo3)

PFDsis = Probability of failure on demand of the SIS logic solver

PFDfce = Probability of failure on demand of the final control elements.

Since there are two final control elements arranged in series, the PFDfce = sum of the PFDs of the final control elements (Relay PFD and MCC PFD).

PFDce = PFDr + PFDmcc (Eq 8)

Where:

PFDr = Probability of failure on demand of the two relays voting as 1oo2 to shutdown gas compressor.

PFDmcc = Probability of failure on demand of the MCC to shutdown the gas compressor

For this example, the following selections are made to model the performance of the interlock.

Model for Sensors (2oo3)

The sensors are considered repairable as one sensor can be replaced while the other two function and provide continuous functionality of the interlock. There are three sensors that will be used in a 2oo3 voting system. From Appendix A, using Equation A-7, the model for the sensors becomes:

$$\text{PFDsavg} = 3\times\left[\frac{(1-DCs)\times(1-\beta s)\times\lambda^{Ds}\times TIs}{2}+\frac{DCs\times(1-\beta s)\times\lambda^{Ds}\times DIs}{2}+(1-\beta s)\times\lambda^{Ds}\times MTTRs\right]^2 +$$

$$\left[\frac{(1-DCs)\times\beta s\times\lambda^{Ds}\times TIs}{2}+\frac{DCs\times\beta s\times\lambda^{D}\times DIs}{2}+\beta s\times\lambda^{Ds}\times MTTRs\right] \qquad \text{(Eq 9)}$$

Where:

$PFD_{savg}$ is the average probability of failure on demand of the sensors

$DC_s$ is the diagnostic coverage for sensor failure

$DI_s$ is the diagnostic interval for the sensors

$MTTR_s$ is the mean time to restore the sensors to functionality given a sensor failure

$TI_s$ is the test interval for the sensors

$\beta_s$ is the common cause failure parameter

$\lambda^{Ds}$ is the failure rate to a dangerous condition for the sensors

MTTR is the mean time to restore the system from the time that failure occurs

Model for SIS Logic Solver

Use a fixed probability of failure on demand as specified by the vendor. Unless detailed design information is provided by the SIS logic solver vendor, this will be the normal default condition for most studies. For this problem a typical PFD of $1.30 \times 10^{-4}$ was selected to represent the SIS Logic Solver.

Model for Final Control Elements

There are two separate paths to shutdown the gas compressor. First is by the SIS logic solver commanding the MCC to shutdown power to the the gas compressor motor. The second is for the SIS logic solver to issue a shutdown command to two interposing relays (R1 and R2) which will cause the power to the gas compressor motor to stop. Two different models are needed to the final control elements. These two subsystems are considered non-repairable in this analysis,

Model for Interposing Relays (1oo2)

There are two interposing relays (R1 and R2 in the interlock). From Appendix A, use Equation A-25 setting parameters DC=DI=MTTR=0 for the non-repairable interposing relays, the model becomes:

$$\text{PFDravg} = \left[\frac{(1-\beta r)\times\lambda^{Dr}\times TIr}{2}+\right]^2+\left[\frac{\beta r\times\lambda^{Dr}\times TIr}{2}+\right] \qquad \text{(Eq 10)}$$

Where:

$PFD_{ravg}$ is the average probability of failure on demand of the relays voting 1oo2 to shutoff the gas compressor.

$TI_r$ is the test interval for the relays

$\beta_r$ is the common cause failure parameter

$\lambda^{Dr}$ is the failure rate to a dangerous condition for the relays

Model for MCC (1oo1)

From Appendix A and using equation A-1 for the non-repairable MCC, the model becomes:

$$\boldsymbol{PFDmccavg} = \frac{\lambda^{Dmcc} * \mathbf{TImcc}}{2} \qquad \text{(Eq 11)}$$

Where:

$PFD_{mccavg}$ is the average probability of failure on demand of the MCC to shutoff the gas compressor.

$TI_{mcc}$ is the test interval for the MCC

$\lambda^{Dmcc}$is the failure rate to a dangerous condition for the MCC

Data

The calculation of the PFD of the interlock requires a set of data to be used to represent the system. Tables 5, 6, and 7 are taken from the Freeman-Summers paper [Ref. 3] and presents the data used to represent the interlock system. Note that these data were originally taken from generic data sources and do not represent any particular device or system.

**Table 5. Uncertainty Data for Level Sensors Used in Example Interlock**
**(All variable probability distributions assumed to be triangularly distributed)**

| Variable | Min | Mode | Max | Mean | Variance |
|---|---|---|---|---|---|
| $\lambda^{Ds}$- Fail Dangerous Rate | $2.84 \times 10^{-3}$ $Yr^{-1}$ | $5 \times 10^{-3}$ $Yr^{-1}$ | $8.5 \times 10^{-3}$ $Yr^{-1}$ | $5.45 \times 10^{-3}$ $Yr^{-1}$ | $1.36 \times 10^{-6}$ $Yr^{-1}$ |
| $DC_s$ – Diagnostic Coverage * | 0.8 | 0.9 | 0.99 | 0.897 | $1.51 \times 10^{-3}$ |
| $DI_s$ – Diagnostic Interval ** | $5.71 \times 10^{-5}$ | $5.71 \times 10^{-5}$ | $5.71 \times 10^{-5}$ | $5.71 \times 10^{-5}$ | 0 |
| $TI_s$ – Test Interval | 4 Yr | 5 Yr | 6 Yr | 5 Yr | $5.56 \times 10^{-2}$ $Yr^2$ |
| $\beta_s$- Common Cause Failure Fraction * | 0 | 0.02 | 0.1 | 0.04 | $4.67 \times 10^{-4}$ |
| $MTTR_s$ - Mean Time to Restore *** | $1.37 \times 10^{-3}$ Yr | $8.22 \times 10^{-3}$ Yr | $1.92 \times 10^{-2}$ Yr | $9.59 \times 10^{-3}$ Yr | $1.34 \times 10^{-5}$ $Yr^2$ |

* Unit less

** 0.5 hours, assumed to be deterministic

***Min of 12 hr, Mode of 72 hours, Max of 168 hrs

| Table 6. Uncertainty Data for Relays Used in Example Interlock<br>(All variable probability distributions assumed to be triangularly distributed) | | | | | |
|---|---|---|---|---|---|
| **Variable** | **Min** | **Mode** | **Max** | **Mean** | **Variance** |
| $\lambda^{Dr}$- Fail Dangerous Rate | $8.76x10^{-9}$ $Yr^{-1}$ | $2.00x10^{-3}$ $Yr^{-1}$ | $4.73x10^{-2}$ $Yr^{-1}$ | $1.64x10^{-2}$ $Yr^{-1}$ | $1.19x10^{-4}$ $Yr^{-2}$ |
| $DC_r$ – Diagnostic Coverage * | NA | NA | NA | NA | NA |
| $DI_r$ – Diagnostic Interval* | NA | NA | NA | NA | NA |
| $TI_r$ – Test Interval | 1 Yr | 1 Yr | 2 Yr | 1.33 Yr | 0.056 $Yr^2$ |
| $\beta_r$- Common Cause Failure Fraction | 0 | 0.02 | 0.1 | 0.04 | $4.67x10^{-4}$ |
| $MTTR_r$ - Mean Time to Restore* | NA | NA | NA | NA | NA |

* Not used in relay model

| Table 7. Uncertainty Data for MCC Used in Example Interlock<br>(All variable probability distributions assumed to be triangularly distributed) | | | | | |
|---|---|---|---|---|---|
| **Variable** | **Min** | **Mode** | **Max** | **Mean** | **Variance** |
| $\lambda^{Dr}$- Fail Dangerous Rate | $1.74x10^{-4}$ $Yr^{-1}$ | $1.31x10^{-3}$ $Yr^{-1}$ | $3.00x10^{-2}$ $Yr^{-1}$ | $1.05x10^{-2}$ $Yr^{-1}$ | $4.76x10^{-5}$ $Yr^{-2}$ |
| $DC_r$ – Diagnostic Coverage* | NA | NA | NA | NA | NA |
| $DI_r$ – Diagnostic Interval* | NA | NA | NA | NA | NA |
| $TI_r$ – Test Interval | 1 Yr | 1 Yr | 2 Yr | 1.33 Yr | $5.6x10^{-2}$ $Yr^2$ |
| $\beta_r$- Common Cause Failure Fraction* | NA | NA | NA | NA | NA |
| $MTTR_r$ - Mean Time to Restore* | NA | NA | NA | NA | NA |

* Not used in MCC model

Uncertainty Analysis Results

Using the VCA method to estimate the PFD of the interlock design yields the results shown in Table 8.

**Table 8. Results of VCA on Example Interlock Design**

| Subsystem | Mean PFD | Variance of PFD |
|---|---|---|
| Sensors | 6.41E-5 | 1.7657E-9 |
| Logic Solver | 1.34E-4 | 0 |
| Relays | 5.46E-4 | 2.0333E-7 |
| MCC | 6.98E-3 | 4.52E-5 |
| **Totals** | **7.73E-3** | **4.54E-5** |

Note: Calculations are carried to an adequate number of places to allow for the resulting PFD to be determined. Rounding to significant number of figures is done at the end.

The mean PFD of 0.00773 implies a mean risk reduction factor of:

RRF = 1/PFD = 1/0.00773 = 129 (Eq 11)

The variance of 4.54E-5 implies a standard deviation in the PFD of

Std Dev = $(\text{Variance})^{1/2}$ = 0.006737 (Eq 12)

The corresponding 95% level of confidence in the PFD:

X95% = 1.645 std Dev + Mean PFD (Eq 13)

X95% = 1.645 (0.006737) + 0.00773= 0.01881 (Eq 14)

The RRF for 95% certain PFD is

RRF95% = 1/0.01881 => 53 (Eq 15)

This is essentially the same result previously reported by Freeman and Summers [Ref. 3] using either Monte Carlo Simulation or numerical approximation methods for the VCA sensitivities. The calculated PFD at the 95% level of 0.01881 indicates that there is only a 5% chance that the interlock will provide an RRF worse that 53. Since this level is only SIL-1 capable and not SIL-2 capable as desired by management, a revised design will be needed to achieve the desired risk reduction or a different protective measure will be needed to control risk.

**Acknowledgement**

This paper is an expanded version of a paper entitled ''General Method for Uncertainty Evaluation of Safety Integrity Level (SIL) Calculations'' presented at the Mary Kay O'Connor 2016 Process Safety Symposium, October 26, 2016, College Station, TX. The author wishes to thank those individuals who gave comments and suggestions for improving this article.

**References**

1. International Society of Automation, Functional Safety: Safety Instrumented Systems for the Process Sector—Parts 1, 2, and 3, ANSI/ISA 84.00.01, 2004 [USA implementation of IEC 61511].

2. International Society of Automation, *Safety Integrity Level (SIL) Verification of Safety Instrumented Functions*, ISA TR84.00.02, Research Triangle Park, NC. (2015).

3. Freeman, R. And Summers, A, "Evaluation of Uncertainty in Safety Integrity Level Calculations," *Process Safety Progress*, Vol 35, No 4, pp 341-348, December 2016. Published Wiley Online Library (wileyonlinelibrary.com). DOI 10.1002/prs.11805, 23 January 2016

4. Center for Chemical Process Safety, *Layer of Protection Analysis – Simplified Process Risk Assessment*, American Institute of Chemical Engineers, New York, 2001

5. Edward B. Haugen, *Probabilistic Approaches to Design*, John Wiley, New York, 1968

6. R.A. Freeman, "Quantifying LOPA Uncertainty," *Process Safety Progress*, Vol 31, No. 3, pp 240-247, 2012

7. R. A. Freeman, "Simplified Uncertainty Analysis of Layer of Protection Analysis Results," *Process Safety Progress*, Vol. 32, No 4, pp 351-360, December 2013, Published online in Wiley Online Library (wileyonlinelibrary.com) DOI 10.1002/ prs11585.pdf, 3 June 2013

8. David Vose, *Risk Analysis – A Quantitative Guide*, 3rd Edition, John Wiley, 2008

9. Paul L. Meyer, *Introductory Probability and Statistical Applications*, John Wiley, New York, 1972

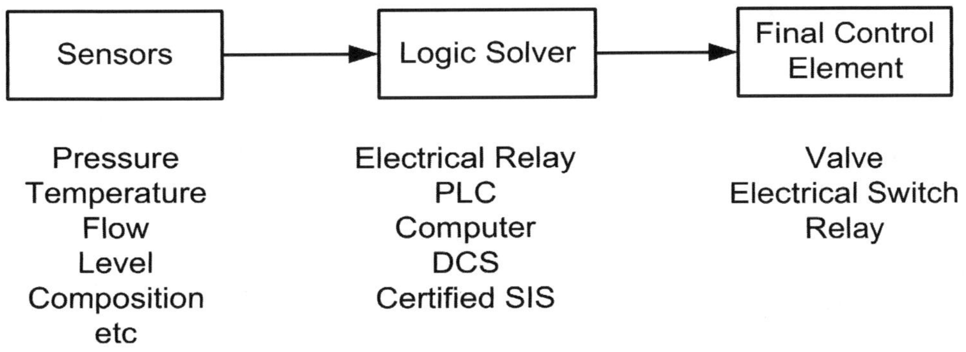


**Figure 1. Typical Components of an SIS Interlock**

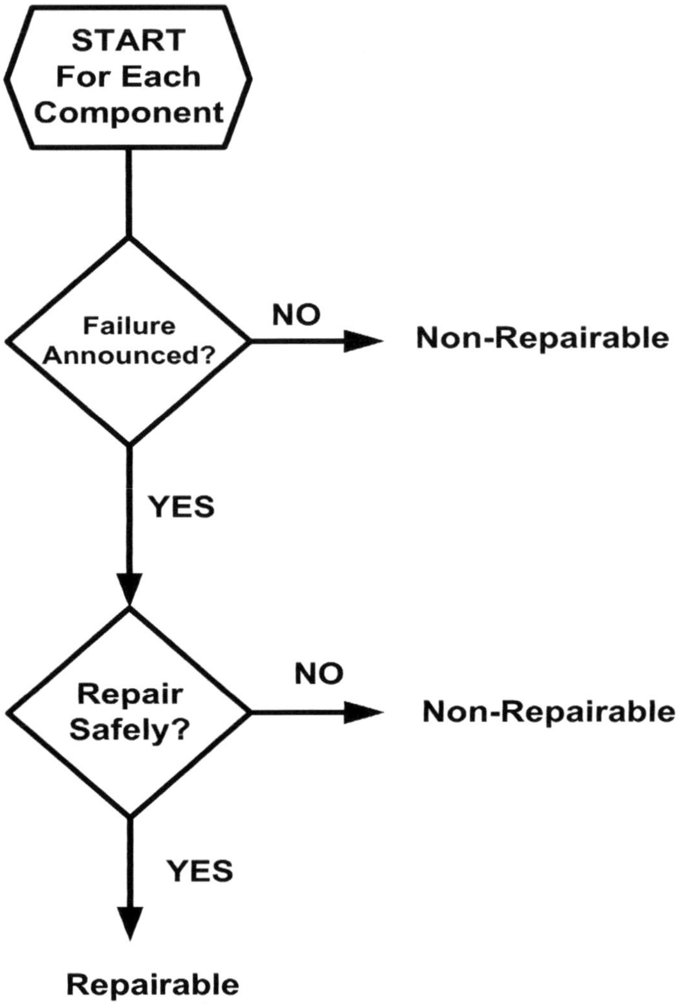


**Figure 2. Decision Tree for Determination of Repairable or Non- Repairable**

## Figure 3. P&ID Sketch for Example Interlock

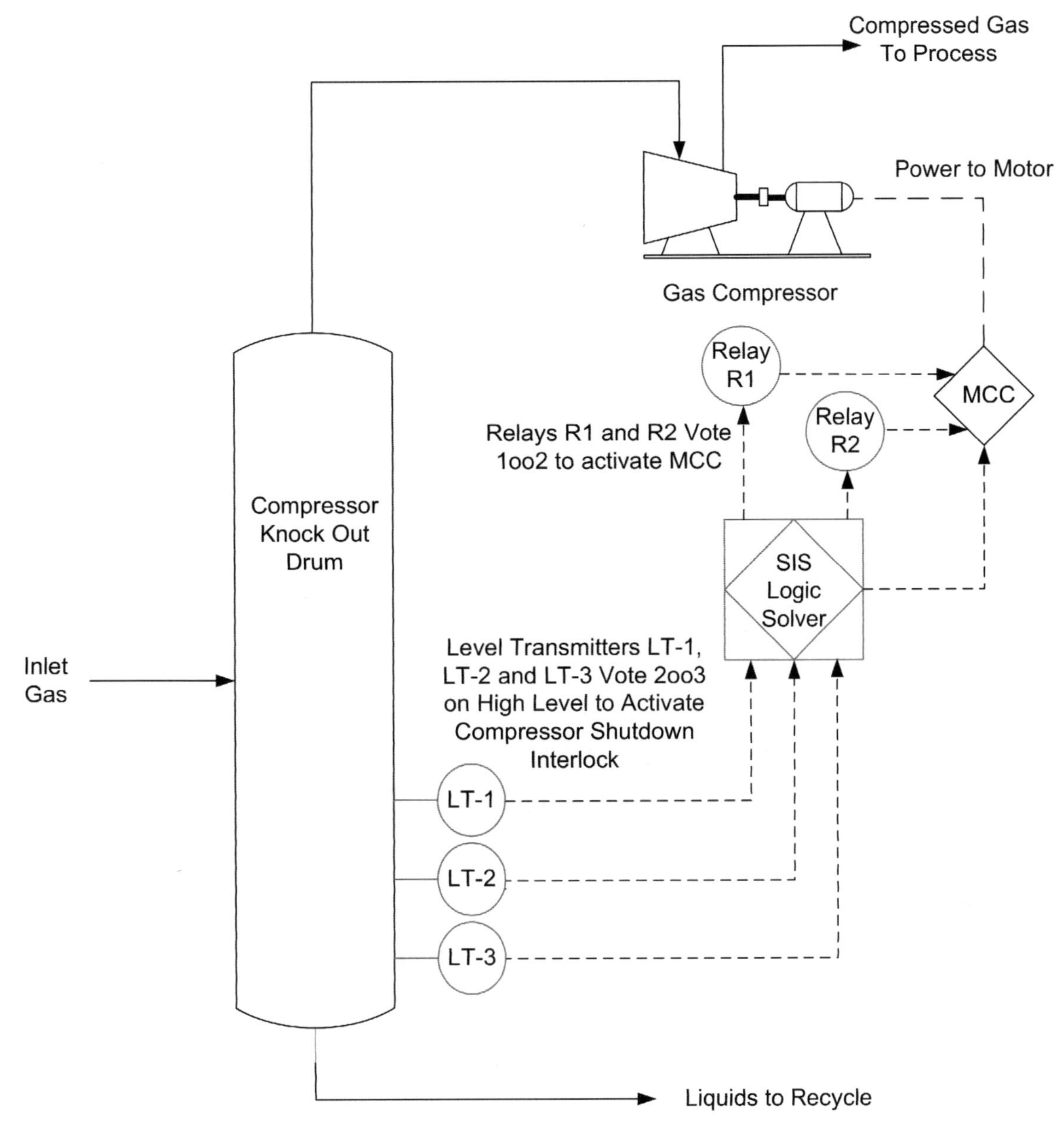

## APPENDIX A -- DEVELOPMENT OF UNCERTAINTY EQUATIONS

This appendix develops the equations needed to evaluate the uncertainty in the SIL PFD as computed using the methods outlined in the ISA Technical Report [Ref. 2]. This appendix is divided into two sections:

- Non-Repairable Systems
- Repairable Systems

The uncertainty equations for other configurations not covered in the ISA report are developed in Appendix B.

### NON-REPAIRABLE SYSTEMS

For systems that are considered non-repairable, the simplified equations of Table A.1 (taken from the ISA Technical Report [Ref. 2]) are used for the analysis.

**Table A.1 Simplified $PFD_{avg}$ formulas for Non-Repairable System without considering CCF, Diagnostics or MTTR.**

| Configuration | Function | PFD based on "Average before" failure rate | Equation Number |
|---|---|---|---|
| 1oo1 | $F_1$ = | $\left[\lambda^D \times \frac{TI}{2}\right]$ | A-1 |
| 1oo2 | $F_2$ = | $\left[(\lambda^D)^2 \times \frac{TI^2}{4}\right]$ | A- 2 |
| 1oo3 | $F_3$ = | $\left[(\lambda^D)^3 \times \frac{TI^3}{8}\right]$ | A-3 |
| 2oo2 | $F_4$ = | $\left[\lambda^D \times TI\right]$ | A-4 |
| 2oo3 | F5 = | $\frac{3}{4} \times \left[(\lambda^D)^2 \times TI^2\right]$ | A-5 |
| 3oo3 | F6 = | $3 \times \left[\lambda^D \times \frac{TI}{2}\right]$ | A-6 |

Where TI is the proof test interval, $\lambda^D$ is the dangerous failure rate.

The expected value of the probability of failure on demand is found by calculating the PFD using the expected value (mean) of the random variables.

Let y a function of some random variables, $x_i$, as:

$y = F(x_i)$ (Eq A-7)

The mean of y can be estimated using the following approximation:

$E(y) = F[E(x_i)]$ (Eq A-8)

Where:

$E(y)$ = expected value of random variable y = mean of y

$E(x_i)$ = expected value of random variable $x_i$ = mean of $x_i$

For the configurations defined in Table A.1, the corresponding expected PFD are presented in Table A.2.

**Table A.2 Expected Value of $PFD_{avg}$ formulas for Non-Repairable System without considering CCF, Diagnostics or MTTR.**

| Configuration | Function | Expected PFD based on "Average before" failure rate | Equation Number |
|---|---|---|---|
| 1oo1 | $E(F_1) =$ | $\left[E(\lambda^D)\times\frac{E(TI)}{2}\right]$ | A-9 |
| 1oo2 | $E(F_2) =$ | $\left[(E(\lambda^D))^2\times\frac{E(TI)^2}{4}\right]$ | A- 10 |
| 1oo3 | $E(F_3) =$ | $\left[(E(\lambda^D))^3\times\frac{E(TI)^3}{8}\right]$ | A-11 |
| 2oo2 | $E(F_4) =$ | $\left[E(\lambda^D)\times E(TI)\right]$ | A-12 |
| 2oo3 | $E(F_5) =$ | $\frac{3}{4}\times\left[(E(\lambda^D))^2\times E(TI)^2\right]$ | A-13 |
| 3oo3 | $E(F_6) =$ | $3\times\left[E(\lambda^D)\times\frac{E(TI)}{2}\right]$ | A-14 |

Where E(x) is the expected value of random variable x.

The variance of the PFD is found by taking the weighted sum of the variance of the random variables. The weighting function is the sensitivity of the PFD with respect to random variable. In mathematical terms the variance of a function may be found in terms of the variance of the random variables as:

Once again let

$$y = F(x_i) \qquad \text{(Eq A-15)}$$

The variance of Y can likewise be estimated as:

$$V(y) = \sum_{i=1}^{n} \left[\frac{\partial y}{\partial x_i}\right]^2 V(x_i) \qquad \text{(Eq A-16)}$$

Where:

$V(y)$ = variance of random variable y as defined above

$V(x_i)$ = variance of random variable $x_i$ as defined above

The sensitivity of y with respect to a random variable $x_i$ is:

$$\text{Sensitivity of y with respect to } x_i = \left[\frac{\partial y}{\partial x_i}\right] \qquad \text{(Eq A-17)}$$

For the configurations defined in Table A.1, the corresponding variance of the PFD are presented in Table A.3.

| Table A.3 Variance of $PFD_{avg}$ formulas for Non-Repairable System without considering CCF, Diagnostics or MTTR. | | | |
|---|---|---|---|
| **Configuration** | **Function** | **Variance in Function $F_i$ based on "Average before" failure rate** | **Equation Number** |
| 1oo1 | $V(F_1) =$ | $\left[\frac{E(TI)}{2}\right]^2 V(\lambda^D) + \left[\frac{E(\lambda^D)}{2}\right]^2 V(TI)$ | A-18 |
| 1oo2 | $V(F_2) =$ | $\left[E(\lambda^D)^2 \frac{E(TI)^2}{2}\right]^2 V(\lambda^D) + \left[E(\lambda^D)^2 \frac{E(TI)}{2}\right]^2 V(TI)$ | A- 19 |
| 1oo3 | $V(F_3) =$ | $\left[3\times E(\lambda^D)^2 \frac{E(TI)^3}{2}\right]^2 V(\lambda^D) + \left[\frac{3}{8} E(\lambda^D)^3 E(TI)^2\right]^2 V(TI)$ | A-20 |
| 2oo2 | $V(F_4) =$ | $E(TI)^2 V(\lambda^D) + E(\lambda^D)^2 \times V(TI)$ | A-21 |
| 2oo3 | $V(F_5) =$ | $\left[\frac{3}{2}\times E(\lambda^D)\times E(TI)^2\right]^2 V(\lambda^D) + \left[\frac{3}{2}\times E(\lambda^D)^2 \times E(TI)\right]^2 V(TI)$ | A-22 |
| 3oo3 | $V(F_6) =$ | $\left[\frac{3}{2}\times E(TI)\right]^2 V(\lambda^D) + \left[\frac{3}{2}\times E(\lambda^D)\right]^2 V(TI)$ | A-23 |

Note that for the calculation of the variance functions, V(Fi), the random variables are evaluated at the expected value (mean).

## REPAIRABLE SYSTEMS

For systems that are considered repairable, the simplified equations of Table A.4 are used for the analysis. The expected value of the probability of failure on demand (PFD) is found by substituting the expected value of each of the random variables into the corresponding equation. The expected value of the PFD for each of the system configurations is presented in Table A.5

**Table A.4 Simplified $PFD_{avg}$ Formulas for Repairable System considering CCF, Diagnostics and MTTR**

| Configuration | Function | PFD based on "Average before" failure rate | Equation Number |
|---|---|---|---|
| 1oo1 | $F_7 =$ | $\left[\frac{(1-DC)\times\lambda^D\times TI}{2}+\frac{DC\times\lambda^D\times DI}{2}+\lambda^D\times MTTR\right]$ | A-24 |
| 1oo2 | $F_8 =$ | $\left[\frac{(1-DC)\times(1-\beta)\times\lambda^D\times TI}{2}+\frac{DC\times(1-\beta)\times\lambda^D\times DI}{2}+(1-\beta)\times\lambda^D\times MTTR\right]^2+\left[\frac{(1-DC)\times\beta\times\lambda^D\times TI}{2}+\frac{DC\times\beta\times\lambda^D\times DI}{2}+\beta\times\lambda^D\times MTTR\right]$ | A- 25 |
| 1oo3 | $F_9 =$ | $\left[\frac{(1-DC)\times(1-\beta)\times\lambda^D\times TI}{2}+\frac{DC\times(1-\beta)\times\lambda^D\times DI}{2}+(1-\beta)\times\lambda^D\times MTTR\right]^3+\left[\frac{(1-DC)\times\beta\times\lambda^D\times TI}{2}+\frac{DC\times\beta\times\lambda^D\times DI}{2}+\beta\times\lambda^D\times MTTR\right]$ | A-26 |
| 2oo2 | $F_{10} =$ | $2\times\left[\frac{(1-DC)\times\lambda^D\times TI}{2}+\frac{DC\times\lambda^D\times DI}{2}+\lambda^D\times MTTR\right]$ | A-27 |

| Table A.4 Simplified $PFD_{avg}$ Formulas for Repairable System considering CCF, Diagnostics and MTTR | | | |
|---|---|---|---|
| **Configuration** | **Function** | **PFD based on "Average before" failure rate** | **Equation Number** |
| 2oo3 | $F_{11}$ = | $3\times\left[\frac{(1-DC)\times(1-\beta)\times\lambda^D\times TI}{2}+\frac{DC\times(1-\beta)\times\lambda^D\times DI}{2}+(1-\beta)\times\lambda^D\times MTTR\right]^2+\left[\frac{(1-DC)\times\beta\times\lambda^D\times TI}{2}+\frac{DC\times\beta\times\lambda^D\times DI}{2}+\beta\times\lambda^D\times MTTR\right]$ | A-28 |
| 3oo3 | $F_{12}$ = | $3\times\left[\frac{(1-DC)\times\lambda^D\times TI}{2}+\frac{DC\times\lambda^D\times DI}{2}+\lambda^D\times MTTR\right]$ | A-29 |

Where:

DC is the diagnostic coverage;

DI is the diagnostic interval;

TI is the proof test interval,

$\lambda^D$ is the dangerous failure rate;

MTTR is the mean time to restore the system to operation

$\beta$ is the common cause failure parameter that is always is between 0 and 1

| Table A.5 Expected PFD Formulas for Repairable System considering CCF, Diagnostics and MTTR | | | |
|---|---|---|---|
| **Configuration** | **Function** | **PFD based on "Average before" failure rate** | **Equation Number** |
| 1oo1 | E(F$_7$) = | $\left[\frac{(1-E(DC))\times E(\lambda^D)\times E(TI))}{2}+\frac{E(DC)\times E(\lambda^D)\times E(DI)}{2} + E(\lambda^D)\times E(MTTR)\right]$ | A-30 |
| 1oo2 | E(F$_8$) = | $\left[\frac{(1-E(DC))\times(1-E(\beta))\times E(\lambda^D)\times E(TI)}{2}+ \frac{E(DC)\times(1-E(\beta))\times E(\lambda^D)\times E(DI)}{2}+ (1-E(\beta))\times E(\lambda^D)\times E(MTTR)\right]^2 + \left[\frac{(1-E(DC))\times E(\beta)\times E(\lambda^D)\times E(TI)}{2}+ \frac{E(DC)\times E(\beta)\times E(\lambda^D)\times E(DI}{2})+ E(\beta)\times E(\lambda^D)\times E(MTTR)\right]$ | A- 31 |
| 1oo3 | E(F$_9$) = | $\left[\frac{(1-E(DC))\times(1-E(\beta))\times E(\lambda^D)\times E(TI)}{2}+ \frac{E(DC)\times(1-E(\beta))\times E(\lambda^D)\times E(DI)}{2}+ (1-E(\beta))\times E(\lambda^D)\times E(MTTR)\right]^3 + \left[\frac{(1-E(DC))\times E(\beta)\times E(\lambda^D)\times E(TI)}{2}+ \frac{E(DC)\times E(\beta)\times E(\lambda^D)\times E(DI)}{2}+ E(\beta)\times E(\lambda^D)\times E(MTTR)\right]$ | A-32 |

| Table A.5 Expected PFD Formulas for Repairable System considering CCF, Diagnostics and MTTR | | | |
|---|---|---|---|
| **Configuration** | **Function** | **PFD based on "Average before" failure rate** | **Equation Number** |
| 2oo2 | $E(F_{10})$ = | $2\times\left[\begin{array}{l}\frac{(1-E(DC))\times E(\lambda^D)\times E(TI))}{2}+\\ \frac{E(DC)\times E(\lambda^D)\times E(DI)}{2}\\ +E(\lambda^D)\times E(MTTR)\end{array}\right]$ | A-33 |
| 2oo3 | $E(F_{11})$ = | $3\times\left[\begin{array}{l}\frac{(1-E(DC))\times(1-E(\beta))\times E(\lambda^D)\times E(TI)}{2}+\\ \frac{E(DC)\times(1-E(\beta))\times E(\lambda^D)\times E(DI)}{2}+\\ (1-E(\beta))\times E(\lambda^D)\times E(MTTR)\end{array}\right]^2+$ $\left[\begin{array}{l}\frac{(1-E(DC))\times E(\beta)\times E(\lambda^D)\times E(TI)}{2}+\\ \frac{E(DC)\times E(\beta)\times E(\lambda^D)\times E(DI)}{2}+\\ E(\beta)\times E(\lambda^D)\times E(MTTR)\end{array}\right]$ | A-34 |
| 3oo3 | $E(F_{12})$ = | $3\times\left[\begin{array}{l}\frac{(1-E(DC))\times E(\lambda^D)\times E(TI))}{2}+\\ \frac{E(DC)\times E(\lambda^D)\times E(DI)}{2}\\ +E(\lambda^D)\times E(MTTR)\end{array}\right]$ | A-35 |

The variance of the PFD of the repairable systems is found in the same manner as that in the non-repairable cases. Once again let:

$$y = F(x_i) \quad \text{(Eq A-36)}$$

The variance of Y can likewise be estimated as:

$$V(y) = \sum_{i=1}^{n} \left[\frac{\partial y}{\partial x_i}\right]^2 V(x_i) \quad \text{(Eq A-37)}$$

Where:

V(y) = variance of random variable y as defined above

$V(x_i)$ = variance of random variable $x_i$ as defined above

The sensitivity of y with respect to a random variable $x_i$ is:

In the case of the repairable systems, there are several more potentially uncertain or random variables:

- DC is the diagnostic coverage;
- DI is the diagnostic interval;
- TI is the proof test interval,
- $\lambda^D$ is the dangerous failure rate;
- MTTR is the mean time to restore the system to operation
- $\beta$ is the common cause failure parameter that is always is between 0 and 1

We must evaluate the partial derivative of the PFD function with respect to each and then combine them using equation A-37. The partial derivatives of function F7 with respect to each random variable are found as follows

$$\frac{\partial F_7}{\partial DC} = \frac{-\lambda^D \times TI}{2} + \frac{\lambda^D \times DI}{2} \quad \text{(Eq A-38)}$$

$$\frac{\partial F_7}{\partial \lambda^D} = \frac{(1-DC)\times TI}{2} + \frac{DC \times DI}{2} + MTTR \quad \text{(Eq A-39)}$$

$$\frac{\partial F_7}{\partial TI} = \frac{(1-DC)\times \lambda^D}{2} \quad \text{(Eq A-40)}$$

$$\frac{\partial F_7}{\partial DI} = \frac{DC \times \lambda^D}{2} \quad \text{(Eq A-41)}$$

$$\frac{\partial F_7}{\partial MTTR} = \lambda^D \qquad \text{(Eq A-42)}$$

We can now compute the variance of function F7 as:

$$V\left(F_7\right) = \sum_{i=1}^{n} \left[\frac{\partial F_7}{\partial x_i}\right]^2 V(x_i) \qquad \text{(Eq A-43)}$$

$$V\left(F_7\right) = \left[\frac{-\lambda^D \times TI}{2} + \frac{\lambda^D \times DI}{2}\right]^2 V(DC) + \left[\frac{(1-DC)\times TI}{2} + \frac{DC \times DI}{2} + MTTR\right]^2 V(\lambda^D) +$$

$$\left[\frac{(1-DC)\times \lambda^D}{2}\right]^2 V(TI) + \left[\frac{DC \times \lambda^D}{2}\right]^2 (DI) + \left[\lambda^D\right]^2 V(MTTR) \qquad \text{(Eq A-44)}$$

Equation 44 represents the variance of the probability of failure on demand of the 1001 configuration when the system is repairable. We note the following relationships between the 1001, 2oo2 and 3oo3 configurations:

$F_{10} = 2\ F_7$ (Eq A-45)

$F_{12} = 3\ F_7$ (Eq A-46)

We can directly determine the variance of functions $F_{10}$ and $F_{12}$ from the properties of the variance operator, V(X).

$V(F_{10}) = V(2\ F_7)$ (Eq A-47)

$V(F_{10}) = 4\ V(F_7)$ (Eq A-48)

$V(F_{12}) = V(3\ F_7)$ (Eq A-49)

$V(F_{12}) = 9\ V(F_7)$ (Eq A-50)

The determination of the variance of functions $F_8$, $F_{10}$, and $F_{11}$ is done in a similar manner. To simplify the presentation of the derivation of variance of functions $F_8$, $F_{10}$, and $F_{11}$, we introduce two functions H and Q as:

$$H = \left[\frac{(1-DC)\times(1-\beta)\times\lambda^D \times TI}{2} + \frac{DC\times(1-\beta)\times\lambda^D \times DI}{2} + (1-\beta)\times\lambda^D \times MTTR\right] \qquad \text{(Eq A-51)}$$

$$Q = \left[\frac{(1-DC)\times\beta\times\lambda^D \times TI}{2} + \frac{DC\times\beta\times\lambda^D \times DI}{2} + \beta\times\lambda^D \times MTTR\right] \qquad \text{(Eq A-52)}$$

We note the following relationships:

$F_8 = H^2 + Q$ (Eq A-53)

$F_9 = H^3 + Q$ (Eq A-54)

$F_{11} = 3\,H^2 + Q$ (Eq A-55)

The variance of $F_8$, $F_9$ and $F_{11}$ are found as:

$V(F_8) = V(H^2 + Q)$ (Eq A-56)

$V(F_9) = V(H^3 + Q)$ (Eq A-57)

$V(F_{11}) = V(3\,H^2 + Q)$ (Eq A-58)

We will need the sensitivity of $F_8$, $F_9$ and $F_{11}$ to each of the potential random variables.

$$\frac{\partial F_8}{\partial x} = \frac{\partial(H^2 + Q)}{\partial x} = \frac{\partial(H^2)}{\partial x} + \frac{\partial(Q)}{\partial x} \quad \text{(Eq A-59)}$$

Likewise

$$\frac{\partial F_9}{\partial x} = \frac{\partial(H^3 + Q)}{\partial x} = \frac{\partial(H^3)}{\partial x} + \frac{\partial(Q)}{\partial x} \quad \text{(Eq A-60)}$$

$$\frac{\partial F_{11}}{\partial x} = \frac{\partial 3(H^2 + Q)}{\partial x} = 3\frac{\partial(H^2)}{\partial x} + 3\frac{\partial(Q)}{\partial x} \quad \text{(Eq A-61)}$$

Using the chain rule for partial derivatives, we now write:

$$\frac{\partial F_8}{\partial x} = 2H\frac{\partial(H)}{\partial x} + \frac{\partial(Q)}{\partial x} \quad \text{(Eq A-62)}$$

$$\frac{\partial F_9}{\partial x} = 3H^2\,\frac{\partial(H)}{\partial x} + \frac{\partial(Q)}{\partial x} \quad \text{(Eq A-63)}$$

$$\frac{\partial F_{11}}{\partial x} = 6H\frac{\partial(H)}{\partial x} + \frac{\partial(Q)}{\partial x} \quad \text{(Eq A-64)}$$

We now evaluate the derivatives of function H with respect to each potential random variables:

- DC is the diagnostic coverage;
- DI is the diagnostic interval;

- TI is the proof test interval,
- $\lambda^D$ is the dangerous failure rate;
- MTTR is the mean time to restore the system to operation
- $\beta$ is the common cause failure parameter that is always is between 0 and 1

$$\frac{\partial(\mathrm{H})}{\partial DC} = \frac{-(1-\beta)\times\lambda^D\times TI}{2} + \frac{(1-\beta)\times\lambda^D\times DI}{2} \qquad \text{(Eq A-65)}$$

$$\frac{\partial(\mathrm{H})}{\partial TI} = \frac{(1-DC)\times(1-\beta)\times\lambda^D}{2} \qquad \text{(Eq A-66)}$$

$$\frac{\partial(\mathrm{H})}{\partial\lambda^D} = \left[\begin{array}{l}\dfrac{(1-DC)\times(1-\beta)\times TI}{2} + \\ \dfrac{DC\times(1-\beta)\times DI}{2} + (1-\beta)\times MTTR\end{array}\right] \qquad \text{(Eq A-67)}$$

$$\frac{\partial(\mathrm{H})}{\partial MTTR} = (1-\beta)\times\lambda^D \qquad \text{(Eq A-68)}$$

$$\frac{\partial(\mathrm{H})}{\partial\beta} = \left[\begin{array}{l}\dfrac{(1-DC)\times(-1)\times\lambda^D\times TI}{2} + \\ \dfrac{DC\times(-1)\times\lambda^D\times DI}{2} + (-1)\times\lambda^D\times MTTR\end{array}\right] \qquad \text{(Eq A-69)}$$

$$\frac{\partial(\mathrm{H})}{\partial DI} = \frac{DC\times(1-\beta)\times\lambda^D}{2} \qquad \text{(Eq A-70)}$$

In a similar manner we evaluate the partial derivatives of function Q with respect to each potential random variables.

$$Q = \left[\frac{(1-DC)\times\beta\times\lambda^{D}\times TI}{2} + \frac{DC\times\beta\times\lambda^{D}\times DI}{2} + \beta\times\lambda^{D}\times MTTR\right] \quad \text{(Eq A-71)}$$

- DC is the diagnostic coverage;
- DI is the diagnostic interval;
- TI is the proof test interval,
- $\lambda^D$ is the dangerous failure rate;
- MTTR is the mean time to restore the system to operation
- $\beta$ is the common cause failure parameter that is always is between 0 and 1

$$\frac{\partial(Q)}{\partial DC} = \frac{(-1)\times\beta\times\lambda^{D}\times TI}{2} + \frac{\beta\times\lambda^{D}\times DI}{2} \quad \text{(Eq A-72)}$$

$$\frac{\partial(Q)}{\partial TI} = \frac{(1-DC)\times\beta\times\lambda^{D}}{2} \quad \text{(Eq A-73)}$$

$$\frac{\partial(Q)}{\partial\lambda^{D}} = \left[\frac{(1-DC)\times\beta\times TI}{2} + \frac{DC\times\beta\times DI}{2} + \beta\times MTTR\right] \quad \text{(Eq A-74)}$$

$$\frac{\partial(Q)}{\partial MTTR} = \beta\times\lambda^{D}\times MTTR \quad \text{(Eq A-75)}$$

$$\frac{\partial(Q)}{\partial\beta} = \left[\frac{(1-DC)\times\lambda^{D}\times TI}{2} + \frac{DC\times\lambda^{D}\times DI}{2} + \lambda^{D}\times MTTR\right] \quad \text{(Eq A-76)}$$

$$\frac{\partial(Q)}{\partial DI} = \frac{DC\times\beta\times\lambda^{D}}{2} \quad \text{(Eq A-77)}$$

We can now determine the sensitivity of the functions F8, F9 , F11 to the random variables of interest. Restating equation A-62 for sensitivity of F8 with respect to the random variables.

$$\frac{\partial F_8}{\partial x} = 2H\frac{\partial(H)}{\partial x} + \frac{\partial(Q)}{\partial x} \quad \text{(Eq A-78)}$$

For $F_8$ we find

$$\frac{\partial F_8}{\partial x} = 2\text{H}\frac{\partial(\text{H})}{\partial x} + \frac{\partial(\text{Q})}{\partial x} \quad \text{(Eq A-80)}$$

$$\frac{\partial F_8}{\partial DC} = 2\left[\frac{(1-DC)\times(1-\beta)\times\lambda^D\times TI}{2} + \frac{DC\times(1-\beta)\times\lambda^D\times DI}{2} + (1-\beta)\times\lambda^D\times MTTR\right]\times$$

$$\left[\frac{-(1-\beta)\times\lambda^D\times TI}{2} + \frac{(1-\beta)\times\lambda^D\times DI}{2}\right] +$$

$$\frac{(-1)\times\beta\times\lambda^D\times TI}{2} + \frac{\beta\times\lambda^D\times DI}{2} \quad \text{(Eq A-79)}$$

$$\frac{\partial F_8}{\partial TI} = 2\left[\frac{(1-DC)\times(1-\beta)\times\lambda^D\times TI}{2} + \frac{DC\times(1-\beta)\times\lambda^D\times DI}{2} + (1-\beta)\times\lambda^D\times MTTR\right]\times\left[\frac{(1-DC)\times(1-\beta)\times\lambda^D}{2}\right] +$$

$$\frac{(1-DC)\times\beta\times\lambda^D}{2} \quad \text{(Eq A-81)}$$

$$\frac{\partial F_8}{\partial MTTR} = 2\left[\frac{(1-DC)\times(1-\beta)\times\lambda^D\times TI}{2} + \frac{DC\times(1-\beta)\times\lambda^D\times DI}{2} + (1-\beta)\times\lambda^D\times MTTR\right]\times$$

$$\left[(1-\beta)\times\lambda^D\right] + \beta\times\lambda^D\times MTTR \quad \text{(Eq A-82)}$$

$$\frac{\partial F_8}{\partial \beta} = 2\left[\frac{(1-DC)\times(1-\beta)\times\lambda^D\times TI}{2} + \frac{DC\times(1-\beta)\times\lambda^D\times DI}{2} + (1-\beta)\times\lambda^D\times MTTR\right]\times$$

$$\left[\left[\frac{(1-DC)\times(-1)\times\lambda^D\times TI}{2} + \frac{DC\times(-1)\times\lambda^D\times DI}{2} + (-1)\times\lambda^D\times MTTR\right]\right] +$$

$$\left[\frac{(1-DC)\times\lambda^D\times TI}{2} + \frac{DC\times\lambda^D\times DI}{2} + \lambda^D\times MTTR\right] \qquad \text{(Eq A-83)}$$

$$\frac{\partial F_8}{\partial DI} = 2\left[\frac{(1-DC)\times(1-\beta)\times\lambda^D\times TI}{2} + \frac{DC\times(1-\beta)\times\lambda^D\times DI}{2} + (1-\beta)\times\lambda^D\times MTTR\right]\times\left[\frac{DC\times(1-\beta)\times\lambda^D}{2}\right] +$$

$$\frac{DC\times\beta\times\lambda^D}{2} \qquad \text{(Eq A-84)}$$

$$\left[\frac{\partial F_8}{\partial \lambda^D}\right] = 2\left[\frac{(1-DC)\times(1-\beta)\times\lambda^D\times TI}{2} + \frac{DC\times(1-\beta)\times\lambda^D\times DI}{2} + (1-\beta)\times\lambda^D\times MTTR\right]\times\left[\frac{(1-DC)\times(1-\beta)\times TI}{2} + \frac{DC\times(1-\beta)\times DI}{2} + (1-\beta)\times MTTR\right] +$$

$$\left[\frac{(1-DC)\times\beta\times TI}{2} + \frac{DC\times\beta\times DI}{2} + \beta\times MTTR\right] \qquad \text{(Eq A-85)}$$

Using the sensitivities for F8 calculated above, the variance in F8 due to uncertain or random variables is found as:

$$V(F_8) = \left[\frac{\partial F_8}{\partial DC}\right]^2 V(DC) + \left[\frac{\partial F_8}{\partial TI}\right]^2 V(TI) + \left[\frac{\partial F_8}{\partial \lambda^D}\right]^2 V(\lambda^D) +$$

$$\left[\frac{\partial F_8}{\partial MTTR}\right]^2 V(MTTR) + \left[\frac{\partial F_8}{\partial \beta}\right]^2 V(\beta) + \left[\frac{\partial F_8}{\partial DI}\right]^2 V(DI) \qquad \text{(Eq A-86)}$$

Using the equation A-63, we now determine the variance of $F_9$.

$$\frac{\partial F_9}{\partial x} = 3H^2 \frac{\partial(H)}{\partial x} + \frac{\partial(Q)}{\partial x} \qquad \text{(Eq A-87)}$$

$$\frac{\partial F_9}{\partial DC} = 3\left[\frac{(1-DC)\times(1-\beta)\times\lambda^D\times TI}{2} + \frac{DC\times(1-\beta)\times\lambda^D\times DI}{2} + (1-\beta)\times\lambda^D\times MTTR\right]^2 \times \left[\frac{-(1-\beta)\times\lambda^D\times TI}{2} + \frac{(1-\beta)\times\lambda^D\times DI}{2}\right] + \frac{(-1)\times\beta\times\lambda^D\times TI}{2} + \frac{\beta\times\lambda^D\times DI}{2} \qquad \text{(Eq A-88)}$$

$$\frac{\partial F_9}{\partial TI} = 3\left[\frac{(1-DC)\times(1-\beta)\times\lambda^D\times TI}{2} + \frac{DC\times(1-\beta)\times\lambda^D\times DI}{2} + (1-\beta)\times\lambda^D\times MTTR\right]^2 \times \left[\frac{(1-DC)\times(1-\beta)\times\lambda^D}{2}\right] + \frac{(1-DC)\times\beta\times\lambda^D}{2} \quad \text{(Eq A-89)}$$

$$\frac{\partial F_9}{\partial MTTR} = 3\left[\frac{(1-DC)\times(1-\beta)\times\lambda^D\times TI}{2} + \frac{DC\times(1-\beta)\times\lambda^D\times DI}{2} + (1-\beta)\times\lambda^D\times MTTR\right]^2 \times \left[(1-\beta)\times\lambda^D\right] + \beta\times\lambda^D\times MTTR \quad \text{(Eq A-90)}$$

$$\frac{\partial F_9}{\partial \beta} = 3\left[\frac{(1-DC)\times(1-\beta)\times\lambda^D\times TI}{2} + \frac{DC\times(1-\beta)\times\lambda^D\times DI}{2} + (1-\beta)\times\lambda^D\times MTTR\right]^2 \times \left[\left[\frac{(1-DC)\times(-1)\times\lambda^D\times TI}{2} + \frac{DC\times(-1)\times\lambda^D\times DI}{2} + (-1)\times\lambda^D\times MTTR\right]\right] + \left[\frac{(1-DC)\times\lambda^D\times TI}{2} + \frac{DC\times\lambda^D\times DI}{2} + \lambda^D\times MTTR\right] \quad \text{(Eq A-91)}$$

$$\frac{\partial F_9}{\partial DI} = 3\left[\frac{(1-DC)\times(1-\beta)\times\lambda^D\times TI}{2} + \frac{DC\times(1-\beta)\times\lambda^D\times DI}{2} + (1-\beta)\times\lambda^D\times MTTR\right]^2 \times \left[\frac{DC\times(1-\beta)\times\lambda^D}{2}\right] + \frac{DC\times\beta\times\lambda^D}{2} \quad \text{(Eq A-92)}$$

$$\left[\frac{\partial F_9}{\partial\lambda^D}\right] = 3 \times \left[\frac{(1-DC)\times(1-\beta)\times\lambda^D\times TI}{2} + \frac{DC\times(1-\beta)\times\lambda^D\times DI}{2} + (1-\beta)\times\lambda^D\times MTTR\right]^2 \times \left[\frac{(1-DC)\times(1-\beta)\times TI}{2} + \frac{DC\times(1-\beta)\times DI}{2} + (1-\beta)\times MTTR\right] + \left[\frac{(1-DC)\times\beta\times TI}{2} + \frac{DC\times\beta\times DI}{2} + \beta\times MTTR\right] \quad \text{(Eq A-93)}$$

Using the sensitivities for $F_9$ calculated above, the variance in $F_9$ due to uncertain or random variables is found as:

$$V(F_9) = \left[\frac{\partial F_9}{\partial DC}\right]^2 V(DC) + \left[\frac{\partial F_9}{\partial TI}\right]^2 V(TI) + \left[\frac{\partial F_9}{\partial\lambda^D}\right]^2 V(\lambda^D) + \left[\frac{\partial F_9}{\partial MTTR}\right]^2 V(MTTR) + \left[\frac{\partial F_9}{\partial\beta}\right]^2 V(\beta) + \left[\frac{\partial F_9}{\partial DI}\right]^2 V(DI) \quad \text{(Eq A-94)}$$

Finally, in a similar manner the variance in $F_{11}$ is computed as:

$$\frac{\partial F_{11}}{\partial x} = 6\text{H}\frac{\partial(\text{H})}{\partial x} + \frac{\partial(\text{Q})}{\partial x} \qquad \text{(Eq A-95)}$$

$$\frac{\partial F_{11}}{\partial DC} = 6\left[\frac{(1-DC)\times(1-\beta)\times\lambda^D\times TI}{2} + \frac{DC\times(1-\beta)\times\lambda^D\times DI}{2} + (1-\beta)\times\lambda^D\times MTTR\right]\times$$
$$\left[\frac{-(1-\beta)\times\lambda^D\times TI}{2} + \frac{(1-\beta)\times\lambda^D\times DI}{2}\right] +$$
$$\frac{(-1)\times\beta\times\lambda^D\times TI}{2} + \frac{\beta\times\lambda^D\times DI}{2} \qquad \text{(Eq A-96)}$$

$$\frac{\partial F_{11}}{\partial TI} = 6\left[\frac{(1-DC)\times(1-\beta)\times\lambda^D\times TI}{2} + \frac{DC\times(1-\beta)\times\lambda^D\times DI}{2} + (1-\beta)\times\lambda^D\times MTTR\right]\times\left[\frac{(1-DC)\times(1-\beta)\times\lambda^D}{2}\right] +$$
$$\frac{(1-DC)\times\beta\times\lambda^D}{2} \qquad \text{(Eq A-97)}$$

$$\frac{\partial F_{11}}{\partial MTTR} = 6\left[\frac{(1-DC)\times(1-\beta)\times\lambda^D\times TI}{2} + \frac{DC\times(1-\beta)\times\lambda^D\times DI}{2} + (1-\beta)\times\lambda^D\times MTTR\right]\times\left[(1-\beta)\times\lambda^D\right] +$$
$$\beta\times\lambda^D\times MTTR \qquad \text{(Eq A-98)}$$

$$\frac{\partial F_{11}}{\partial \beta}=6\left[\frac{(1-DC)\times(1-\beta)\times\lambda^{D}\times TI}{2}+\frac{DC\times(1-\beta)\times\lambda^{D}\times DI}{2}+(1-\beta)\times\lambda^{D}\times MTTR\right]\times$$
$$\left[\left[\frac{(1-DC)\times(-1)\times\lambda^{D}\times TI}{2}+\frac{DC\times(-1)\times\lambda^{D}\times DI}{2}+(-1)\times\lambda^{D}\times MTTR\right]\right]+$$
$$\left[\frac{(1-DC)\times\lambda^{D}\times TI}{2}+\frac{DC\times\lambda^{D}\times DI}{2}+\lambda^{D}\times MTTR\right] \qquad \text{(Eq A-99)}$$

$$\frac{\partial F_{11}}{\partial DI}=6\left[\frac{(1-DC)\times(1-\beta)\times\lambda^{D}\times TI}{2}+\frac{DC\times(1-\beta)\times\lambda^{D}\times DI}{2}+(1-\beta)\times\lambda^{D}\times MTTR\right]\times\left[\frac{DC\times(1-\beta)\times\lambda^{D}}{2}\right]+$$
$$\frac{DC\times\beta\times\lambda^{D}}{2} \qquad \text{(Eq A-100)}$$

$$\left[\frac{\partial F_{11}}{\partial \lambda^{D}}\right]=6\left[\frac{(1-DC)\times(1-\beta)\times\lambda^{D}\times TI}{2}+\frac{DC\times(1-\beta)\times\lambda^{D}\times DI}{2}+(1-\beta)\times\lambda^{D}\times MTTR\right]\times$$
$$\left[\frac{(1-DC)\times(1-\beta)\times TI}{2}+\frac{DC\times(1-\beta)\times DI}{2}+(1-\beta)\times MTTR\right]+$$
$$\left[\frac{(1-DC)\times\beta\times TI}{2}+\frac{DC\times\beta\times DI}{2}+\beta\times MTTR\right] \qquad \text{(Eq A-101)}$$

Using the sensitivities for $F_{11}$ calculated above, the variance in $F_{11}$ due to uncertain or random variables is found as:

$$V(F_{11}) = \left[\frac{\partial F_{11}}{\partial DC}\right]^2 V(DC) + \left[\frac{\partial F_{11}}{\partial TI}\right]^2 V(TI) + \left[\frac{\partial F_{11}}{\partial \lambda^D}\right]^2 V(\lambda^D) +$$

$$\left[\frac{\partial F_{11}}{\partial MTTR}\right]^2 V(MTTR) + \left[\frac{\partial F_{11}}{\partial \beta}\right]^2 V(\beta) + \left[\frac{\partial F_{11}}{\partial DI}\right]^2 V(DI) \qquad \text{(Eq A-102)}$$

**PROCEDURE FOR DETERMINATION OF VARIANCE FOR REPAIRABLE 1oo2, 1oo3 AND 2oo3 SYSTEMS**

Based on the above derivations we may now write the formulas for the variance of various configurations of repairable equipment.

Step 1. Determine the configuration of equipment to be studied.

Step 2. Evaluate Function H using the expected value of all random or uncertain variables, equation A-51

Step 3. Evaluate Function Q using the expected value of all random or uncertain variables, equation A-52

Step 4. Evaluate the sensitivity of function H with respect to the random variables using equations A-65 thru A-70.

Step 5. Evaluate the sensitivity of function Q with respect to the random variables using equations A-72 thru A-77.

Step 6. Evaluate the variance as:

a. For 1oo2 systems use equation A-86 for $V(F_8)$
b. For 1oo3 systems use equation A-94 for $V(F_9)$
c. For 2oo3 systems use equation A-102 for $V(F_{11})$

The end result of the above calculations is the determination of the mean and variance of the configuration used in the sensor, logic solver or final control element systems. Table 4 of the main body of this paper presents a cross reference roadmap for the determination of the mean and variance of various of the probability of failure on demand (PFD) of various hardware configurations. The mean and variance for the particular configuration is then returned to the overall calculations of the PFD of the interlock.

# APPENDIX B -- UNCERTAINTY EQUATIONS FOR CONFIGURATIONS NOT COVERED IN ISA TR 84.00.02

The ISA Technical report [Ref. 2] does not define the calculation equations for the PFDavg for non-repairable systems where duplicated devices create the potential for common cause failures. This appendix derives the equations needed to evaluate the uncertainty for non-repairable systems with the potential for common cause failures.

To derive the needed functions for the non-repairable systems with common cause failure (CCF) potential, we start with the equations for the repairable system with common cause failure allowed (Table A.4). First we note that a system composed of only one set of components (1oo1) without duplication has no potential for a common cause failure. In addition, a review of the equations of Table 2 finds that systems 1oo1, 2oo2 and 3oo3 do not contain the common cause failure (CCF) parameter $\beta$. Therefore, systems 1oo1, 2oo2, and 3oo3 can be eliminated from consideration as common cause failures cannot lead to failure of the specified interlocks.

For the other situations [1oo2, 1oo3 and 2oo3] we note that no repair will be allowed during the mission of the non-repairable system. This observation implies that the diagnostic coverage (DC), the diagnostic interval (DI) and the mean time for restoration (MTTR) are all zero (0). We will set DC = DI = MTTR = 0

Substitution of the zero value for DC, DI and MTTR into the equations of Table A.4 yields the non-repairable system PFD equations where CCF is allowed. Table B.1 presents the results of these algebraic substitutions.

**Table B.1 Simplified $PFD_{avg}$ formulas for Non-Repairable System Allowing for CCF**

| **Configuration** | **Function** | **PFD based on "Average before" failure rate** | **Equation Number** |
|---|---|---|---|
| 1oo2 | $F_{13}$ = | $\left[\frac{(1-\beta)\times\lambda^D\times TI}{2}\right]^2+\left[\frac{\beta\times\lambda^D\times TI}{2}\right]$ | B-1 |
| 1oo3 | $F_{14}$ = | $\left[\frac{(1-\beta)\times\lambda^D\times TI}{2}\right]^3+\left[\frac{\beta\times\lambda^D\times TI}{2}\right]$ | B-2 |
| 2oo3 | $F_{15}$ = | $3\times\left[\frac{(1-\beta)\times\lambda^D\times TI}{2}\right]^2+\left[\frac{\beta\times\lambda^D\times TI}{2}\right]$ | B-3 |

Where TI is the proof test interval, $\lambda^D$ is the dangerous failure rate.

The expected value of the probability of failure on demand is found by calculating the PFD using the expected value (mean) of the random variables.

Let y a function of some random variables, $x_i$, as:

$$y = F(x_i) \qquad \text{(Eq B-4)}$$

The mean of y can be estimated using the following approximation:

$$E(y) = F[E(x_i)] \qquad \text{(Eq B-5)}$$

Where:

$E(y)$ = expected value of random variable y = mean of y

$E(x_i)$ = expected value of random variable $x_i$ = mean of $x_i$

For the configurations defined in Table A.1, the corresponding expected PFD are presented in Table B.2.

| Table B.2 Expected Value of $PFD_{avg}$ formulas for Non-Repairable System Allowing CCF | | | |
|---|---|---|---|
| **Configuration** | **Function** | **Expected PFD based on "Average before" failure rate** | **Equation Number** |
| 1oo2 | $E(F_{13}) =$ | $\left[\frac{(1-E(\beta))\times E(\lambda^D)\times E(TI)}{2}+\right]^2 + \left[\frac{E(\beta)\times E(\lambda^D)\times E(TI)}{2}\right]$ | B-6 |
| 1oo3 | $E(F_{14}) =$ | $\left[\frac{(1-E(\beta))\times E(\lambda^D)\times E(TI)}{2}+\right]^3 + \left[\frac{E(\beta)\times E(\lambda^D)\times E(TI)}{2}\right]$ | B-7 |
| 2oo3 | $E(F_{15}) =$ | $3\times\left[\frac{(1-E(\beta))\times E(\lambda^D)\times E(TI)}{2}+\right]^2 + \left[\frac{E(\beta)\times E(\lambda^D)\times E(TI)}{2}\right]$ | B-8 |

Where E(x) is the expected value of random variable x.

The variance of the PFD is found by taking the weighted sum of the variance of the random variables. The weighting function is the sensitivity of the PFD with respect to random variable. In mathematical terms the variance of a function may be found in terms of the variance of the random variables as:

Once again let

$$y = F(x_i) \quad \text{(Eq B-9)}$$

The variance of Y can likewise be estimated as:

$$V(y) = \sum_{i=1}^{n} \left[\frac{\partial y}{\partial x_i}\right]^2 V(x_i) \qquad \text{(Eq B-10)}$$

Where:

V(y) = variance of random variable y as defined above

$V(x_i)$ = variance of random variable $x_i$ as defined above

The sensitivity of y with respect to a random variable $x_i$ is:

$$\text{Sensitivity of y with respect to } x_i = \left[\frac{\partial y}{\partial x_i}\right] \qquad \text{(Eq B-11)}$$

The variance of the PFD for the repairable cases 1oo2, 1oo3 and 2oo3 are presented in equations of Appendix A as:

- 1oo2, the variance is specified by equation A-86
- 1oo3, the variance is specified by equation A-94
- 2oo3, the variance is specified by equation A-102

These three equations are the basis upon which the equations for the non-repairable system with common cause failure (CCF) are developed.

## Variance of PFD for Non-Repairable 1oo2 with CCF

To derive the variance for $F_{13}$ we start with the general equation for the variance as:

$$V(F_{13}) = \left[\frac{\partial F_{13}}{\partial DC}\right]^2 V(DC) + \left[\frac{\partial F_{13}}{\partial TI}\right]^2 V(TI) + \left[\frac{\partial F_{13}}{\partial \lambda^D}\right]^2 V(\lambda^D) +$$

$$\left[\frac{\partial F_{13}}{\partial MTTR}\right]^2 V(MTTR) + \left[\frac{\partial F_{13}}{\partial \beta}\right]^2 V(\beta) + \left[\frac{\partial F_{13}}{\partial DI}\right]^2 V(DI) \qquad \text{(Eq B-12)}$$

The variance of a constant is zero. Thus, setting DC=DI=MTTR =0 results in the variance of these variables as becoming zero.

$$V(DC) = V(DI) = V(MTTR) = 0 \qquad \text{(Eq B-13)}$$

Thus, the variance for the 1oo2 case becomes:

$$V(F_{13}) = \left[\frac{\partial F_{13}}{\partial TI}\right]^2 V(TI) + \left[\frac{\partial F_{13}}{\partial \lambda^D}\right]^2 V(\lambda^D) + \left[\frac{\partial F_{13}}{\partial \beta}\right]^2 V(\beta) \qquad \text{(Eq B-14)}$$

The evaluation of the sensitive's of the PFD may now be taken from appendix A equations as:

$\left[\frac{\partial F_{13}}{\partial TI}\right]$ from equation A-81

$\left[\frac{\partial F_{13}}{\partial \lambda^D}\right]$ from equation A-85

$\left[\frac{\partial F_{13}}{\partial \beta}\right]$ from equation A-83

Substituting DC=DI=MTTR=0 into these sensitivity equations yields:

$$\frac{\partial F_{13}}{\partial TI} = 2\left[\frac{(1-\beta)\times\lambda^D\times TI}{2} + \right]\times\left[\frac{(1-\beta)\times\lambda^D}{2}\right] + \frac{\beta\times\lambda^D}{2} \qquad \text{(Eq B-15)}$$

$$\frac{\partial F_{13}}{\partial \lambda^D} = 2\left[\frac{(1-\beta)\times\lambda^D\times TI}{2} + \right]\times\left[\frac{(1-\beta)\times TI}{2}\right] + \left[\frac{\beta\times TI}{2}\right] \qquad \text{(Eq B-16)}$$

$$\frac{\partial F_{13}}{\partial \beta} = 2\left[\frac{(1-\beta)\times\lambda^D\times TI}{2}\right]\times\left[\frac{(-1)\times\lambda^D\times TI}{2}\right] + \left[\frac{\lambda^D\times TI}{2}\right] \qquad \text{(Eq B-17)}$$

**Variance of PFD for Non-Repairable 1oo3 with CCF**

To derive the variance for $F_{14}$ we start with the general equation for the variance as:

$$V(F_{14}) = \left[\frac{\partial F_{14}}{\partial DC}\right]^2 V(DC) + \left[\frac{\partial F_{14}}{\partial TI}\right]^2 V(TI) + \left[\frac{\partial F_{14}}{\partial \lambda^D}\right]^2 V(\lambda^D) +$$

$$\left[\frac{\partial F_{14}}{\partial MTTR}\right]^2 \text{V(MTTR)} + \left[\frac{\partial F_{14}}{\partial \beta}\right]^2 \text{V}(\beta) + \left[\frac{\partial F_{14}}{\partial DI}\right]^2 \text{V(DI)} \quad \text{(Eq B-18)}$$

The variance of a constant is zero. Thus, setting DC=DI=MTTR =0 results in the variance of these variables as becoming zero.

$$\text{V(DC)} = \text{V(DI)} = \text{V(MTTR)} = 0 \quad \text{(Eq B-19)}$$

Thus, the variance for the 1oo3 case becomes:

$$\text{V}(F_{14}) = \left[\frac{\partial F_{14}}{\partial TI}\right]^2 \text{V(TI)} + \left[\frac{\partial F_{14}}{\partial \lambda^D}\right]^2 \text{V}(\lambda^D) + \left[\frac{\partial F_{14}}{\partial \beta}\right]^2 \text{V}(\beta) \quad \text{(Eq B-20)}$$

The evaluation of the sensitive's of the PFD may now be taken from appendix A equations as:

$\left[\frac{\partial F_{14}}{\partial TI}\right]$ from equation A-89

$\left[\frac{\partial F_{14}}{\partial \lambda^D}\right]$ from equation A-93

$\left[\frac{\partial F_{14}}{\partial \beta}\right]$ from equation A-91

Substituting DC=DI=MTTR=0 into these sensitivity equations yields:

$$\frac{\partial F_{14}}{\partial TI} = 3\left[\frac{(1-\beta)\times\lambda^D\times TI}{2}\right]^2 \times \left[\frac{(1-\beta)\times\lambda^D}{2}\right] + \frac{\beta\times\lambda^D}{2} \quad \text{(Eq B-21)}$$

$$\frac{\partial F_{14}}{\partial \lambda^D} = 3\left[\frac{(1-\beta)\times\lambda^D\times TI}{2}\right]^2 \times \left[\frac{(1-\beta)\times TI}{2}\right] + \left[\frac{\beta\times TI}{2}\right] \quad \text{(Eq B-22)}$$

$$\frac{\partial F_{14}}{\partial \beta} = 3\left[\frac{(1-\beta)\times\lambda^D\times TI}{2}\right]^2 \times \left[\frac{(-1)\times\lambda^D\times TI}{2}\right] + \left[\frac{\lambda^D\times TI}{2}\right] \qquad \text{(Eq B-23)}$$

## Variance of PFD for Non-Repairable 2oo3 with CCF

To derive the variance for $F_{15}$ we start with the general equation for the variance as:

$$V(F_{15}) = \left[\frac{\partial F_{15}}{\partial DC}\right]^2 V(DC) + \left[\frac{\partial F_{15}}{\partial TI}\right]^2 V(TI) + \left[\frac{\partial F_{15}}{\partial \lambda^D}\right]^2 V(\lambda^D) +$$

$$\left[\frac{\partial F_{15}}{\partial MTTR}\right]^2 V(MTTR) + \left[\frac{\partial F_{15}}{\partial \beta}\right]^2 V(\beta) + \left[\frac{\partial F_{15}}{\partial DI}\right]^2 V(DI) \qquad \text{(Eq B-24)}$$

The variance of a constant is zero. Thus, setting DC=DI=MTTR =0 results in the variance of these variables as becoming zero.

$$V(DC) = V(DI) = V(MTTR) = 0 \qquad \text{(Eq B-25)}$$

Thus, the variance for the 2oo3 case becomes:

$$V(F_{15}) = \left[\frac{\partial F_{15}}{\partial TI}\right]^2 V(TI) + \left[\frac{\partial F_{15}}{\partial \lambda^D}\right]^2 V(\lambda^D) + \left[\frac{\partial F_{15}}{\partial \beta}\right]^2 V(\beta) \qquad \text{(Eq B-26)}$$

The evaluation of the sensitive's of the PFD may now be taken from appendix A equations as:

$\left[\frac{\partial F_{15}}{\partial TI}\right]$ from equation A-97

$\left[\frac{\partial F_{15}}{\partial \lambda^D}\right]$ from equation A-101

$\left[\frac{\partial F_{15}}{\partial \beta}\right]$ from equation A-99

Substituting DC=DI=MTTR=0 into these sensitivity equations yields:

$$\frac{\partial F_{15}}{\partial TI} = 6\left[\frac{(1-\beta)\times\lambda^D\times TI}{2} + \right] \times \left[\frac{(1-\beta)\times\lambda^D}{2}\right] + \frac{\beta\times\lambda^D}{2} \quad \text{(Eq B-27)}$$

$$\frac{\partial F_{15}}{\partial \lambda^D} = 6\left[\frac{(1-\beta)\times\lambda^D\times TI}{2}\right] \times \left[\frac{(1-\beta)\times TI}{2}\right] + \left[\frac{\beta\times TI}{2}\right] \quad \text{(Eq B-28)}$$

$$\frac{\partial F_{15}}{\partial \beta} = 6\left[\frac{(1-\beta)\times\lambda^D\times TI}{2}\right] \times \left[\frac{(-1)\times\lambda^D\times TI}{2}\right] + \left[\frac{\lambda^D\times TI}{2}\right] \quad \text{(Eq B-29)}$$

The evaluation of the functions for the variance are always done using the expected value of the variable. For example when the variable TI appears in an equation, the numerical evaluation of the function would be using E(TI) where E(TI) is the expected value of the random variable TI.

# Comparison of Batch versus Continuous Process in Pharmaceutical Industry Based on Safety Consideration

**Shiqi Chen**
**Mary Kay O'Connor Process Safety Center**
**Artie McFerrin Department of Chemical Engineering**
**Texas A&M University**
**College Station, Texas 77843-3122, USA Email address**
**shiqi_chen@tamu.edu**

**Yogesh Koirala, M. Sam Mannan**
**Mary Kay O'Connor Process Safety Center**
**Artie McFerrin Department of Chemical Engineering**
**Texas A&M University**
**College Station, Texas 77843-3122, USA Email address**
**ykoirala@tamu.edu, mannan@tamu.edu**

Prepared for Presentation at
American Institute of Chemical Engineers
2017 Spring Meeting and 13th Global Congress on Process Safety
San Antonio, TX
March 26 – 29, 2017

# Comparison of Batch versus Continuous Process in Pharmaceutical Industry Based on Safety Consideration

**Shiqi Chen**
**Mary Kay O'Connor Process Safety Center**
**Artie McFerrin Department of Chemical Engineering**
**Texas A&M University**
**College Station, Texas 77843-3122, USA Email address**
**shiqi_chen@tamu.edu**

**Yogesh Koirala, M. Sam Mannan**
**Mary Kay O'Connor Process Safety Center**
**Artie McFerrin Department of Chemical Engineering**

**Keywords:** Continuous Process, Hazard Identification, Pharmaceutical Manufacturing

## Abstract

As opposed to the petro-chemical and bulk chemical industry, where continuous processes are widely applied, the pharmaceutical industry still primarily relies on traditional batch process due to the complexity of product, multi-step operation and low-volume production. Considering these conditions, the versatile batch process is more appropriate. Nowadays, driven by the contradiction between increasing demand for drugs and inefficient batch production mode, there is a trend in pharmaceutical industry, that is the transformation from traditional batch process to novel continuous process. Related projects and research that is aimed at analyzing this transition are conducted in worldwide, and the scale of these studies ranges from lab-scale reactions to overall arrangement of the factories.

Although continuous pharmaceutical process is thriving, the safety issue in this field is not promoted at the same time. Since the continuous pharmaceutical process is a novel technology, little information can be provided to evaluate its safety level. Moreover, process conditions are usually intensified for continuous process comparing to batch process. It also may bring potential risks and make continuous manufacturing inappropriate for some of pharmaceutical productions.

This research provides a comprehensive comparison for batch versus continuous pharmaceutical process by application of Dow's Fire and Explosion Index. In addition to this conventional safety evaluation, influences from production efficiency and specialties in pharmaceutical production are integrated into the comparison. Production of 2-methyl benzimidazole and peracetic acid via batch and continuous processes are conducted in this research. Via these integrative and systematic studies, safety levels of batch and continuous processes is analyzed and compared. The ways in which process conditions, production efficiency, and other requirements influence safety level for different production modes are illustrated.

# 1 Introduction

Continuous process is widely applied in petro-chemical and bulk chemical industry for its high production rate, automated operation and saving in cost[1]. In contrast, the pharmaceutical industry still primarily relies on traditional batch process. This aspect stems from intrinsic limitations in active pharmaceutical ingredients (APIs) production. Complexity of medical substances means multistep synthesis. An average of 8.1 steps is needed for one API synthesis[2]. Another limitation is production volume. Demand for drugs are not comparable with most of the bulk chemicals. These limitations make batch process dominate pharmaceutical industry for a long time.

However, due to the gradual expansion of the pharmaceutical market, batch process is behind the time because of its low production rate caused by low yield and inefficient operation mode. Driven by this contradiction between increasing demand for drugs and inefficient production mode, many pharmaceutical manufacturers are trying to shift from a traditional batch process to a novel continuous process. Experimental and industrial trials that are aimed at this transitioning have been done and exhibit the practicability of this new production mode. Several companies such as Lonza, Phoenix Chemical, and Corning make significant progress in this area[3-5]. Novel continuous process exhibits great potential for the pharmaceutical industry.

"Which process should be applied?" and "Is there enough benefit to move forward?" are the questions that are raised when an innovative process is introduced. To answer these questions, comprehensive comparison between batch process and continuous process is required.

This comparison should be specific in the early stages. For example, for a given substance, based on its intrinsic properties and demand in market, reasonable comparison is required to make a selection between batch and continuous process. After composing a conclusive summary that is based on number of different specific cases, the result of these comparisons can be modified and applied to generate a practical process screening methodology.

For the questions mentioned above, economic benefits, safety and environmental impacts should all be considered to get a reliable answer. In the batch versus continuous process comparisons, related economic analysis and environmental analysis have been done by many researchers[6-8]. However, there remains gaps in research and understanding of safety considerations in comparison of batch versus continuous process. Therefore, the need arises for related comparison within safety consideration and systematic methodology for safety evaluation.

In the following sessions, the development in continuous pharmaceutical process, the previous studies in process screening, selecting and comparing, and concept of inherent safety assessment are discussed. The statement of research objective, the description of methodology and results of case studies are illuminated as well.

# 2 Literature Survey

To decide whether to head to novel continuous process or not, evaluations of batch and continuous process are in demand for pharmaceutical companies.

In preliminary stage of the evaluation, fundamental information about reaction kinetics, heat transmission and mass transfer is collected and analyzed. The objective of this investigation is looking into the technical feasibility of chosen reaction. After these basic assessments for intrinsic properties of reaction, far-reaching analyze in economic, environmental and safety aspect can be achieved. In this more comprehensive comparison based on overall picture, more factors such as expenditure, wastewater treatment and loss control measure should be covered. Among these factors, which factor should be selected as criterion, how these factors interact, and what's the correlation between these factors and comparison results are questions require to be answered.

To answer these questions, some researchers put efforts in individual comparisons of manufacturing for specific active pharmaceutical ingredient (API). Most of these studies focus on economic advantage of continuous process. Spencer D. Schaber et al.[12] estimated the capital, operating and the overall cost for production of a given API from synthesis to tablet formulation by integrated continuous and batch pharmaceutical manufacturing respectively. Based upon the equal annual yield, the result shows that the capital expenditures for continuous production could be markedly lower than those of batch mode. Differences in operating and overall cost are not such significant but potential saving by material handling and labor in continuous process is explicitly illustrated in the study. Hessel, V. et al. compared continuous vs. batch pharmaceutical manufacturing for production of ibuprofen and artemisinin on the standpoint of cost and environmental impact. Capital expenditure and operating expenditure savings are up to 57.0% and 51.6% respectively for ibuprofen, while capital expenditure and operating expenditure savings are up to 19.6% and 29.3% respectively for artemisinin[6, 8]. These study shows the potential economic benefits in application of continuous pharmaceutical processes.

In addition to process comparison, some researchers developed general criterions or created systematic methodologies for process selection. Dominique M. Roberge et al.[13] classified reactions used in the fine and pharmaceutical industry into 4 categories according to kinetic properties. Based on this classification, the very fast reactions (reactions time less than 1s) and the fast reactions (reactions occurring from 1s to10min) would benefit the most from continuous process. These two types of reactions occupy 44% of all reactions investigated. Ryan L. Harman and Jonathan P. McMullen[14] developed a decision roadmap of flow process for use in lab-scale. Considering objectives of the study, the intrinsic properties of the chemistry and requirements for mixing, heat transferring, related questions are asked at the various nodes in this flow map to get the final choice. Gary S. Calabrese and Sergio Pissavini[15] established an conceptual methodology to evaluate applicability of continuous reactor in the form of reaction properties test. They measured applicability of flow reactor via series of questions. These questions covered issues in reaction properties and some requirements on process. Three "zones" that tagged with different level of applicability are defined. Different questions are included for each "zone". For any given reaction, the more answers with "yes" in the "zone", the higher possibility of this reaction to be at the level of applicability that the "zone" stands for.

## 3 Problem Statement

Although advantages of continuous process in pharmaceutical manufacturing are illustrated by many studies, it is still too early to claim that continuous process is the right direction to go forward. Several gaps in the literature survey are required to be focused on.

Firstly, most of process selection methodologies are confined to analyzing separate properties of reaction taken in process, rather than evaluation in a systematic way. For example, small inventory is usually identified as one of significant advantage of continuous process, however, it is possible that to achieve similar yield in the corresponding batch process, the reaction in continuous process requires more intensified conditions like higher temperature, higher pressure, and higher concentration of hazardous chemicals. Selection and comparison by independent and individual parameters without integration is not enough.

Secondly, existing comparisons between batch and continuous processes mainly focus on economic aspect, while lacking combination of safety issue and other important aspects from real production. Although the driving force for the application of this novel process is the economic interests, safety issues directly determine its feasibility. A choice between batch and continuous process needs to be made based on business performance within acceptable safety levels. Those comparisons that merely consider the profits are not enough.

Thirdly, the existing studies are limited to fixed production system. Potential problems in a different scale are not considered. In the pharmaceutical industry, gradually increasing production scales is required which is different from the bulk chemical industry. Hence, safety of process conditions needs to be guaranteed for all scale production. Comparison only based on fixed scale is not comprehensive.

Therefore, the primary objective of this study is to provide a comprehensive and systematic comparison between batch and continuous process used in pharmaceutical industry. Then on the basis of this analysis, this study will evaluate how main factors considered in Dow F&EI impact the safety level for these two different processes. Finally, this study will provide support to further improve existing methodology for process screening and selection for pharmaceutical industry.

It is worth mentioning that there are other risks such as chemical exposure for pharmaceutical manufacturing, however, this study mainly focuses on potential fire, explosion and reactivity incident. In addition, comparison in this study focuses on intrinsic properties of the reaction and process which are the major hazards of the reaction system. Therefore, other factors like loss control measures and emergency equipment are supposed to be effective for both batch and continuous systems.

# 4 Methodology

To measure the realistic fire, explosion and reactivity potential of batch and continuous process, Dow's Fire and Explosion Index (F&EI) methodology is applied as hazard index in this study, since it is the most widely used hazard index that recognized by the chemical industry[18].

From the first edition of Dow's Fire & Explosion Index Hazard Classification Guide in 1964, this index has been continuously improving though the past 50 years. The last version is the seventh edition that published in 1994. The purposes of this index include forecasting quantitative damage of potential explosion, reactivity incidents and fire; determining hazardous equipment which is possible to give rise to incidents in the system; conveying this F&EI risk potential to related managements and organizations. Since the aim of this study is to identify major hazard that affect selection, application of Dow F&EI will be mainly focus on its function of hazard identification.

In the Dow's Fire & Explosion Index Hazard Classification Guide, economic analysis is confined to financial loss after an incident. Correlation between safety and economic gain from production is not built. Actually, this kind of connection could be established via yield of reactions and geometry and inventory of units, since these factors take important role in both safety assessment and process gain. In the current study, this correlation between safety and profit is constructed on the foundation of Dow's Fire & Explosion Index.

## 5 Case Study I: Benzimidazole Derivative Synthesis

### *5.1 overview*

Benzimidazole derivative is an important type of intermediates in pharmaceutical industry. It is widely exhibited in a variety of pharmaceutical substance, such as proton pump inhibitors, antimicrobials, antivirals, anticancer, anti-inflammatory, hormone modulators as well as depressants, antidiabetics and so on[22]. Famous drug with benzimidazole nucleus includes Omeprazole and Albendazole. Omeprazole is used to treat gastroesophageal reflux disease and peptic ulcer disease[23], and Albendazole is an effective drug in treatment of various parasitic worm infestations[24].

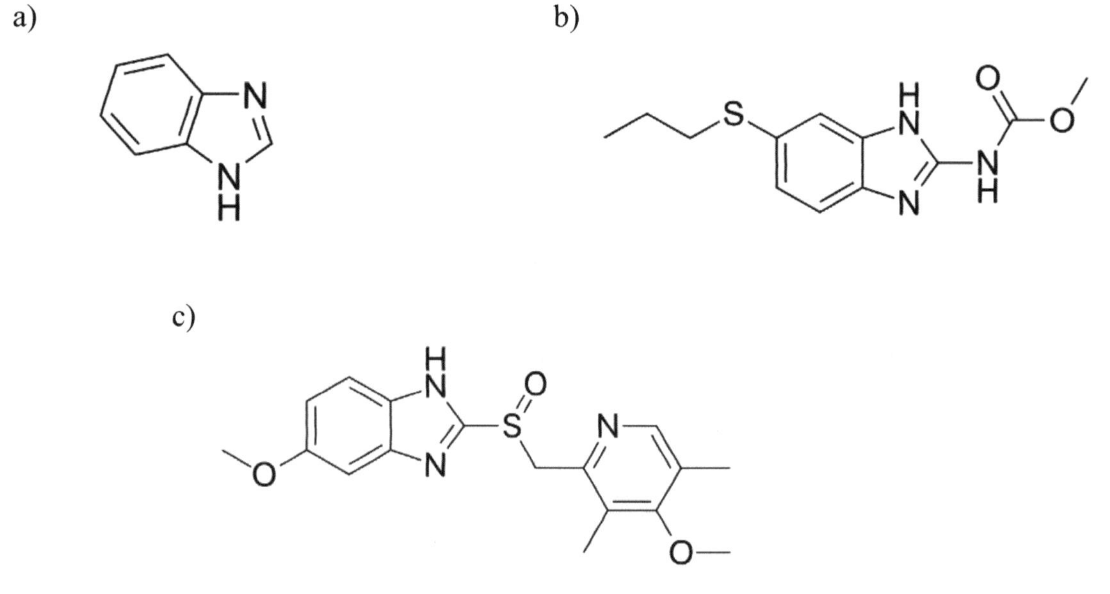


**Figure 1. a) Molecular Structure of Benzimidazole, b) Molecular Structure of Albendazole, c) Molecular Structure of Omeprazole**

2-methyl benzimidazole is a pharmaceutical-potential substance that belongs to this benzomidazole ring system. The condensation of o-phenylenediamines with carboxylic acids is one of the most common synthesis route for 2-methyl benzimidazole[25]. Conditions and production comparisons of o-phenylenediamines condensation by batch vs. continuous processes are showed in Table 1. These fundamental parameters are used for safety evaluation by Dow's F&EI in the following study.

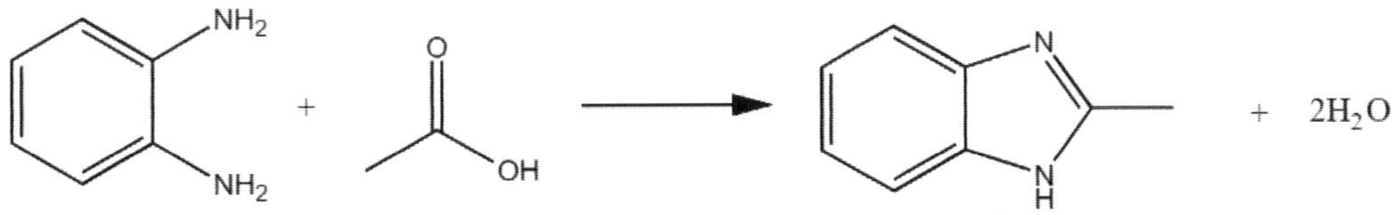

**Figure 2. Synthesis of 2-methyl Benzimidazole by Condensation of o-phenylenediamine**

**Table 1. Production Conditions of Batch and Continuous Processes for Synthesis of 2-methyl benzimidazole**

| | Batch Process | Continuous Process |
|---|---|---|
| Temperature (℃) | 200 | 270 |
| Pressure (psi) | 145 | 1885 |
| Reaction Time (min) | 5 | 0.5 |
| Yield (%) | 98 | 94 |
| Substrate Concentration (M) | 5 | 1 |
| Reaction Volume (ml) | 1000 | 4 |

In order to concentrate on study in intrinsic differences between two processes, and simplify the hazard identification in F&EI, several assumptions are adopted:

1. The reaction mass is perfect mixed in reactors;
2. Adjunct cooling systems in both processes are in perfect performances;
3. Emergency equipment is same and in perfect performances in both processes;
4. Scalabilities of both processes are good enough that reaction conditions remain same in all scales.

### *5.2. Safety Comparison Result*

The comparison is conducted based on material properties and process features. Related explanation is stated below.

5.2.1. comparison on material

In this study, impacts from intrinsic properties of reactants are same for both processes. Material factors of all substances are listed in Table 2. According to Dow's Fire and Explosion Index Hazard Classification Guide, impact from the most hazardous material dominates the material factor of the whole process, therefore only impact from acetic acid is considered for both processes. In addition, although temperatures and composition are different in two processes and may lead to different chemical properties, these features are neglected in this part and contained in process comparison.

**Table 2. Material Factor for Batch and Continuous Reactor**

| | Batch Reactor | Continuous Reactor |
|---|---|---|
| o-phenylenediamine | 10 | 10 |
| 2-methylbenzimidazole | 1 | 1 |
| Acetic Acid | 14 | 14 |

5.2.2. comparison on process

First of all, it's worth mentioning that albeit lots of distinctions exist between batch and continuous processes, comparisons in this study are confined to synthesis in reactor on account of its core role in the whole process: operation mode of reactor – continuously or intermittently – determines whether the process is continuous or not from the start. Therefore, the reactor is selected as the pertinent process unit without conventional unit screening procedure.

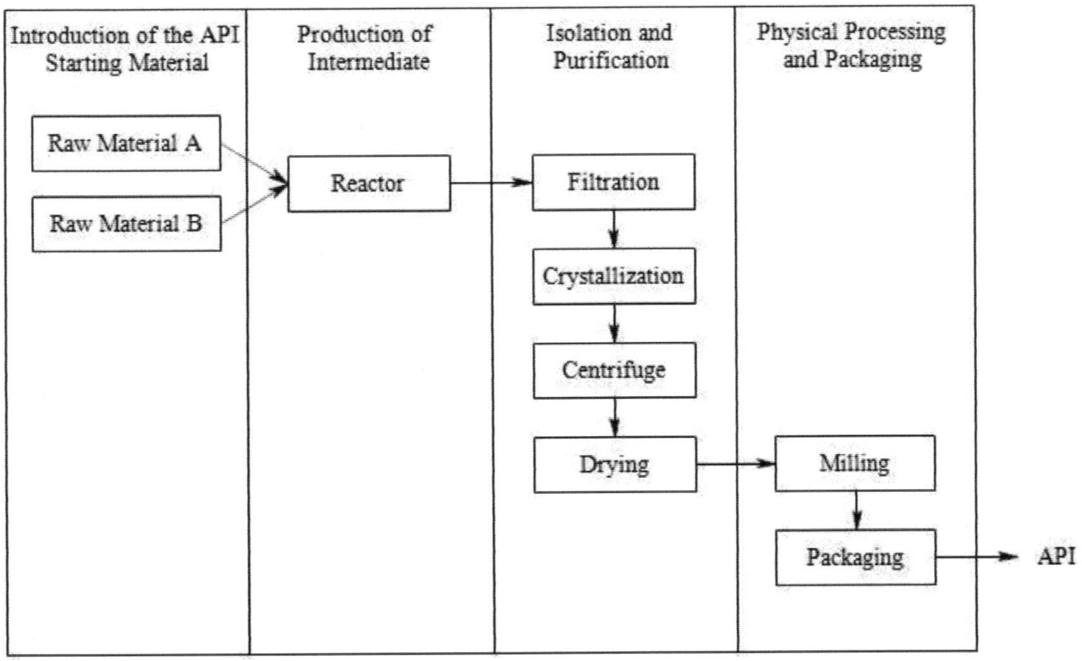


**Figure 3. Procedure for Pharmaceutical Production**

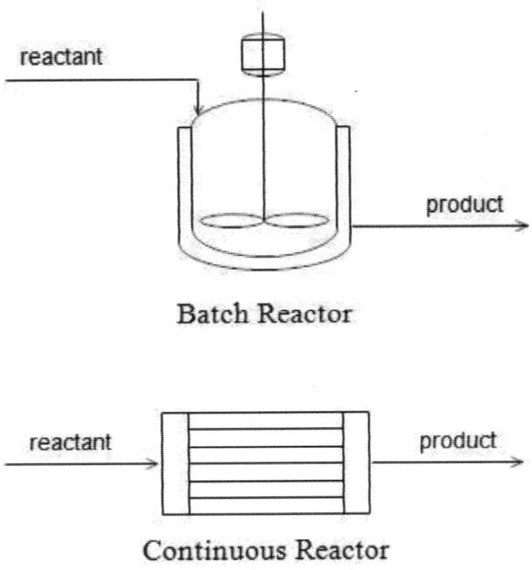


**Figure 4. Differences in Structure between Batch and Continuous Reactors**

On the foundation of safety evaluation by Dow's Fire & Explosion Index, quantitative judges between batch and continuous process are illustrated through comparison in sensitivity and productivity within safety consideration.

This sensitivity analysis focus on correlation between safety level and inventory of reactor. Inventory of unit plays an important role in the safety. In addition, this sensitivity analysis can also provide supportive information for scaling up procedure.

Inventory of unit plays an important part in the safety via quantity of flammable or unstable material. This potential is considered in both the General Process Hazards (F1) and the Special Process Hazards (F2) according to Dow's F&EI Hazard Classification Guide. In general, more the flammable or unstable material, higher the risk of fire and explosion within the process. For different unit (process unit or storage unit) and different phase (liquid, gas or solid), different correlation formula between inventory and penalty is given in the Special Process Hazards (F2) part of the Dow's F&EI Guide. Moreover, other issues such as whether these materials are processed indoor or outdoor, what's the temperature to handle these materials influences the safety level as well. These questions are included in the General Process Hazards (F1) part.

In this sensitivity analysis, F&EI that combines both the general and the special process hazards is expressed as a dependent variable that changes with the variation of total inventory of reactor. Other operation conditions remain the same for different inventory. Correlations between the F&EI and material inventory for batch and continuous process are shown in the Figure 5.

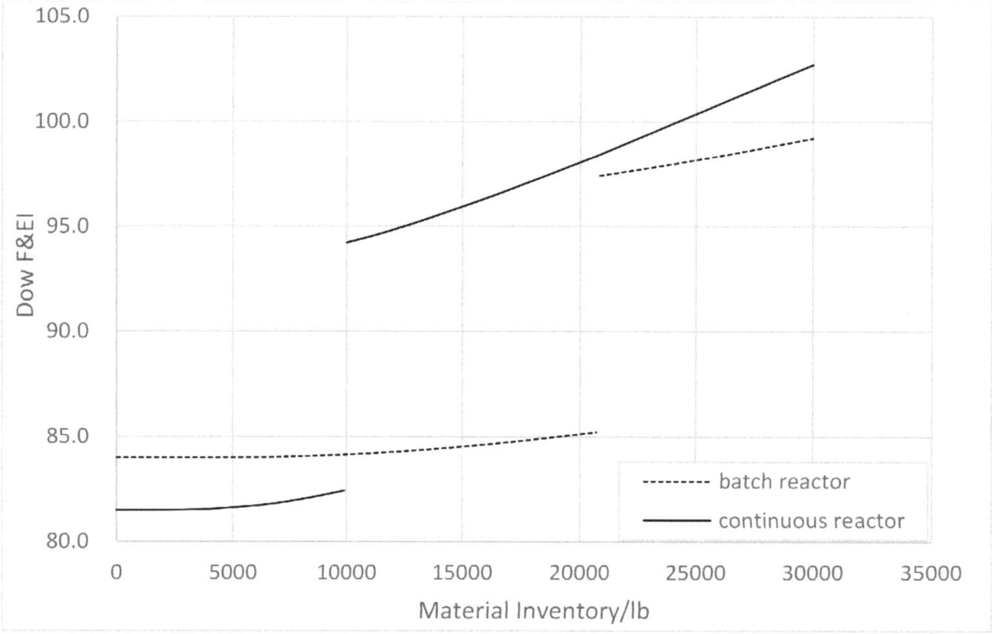


**Figure 5. Analysis of Sensitivity to Inventory for Batch and Continuous Reactors Based on 2-methyl Benzimidazole Synthesis**

for 0 lb < Inventory < 10000 lb:

F&EI of batch reactor is slightly greater than continuous reactor. This difference mainly comes from difference in usage of stirrer and reaction pressure. Due to requirement for homogeneous mixture, stirrer is necessary in batch reactor, while such a stirrer is not needed (or not allowed) in a continuous tubular reactor. Thus, stirrer, as a hazardous rotating equipment, adds extra penalty to batch process. As for difference in pressure, higher penalty is given to continuous process for its intensified condition. From the result, influence from rotating equipment is greater than that from pressure in this case.

Another observation is that, the magnitude of the difference in the F&EI between two reactors decreases with the growing inventory. This trend stems from the increasing impact from quantity of hazardous material. When inventory approaches zero, difference in F&EI is mainly caused by stirrer and distinct pressure aforementioned. As inventory increases, influence from unstable material extends. This influence depends on composition of mixture, or more specifically, concentration of acetic acid that defines the Material Factor (MF). Based on information of reactions, molar concentrations of acetic acid in batch and continuous reactor are 8.31M and 15.66M respectively. Thus, for similar total inventory, penalty from hazardous material in continuous reactor is greater than batch reactor.

when 10000 lb < Inventory < 20833 lb:

F&EI of continuous reactor is signally higher than batch reactor, and this difference grows along with the increasing inventory. In this range, part of influences come from rotating equipment, pressure and increasing inventory that mentioned above. More importantly, quantity of flammable and unstable liquids handled above boiling point exceeds threshold of 10,000lb for continuous reactor, therefore give a vertical rise to F&EI for continuous reactor, as shown in the Figure 5. This threshold is illustrated clearly in Dow's F&EI Hazard Classification Guide: when quantity of flammable and unstable liquids above its boiling point is less than 10,000 lb, a penalty of 0.60 is determined. A penalty of 0.90 is given when this liquids quantity above its boiling point is greater than 10,000 lb. In this case study, operation temperature of continuous reactor (270℃) is higher than boiling point of both o-phenylenediamine (257℃) and acetic acid (118℃), while operation temperature of batch reactor (200℃) merely higher than boiling point of acetic acid (118℃). Therefore, for inventory in the range of 10,000lb to 20833lb, total mass in continuous reactor is defined as unsafe, while merely acetic acid mass is defined as unsafe in the batch reactor. This divergence leads to different penalty in this range of inventory.

when 20833 lb < Inventory < 30000 lb:

F&EI of batch reactor drastically increases. However, F&EI of continuous reactor is still slightly greater than batch reactor, and the magnitude of this difference increases with the growing inventory.

This vertical rise in the F&EI of batch reactor is also caused by critical quantity of material above the boiling point. When total quantity in batch reactor goes beyond 20,833lb, quantity of flammable and unstable substance - acetic acid - also goes beyond threshold of 10,000 lb. Thus, penalty from material above boiling point is similar for both of processes.

The higher F&EI of continuous reactor originated from quantity of flammable and unstable material in this range of inventory. Since higher concentration of acetic acid is applied in continuous reactor, more proportion of substance is identified as unsafe for continuous reactor, thus higher penalty is given. This explanation is similar to that in range of low inventory (< 10,000 lb). However, unlike appearance in the low inventory range, influence from these flammable and unstable materials exceeds impact from pressure and stirrer. Therefore, continuous reactor becomes more dangerous for this range of total inventory.

It's worth mentioning that higher inventory is not considered here. Upper limit of inventory (30,000 lb) is set based on the general highest inventory of reactor used in pharmaceutical manufacturing.

Degree of hazard for batch versus continuous reactor used in this case study is shown in Table 3. In the range of inventory investigated, batch reactor is identified as moderate hazardous for a broader range of inventory, while continuous reactor is identified as intermediate hazardous for a broader range of inventory. Even so, It's worth mentioning that the F&EI values for both reactors are in a relative safe range (F&EI > 128 indicates heavy even severe hazard).

**Table 3. Degree of Hazard for Batch and Continuous Reactors**

| Material Inventory X | Batch Reactor | Continuous Reactor |
|---|---|---|
| X < 15,152 lb | Moderate | Moderate |
| 15,152 lb < X < 20,833 lb | Moderate | Intermediate |
| 20,833 lb < X < 30,000 lb | Intermediate | Intermediate |

In summary, F&EI for both batch and continuous reactors increase as material inventory increases. Moreover, F&EI of continuous reactor is more sensitive to change of inventory comparing to batch reactor. This higher sensitivity caused by comprehensive influence from higher concentration of unstable and flammable material, higher operation temperature and higher pressure it applied.

Evaluation in this part mainly focuses on safety of batch versus continuous process base on productivity considerations. Emphasis on productivity scale of this study primary comes from flexibility required for pharmaceutical production. Pharmaceutical industry is highly supply driven. Requirement on production rate is changed according to demand of market all the time. Therefore, better understanding on correlation between safety and productivity provides practicable support for research in process selection.

To approach this target, safety level is evaluated for the process at the same productivity capacity. In this part, F&EI is regarded as indicator for safety evaluation as above. Annual production is set as indicator for productivity. Related formula is shown below.

General calculations for productivity capacity are:

$$\text{Annual Production} = \text{Product Yield} \times \text{Annual Operation Time}$$

$$\text{Product Yield} = \text{Mass Flow of Reagent} \times \text{Conversion}$$

Mass flow is supposed to be proportional to reactor volume as set in the laboratory trial.

According to definition of fine chemical, annual production of fine chemical is limited to 1,000 tons per year[26]. Therefore, 1,000 tons per year is set as the upper limit in this evaluation.

As for product yield and annual operation time, the former is provided in Table 1, and expression for the latter is formed referring to the normal operation time[27].

Annual operation time for batch reactor is:

$$\text{Annual Operaton Time} = 300\ \text{day/yr} \times 24\ \text{hr/day}$$

Annual operation time for continuous reactor is:

$$\text{Annual Operation Time} = 8000\ \text{hr/yr}$$

Normally, in pharmaceutical industry, multi products are produced in one production line. Therefore, the same product cannot occupy the producing line for the whole year. In this study, 2-methyl benzimidazole is assumed to be produced in 2 months of one year.

In addition, circulating time in batch reactor need to be considered. According to the literature[11], the ratio of reaction time to total operation time for this reaction is shown below. Excepted reaction, other operations include filling, cleaning and cooling and other additional work.

$$\frac{\text{Reaction Time}}{\text{Total Operation Time}} = 0.185$$

Based on these formulas and assumptions, correlation between annual productions and F&EI is shown in the Figure 6.

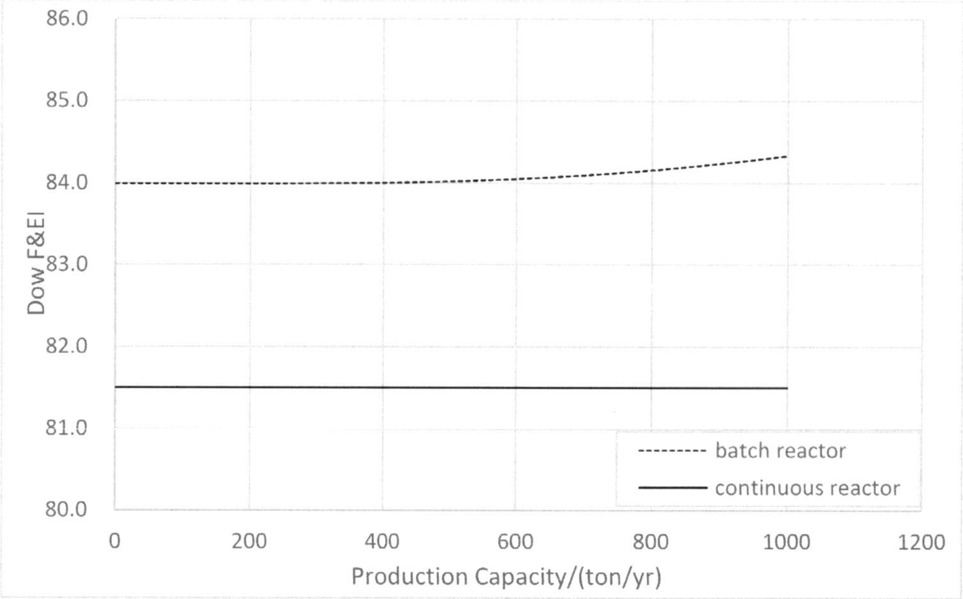


**Figure 6. Analysis of Safety Based on Production for Batch and Continuous Reactors Based on 2-methyl Benzimidazole Synthesis**

According to the Figure 6, F&EI value is constant in the range of practicable annual production (0 ~ 1000 tons per year). Plus, the F&EI for both reactors corresponds to a moderate level of hazard (61 < F&EI < 96).

This low F&EI resulted from relatively low demand for production. According to the sensitivity analysis, F&EI for both batch and continuous reactor increase as material inventory increases. Both processes could be intermediate hazardous once its inventory is higher than a specific value. However, these values are unapproachable in practical pharmaceutical manufacturing. Since demand for fine chemical, always less than 1,000 tons per year, confines scale of production.

Another observation is that, the F&EI of continuous reactor is always lower than that of batch reactor for same production. Plus, slight rise appears for batch reactor when production capacity exceeds 500 tons per year. These appearances predicate consistent superiority in safety for continuous reactor than batch reactor. Explanation is stated below:

Firstly, as stated in the sensitivity analysis, settled elements, such as operation pressure, temperature and use of rotating equipment, take major account for the difference at small-scale production. Advantages of continuous process in these aspects determine its superiority in small-scale pharmaceutical production.

Secondly, continuous reactor possesses higher space-time yield compared to batch reactor. This advantage stems from the intrinsic property of continuous operation mode - the low residence time and high flow throughput. That is to say, it costs fewer time and lower inventory for continuous reactor to produce same amount of product than batch reactor. These benefits in production efficiency influence safety level indirectly but significantly: Smaller inventory means fewer unstable materials in process, and shorter operation time means less chance to fail. This point is not considered in separate safety evaluation, but can be clearly illustrated by correlation between production capacity and F&EI. In this case, when demand on production growing, demand on inventory grows faster for batch reactor than continuous reactor. This faster rise in inventory lead to faster rise in amount of unstable material in process so as to the faster rise in F&EI value. It explains why slight rise appears for batch reactor when production capacity exceeds 500 tons per year.

In fact, the comparable batch and continuous reactor are rarely at the same level of volume. Referring to literature, for commercial production of fine chemical, the volumes of continuous reactor are usually in a range of 0.1 to 3 liter, while the volume of batch reactor is usually around the range of 2 to 10 cubic meters[6]. For this case, based on assumption of same annual production, ratio of batch reactor inventory and continuous reactor inventory is as much as one hundred.

## 6 Case Study II: Oxidation with Hydrogen Peroxide

### *6.1. overview*

Hydrogen peroxide is one kind of widely used oxidation agent in pharmaceutical production. Typical reaction involving hydrogen peroxide includes Baeyer–Villiger oxidations for synthesis of lactone[9], N-oxidation of alkyl pyridine for synthesis of alkyl pyridine N-oxidant[28] and benzylic hydro-peroxide rearrangement for synthesis of Brivanib alaninate[29].

This common oxidation agent is also used in production of peracetic acid. Peracetic acid is produced via oxidation of acetic acid by hydrogen peroxide with sulfuric acid as the catalyst. Conditions for peracetic acid production by batch vs. continuous process are showed in Table 4[30].

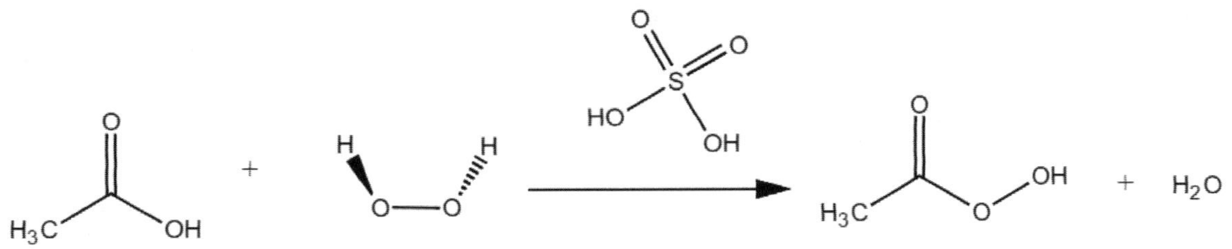


**Figure 7. Synthesis of Peracetic Acid by Oxidation of Acetic Acid**

**Table 4. Production Conditions of Batch and Continuous Process for Synthesis of Peracetic Acid**

| | Batch Process | Continuous Process |
|---|---|---|
| Temperature (℃ ) | 60 | 80 |
| Pressure (psi) | 14.5 | 14.5 |
| Reaction Time (min) | 30 | 15 |
| Yield (kg/h) | 170 | 20 |
| Acetic Acid Concentration (M) | 7.3 | 7.3 |
| Reaction Volume (L) | 4000 | 10 |

Assumption for this case study is listed below:

1. The reaction mass is perfect mixed in reactors.
2. Since the temperature for decomposition of hydrogen peroxide in peracetic acid synthesis reaction is higher than 120 ℃ [31], which is larger than operation temperature for both processes in the case study, decomposition of hydrogen peroxide is assumed to be inexistent.
3. Sulfuric acid is dissociated completely to $H^+$ and $HSO_4^-$, in addition, concentration of $H^+$ is assumed to be constant of 0.67 mol/L during the reactions for both processes.
4. Influence from sulfuric acid is not considered in safety evaluation.
5. Emergency equipment and cooling system are perfectly performed for both of processes.
6. Scalabilities of both processes are good enough that reaction conditions remain same in all scales.

### *6.2. safety comparison result*

The comparison is conducted based on material property and process feature. Related explanations are stated in the following paragraph.

#### 6.2.1. comparison on material

In this case, impact from material within the reactor is same for both of processes. Material factors for reactants involved are listed in Table 5. According to Dow's Fire and Explosion Index Hazard Classification Guide, influence from the most hazardous material dominates the material factor. Although material factor values are same for hydrogen peroxide and peracetic acid, since concentration of hydrogen peroxide is higher than peracetic acid, hydrogen peroxide is chosen to be the dominant material. Therefore, only impact from hydrogen peroxide is considered for both processes. Change on substances' property from temperatures are neglected in this part and reconsidered in process comparison.

**Table 5. Material Factor for Batch and Continuous Reactor**

| | Batch Reactor | Continuous Reactor |
|---|---|---|
| Acetic Acid | 14 | 14 |
| Hydrogen Peroxide | 24 | 24 |
| Peracetic Acid | 24 | 24 |

6.2.2. comparison on process

The major difference between these two production processes come from distinctions in reaction conditions and reactor constructions. These differences are shown in Table 4. Based on safety evaluation by Dow's Fire & Explosion Index, quantitative judges between batch and continuous process are illustrated through comparison in sensitivity and productivity within safety consideration.

Procedure for sensitivity comparison is the same as the first case study. Correlations between the F&EI and material inventory for batch and continuous process are shown in the Figure 8.

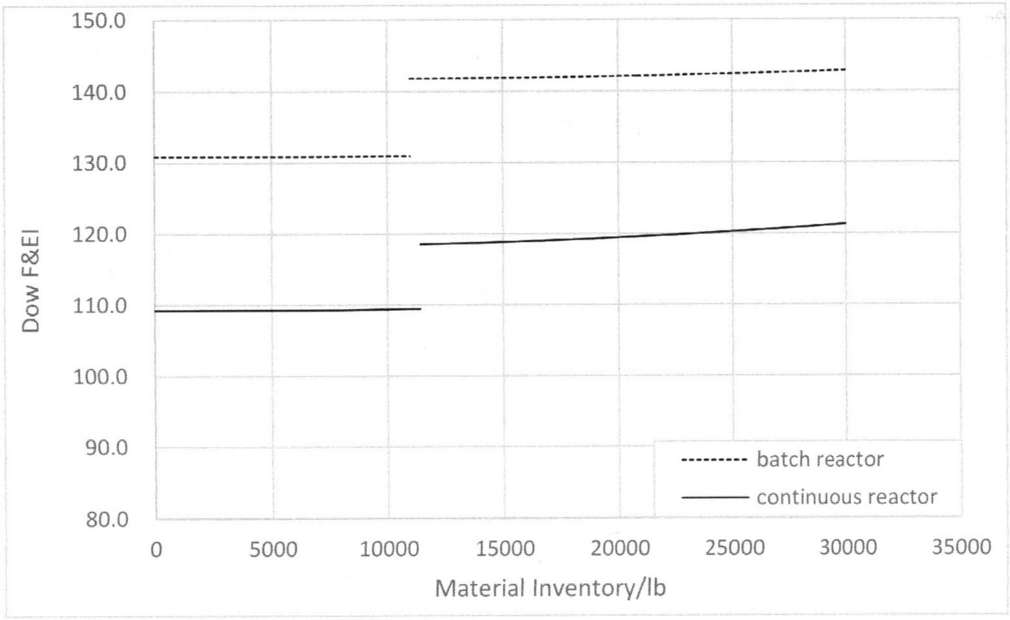


**Figure 8. Analysis of Sensitivity to Inventory for Batch and Continuous Reactors Based on Peracetic Acid Synthesis**

for 0 lb < Inventory < 11000 lb:

F&EI of batch reactor is larger than continuous reactor. This difference is almost caused by usage of stirrer in batch reactor. Application of such a rotating equipment increases penalty for batch process.

when 11000 lb < Inventory < 11430 lb:

F&EI of batch reactor is still higher than continuous reactor, but this difference grows drastically due to the sharp transition in F&EI of batch reactor. This transitions are caused by quantity of flammable liquids that handled above flashing point exceeds threshold of 10,000lb. This threshold is illustrated clearly in Dow's F&EI Hazard Classification Guide: when quantity of flammable liquids above their flashing point is less than 10,000 lb, a penalty of 0.30 is determined. A penalty of 0.45 will be given when this quantity of flammable liquids above their flashing point is larger than 10,000 lb. For peracetic acid production, operation temperature for batch reactor (60℃ ) and continuous reactor (80℃ ) is higher than flashing point of both acetic acid (39℃ ) and peracetic acid(40.5℃ ). Therefore, the value of inventory for higher penalty is determined by compositions of peracetic acid and acetic acid in reactors. Since concentrations of these two materials is slightly higher for batch comparing to continuous reactor, batch reactor approaches this transition in F&EI at a smaller total inventory.

when 11430 lb < Inventory < 30000 lb:

F&EI of continuous reactor exceed the transition caused by quantity of unsafe material above flashing point. However, it is still lower than F&EI of batch reactor.

Compared to the first case study, value of slope for both processes is similar low. It results in the consistent superiority of continuous reactor in safety level compared to batch reactor. This similarity in slopes comes from similarity in operation conditions. Differences only appear in operation temperature and hydrogen peroxide concentration. These differences are not such significant that lead to huge divergence between slopes. Thus, difference in F&EI value between two reactors does not change along with increasing inventory.

Degree of hazard of this case study is exhibited in Table 6. Although marked transitions appears for both reactor, safety levels for two reactors do not change based on scaling up of reactor size. The safety levels are majorly decided by settled parameters, such as operation pressure and usage of stirrer. These parameters do not change with inventory as variable. Hence, inventory does not play a significant role in judgement on degree of hazard.

**Table 6. Degree of Hazard for Batch and Continuous Reactors**

| Reactor Type | Batch Reactor | Continuous Reactor |
|---|---|---|
| Degree of Hazard | Heavy | Intermediate |

In sum, F&EI for both batch and continuous reactor increase as material inventory increases due to the increases of quantity above flashing point. However, since operation conditions are relatively mild and similar for these two reactors, both reactor do not exhibit high sensitivity when inventory is growing gradually.

For evaluation based upon productivity, assumption for practical production are set as following:

1. Confined by the conversion of this reaction route and volume of reactor, upper limit for annual production is settled to be 580 tons per year.
2. Same as the first case study, referring to normal operation time for this two operation mode[27], annual operation time for batch reactor is 8760 hour per day, and annual operation time for continuous reactor is 8000 hour per day.
3. Peracetic acid is assumed to be produced for 2 months in one year.
4. According to the literature[30, 32], the ratio of reaction time to total operation time for this reaction is 0.143

Based upon these formulas and assumptions, correlation between annual productions and F&EI is shown in the Figure 9.

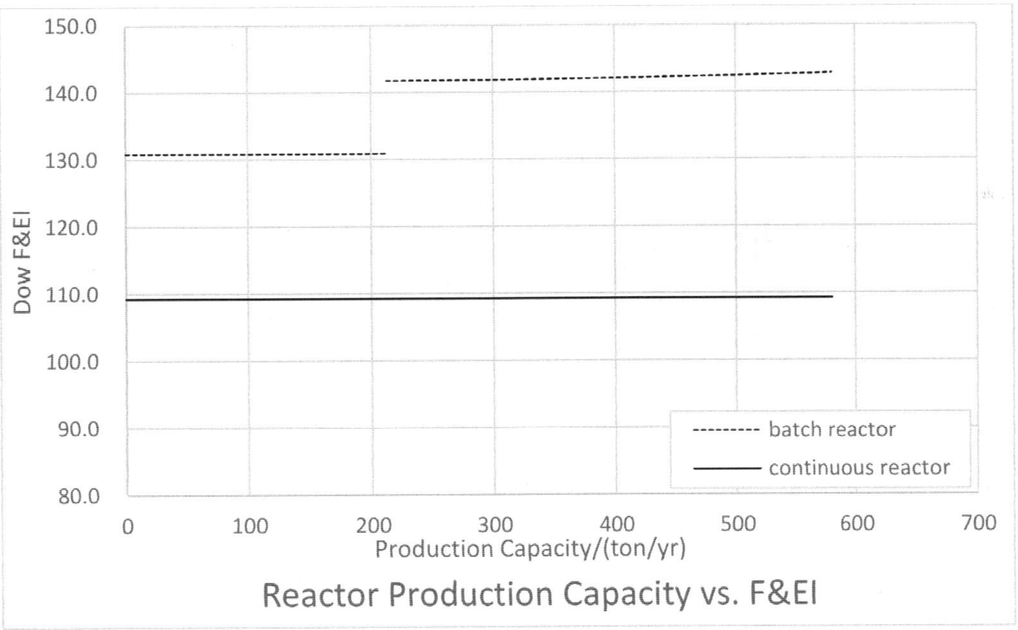


**Figure 9. Analysis of Safety Based on Production for Batch and Continuous Reactors Based on Peracetic Acid Synthesis**

According to the Figure 9, F&EI is stabilized at 109 in the range of practicable annual production for continuous reactor. For batch reactor, the F&EI value rise drastically when the annual production is higher than 220 tons per year.

For the batch production mode, relatively low production efficiency leads to substantial increase in reactor volume when demand on production grows. This increase in reactor volume is so significant that exceed the threshold for large quantity of flammable and unstable material above flashing point. Therefore, the F&EI value jump from 131 to 143 when the annual production is higher than 220 tons per year.

For continuous production mode, high production efficiency causes only a few change in volume hence only few change in F&EI in as well. This difference in F&EI between batch and continuous reactor means that, to produce the same amount of peracetic acid in the given period, the volume of continuous reactor is much smaller than batch reactor.

# 7 Result Summary

Figure 10 represents the comparisons of two case studies. The X-axis represents two different case studies, and the Y-axis represents the average Dow's F&EI value which can be used to define the degree of hazard.

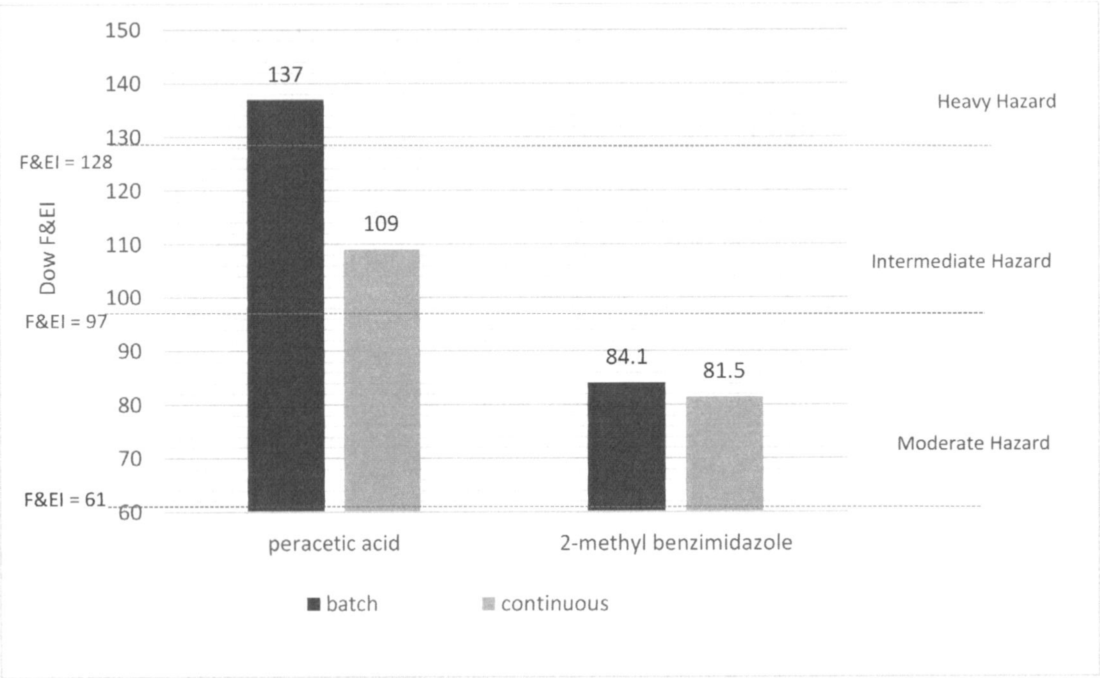


**Figure 10. Comparison of Safety Level by Batch and Continuous Productions for Two Case Studies**

Regardless of the reactor type, the safety level is predetermined by properties of the synthesis reaction. Although distinctions in F&EI appear when using different reactors for both cases, peracetic acid production is generally more hazardous than 2-methyl benzimidazole production, no matter which reactor it uses. This higher level of danger comes from the intrinsic properties, which consist of the flammability and reactivity of the chemicals involved in the synthesis of peracetic acid.

Taking batch and continuous reactors into consideration, distinction in F&EI comes from the differences in operation conditions and production efficiency. These varying factors such as temperature, pressure, and yield do not influence F&EI as individual factors. Instead, they are correlated and impact the F&EI as an ensemble. Difference in these ensemble decides the variance in F&EI value between reactors, then determines to which extent the selection of reactor can make a difference. For this study, the variance in F&EI value between batch and continuous reactor is greater for peracetic acid production (variance in F&EI = 28) than 2-methyl benzimidazole production (variance in F&EI = 2.6). Therefore, selecting a reactor is more important for peracetic acid comparing to 2-methyl benzimidazole production.

The importance of choosing a reactor is different for the two cases, the F&EI value for a continuous reactor is lower than a batch reactor for both of these cases. This consistency exhibits the superiority of a continuous reactor in safety.

# 8 Conclusion

A comprehensive comparison between the batch and continuous process used in the pharmaceutical industry based upon Dow's F&EI is made in this study. Two typical pharmaceutical productions, the production of 2-methyl benzimidazole and the production of peracetic acid are analyzed. Safety levels for batch and continuous reactors are compared based on both inventory and annual production being used as variables. The conclusions on the foundation of these two case studies are stated below:

1. The comparison of batch versus continuous processes in the pharmaceutical industry is complex. It's determined by reaction properties, operation conditions, structures of equipment, production efficiency, scale of production, and many other factors. To get a conclusive comparison between these two production modes, more systematic studies based on the integrated batch and continuous process need to be conducted for safety consideration.

2. Just focusing on the reactor in which part of API synthesis occurs, the continuous reactor is more sensitive to inventory in general. The magnitude of this sensitivity is determined by operation conditions such as operation temperature, operation pressure and concentration of hazardous material. This sensitivity to inventory might lead to problems for continuous production in a relatively large scale.

3. However, when considering practical pharmaceutical production, which is always in a small scale (no larger than 1000 tons per year), hazardous large-volume continuous reactor is not necessary. In addition, the space time yield of continuous production mode is usually higher than the batch production mode, application of a continuous reactor is relatively safer when compared to batch reactors.

# 9 References

1. Plumb, K., Continuous processing in the pharmaceutical industry: changing the mind set. *Chemical Engineering Research and Design* **2005,** *83* (6), 730-738.

2. Laird, T., What Type of Reactions Do Process Chemists Use on Scale? *Organic Process Research & Development* **2006,** *10* (5), 851-852.

3. Braune, S.; Poechlauer, P.; Reintjens, R.; Steinhofer, S.; Winter, M.; Lobet, O.; Guidat, R.; Woehl, P.; Guermeur, C., Selective nitration in a microreactor for pharmaceutical production under cGMP conditions. *Chimica oggi* **2009,** *27* (1), 26-29.

4. Proctor, L. D.; Warr, A. J., Development of a Continuous Process for the Industrial Generation of Diazomethane1. *Organic process research & development* **2002,** *6* (6), 884-892.

5. Roberge, D. M.; Gottsponer, M.; Eyholzer, M.; Kockmann, N., Industrial design, scale-up, and use of microreactors. *Chemistry Today* **2009,** *27* (4), 8-11.

6. Hessel, V.; Vural Gürsel, I.; Wang, Q.; Noel, T.; Lang, J., Potential analysis of smart flow processing and micro process technology for fastening process development: use of chemistry and process design as intensification fields. *Chemical Engineering & Technology* **2012,** *35* (7), 1184-1204.

7. Schaber, S. D.; Gerogiorgis, D. I.; Ramachandran, R.; Evans, J. M.; Barton, P. I.; Trout, B. L., Economic analysis of integrated continuous and batch pharmaceutical manufacturing: a case study. *Industrial & Engineering Chemistry Research* **2011,** *50* (17), 10083-10092.

8. Jolliffe, H. G.; Gerogiorgis, D. I., Plantwide design and economic evaluation of two continuous pharmaceutical manufacturing (CPM) cases: ibuprofen and artemisinin. *Computers & Chemical Engineering* **2016,** *91*, 269-288.

9. Gutmann, B.; Cantillo, D.; Kappe, C. O., Continuous ‐ Flow Technology—A Tool for the Safe Manufacturing of Active Pharmaceutical Ingredients. *Angewandte Chemie International Edition* **2015,** *54* (23), 6688-6728.

10. Wakami, H.; Yoshida, J.-i., Grignard exchange reaction using a microflow system: from bench to pilot plant. *Organic process research & development* **2005,** *9* (6), 787-791.

11. Damm, M.; Glasnov, T. N.; Kappe, C. O., Translating high-temperature microwave chemistry to scalable continuous flow processes. *Organic Process Research & Development* **2009,** *14* (1), 215-224.

12. Schaber, S. D.; Gerogiorgis, D. I.; Ramachandran, R.; Evans, J. M. B.; Barton, P. I.; Trout, B. L., Economic Analysis of Integrated Continuous and Batch Pharmaceutical Manufacturing: A Case Study. *Ind Eng Chem Res* **2011,** *50* (17), 10083-10092.

13. Roberge, D. M.; Ducry, L.; Bieler, N.; Cretton, P.; Zimmermann, B., Microreactor technology: A revolution for the fine chemical and pharmaceutical industries? *Chem Eng Technol* **2005,** *28* (3), 318-323.

14. Hartman, R. L.; McMullen, J. P.; Jensen, K. F., Deciding Whether To Go with the Flow: Evaluating the Merits of Flow Reactors for Synthesis. *Angew Chem Int Edit* **2011,** *50* (33), 7502-7519.

15. Calabrese, G. S.; Pissavini, S., From Batch to Continuous Flow Processing in Chemicals Manufacturing. *Aiche J* **2011,** *57* (4), 828-834.

16. Kletz, T. A., What you don't have, can't leak. *Chemistry and Industry* **1978,** *6*, 287-292.

17. Kletz, T. A., *Plant design for safety: a user-friendly approach.* Taylor & Francis: 1991.

18. Fire, D., Explosion Index Hazard Classification Guide. *AICHE New York* **1994**.

19. Lawrence, D. Quantifying inherent safety of chemical process routes. © Duncan Lawrence, 1996.

20. Heikkilä, A.-M., *Inherent safety in process plant design: an index-based approach.* VTT Technical Research Centre of Finland: 1999.

21. Gentile, M.; Rogers, W.; Mannan, M., Development of a fuzzy logic-based inherent safety index. *Process Safety and Environmental Protection* **2003,** *81* (6), 444-456.

22. Bansal, Y.; Silakari, O., The therapeutic journey of benzimidazoles: a review. *Bioorganic & medicinal chemistry* **2012,** *20* (21), 6208-6236.

23. omeprazole. https://www.drugs.com/monograph/omeprazole.html.

24. albendazole. https://www.drugs.com/monograph/albendazole.html.

25. Patil, A.; Ganguly, S.; Surana, S., A systematic review of benzimidazole derivatives as an antiulcer agent. *Rasayan J Chem* **2008,** *1* (3), 447-460.

26. Patt, R.; Kordsachia, O.; Süttinger, R.; Ohtani, Y.; Hoesch, J.; Ehrler, P.; Eichinger, R.; Holik, H.; Hamm, U.; Rohmann, M., Ullmann's encyclopedia of industrial chemistry. Wiley: 2002.

27. Annual Production. http://www.eng-tips.com/viewthread.cfm?qid=389883.

28. Pineda-Solano, A.; Saenz-Noval, L.; Nayak, S.; Waldram, S.; Papadaki, M.; Mannan, M. S., Inherently safer reactors: Improved efficiency of 3-picoline N-oxidation in the temperature range 110–125° C. *Process Safety and Environmental Protection* **2012,** *90* (5), 404-410.

29. LaPorte, T. L.; Spangler, L.; Hamedi, M.; Lobben, P.; Chan, S. H.; Muslehiddinoglu, J.; Wang, S. S., Development of a Continuous Plug Flow Process for Preparation of a Key Intermediate for Brivanib Alaninate. *Organic Process Research & Development* **2014,** *18* (11), 1492-1502.

30. Ebrahimi, F.; Kolehmainen, E.; Turunen, I., Safety advantages of on-site microprocesses. *Organic Process Research & Development* **2009,** *13* (5), 965-969.

31. Kadla, J. F.; Chang, H., The reactions of peroxides with lignin and lignin model compounds. ACS Publications: 2001.

32. Ebrahimi, F.; Kolehmainen, E.; Oinas, P.; Hietapelto, V.; Turunen, I., Production of unstable percarboxylic acids in a microstructured reactor. *Chemical engineering journal* **2011,** *167* (2), 713-717.

# 478238 Risk Based Decision Making Methods for Evaluating Complex Technologies, a Comparative Analysis on Pipelines and Railroad for Crude Oil Transport

**Abstract:**

Decision making for long-term capital intensive projects involve large number of uncertainties. Traditional risk assessments are based on probabilities which are estimated from limited data sets and provide limited flexibility for complex technologies. Further, the environment and specific conditions under which the data is obtained may be unknown and yield to erroneous extrapolations. The method proposed in this work utilizes a novel method for estimation of the risk based on the net present value of technologies and provides the basis for a decision support tool based on the expected value theorem. This is demonstrated using the case of the safety and economic crisis in onshore crude oil transport in the continental United States. The exponential growth of crude oil production in the continental United States has led to crisis in the mid-stream industry with existing pipeline infrastructure under pressure to service the increased production. This fact is reflected in the 4000% increase in the carloads in a span of five years from 2009 to 2013. This has been compounded by the volatile fluctuation in crude oil price from $100 to $30 over the past few years. This has put enormous constraints on capital decision making. Since railroad infrastructure albeit aged, already exists in many regions, loading and unloading crude oil is relatively easy and their connection to the refining centers along the Gulf of Mexico apart from sites in the east and west coast make it an option to many producers. In this paper a novel decision making model is proposed using the concept of 'Net present value' of a technology to evaluate the option between railroad and as suitable mid-stream alternatives for the transportation of crude oil. The model is built on using a prescriptive decision making technique for the quantification of factors affecting the decision, incorporating also the utility of wealth of the decision maker. Monte-Carlo simulation is further used to obtain results and trends based on variation of initial parameters. This analysis can provide a basis for making decisions given the complex environment of crude oil transport operations and can be extended to a variety of geographies given that they satisfy the requirements of the model. The method can also be further extended to other technology evaluation given appropriate factors can be identified to build the model.

# 479492 Key Modeling Parameters and Process Safety Information for Effective Atmospheric Dispersion Modeling

**Abstract:**

Dispersion modeling is an important part of Quantitative risk assessment (QRA) documentation. Dispersion modeling can be performed using commercially available software. Variety of modeling parameters and Process Safety Information (PSI) data are needed for atmospheric dispersion modeling. Some of the parameters include weather, facility geographical location, flammability and toxicity data of the service fluid, radiation intensity, software tool, and model. Some PSI data needs include process and instrumentation diagrams (PIDs), Process Flow Diagrams (PFDs), relief system evaluation documentation, Heat and Material Balance (H&MB), and equipment data sheets. This paper describes the key parameters and their selection for effective dispersion modeling.

# 479373 Implementation of Hazard Mitigation Solutions: Consequence Vs. Risk-Based Building Specifications

**Abstract:**

Implementation of Hazard Mitigation Solutions: Consequence vs. Risk-Based Building Specifications

Travis Holland, Joshua Bruce-Black

Baker Engineering and Risk Consultants

3330 Oakwell Court, Suite 100

San Antonio, TX 78218

tholland@bakerrisk.com, jbblack@bakerrisk.com

Abstract

There are various strategies to mitigate occupant vulnerability and risk after completion of a consequence based facility siting study (FSS) or quantitative risk assessment (QRA). One potentially cost-effective mitigation strategy is to focus on reducing the occupant vulnerability and risk to personnel by upgrading existing buildings or designing new buildings. A building specification is commonly used to relay a small set of building design parameters based on the results of a consequence based FSS to engineers or contractors. However, with the industry shift towards risk based studies (QRAs), there has been a growing perspective that the information required to develop an appropriate building specification is more complicated given the increased number of variables involved in a QRA. Using the development of a new control room specification as an example, this paper will compare and contrast the steps and benefits/drawbacks involved in developing an appropriate building design basis employing a consequence and risk based approach. This paper will highlight how establishing tolerance criteria, emergency response strategy, personal protective equipment (PPE), and building occupancy early in the project definition stage promotes effective streamlined communication in the specification and reduces the differences between consequence and risk-based building specifications.

13TH GLOBAL CONGRESS ON PROCESS SAFETY

# Generalization of Modeling Along-Wind Dispersion

**Jessica M. Morris**
**Tom Spicer, Ph.D., P.E.**
**Ralph E. Martin Department of Chemical Engineering**
**University of Arkansas, Fayetteville, AR**
**Jmm065@uark.edu**

Prepared for Presentation at
American Institute of Chemical Engineers
2017 Spring Meeting and 13th Global Congress on Process Safety
San Antonio, TX
March 26 – 29, 2017

# Generalization of Modeling Along-Wind Dispersion

**Jessica M. Morris**
**Tom Spicer, Ph.D., P.E.**
**Ralph E. Martin Department of Chemical Engineering**
**University of Arkansas, Fayetteville, AR**
**Jmm065@uark.edu**

**Keywords:** along-wind, dispersion, finite duration, time-varying, atmosphere, releases, modeling

## Abstract

The method to model along-wind dispersion is an important aspect of currently used commercial atmospheric dispersion models when considering time-limited or time-varying releases of toxic or flammable materials. Past wind tunnel experimental data has shown the validity of this approach. Inclusion of the effects of along-wind dispersion has proven to be critically important to modeling available field-scale data with current models for such time-varying releases. Many investigators have considered modeling along-wind dispersion from a theoretical point of view as well as basic parameterization of available field test data. This paper reviews past work and proposes a non-dimensionalized parameterization which shows improved agreement with available field-scale data.

## 1 Introduction

Continued world-wide industrial development has brought the wide spread use of industrial chemicals, many of which are hazardous. Whether released accidentally (or intentionally), air borne hazardous chemicals effect large populations quickly as evidenced by several industrial chemical accidents (e.g., Bhopal[1], Flixborough[2], Seveso[3]). Understanding and modeling hazardous chemical releases is vital to protecting the public. Many researchers have worked to assess these hazards using physical and computational models. Predicting the potential impact of hazardous gas releases in the atmosphere is imperative to properly determine where transport, storage, and use of such chemicals should be located. In the case of an actual loss of containment, understanding where air borne chemicals can travel is important to determine the proper response to an emergency.

While models exist to address these questions, there are opportunities to improve modeling methods particularly for (realistic) finite duration and time varying hazardous gas releases, which depend strongly on parameterizing the effects of along-wind dispersion (the elongation of the cloud in the along-wind direction). Modeling methods for along-wind dispersion are frequently based on established theoretical approaches which provide guidance for analyzing currently available field scale data. This paper considers experimental data and established theoretical methods to consider how the effects of along-wind dispersion can be improved.

# 2 Theoretical Description

## *2.1 Theoretical Models*

Along-wind dispersion can be calculated under many different conditions and assumptions. There have been many established theories accounting for different conditions and assumptions. Scientists choose the along-wind dispersion coefficient as either a function of time or distance. Majority of scientists choose the along-wind dispersion coefficient as a function of the friction velocity. Some of the most prominent theories about along-wind dispersion are discussed below.

## *2.2 Batchelor*

Batchelor (1964)[4] determined the Lagrangian Similarity Hypothesis. This hypothesis relates Eulerian and Lagrangian quantities when deriving theoretical solutions to problems. Applying known properties of fluid flow to the atmosphere, Batchelor was able to establish a generalized equations under neutral stability conditions, with no restraints on maximum height, for center line concentration at ground level and wind speed dependence with height. Batchelor derived many equations that are the foundation for modern theoretical equations commonly utilized. Batchelor's relationship between turbulence in a constant stress region on a boundary layer is a function of the friction velocity and time. This relationship and the Lagrangian Similarity Hypothesis serve as a template for along-wind dispersion equations to come.

Release height can be interpreted differently on its effect on dispersion. Restating Batchelor's similarity theory, "the statistical properties of the velocity of a marked fluid particle at time, $t$, after release at the ground depend only on friction velocity, $u_*$ and $t$." This theory is then extended to the case of an elevated release: "The statistical properties of the velocity of a marked particle at time, $t$, after release at height, $h$, are the same as those of a particle released at the ground (z=0) at the instance $-t_1$, provided $t \gg t_1$ where $t_1$ is expected to be of the order of magnitude of the time-scale of the turbulence at height $h$, that is, of order $h/u_*$". Under Batchelor's similarity theory, after a sufficient amount of time, the height of release can be ignored.

## *2.3 Chatwin*

Chatwin (1968)[5] was the first mathematician to theoretically derive a full equation with numerical values for the along-wind dispersion coefficient. He also derived the shear stress component of the along-wind dispersion equation. Chatwin used Batchelor's similarity theory along with the assumptions of the ability of turbulence to be described as an eddy diffusivity to solve for along-wind dispersion, $\sigma_x$. He makes further assumptions of neutral stability, ground release, and an unbounded boundary layer. The generalized equations for both overall along-wind dispersion, $\sigma_x$ , and wind shear, $\sigma_{xs}$ , by Chatwin are as followed

$$\sigma_x = 0.803\frac{u_* t}{\kappa} \quad (1)$$

$$\sigma_{xs} = 0.596\frac{u_* t}{\kappa} \quad (2)$$

where $u_*$ is the friction velocity and $\kappa$ is the van Karman's constant, approximately equal to 0.41.

## *2.4 Van Ulden*

Van Ulden (1992)[6] derived an equation under slightly different dispersion definitions. Van Ulden made the distinction between absolute and relative dispersion. Wheatley and Chatwin were able to authenticate their equations with the assumption of a hypothetical atmosphere: where no large horizontal eddies exist. Taking into consideration this distinction of absolute and relative dispersion, van Ulden derived a new equation for $\sigma_x$, the total variance of the concentration distribution in the wind direction

$$\sigma_x{}^2 = \sigma_c{}^2 + \sigma_d{}^2 \quad (3)$$

where $\sigma_c$ is the deterministic part (or centroid variance), and $\sigma_d$ is the random part (or diffusive variance). The total longitudinal variance, $\sigma_x$, is dependent on three separate quantities: centroid variance, diffusive variance due to shear effects, and direct dispersion. These three variables were able to be solved for shear, centroid, and diffusive effects respectively as a function of friction velocity and time

$$\sigma_s = 0.5\frac{u_* t}{\kappa} \quad (4)$$

$$\sigma_c \simeq 0.54\frac{u_* t}{\kappa} \quad (5)$$

$$\sigma_d = 0.67\frac{u_* t}{\kappa}. \quad (6)$$

Under the assumption that $\sigma_c \simeq \sigma_s = 0.5\frac{u_* t}{\kappa}$, these variables can be substituted into equation ( 3 ) to get the total longitudinal variance. An exact equation for $\sigma_x$ is obtained:

$$\sigma_x = 0.84\frac{u_* t}{\kappa} \quad (7)$$

Van Ulden not only derived an along-wind dispersion equation under neutral conditions from a ground release in an unbounded boundary layer, but also for an along-wind dispersion equation for very stable conditions. Under very stable conditions, the wind speed profile and eddy diffusivities change. When applying the new constraints for wind speed and eddy diffusivities in a different stability class, the total variance of concentration distribution in the wind direction is shown to be:

$$\sigma_x = \{(\frac{3}{4} - \frac{16}{9\pi})\frac{5u_*{}^3 t^3}{kL}\}^{1/2} \simeq 0.43(\frac{5u_*{}^3 t^3}{kL})^{1/2}. \quad (8)$$

where $L$ is the Monin-Obukov length.

### *2.5 Summary*

The relationship between advection cloud speed, time, and distance to determine the along-wind diffusion coefficient is complex. Changing conditions such as stability class, surface roughness, wind speed, function of distance or time, release height, and other parameters all have an effect on determining the along-wind diffusion coefficient. With the varying conditions, not all the equations are able to be integrated together into an overarching equation meeting all conditions. Different assumptions or criteria result in diverse equations for the same diffusion coefficient. An equation that takes into account all of these parameters is desired.

# 3 Correlation of Field Data by Hanna and Franzese

The 12 field and wind tunnel tests mentioned prior were analyzed by Hanna and Franzese to determine a suitable along-wind dispersion coefficient that resulted in an equation comparable to Chatwin's theoretical equation[5Error! Bookmark not defined.]. Results of the data and the predicted equation are shown below in Figure 1.

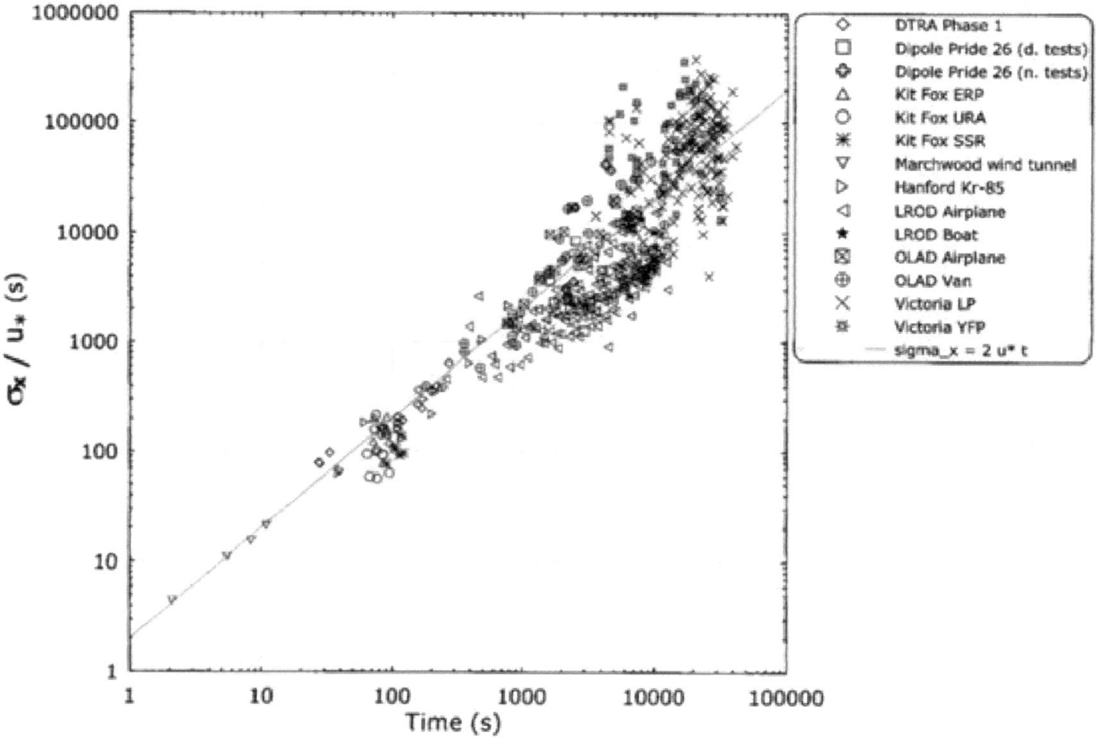


**Figure 1.** Hanna and Franzese results of along-wind dispersion coefficient divided by friction velocity and plotted vs time. Line represents $\sigma_x = 2u_*t$.[8]

# 4 Reconsideration of Field Data Using Dimensionless Groups

Figure 1 shows a correlation of the along-wind dispersion coefficient with the friction velocity as a function of time. While this is linear, their data can be up to a factor of 10 away from the predicted equation. There is also a noticeable trend of data beneath the predicted line when time is less than 10,000 seconds and above the predicted line when time is greater than 10,000 seconds. This is believed to be an effect of the near field vs. the far field. This equation for the along-wind dispersion coefficient does not take into effect the near field and far field which results in a very large portion data over predicted from 1,000-10,000 seconds and a large portion of data under predicted from 10,000 seconds and above. An improved equation to improve the correlation accounting for the error and trends is desired.

A non-dimensionalized equation was created to better predict the along-wind dispersion coefficient. The parameters of the non-dimensionalized equation were based off theory and assumptions regarding trends seen with atmospheric dispersion and along-wind dispersion. The non-dimensionalized equation allows for the along-wind dispersion coefficient to be predicted based on wind speed, distance, and stability class for each test in addition to the previous parameters of time and friction velocity.

The non-dimensionalized equation takes into account multiple theories and trends. Batchelor's Lagrangian Similarity Theory predicts $\sigma_x$ to be a function of friction velocity, $u_*$ , and time, $t$, while ignoring the release height[4]. Van Ulden predicts separate equations for both neutral and stable conditions[6]. Surface roughness has been shown to have an effect on wind speed[7]. The overall non-dimensionalized equation for along-wind dispersion is shown below

$$\frac{\sigma_x}{x} = \left(\frac{ut}{x}\right)^p \left(\frac{u_*}{u}\right)^q C_1 \qquad (9)$$

where $x$ is the distance in meters and p, $q$, and $C_1$ are all dimensionless constants. The $\left(\frac{ut}{x}\right)$ term accounts for the effect of wind speed, $u$, time, $t$, and distance, $x$. The $\left(\frac{u_*}{u}\right)$ term is to take into effect the wind speed, $u$, and friction velocity, $u_*$, which ultimately represents the stability class.

Data was accumulated and analyzed by Stephen Hanna and Pasqual Franzese in 1999 from 11 field tests and 1 wind tunnel test[8]. The accumulation of these 12 tests produced 446 data sets over varying conditions. Other tests were accumulated and stored in the Modeler's Data Archive[9] for future reference. The following tests were analyzed in hopes of a better fit for along-wind dispersion in atmospheric modeling. Of the 446 data sets, only six data points were not used in the analysis due to lack of recorded friction velocity and/or wind speed. Reference heights for recording the wind speed from each of the data sets vary. A constant reference height of 10m was implemented across all data sets for a more standardized data analysis. Data sets used in this analysis are described below.

### *4.1 DTRA Phase* 1

The DTRA Phase 1 tests were carried out from the 9th -26th of September in 1996 at the Dugway Proving Ground on Pad 11 producing 23 ensemble averages[10]. Propylene was instantaneous released from air cannons at ground level. Each air cannon produced 0.36kg of propylene in a period of 0.2-0.4 seconds. DigiPIDs and Ultra-Violet Ion Collectors were both used at distances of 200 m, 300 m, 400 m, 800 m, and 1200 m along a fixed sampling line 1.5m above the ground. Increasing distances resulted in an increased number of air cannons to result in an increased mass of propylene for the increased distance. 1 air cannon was used for distances of 200 m and 300 m, 2 air cannons for 400 m, 4 air cannons for 800 m, and 8 air cannons for 1200 m. Wind speed is referenced at a height of 2m which ranged from 2-7 m/s. The Great Salt Lake Desert produced a surface roughness of less than 1mm.

### *4.2 Dipole Pride*

Dipole Pride tests were grouped into day and night tests from the 4th -20th of November in 1996 at the Department of Energy Nevada Test Site – Yucca flat. $SF_6$ was released from 2 vertically mounted, pressurized cylinders on the back of a 5-ton flatbed truck. These point sources released 8-21.6 kg of mass in a duration of 1-3 seconds[11]. 90 whole air samplers, with 15 minute time averaging, and 6 continuous TGA-4000 $SF_6$ analyzers were placed in 3 lines at distances of 1 km, 10 km and 20 km[12]. Wind speeds were referenced at a height of 10 m resulting in wind speeds of 3-5 m/s during the day and 2-5 m/s during the night. Surface roughness at the Yucca flat was around 32 mm. The 23 total tests produced 10 adequate day releases and 5 adequate night releases.

### *4.3 Kit Fox (ERP, URA, SSR)*

All Kit Fox releases took place at the National Spill Test Facility at the Nevada Test Site. The Equivalent Roughness Pattern (ERP) and Uniform Roughness Array (URA) tests took place from the 24th -31st of August in 1995 whereas the Smooth Surface Rough (SSR) tests took place the 11th -15th of September in 1995.[13] The tests were similar in all conditions other than surface roughness. The Kit Fox ERP tests had a surface roughness of 0.2 m, the Kit Fox URA tests had a surface roughness of 0.02 m, and the Kit Fox SSR tests had a surface roughness of 2 mm[14]. Carbon Dioxide was released from a ground level area source with dimensions of 1.5 m x 1.5 m x 1 m at a rate of 1 m/s resulting in 1.5 kg/s – 4 kg/s being released[15]. Concentrations were measured at distances of 25 m, 50 m, 100 m, and 225 m. 84 high and low infrared concentration sensors, solid state infrared $CO_2$ sensors, and sample bag Nova sensors, were all used to detect concentrations of $CO_2$ with an averaging time of around 20 seconds[16]. Wind speeds were referenced at a height of 2 m which resulted in wind speeds of 2-3 m/s for the ERP and SSR test and 3-4 m/s for the URA tests. All tests took place right before sunset which resulted in lower wind speeds with neutral to stable stability classes. The ERP and SSR tests resulted in 6 sufficient trials each and the URA tests resulted in 9 sufficient trials to be used in this analysis.

### *4.4 Hanford Kr-85*

The Hanford Fr-85[17] tests took place from 18 August to 8 November 1967 at the Battelle Pacific Northwest Laboratory. A vial containing Krypton 85 at ambient pressure was crushed by a guillotine-like device at ground level, instantaneously releasing a 1 m puff of the radioactive gas. Concentration was detected by 64 halogen-quenched Geiger-Muller tubes at a distance of 200 m and an elevation of 1.5 m above the ground. 24 halogen-quenched Geiger-Muller tubes were positioned on 6 towers at a distance of 800 m at elevations of 0.8 m, 4.6 m, 10.7 m and 21.3 m. The time averaging associated with the Geiger-Muller tubes was 4.8 seconds. 8 trials were performed, but only 6 releases remained within the boundary of the array of sensors. The wind speed was taken at a reference height of 1.5 m which resulted in wind speeds of 1-8 m/s. The roughness was around 30 mm due to the sagebrush and steppe grass.

### *4.5 LROD*

The Long Range Overwater Dispersion (LROD)[18] tests took place from 13$^{th}$ -26$^{th}$ of July in 1993 at the Pacific Missile Range Facility in Kauai, Hawaii. A crosswind line source of pressurized, liquid $SF_6$ was released from a high pressure hose from the back of a plane at an elevation of 90 m at a rate of 9.6 g/meter and a speed of 129 m/s over water. A total duration of the release of $SF_6$ was around 13 minutes for the 100 km distance. Concentration was measured at ground level from boats along with planes at various elevations using electron capture detectors (ECD) and TGA-4000 whole air samplers. The averaging time was around 30-120 seconds for the sequential bag samples. Noise associated with the continuous analyzer resulted in limited use. Concentration was detected from the plane by flying parallel to the dissemination line and 45 km south of the sampling line (5 km, 15 km, 30 km, 60 km, and 100 km). The plane, equipped with a continuous $SF_6$ analyzer, would fly to the dissemination line then back to 25 m. It would slowly ascend from 25 m to 2500 m for various heights. Testing over water resulted in a surface roughness of 0.02-0.86 mm. A reference height of 10 m was used for wind speed which resulted in speeds of 3-11 m/s. The LROD tests produced 11 ensemble averages.

### *4.6 OLAD*

The Over Land Along-wind Dispersion (OLAD)[19] tests were similar to the LROD tests, except over land. These tests took place from the 8$^{th}$ -25$^{th}$ of September in 1997 at the Dugway Proving Ground, West Desert Test Center. $SF_6$ was released from both plane and truck line sources. The plane released 100 kg of $SF_6$ at an elevation of 100 m from a $N_2$ pressurized tank at a rate of 33.3 kg/min for a distance of 20 km. The truck released 15 kg of $SF_6$ at an elevation of 3 m from a mounted tank at a rate of 1.5 kg/min for a distance of 10 km. Concentrations were recorded with TGA-4000 whole air samplers, with time averaging of 15 minutes, and 2 continuous electron capture detector (ECD) analyzers at the end of each line at distances of 2 km, 5 km, and 10 km for the truck releases and 10 km, 15 km, and 20 km for the plane releases. Wind speed was measured at a height of 2 m which resulted in speeds of 0.5 – 7 m/s. The surface roughness was measured around 30 mm. These tests resulted in 3 ensemble averages for the aircraft release and 8 surface ensemble averages for the truck release.

### *4.7 Marchwood wind-tunnel*

The Marchwood[20] wind tunnel tests are the only tests that took place in a wind tunnel. A mixture of propane and helium, resulting in neutral buoyancy, was released from a 15 mm ground level point source. The mixture was released for a duration of 0.5 – 12 seconds. 150-300 releases were summarized into 4 ensemble averages. A single FID measured the concentration with a 5 ms sampling interval at locations of 1 m, 3 m, 5 m, and 7 m. Wind speeds ranged from 1-4 m/s at neutral stability class. Surface roughness was measured at 0.3 mm for the wind tunnel.

### *4.8 Oceanside*

The Oceanside[21] tests took place at the shoreline of Oceanside and Del Mar in Southern California. 20 μm particles of green and yellow fluorescent pigment (FP) were released by plane, boat, and land with standardized Dugway Proving Ground equipment. For the plane releases, the DPG model D-1 released 1500 g/min of FP for a duration of 10 minutes at an elevations of 61 m, 122 m, and 152 m over a 40 km distance. For the boat releases, a DPG Mark IV was used to release 110 g/min for a duration of 2 hours at sea level over 40 km. For instantaneous land releases, the DPG Mark IX was used at three different inland locations to release 10-50 g of FP instantaneously at ground level. Concentration of FP were recorded from rotorod samplers, sequential samplers, vertical samplers, 100 m square grids, dense sampling lines, and metronics drum impact samplers. Time averaging ranged from 5 minutes at coastal regions to 15 minutes at inland conditions. Surface roughness ranged from 30-50 m due to the coastal shoreline. The wind speed reference height varied from 8-120 m which resulted in wind speeds ranging from 1-5 m/s.

### *4.9 Fort Wayne*

Fort Wayne[22] releases took place in the Summer and Fall of 1963 in Fort Wayne, Indiana. An aerosol of fluorescent pigment (FP) was released via 2 aircrafts in the form of an elevated, 91-244 m, line source. Dry green and yellow FP were released from the planes via FP dispensers supplied by the Dugway Proving Ground. The amount of release was measured by the difference in weight, before and after. The release was upwind of the 180,000 population urban setting with a river running through the city and agriculture surrounding. Rotorods, Gelman throw-away plastic filter holders, and Gelman paper tape samplers were used for concentration measurements of the FP. The tape provided an averaging time of 5 minutes where the filter holders had an averaging time of 30 minutes. Wind speed was measured at the release height and resulted in speeds ranging from 5-16 m/s.

### *4.10 Short Range Diffusion Experiment Series*

The Short Range Diffusion Experiment Series[23] took place in 1979 in Japan. A 200 m x 200 m vacant lot covered in turf and surrounded by trees was used for the releases. $NO_x$ was released instantaneously from a balloon at an elevation of 1.5 m. Chemiluminescence $NO_x$ meters were used to detect concentration with a sampling time of <2 minutes. Sample lines were chosen at distances of 60 m and 100 m. Wind speeds were referenced at heights of 0.5 m, 3 m, and 9 m which resulted in speeds of 1-2 m/s.

### *4.11 Results and Discussion*

A multivariable regression analysis, also known as multiple regression, is an extension of linear regression[24]. A regression analysis determines the relationships among variables.

With a multivariable regression analysis, the relationships between multiple independent variables can be determined. A regression analysis is a useful tool in understanding the changes of one variable with respect to others. The form of a typical multi-variable regression analysis is shown below

$$Y = a + b_1 x_1 + b_2 x_2 + \cdots + b_n x_n \tag{10}$$

A multi-variable regression analysis can be applied to the non-dimensionalized equation. The non-dimensionalized equation is rearranged with the natural log of each variable to perform the multi-variable regression analysis. This equation is shown below

$$ln\left(\frac{\sigma_x}{x}\right) = p\, ln\left(\frac{ut}{x}\right) + q\, ln\left(\frac{u_*}{u}\right) + C_2 \tag{11}$$

where $C_2$= $\ln(C_1)$. A multivariable regression analysis with known data is used with this equation to determine the constants for p, q, and $C_1$. An analysis is done for the overall data and categorized into stability classes using macros in Excel[25].

Multiple parameters were analyzed for optimal results using the multivariable regression analysis. Data was quantified and analyzed based on individual tests, surface roughness, friction velocity, Monin-Obukhov length, distance, time, near field, and far field.

Data from all 12 experimental data sets referenced above were analyzed with a multi-variable regression analysis to determine appropriate exponents and a constant for an optimum non-dimensionalized equation. While many parameters were originally analyzed for the optimum non-dimensionalized equation, the quantification of near field and far field as a function of time was chosen as the most prominent parameter. The optimum non-dimensionalized equation was parameterized by near field with time less than 10,000 seconds and far field with time greater than 10,000 seconds. These equations are shown below in Table 1. The ratio, α, of the along-wind dispersion coefficient, $\sigma_x$, for the data vs $\sigma_x$ for the predicted non-dimensionalized equation is used for comparison of the difference in equations. A non-parametric confidence interval was determined for the non-Gaussian distribution of $\alpha$.[26] Hanna and Franzese's along-wind dispersion coefficient was analyzed in the same manner in Table 1.

**Table 1. Comparison of along-wind dispersion equations with 90% confidence intervals**

| Model | Equation | 90% Confidence Interval of $\alpha = \frac{\sigma_x\ (predicted)}{\sigma_x\ (data)}$ | Range of Confidence Interval |
|---|---|---|---|
| Hanna and Franzese | $\sigma_x = 2u_* t$ | 1.63 < α < 1.88 | 26% |
| Near Field Non-Dimensionalized | $\frac{\sigma_x}{x} = 1.72\left(\frac{ut}{x}\right)^{1.18}\left(\frac{u_*}{u}\right)^{1.10}$ | 1.16 < α < 1.31 | 16% |
| Far Field Non-Dimensionalized | $\frac{\sigma_x}{x} = 0.05\left(\frac{ut}{x}\right)^{0.67}\left(\frac{u_*}{u}\right)^{-0.27}$ | 0.85 < α < 1.05 | 20% |

From Table 1, the range of the 90% confidence intervals for each of the equations is shown. The non-dimensionalized equations have a tighter correlation of the predicted values. The 90% confidence intervals shows that Hanna's predicted along-wind dispersion equation has a trend over predict the $\sigma_x$ values. An interesting result from the multi-variable regression analysis is the determination of the coefficient for the $\left(\frac{u_*}{u}\right)$ term. The -0.27 exponential value essentially confirms that the stability class has no effect on the along-wind dispersion coefficient after a specific amount of time. A discontinuity is understood between the near field and far field non-dimensionalized equation and is currently being evaluated.

The results of the non-dimensionalized predicted $\sigma_x$ values were divided by the friction velocity and plotted against the original data to determine their relationship. This relationship is shown below in Figure 2.

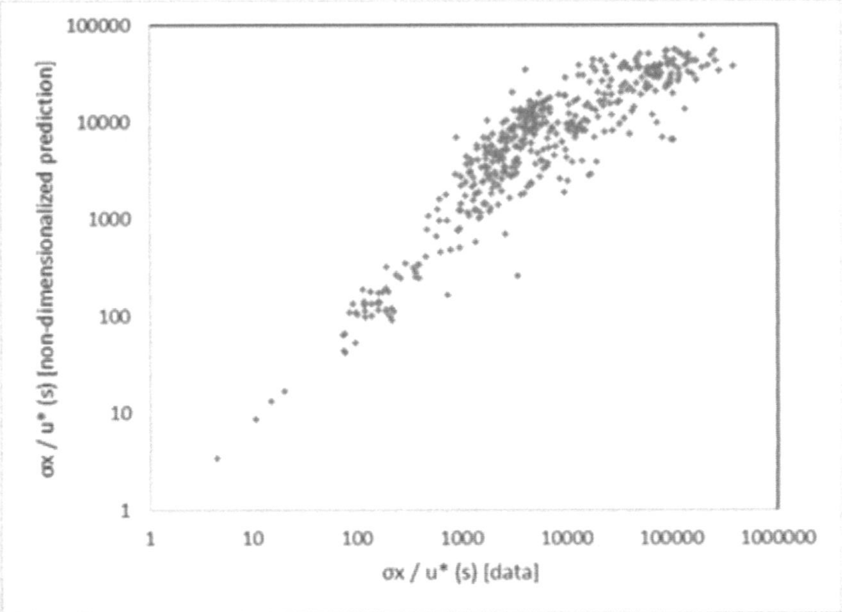


**Figure 2.** Along-wind dispersion coefficient divided by the friction velocity of the non-dimensionalized modeled values vs the original data.

As shown in Figure 2, the correlation between the data and non-dimensionalized predicted values shows a 45° line. This correlates a good fit for the predicted $\sigma_x$ values when compared to the original $\sigma_x$ values.

A visual representation of the non-dimensionalized predicted $\sigma_x$ values, Hanna and Franzese's $\sigma_x$ values, and the original $\sigma_x$ values as a function of time are all shown below in Figure 3.

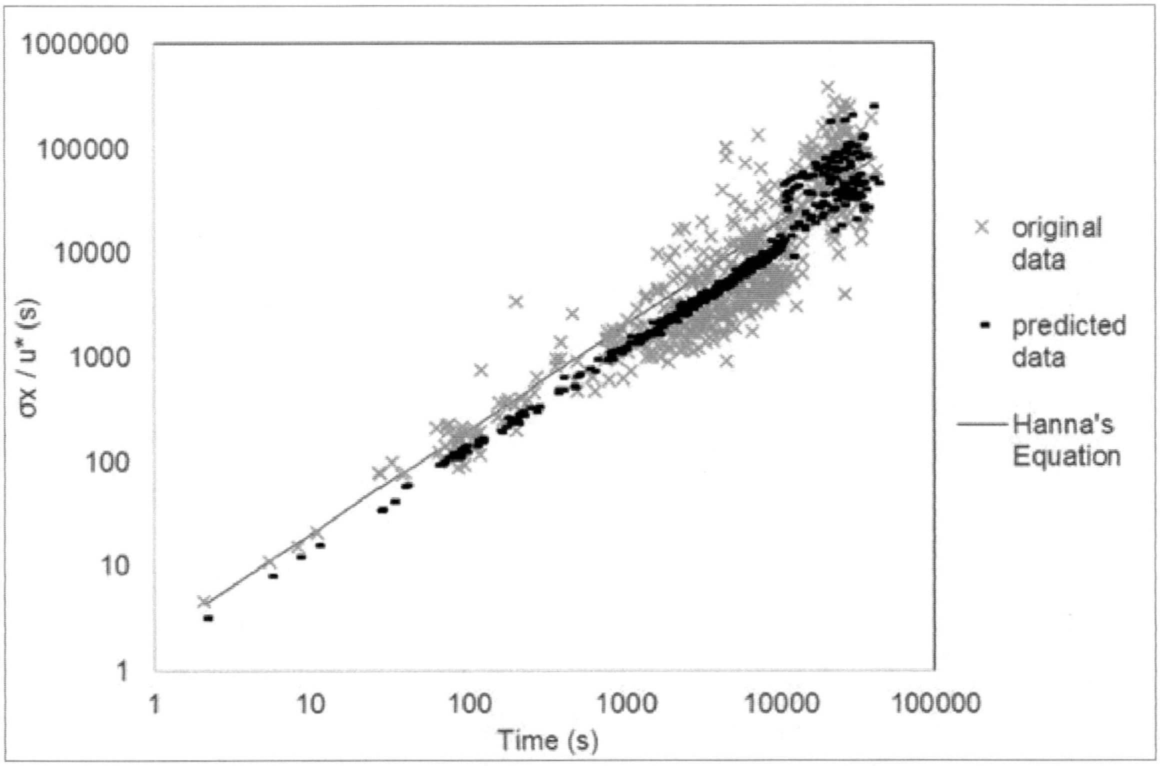


**Figure 3.** Time vs. the along-wind dispersion coefficient divided by the friction velocity as a function of the original data, non-dimensionalized predicted data, and Hanna's predicted data.

From this figure, the predicted values from both Hanna and Franzese's predicted data and the non-dimensionalized predicted data are shown in comparison with the original data. The non-dimensionalized equation takes into account additional parameters, other than time, which results in the spread of $\sigma_x$ values as a function of time and additional parameters in comparison to the single line by the $\sigma_x = 2u_*t$ along-wind dispersion equation. The non-dimensionalized equation also takes into effect the near field and far field. This is clearly seen at times greater than 10,000 seconds where the $\sigma_x$ values are spread over a wider range. Overall, the non-dimensionalized equation for along-wind dispersion proves better than the previous $\sigma_x = 2u_*t$ correlation.

## 5 Summary

Along-wind dispersion is critical when modelling toxic, flammable, or hazardous time-limited or finite-duration releases into the atmosphere. While many models exist from a theoretical standpoint or parameterization of experimental data, there is no overarching correlation to predict the along-wind dispersion coefficient. This paper took into effect established theoretical equations validated with experimental data to determine a better equation to model the effects of along-wind dispersion. A successful non-dimensionalized equation parameterized by the near field and far field was determined and validated against available field-scale data which showed improvement in comparison to previous equations. The non-dimensionalized equation for along-wind dispersion solved for in this paper will improve the existing atmospheric modeling with along-wind dispersion.

## 6 Acknowledgments

We gratefully thank our colleagues from DNV GL, Henk Witlox and Maria Fernandez, who provided insight and expertise that greatly assisted the research. We also thank Steven R. Hanna and Pasquale Franzese for sharing their accumulated data for further analysis.

## 7 References

[1] Boybeyi, Zafer, Sethu Raman, and Paolo Zannetti. "Numerical investigation of possible role of local meteorology in Bhopal gas accident." *Atmospheric Environment* 29.4 (1995): 479-496.
[2] Høiset, S., et al. "Flixborough revisited—an explosion simulation approach." *Journal of hazardous materials* 77.1 (2000): 1-9.
[3] De Marchi, Bruna, Silvio Funtowicz, and Jerome Ravetz. "Seveso: A paradoxical classic disaster." *The long road to recovery: Community responses to industrial disaster* (1996): 86-120.
[4] Batchelor, G. K. "Diffusion from sources in a turbulent boundary layer." *Arch. Mech. Stosow* 3.6 (1964): 661-670.
[5] Chatwin, P. C. "The dispersion of a puff of passive contaminant in the constant stress region." Quarterly Journal of the Royal Meteorological Society 94.401 (1968): 350-360.
[6] Van Ulden, A. P. "A surface-layer similarity model for the dispersion of a skewed passive puff near the ground." *Atmospheric Environment. Part A. General Topics* 26.4 (1992): 681-692.
[7] DNV GL. DNV GL Theory Manual for Unified Dispersion Model. London UK 2015.
[8] Hanna, Steven R., and Pasquale Franzese. "Alongwind dispersion—a simple similarity formula compared with observations at 11 field sites and in one wind tunnel." *Journal of Applied Meteorology* 39.10 (2000): 1700-1714.
[9] Chang, Joseph C., and Steven R. Hanna. "Modelers' Data Archive – A Collection of Atmospheric Transport and Dispersion Data Sets." (2010).
[10] Biltoft, C., Phase I of DSWA transport and dispersion model validation study. DPG-FR-97-058, (1997): 156.

[11] Biltoft, C., Dipole Pride 26: Phase II of DSWA transport and dispersion model validation study. DPG-FR-98-001, (1998): 130.
[12] Watson, T. B, R. E. Keislar, B. Reese, D. H. George, and C. A. Biltoft, The DSWA Dipole Pride 26 field experiment. NOAA Tech. Memo. ERL ARL-225, (1998): 99.
[13] King, S. Bruce. The Kit Fox Field Demonstration Project and Data Set. *International Conference and Workshop on Modeling the Consequences of Accidental Releases of Hazardous Materials.* (1999).
[14] Spicer, T. and Havens, J. Description and Analysis of Atmospheric Dispersion Tests Conducted by EPA. *International Conference and Workshop on Modeling the Consequences of Accidental Releases of Hazardous Materials.* (1999).
[15] Sheesley, D. et al.. Chemical Dispersion Modeling Project, 1995 Kit Fox Field Demonstration DOE Nevada Hazardous Spill Center. Volume I – Experiment Description and Data Processing. (1998).
[16] Hanna, S. R., and K. W. Steinberg, Studies of dense gas dispersion from short-duration transient releases over rough surfaces during stable conditions. *Air Pollution Modeling and Its Applications XI,* Plenum Press, (1996): 481–490.
[17] Nickola, P. W., J. D. Ludwick, and J. V. Ramsdell Jr., An inert gas tracer system for monitoring the real-time history of a diffusing plume or puff. *J. Appl. Meteor.,*9, (1970): 621–626.
[18] Bowers, J. F., G. E. Start, R. G. Carter, T. B. Watson, K. L. Clawson, and T. L. Crawford, Experimental design and results for the Long-Range Overwater Diffusion (LROD) Experiment. DPG/JCP-94/012, (1994): 531.
[19] Biltoft, C., S. D. Turley, T. B. Watson, G. H. Crescenti, and R. G. Carter, Over-Land Atmospheric Dispersion (OLAD) test summary and analysis. WDTC Doc. WDTC/JCP-99/048, (1999): 38.
[20] Robins, A. G., and J. E. Fackrell, An experimental study of the dispersion of short-duration emissions in a turbulent boundary layer. *Air Pollution VI,* C. A. Brebbia, C. F. Ratto, and H. Powers, Eds., WIT Press, (1998): 697–707.
[21] Smith, and B. L. Niemann, Shoreline diffusion program, Oceanside, CA. Vols. I and II. MRI-64-FR-860, (1969): 230 and 456.
[22] Hilst, G. R., and N. E. Bowne, A study of the diffusion of aerosols released from aerial line sources upwind of an urban complex. Vols. I and II. Rep. DA-42-007-AMC-38®, (1966): 231 and 583.
[23] Sato, J., F. Kimura, and T. Yoshikawa, The longitudinal spread of puff in the short range diffusion experiment. *Pap. Meteor. Geophys.,*32, (1981): 155–162.
[24] Draper, Norman Richard, Harry Smith, and Elizabeth Pownell. *Applied regression analysis*. Vol. 3. New York: Wiley, (1966).
[25] Davis, Richard A. “Practical Numerical Methods for Chemical Engineers: Using Excel with VBA”. Lexington, KY, (2015).
[26] Conver, W. J. “Practical Nonparametric Statistics”. Wiley, New York, (1971).

# 478500 Review and Analyze the Effects of Parameters for $H_2s$ Gas Mixture Dispersion Models: Degadis and CFD Models


**Abstract:**

Due to the increasing demands of energy, natural gas explorers and developers are interested in exploring sour gas fields, which have a high concentration of hydrogen sulfide ($H_2S$). Because $H_2S$ is highly corrosive and could reduce the heating values of natural gas, $H_2S$ needs to be removed from natural gas before transfer. Then the removed $H_2S$ gas will be converted to sulfide as the products in the Claus process. The $H_2S$ mixture (*e.g.*, $H_2S$ and $CO_2$) may cause corrosion to pipelines and thus, could lead to $H_2S$ gas mixture release.

$H_2S$ mixture behaves as a heavy gas, which travels down to the ground causing harm to the local population and the environment. In this paper, the DEnse GAs DISpersion (DEGADIS) model and the Computational Fluid Dynamics (CFD) model (*e.g.*, ANSYS Fluent) are implemented for the $H_2S$ mixture simulation, and assumptions in the models are also studied. The effects of topography and meteorological parameters such as atmospheric temperature, wind speed, and humidity are investigated. Finally, the simulation results from the DEGADIS model and the CFD model for flat terrain with constant wind velocity are compared.

The simulation results show that the DEGADIS model, which is suitable for flat terrain and constant wind conditions, can be applied to simple terrain geography and stable atmospheric conditions. Whereas the CFD model, which considers unsteady wind speed and turbulence near the obstacles, can be utilized for complex topography. The series of simulation results can provide guidance for the risk assessment of $H_2S$ mixture releases under different topographies and meteorological conditions.

# 478998 Considerations When Justifying the Purchase of an Operator Training Simulator

**Abstract:**

Highly skilled plant operators ensure higher levels of plant safety, reliability and profitability. An Operator Training Simulator accelerates both the time to competence and raises the competency level of an operator. The ARC Advisory Group recently surveyed companies who have invested in Operator Training Simulators (OTS). Two critical results from the survey are that over sixty five percent of those companies report fewer abnormal situations, fewer trips, and incidence avoidance attributable to the use of OTS. Over sixty percent of these companies report improved safety and reduced risk and lower liability attributable to the use of the OTS.

Many leading companies have embraced the use of Operator Training Simulators across their Enterprise however many facilities and companies have yet to embrace the idea. The most common reason given for not investing in such a solution is that the cost of an OTS cannot be justified. The nuclear industry mandates that a nuclear power plant must have an OTS whereas the American Petroleum Institute only recommends the use of Training Simulators. The Process industries overall will benefit by the greater adoption of training simulators and this paper discusses the different ways that can be used to justify the investment in an OTS.

# 479500 Consequence Analysis of Gas Explosion in Underground Common Utility Tunnels


**Abstract:**

The purpose of this study is to figure out the effects of blast overpressure in gas piping leakage in underground common utility tunnels through the explosion simulation tool FLACS (Flame Acceleration Simulator) and find and apply methods that can reduce the risk.

Common utility tunnels are facilities to jointly bury pipes in underground spaces. Placing gas piping in common utility tunnels has the advantages of reducing the risk of corrosion and preventing the physical damage due to civil engineering works and excavating work on the ground. However, when an explosion has occurred in a common utility tunnel, the consequences could be larger in that it is a closed space. Nevertheless, studies on underground common utility tunnels are currently limited to fire accidents. Therefore, risks due to explosions should be quantitatively studied.

Therefore, in this study, the effects of explosions were analyzed assuming cases where gases leaked from gas piping in underground common utility tunnels exploded. In addition, blast overpressures were evaluated for diverse scenarios regarding ignition sources, leakage locations, and discharge rates and the results were statistically derived. Blast overpressures high enough to cause dangerous situations of the underground common utility tunnels were observed. Methods that can reduce the risk, that is, methods that can reduce the frequency of explosions, and methods for designs of internal structures of common utility tunnels that can mitigate the consequences will be presented.

# 479158 Air Cooler Explosion Analysis Using CFD

**Abstract:**

Air coolers or fin-fans are located in most process plants. They cool process fluids using forced air flow over radiators containing the fluid. Air coolers initially appear to offer many features that would result in a significant explosion source should a flammable cloud reach the cooler. The radiator coils are closely spaced pipes which will accelerate flames. The metal box represents confinement and the turbulent air flow produced by the fans will accelerate flames similar to how congestion produces turbulence ahead of a flame front. Also, the flow of air can draw rich flammables near the ground up into the air cooler.

This study examines the severity of an air cooler explosion using computational fluid dynamics (CFD). The commercial software FLACS is used to examine the parameters mentioned above (congestion, confinement, and turbulence) and their effect on explosion severity. Comparisons between FLACS and the Multi-Energy Method are made and recommendations provided for assessing air coolers in non-CFD facility siting studies.

# Revisions to the QMEFS Vapor Cloud Explosion Model

**Jeffrey D. Marx and Benjamin R. Ishii**
**Quest Consultants Inc.®**
**908 26th Avenue N.W.**
**Norman, OK 73069**
**jdm@questconsult.com / bri@questconsult.com**

Prepared for Presentation at
American Institute of Chemical Engineers
2017 Spring Meeting and 13th Global Congress on Process Safety
San Antonio, TX
March 26 – 29, 2017

# Revisions to the QMEFS Vapor Cloud Explosion Model

**Jeffrey D. Marx and Benjamin R. Ishii**
**Quest Consultants Inc.®**
**908 26th Avenue N.W.**
**Norman, OK 73069**
**jdm@questconsult.com / bri@questconsult.com**

**Keywords:** vapor cloud explosion (VCE), Baker-Strehlow-Tang (BST) model, estimating flame speeds

## Abstract

The Quest model for estimating flame speeds (QMEFS) was introduced in 2007, has since been published, and has been used by Quest and other companies to provide a detailed method for vapor cloud explosion (VCE) predictions. QMEFS provides a set of parameters that predict a flame speed, which then is used with the Baker-Strehlow-Tang (BST) blast curves to predict overpressure and impulse. The original intent of the QMEFS model was to provide better characterization of confined or congested areas, as well as fuel reactivities, which fit between the low, medium, and high categories offered by the BST model, so that the large differences in predictions between categories could be eliminated. In this way it provided a more refined system for predicting the consequences of a VCE, while remaining anchored within the flame speeds provided by the BST matrix.

In the years since QMEFS' publication, Baker has added to the BST flame speed matrix. In addition, use of the model has highlighted some of the areas where the model could be improved. This paper shows several ways in which the model has been enhanced to encompass a wider range of conditions than are covered by the BST model.

## 1 Introduction

Any release of a flammable fluid in a petrochemical facility has the potential to generate a flammable vapor cloud that, if ignited, could produce a vapor cloud explosion (VCE). Most VCEs that generate damaging levels of overpressure originate in what is called a potential explosion site (PES), where confinement and congestion contribute to flame acceleration and the generation of overpressure. When flammable vapors enter the PES and are ignited, the possibility of human injury/death, asset damage, or event escalation becomes a concern. The concern for human injury or death is most often addressed in the form of a building siting study. Because people are somewhat less likely to be injured or killed when outside, as compared to when inside a building, the siting study focuses on the potential impacts to buildings within and around petrochemical facilities. It then becomes the task of process safety professionals to estimate the potential for VCE events, and their resulting overpressure impacts on buildings.

The Baker-Strehlow-Tang (BST) model and the TNO Multi-energy (ME) model use blast curves to estimate the overpressure impacts generated by an exploding vapor cloud. Both approaches require the user to estimate the strength of the explosion as a function of the reactivity of the flammable material and the degree of confinement or congestion present in the cloud. This information is then used to determine which strength curve (in the case of the TNO model) or flame speed curve (in the case of the BST model) is used to calculate the overpressure of the explosion as a function of distance from the center of the explosion. It is not clear, however, how best to choose the correct curve when modeling an explosion.

There have been several papers published regarding blast curve selection for the ME model. These methods are generally presented as a qualitative ruleset which acts as an index into which of the 10 curves should be used. The BST model provides a simple set of guidelines that result in a selection of a flame speed, which corresponds to an overpressure vs. distance relationship. Three parameters are used to determine the flame speed: reactivity of the flammable gas, the degree of confinement of the flammable cloud, and the degree of obstruction due to obstacles within the flammable cloud. These parameters are represented by a set of what are effectively low, medium, and high choices for each parameter.

While the BST model provides reasonable guidance on choosing the explosion parameters, there are still areas that could be improved. One problem is that the BST flame speed matrix includes large jumps in flame speed between categories of confinement, reactivity, or congestion. This was first addressed by Baker in a 1998 paper [1] where a new confinement category, 2½-D, was added for those situations that were more confined than the 3-D case, but less so than the 2-D case. The flame speeds used for the 2½-D confinement were simply the arithmetic average of the flame speeds for the 2-D and 3-D cases for a given reactivity and congestion class. Even with this extension, large discontinuities remained among the categories for flame speed according to the prescribed methodology. These discontinuities drive the need for a new method to determine flame speed based upon quantifiable properties of the flammable gas and its surroundings that varies smoothly across the range of conditions that are found in actual process plants.

The creation of the QMEFS model [2] improved upon the basic BST scheme by combining it with the GAMES correlations [3] and providing a numerical resolution to the BST matrix. Integration of the GAMES model allows the strength of the explosion to be defined by numeric input parameters and avoids the low/medium/high categorizations.

## 2 Parameters Affecting Flame Speed

With the incorporation of the GAMES correlations, the resulting explosion flame speed is characterized by five parameters:

- Laminar burning velocity (LBV) is used as the characteristic property for fuel reactivity

- Volume blockage ratio (VBR) and average obstacle diameter define the level of congestion
- Confinement is characterized as the number of confining planes, or walls, that are available to confine the expanding gases.
- Flame run-up distance is used to characterize the flame acceleration path

The last part of explosion characterization is the explosive energy of the fuel/air mixture, which is defined by the gas concentration, the combustion energy of the fuel, and the volume occupied by the fuel/air mixture.

## *2.1 Reactivity*

Fuel reactivity is a measure of the propensity of the flame front to accelerate and create overpressures or potentially undergo a deflagration-to-detonation transition (DDT). The BST model originally classified reactivity as high, medium, or low based on the fuel in the fuel-air mixture. The original rule set [1] defined propylene oxide, ethylene oxide, and hydrogen (or materials with a 0.8 m/s LBV or larger) as high reactivity; methane and carbon monoxide were classified as low reactivity, and everything else as medium reactivity. According to the current BST model, materials having a laminar flame speed greater than 0.75 m/s are considered high reactivity (to accommodate ethylene being high reactivity) while those having a laminar flame speed below 0.4 m/s are considered low reactivity [1], although some BST publications list the low boundary as 0.45 m/s. In 2016, the BST categories were expanded to include a very-low category for fuels like ammonia, which has an LBV in the range of 0.07 m/s to 0.15 m/s [4].

In the experimental studies that form the basis of the BST model, ammonia, methane, propane, and ethylene were used to define the very low, low, medium, and high reactivity categories. These categories produce confusion due to how they create boundaries between the BST categories. For example, the LBV of ethylene, depending on the cited reference, ranges between 0.64 and 0.83 m/s. With the current definition of 0.75 m/s as the high reactivity boundary, this makes ethylene a "borderline" fuel – one that may either be defined as medium or high reactivity. However, ethylene is explicitly categorized as a high reactivity material, as their experimental work used ethylene as the high reactivity material [5]. This then begs the question: what about ethylene oxide (LBV 1.0 m/s) or acetylene (LBV 1.6 m/s)? These two materials are clearly more reactive than ethylene, but are still categorized as high reactivity. The BST model effectively says that an ethylene explosion will be as severe as an acetylene explosion, when all other parameters are held constant.

## *2.2 Congestion*

In the original BST model, low obstacle density was defined as having an area blockage ratio (ABR) of less than 10%, while high obstacle density was defined as an ABR of 40% or greater, and everything in between is considered medium [6]. Later BST guidelines have used the volume blockage ratio (VBR) as the parameter for classifying obstruction [5]. In practice both blockage ratios will affect how fast a flame accelerates, but, for a uniform obstacle field, the two are related simply by the pitch-to-diameter ratio of the obstacles. If the obstacles included in an ABR are not repeated quasi-uniformly throughout the obstacle field, then their effect would be more accurately portrayed by VBR.

The classifications of low, medium, and high are often too coarse for proper assignment of congestion to a particular PES. In the experimental programs that helped to define the BST model, VBRs of 1.5%, 4.3%, and 5.7% were used to represent the low, medium, and high categories. As with other VCE parameters, the differences between these categories leaves significant room for variation between the categories. In addition, no definition of 0% VBR is provided, nor is any relationship to VBRs higher than 5.7%.

## *2.3 Confinement*

The effect of the confinement of the flammable cloud is taken into account by determining the number of dimensions in which the burning gas may expand. 3-D expansion allows the burning cloud to expand freely in all directions and results in the slowest flame acceleration and lowest overpressures. 2-D expansion, such as a flame between two flat plates, generates higher overpressures because the combustion gases have fewer directions in which to expand resulting is a higher flame acceleration.

Confinement is included in the BST model by identifying the number of dimensions that are available to the products of combustion for expansion. Expansion into free space (except for the plane of the ground) is considered 3-D expansion, expansion between two parallel planes is considered 2-D expansion, and expansion in a pipe is considered 1-D expansion. The 1-D case was removed in the more recent publications discussing the BST model [5]. To handle the case of a frangible or partially-confining plane, such as a very closely spaced pipe rack, the BST model added a 2.5-D classification which simply averaged the 2-D and 3-D flame speed results [5].

While the classification is fairly straightforward, especially after the 2.5-D category was added, some situations are still difficult to define. These include partial walls, pieces of equipment that may form more confinement than congestion, or surfaces that are porous such as grated decks.

## *2.4 Other Factors*

Several researchers [5, 7, 8] have acknowledged that the overall dimensions of the flammable vapor cloud before ignition directly affect the final flame speed and consequent overpressures of the vapor cloud explosion. Since a flame will accelerate until it reaches a maximum sustainable value (or undergoes a deflagration to detonation transition, a DDT), the maximum dimension of a flammable cloud is expected to be an important variable in the creation of overpressure. As a point of reference, the length available for flame acceleration in the more recent BST tests was 15 meters [5].

In addition, research [7, 8] suggests that the scale of obstacles within the congested area also affects the maximum flame speed that may be achieved. As a flame front moves around and past obstacles, the turbulence that is generated can increase the flame surface area and accelerate combustion. The characteristic obstacle diameter for the BST tests was 2 inches (0.0508 meters); other test series (including the MERGE/EMERGE tests, varied the average obstacle diameter).

### *2.5 Parameters Comparison*

The BST parameters discussed above are listed in Table 1, along with the corresponding numerical values that are equivalent to those applied in the QMEFS model. When the numerical values in Table 1 are applied in the QMEFS model, the results will match the predictions of the BST model.

**Table 1. BST Flame Speed Parameters**

| BST Parameter | BST Category | Model Parameter | Corresponding Parameter Value |
|---|---|---|---|
| Reactivity | Very-Low (e.g., Ammonia) | $S_l$ | 0.08 m/s |
| | Low (e.g., Methane) | | 0.37 m/s |
| | Medium (e.g., Propane) | | 0.43 m/s |
| | High (e.g., Ethylene) | | 0.75 m/s |
| Obstacle Density | Low | *VBR* | 0.015 |
| | Medium | | 0.043 |
| | High | | 0.057 |
| Expansion | 3-D | *nPlanes* | 1 |
| | 2.5-D | | 1.5 |
| | 2-D | | 2 |

## 3 Model for Estimating Flame Speed

The QMEFS model was created to overcome some of the limitations described above. By incorporation of the GAMES correlations and the overpressure/impulse curves from the BST method, a model that provides the capability for more detailed descriptions of explosion scenarios has been created. Thus, QMEFS incorporates the modeling methodologies and experimental data sets that have been used in both the BST model and in the GAMES project.

### *3.1 Fundamental Correlations*

The fundamental correlation is based on correlations developed from the GAMES project [3]. These correlations related overpressure to parameters based on the MERGE and EMERGE experiments [7,8]. These correlations can be combined with acoustic theory to relate flame speed to explosion parameters [2] to produce the equations below.

For no confining surfaces other than the ground (3D expansion):

$$2.4 * \frac{M_f^2}{1+M_f} * p_0 = 0.84 * \left(\frac{VBR*L_f}{D}\right)^{2.75} * S_l^{2.7} * D^{0.7} \tag{1}$$

For confinement between two flat planes (2D expansion):

$$2.4 * \frac{M_f^2}{1+M_f} * p_0 = 3.38 * \left(\frac{VBR*L_f}{D}\right)^{2.25} * S_l^{2.7} * D^{0.7} \tag{2}$$

where:

- $VBR$ = volume blockage ratio
- $L_f$ = maximum distance a flame may propagate in the obstructed region (i.e., the run-up distance), m
- $D$ = average obstacle diameter, m
- $S_l$ = laminar burning velocity (LBV) of the flammable gas, m/s
- $p_0$ = the ambient pressure, bara
- $M_f$ = the flame speed relative to a fixed observer, expressed as a Mach number

To correct the flame speed to account for confinement in spaces other than one or two planes, a curve fit of the ratio of burned cloud to unburned cloud radii versus the number of confining planes was made. A curve fit was generated with an expansion ratio ( $\alpha$ ) equal to 7, a typical expansion ratio for common hydrocarbon fuels in air. This is also the value assumed in the derivation of portions of the BST model [9]. The equation resulting from this curve fit is:

$$NPF = \frac{1}{0.5035 - 0.0757 \cdot nPlanes} - 1.338 \tag{3}$$

where:

- $NPF$ = flame speed correction factor
- $nPlanes$ = number of confining planes

For true 3-D spherical expansion, $nPlanes = 0$. For a typical explosion at grade level, $nPlanes = 1$. For 2-D expansion, such as a flame propagating between a floor and a ceiling, $nPlanes = 2$. For 1-D expansion, such as a flame front in a pipe, $nPlanes = 4$. Planes that are not completely solid or rigid may be accounted for using a fraction of a plane.

The flame speed correction factor, *NPF*, does not fully account for increases in flame speed as the number of confining planes increases. This is accomplished by applying the following factor to the flame speed when *nPlanes* is equal to 2:

$$\beta = \frac{M_{f,2D}}{M_{f,3D} \cdot NPF(nPlanes = 2)} \tag{4}$$

This can be implemented for nPlanes =1 to nPlanes = 4:

$nPlanes = 1$

$$M_f = M_{f\,3D} * NPF \tag{5}$$

$1 > nPlanes < 2$

$$M_f = M_{f\,3D} * NPF * \beta \tag{6}$$

$nPlanes >= 2$

$$M_f = M_{f\,3D} * NPF * [1 + (\beta - 1) * (nPlanes - 1)] \tag{7}$$

### *3.2 Flame Speed Limits*

The equations above are implemented in the QMEFS model to define the flame speed ($M_f$) and then coupled with the latest BST flame speed matrix [4]. One limitation to the GAMES correlations is that they can predict unrealistic flame speeds. This is a direct result of the correlations fitting the data within the scope of the MERGE and EMERGE tests, and not being designed to extend beyond that scope. By coupling with the BST model, this problem is partially resolved because the BST matrix provides a set of upper limits to flame speeds (and by extension, the explosion strength).

Upon examination, it can be seen the GAMES correlations will predict higher and higher flame speed values as parameters such as $L_f$ or $S_l$, increase. In reality, for a given obstacle configuration, geometry, and flammable gas mixture, there is a limit to the flame speed that can be achieved. This flame speed may be either subsonic or the cloud may undergo a deflagration-to-detonation transition (DDT) in which the flame front becomes a detonation propagating at roughly the Chapman-Jouguet (C-J) detonation velocity for the flammable mixture. For most common flammable hydrocarbons in air, this speed is roughly 1800 m/s or $M_f$= 5.2. For purposes of this model, any flame that accelerates to a calculated velocity greater than $M_f$ = 3.0 is considered to have undergone DDT [10, 11] and the flame speed is set to the C-J velocity, $M_f$= 5.2.

The flame speeds suggested by the latest BST model are applied as the upper limit flame speeds to be used in this model, as they have been "scaled up" to account for industrial scale flammable clouds [5] and account for DDTs.

The result of these correlations is a model that estimates the flame speed based on $VBR$, $D$, $S_l$, $L_f$ and the number of confining planes. The resulting flame speed is compared to the published BST model, which provides a matrix of flame speeds as a function of reactivity, obstacle density, and flame expansion. The BST matrix values are used as maximum flame speed values for the set of parameters given in Table 1. Values between the BST matrix elements are calculated by linear interpolation (as was done for Baker's 2½-D flame expansion category). This methodology provides an extended, systematic approach for estimating flame speeds resulting from the combustion of a flammable cloud in an obstructed and/or confined region. Predicted flame speeds are used with the existing BST blast curves to produce estimates of overpressure at a distance from the explosion source.

### 3.3 *Expansion of the BST Matrix*

The range of the QMEFS model is limited by the upper and lower bounds of the BST matrix elements for congestion, confinement, and fuel reactivity. Through application of this model to various petrochemical projects, the need to expand the matrix became obvious. As described above, the model is built to provide predictions for the *nPlanes* (confinement) parameter in the range of 1.0 to 4.0 (the model is not validated nor trusted above this value). Values of average obstacle diameter and flame run-up distance are built into the GAMES correlations, so naturally are accounted for within reasonable limits. This leaves the congestion (VBR) and fuel reactivity parameters as places to potentially expand the model.

The intent of model expansion is to widen the range of congestion and reactivity inputs so that it removes questions about its use in the "below low" and "above high" regions. With proper expansion of the model described above, the predicted flame speed following ignition of a flammable cloud may then be used with the BST blast curves to estimate the overpressure impacts in the area surrounding a vapor cloud explosion.

To accomplish this expansion, additional categories were added to the BST matrix for congestion and reactivity. This follows Baker's own expansion into the **very-low reactivity** category when ammonia explosion experimental results were published [4].

For a **very-high reactivity** category, hydrogen was assumed as the characteristic fuel, and represents a true upper bound for LBV. Expansion to the very-high category, for parts of the matrix with high flame speeds is simple because DDTs are expected in these conditions already. For other areas, published data was required for the basis of the expansion.

In the case of the **very-low congestion** category, open air (3-D flame expansion) explosion tests were performed by Harris and Wickens [12] as well as Sato and Iwabuchi [13] for a variety of materials. The **very-high congestion** category, several of the BFETS series of tests were used [14] due to their use of a VBR in the 10% range.

The expanded matrix applied in the new QMEFS model is presented in Table 2. The latest BST flame speed matrix values are outlined by bold lines in Table 2. Values outside these areas have been added as part of the QMEFS expansion.

While a 1-D confinement is not expected to be applied in large-scale industrial scenarios, the *nPlanes* = 5 (1-D) confinement category is included in this model to provide an interpolation point beyond *nPlanes* = 2 (2-D flame expansion).

**Table 2. Expanded Flame Speed Matrix**

| Confinement | Reactivity | Congestion | | | | |
|---|---|---|---|---|---|---|
| | | Very-Low | Low | Medium | High | Very-High |
| 1D | Very-High | DDT | DDT | DDT | DDT | DDT |
| | High | DDT | DDT | DDT | DDT | DDT |
| | Medium | 1.029 | 1.029 | 1.765 | 2.265 | DDT |
| | Low | 0.5 | 0.5 * | 1.029 | 2 * | DDT |
| | Very-Low | 0.023 | 0.023 | 0.124 | 0.168 | 0.303 |
| 2D | Very-High | DDT | DDT | DDT | DDT | DDT |
| | High | 0.588 | 0.588 | DDT | DDT | DDT |
| | Medium | 0.47 | 0.47 | 0.66 | 1.6 | DDT |
| | Low | 0.079 | 0.079 | 0.47 | 0.66 | 1.244 |
| | Very-Low | 0.008 | 0.008 | 0.049 | 0.069 | 0.130 |
| 3D | Very-High | 0.5 | DDT | DDT | DDT | DDT |
| | High | 0.06 | 0.36 | DDT | DDT | DDT |
| | Medium | 0.03 | 0.11 | 0.44 | 0.5 | 1.084 |
| | Low | 0.025 | 0.026 | 0.23 | 0.34 | 0.678 |
| | Very-Low | 0.003 | 0.003 | 0.024 | 0.036 | 0.073 |

* Values adjusted from BST publication for model consistency

# 4 Application

It is helpful to test the QMEFS model outside of the published BST matrix, in the areas where the QMEFS matrix has been expanded. Large scale hydrogen experiments were performed in a hemispherical test rig with no congestion. The test volume was 300 $m^3$ with a base of 10.4 meters diameter [13] and center ignition. The tests were modeled with the latest BST model and the expanded BST model. The model parameters listed in Table 3 were applied to these calculations. Figure 1 presents the overpressure in relation to distance from center of the fuel-air mixture, for the test results and both BST and QMEFS predictions. The BST model does not have a category for a hydrogen vapor cloud with no congestion, thus the scenario falls into the high fuel reactivity and low congestion portion of the matrix. The result is an underprediction of the overpressure. The BST high fuel reactivity works well for fuel such as ethylene (LBV = 0.76 m/s), but will often underpredict the impacts of a hydrogen (LBV = 3 m/s) explosion. The expanded matrix allows a higher fuel reactivity category and a lower congestion category, and consequently a better prediction of the explosion impacts. As seen in Figure 1, the QMEFS result is a slight overprediction of the overpressure.

**Table 3. Modeling Parameters for an Unconfined Hydrogen Explosion**

| Parameter | BST | Expanded QMEFS |
|---|---|---|
| Volume [$m^3$] | 300 | 300 |
| *nPlanes* | 1 | 1 |
| Reactivity | High | Very-High |
| VBR | N/A | 0 |
| Congestion | low | N/A |
| Run-up distance [m] | N/A | 5.2 |

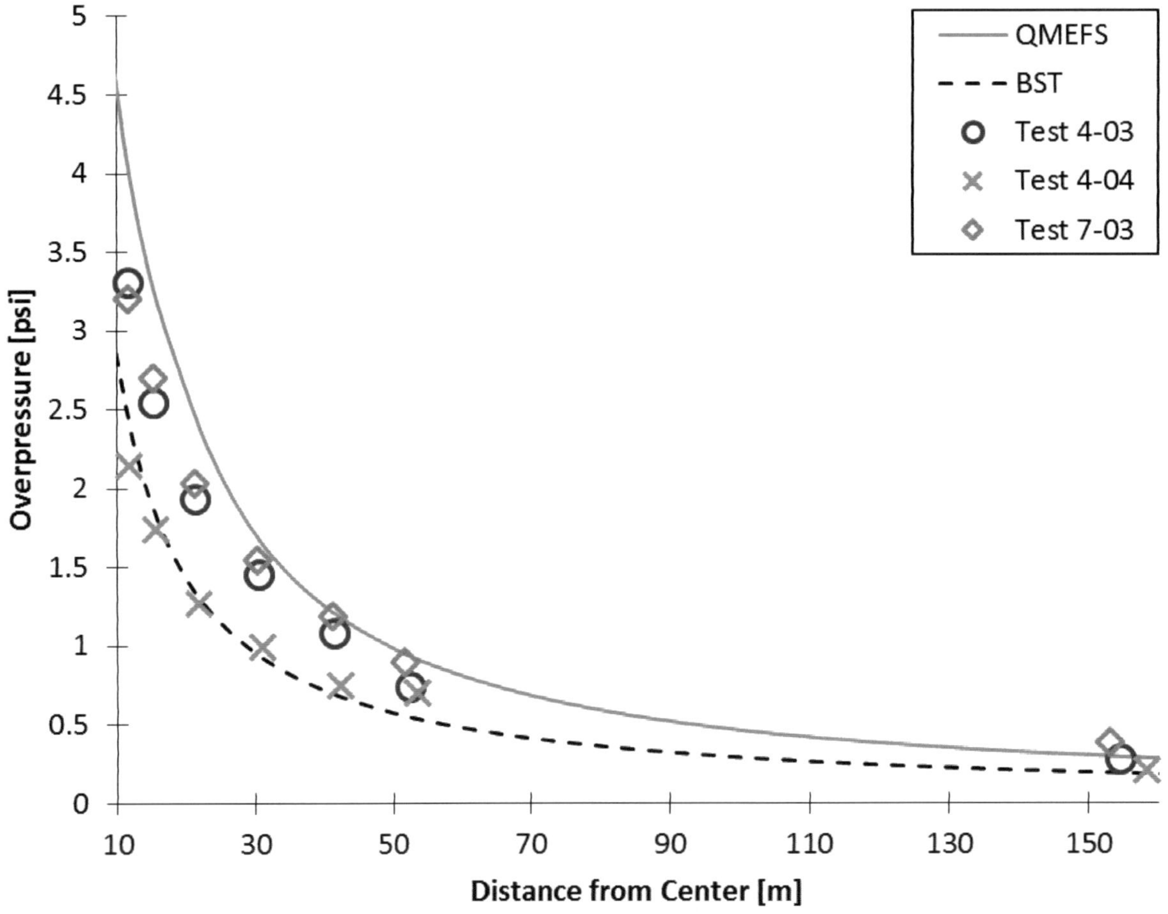


**Figure 1. Overpressure vs. Distance for Open-Air Hydrogen Tests**

As discussed in the paper, the distance that is available for the flame to accelerate, i.e. the run-up distance, can have a significant effect on the flame speed attained in a flammable gas cloud. The flame speeds presented in the original Baker-Strehlow (BS) model [6] were generated in a test rig whose largest dimension was less than 6 feet (1.8 meters) [5]. The flame speeds presented in the 2005 BST paper are based on tests where the largest dimension was 48 feet (14.6 meters) with the published flame speed results "scaled up" to account for the maximum size of a typical industrial plant [5]. One test of the QMEFS model is to determine the flame speed versus run-up distance for the test configurations used in the newest BST model when specific parameters are varied.

Figure 2 shows the results for the 3-D flame expansion (*nPlanes* = 1), medium reactivity ($S_l$ = 0.43 m/s) case. The curves plotted in Figure 2 show six VBR values – the high (5.7%), medium (4.3%), and low (1.5%) congestion values used in the BST experiments, one intermediate value to show how the model behaves as the VBR is adjusted, a fifth value (1%) below the BST low category to demonstrate the behavior in the "less-than-low" congestion range, and the sixth value (10%) above the BST high category to demonstrate the behavior in the "higher-than-high" congestion range.

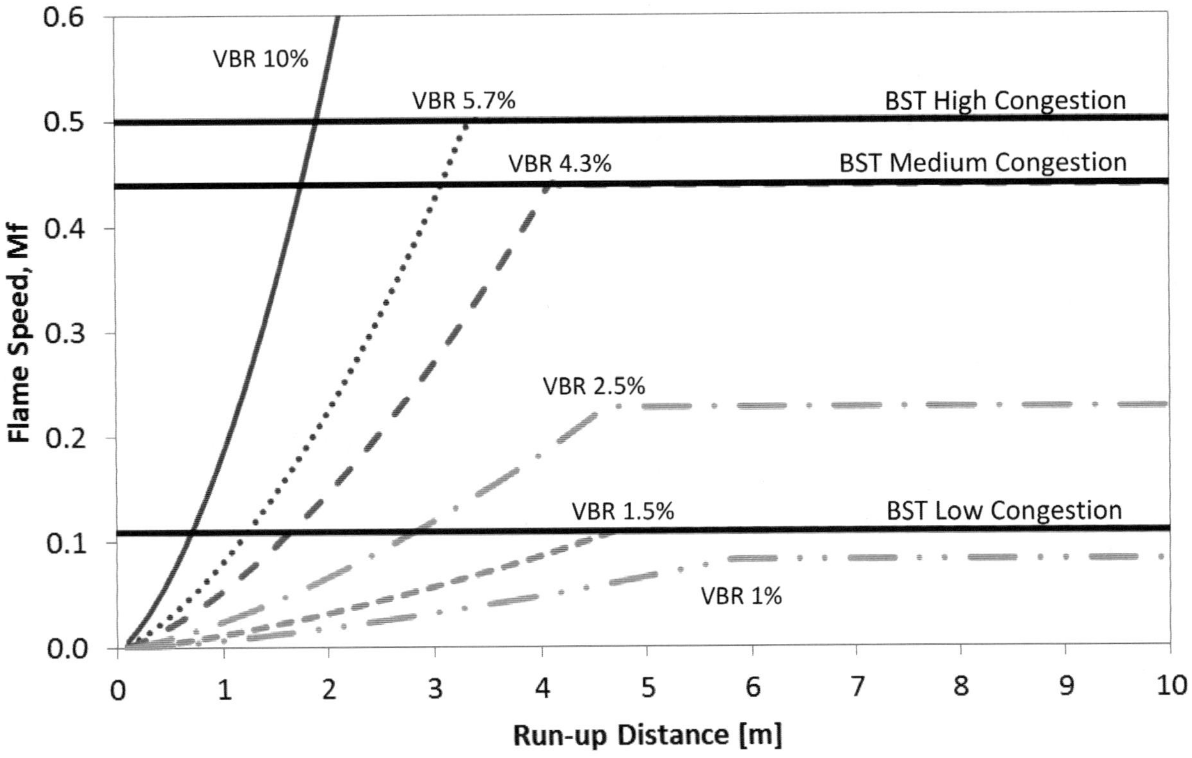


**Figure 2. Flame Speed vs. Run-Up Distance; 3-D, Propane, D = 2"**

Figure 3 demonstrates how flame speed varies with *nPlanes* and LBV. The basis of the figure is an explosion involving a medium VBR (0.043) and a 4-inch average obstacle diameter. The *nPlanes* parameter is varied between 1.0 (3-D flame expansion) and 4.0 (the maximum value recommend for use in the QMEFS model). The curves plotted in Figure 3 show five materials with various LBV values – ammonia (0.08 m/s), methane (0.37 m/s), propane (0.43 m/s), ethylene (0.76

m/s) and ethylene oxide (0.91 m/s). In the BST model ethylene oxide and ethylene are grouped together into the high category, but these two fuels have different burning properties as can be seen with their LBV values. Thus, by accounting for the LBV of the fuel, the new model provides a finer resolution for flame speed. In addition, as seen in Figure 3, the QMEFS model provides more resolution for the flame expansion parameter through the use of the *nPlanes* variable.

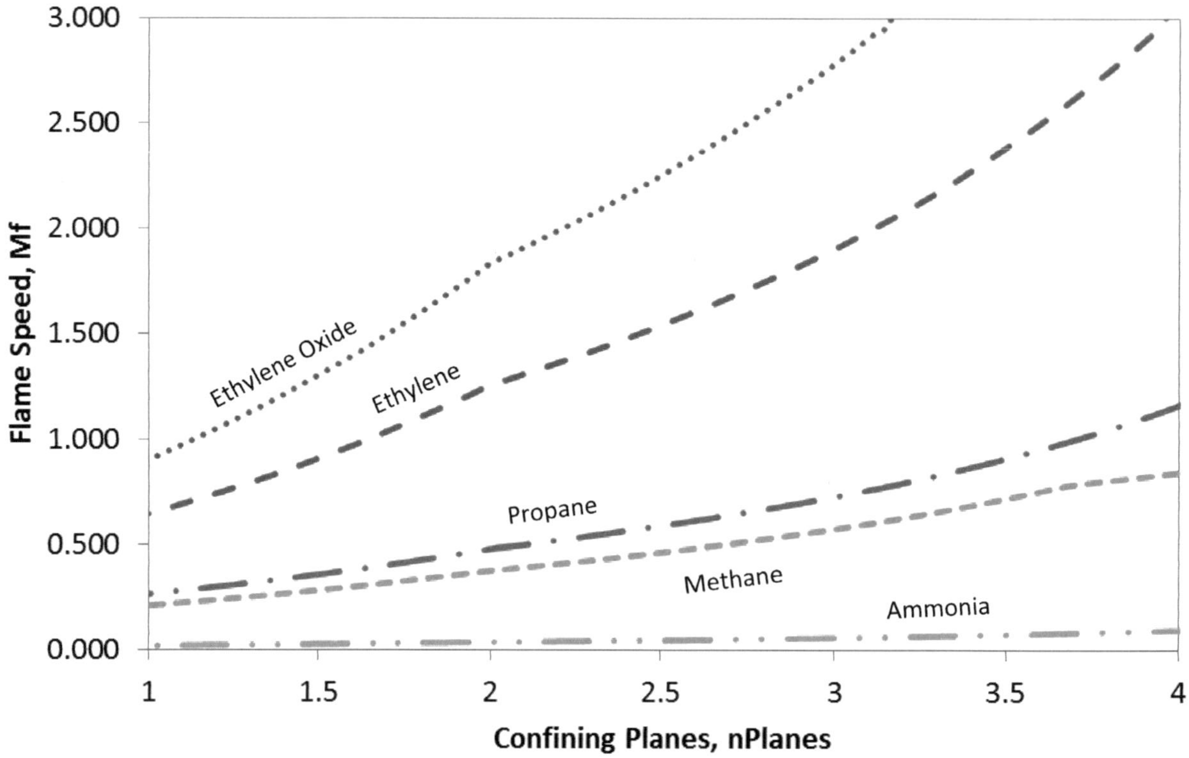


**Figure 3. Flame Speed vs. Confining Planes; Medium VBR, D = 4"**

# 5 Conclusions

The model described in this paper provides an improved method to describe the characteristics of a vapor cloud explosion over a wide range of conditions. It combines elements of the work conducted by Baker to develop the BST model, as well as the European MERGE/EMERGE projects; thus incorporating data from the two largest, modern vapor cloud explosion test projects.

Because the BST categories are resolved into numeric ranges, a higher resolution solution space and a clearer set of measureable inputs are achieved. With the ability to model the reactivity of a flammable gas cloud based on laminar flame speed instead of a low, medium, or high classification, the model is able to more accurately describe a wide range of flammable gases and mixtures. Expansion of the constraining flame speed categories to very-low and very-high then provides an even wider range of possibilities.

With the QMEFS matrix expanded to congestion (VBR) categories of very-low and very-high, the model provides a wider range of predictive capability for explosion scenarios. These expansions and enhancements are still confined by the flame speeds in the BST system, and are applied within the set of curves for overpressure and impulse. Thus, with the QMEFS model presented here, process safety experts have a more versatile tool to predict VCE impacts in petrochemical plants.

## 6 References


[1] Baker, Q.A., C.M. Doolittle, G.A., Fitzgerald, and M.J.Tang, "Recent Developments in the Baker-Strehlow VCE Analysis Methodology," *Process Safety Progress*, v.17, No.4 (1998).

[2] Melton. T.A. and Marx J.D., "Estimating Flame Speeds for Use with BST Blast Curves," Process Safety Progress, v. 28, March, (2009).

[3] Mercx, W.P.M., A.C. van den Berg, and D. van Leeuwen, A*pplication of correlations to quantify the source strength of vapour cloud explosions in realistic situations*. Final report for the project: 'GAMES', TNO report PML 1998-C53, Rijswijk, The Netherlands (1998)

[4] Thomas, K.J., Malik. D.R., Downes, A.M., Rodges, S., "Ammonia Vapor Cloud Explosion Testing and Implications for Facility Siting Studies, Consequence Analyses and Risk Assessments," *50th Loss Prevention Symposium*, AIChE, April, (2016).

[5] Pierorazio, A.J., J.K.Thomas, Q.A.Baker, and D.E.Ketchum, "An Update to the Baker-Strehlow-Tang Vapor Cloud Explosion Prediction Methodology Flame Speed Table," *Process Safety Progress*, v.24, no.1 (2005).

[6] Baker, Q.A., M.J. Tang, E. Scheier, and G.J. Silva, "Vapor Cloud Explosion Analysis," *28th Loss Prevention Symposium*, AIChE, April, (1994).

[7] Mercx, W.P.M., editor, *Extended Modelling and Experimental Research into Gas Explosions*, Final summary report of the project EMERGE, CEC contract EV5VCT930274, European Commission, Directorate General XII, Brussels, Belgium (1997).

[8] Mercx, W.P.M., A.C. van den Berg, C.J. Hayhurst, N.J. Robertson, and K.C. Moran (2000), *Developments in Vapour Cloud Explosion Blast Modeling*, Journal of Hazardous Materials 71 (2000) 301-319.

[9] Tang, M.J. and Q.A.Baker, "A New Set of Blast Curves from Vapor Cloud Explosion," *Process Safety Progress*, v.18, no.4 (1999).

[10] Oran, E.S. and V.M. Gamezo, "Origins of the deflagration-to-detonation transition in gas-phase combustion," *Combustion and Flame*, 148 (2007).

[11] Vasil'ev, A.A., "Estimation of Critical Conditions for the Detonation-to-Deflagration Transition," *Combustion, Explosion, and Shock Waves*, v.42, no. 2 (2006).

[12] Harris, R.J. Wickens, M.J., "Understanding Vapor Cloud Explosions – An Experimental Study," The Institute of Gas Engineers 55th Autumn Meeting, (1989).

[13] Sato, Y., Iwabuchi, H., Groethe, M., Merilo, E., Chiba, S., "Experiments on Hydrogen Deflagration," Journal of Power Sources v.159 (2006).

[14] Health & Safety Executive, "Explosions in Full Scale Offshore Module Geometries Main Report," Offshore Technology Report – OTO 1999 043. Prepared by BG Technology, April, (2000).

## 479444 Do We Need a Quantitative Risk Assessment (QRA) If My Consequence Analysis Facility Siting Presents a Problem?

**Abstract:**

There are three steps that a regular facility siting study (FSS) can go through. The first step is the basic separation table approach descrived in API 752. The second step is the consequence analysis facility siting and the third and final one is the quantitative risk assessment. It turns out, however, that the vast majority of the consequence analysis FSS studies quickly demonstrate that the impact to buildings can be more significant that what have been defined as acceptable. For such cases, the analysis normally turns into the QRA, exceedance curves, building structural assessment, etc. The idea of this paper is to cover a intermediate step in early stages to use computational fluid dynamics (CFD) technique in the early stage of the analysis without compromising the time schedule and the budget, this would include dispersion, radiation and explosion modeling. The more accurate results obtained with CFD modelling lead to cloud sizes which are mainly less conservative than traditional methods. The objective of this paper is to go a setp beyond a previous analysis for the dispersion using CFD, but rather make a full evaluation of a FSS using CFD.

# Proposed Method for Investigating the Reactivity of Chlorine with Environmental Materials under Relevant and Controlled Conditions

**Audrey Feuvrier, PhD.**
**Ralph E. Martin Department of Chemical Engineering**
**University of Arkansas, Fayetteville, AR**
**audrey.feuvrier@uark.edu**

**Tom Spicer, Ph.D., P.E.**
**Ralph E. Martin Department of Chemical Engineering**
**University of Arkansas, Fayetteville, AR**
**tos@uark.edu**

Prepared for Poster Presentation at
American Institute of Chemical Engineers
2017 Spring Meeting and 13th Global Congress on Process Safety
San Antonio, TX
March 26 – 29, 2017

# Proposed Method for Investigating the Reactivity of Chlorine with Environmental Materials under Relevant and Controlled Conditions

*Audrey Feuvrier, PhD.*
*Ralph E. Martin Department of Chemical Engineering*
*University of Arkansas, Fayetteville, AR*
*audrey.feuvrier@uark.edu*

*Tom Spicer, Ph.D., P.E.*
*Ralph E. Martin Department of Chemical Engineering*
*University of Arkansas, Fayetteville, AR*

**Keywords:** Chlorine reactivity, chlorine release, dry deposition, Jack Rabbit II

## Abstract

Many common toxic industrial chemicals (TICs) such as chlorine and ammonia are very reactive with commonly encountered materials in the environment. The amount of TIC material reacted in the environment is an important factor in the impact of a release. Typically characterized with a deposition rate in atmospheric dispersion models, the reaction rate and total deposition quantities have predominately been studied in field scale experiments. Laboratory experiments have also been undertaken, but past investigations have inadequately quantified the available turbulence which can importantly impact surface reaction rates. The state of the art description of surface deposition represents the process as a series of resistances including the effect of atmospheric turbulence and surface reaction(s) depending on the surface material (substrate). Sponsored by the Chemical Security Analysis Center (CSAC) of the U.S. Department of Homeland Security and Transport Canada, an experimental program is underway to test potential substrates under known inlet gas concentrations with levels of turbulence that are comparable to those of the atmosphere. The objective is to develop a test methodology for subjecting a substrate to a reactive gas flow of known TIH concentration with turbulence levels consistent with those in the atmosphere. With data from repeat experiments, deposition of a reactive TIH chemical can be quantified with the proposed experimental protocol.

## 1 Introduction

Estimating the quantity of a Toxic Industrial Chemical (TIC) removed from the environment through chemical processes is critical to proper consequence assessment of accidental (or intentional) releases, particularly in urban areas [1]. TIC removal mechanisms play an important role in the rational assessment of risk, as these phenomena may significantly and quickly reduce the TIC concentration near a release [2,3], particularly for a reactive chemical such as chlorine. Typically characterized as dry deposition, reactions of chlorine with the environment was not included in the computational models used to predict the consequences of large releases of chlorine in the past [2], but new efforts are underway to characterize chlorine reactivity and incorporate such effects in consequence and risk assessments.

While species like sulfur dioxide and iodine have been the subject of extensive experimental studies [3], studies about deposition involving chlorine are scarce and show a great deal of variability. The high reactivity of chlorine as well as its toxicity make it difficult to study in a laboratory environment, and consequently, many of the studies are conducted as field studies [4]–[6], which restricts test conditions to a limited set of atmospheric conditions. In one set of laboratory experiments, Hill [7] studied the effect of deposition of several TICs on an alfalfa canopy in a chamber where humidity, light, and wind speed were controlled. He showed that chlorine deposition is rapid on alfalfa and that chamber air velocity has a significant effect on the chlorine uptake, but the level of turbulence in the chamber is unknown. Other experimental studies gave promising results such as [8]–[10], but these studies involved no other movement of the flow than the one produced by the mixing device or the circuit used to monitor the concentration level. The state of the art description of surface deposition, [11], represents the process as a series of resistances including the effect of atmospheric turbulence and surface reaction(s) depending on the surface material (substrate).

In this context, an experimental program, sponsored by the Chemical Security Analysis Center (CSAC) of the U.S. Department of Homeland Security and Transport Canada is underway to test potential substrates in a controlled environment. The project includes the design and characterization of a mixing chamber as well as development of a protocol for the measurement of chlorine removal rate for different samples of soil and vegetation. Concentration, velocity, and turbulence measurements will be made in the working section to confirm that the flow is uniform and consistent. It is also desired that the flow in the working section of the apparatus is consistent with the atmospheric flow with target turbulence intensities on the order of 10%. Working section flow speed will be variable, with a maximum in the section of 8 m/s, and expected velocities between 0.5 to 3 m/s. Passive flow conditioning will be used to eliminate non-uniformities in the flow. Recognizing that chlorine gas concentrations as high as 10% or 100,000 ppm may be used, apparatus materials of construction will be carefully chosen to resist damage and also to not remove significant amounts of chlorine during a test. Finally, to limit the hazards involved with the use of chlorine at high concentration level, the apparatus will need to fit under a 6 ft long (about 1.8 m) fume hood.

In the following, the proposed designs for the apparatus and its main elements are introduced. The choice of the mixer, fan, and materials will be discussed. In the second part of the article, the future protocol of measurement will be presented. Focus will be put on the means of measurement used and the characterization of the chamber. The steps to determine chlorine removal rates for several samples types under different environmental conditions will be considered.

## 2 Proposed design

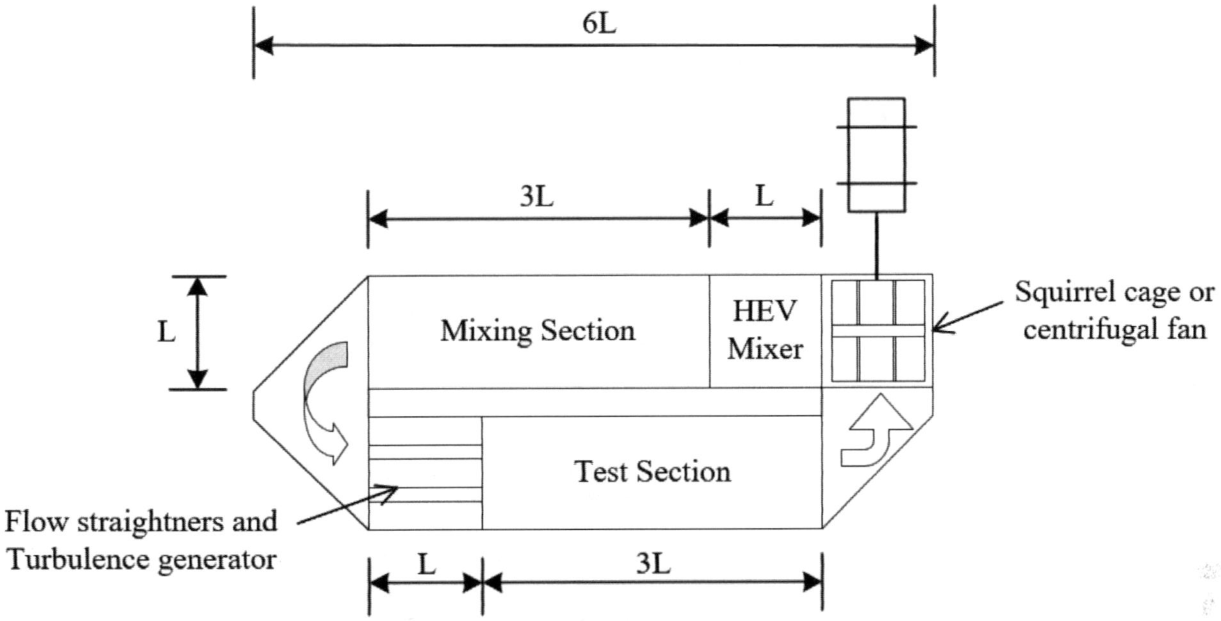


**Figure 1. Preliminary apparatus design**

Figure 2 shows the preliminary design of the apparatus following the criteria laid down previously using a simple closed return design. The test or working section is placed under the mixing section to facilitate access to the test section through a lab hood sash. The characteristic length, L, chosen to fit in a standard 6 ft hood, will be between 0.25 and 0.3 m, giving a total volume for the apparatus of between 0.16 and 0.28 $m^3$. A standard lecture bottle contains 454 g of chlorine, which would be sufficient to provide chlorine for 8 to 14 tests depending on the length L based on a maximum concentration of 100,000 ppm. The use of lecture bottles limits the quantity of chlorine required in the lab during the experiments and effectively eliminates the hazard associated with the storage of large quantities of chlorine on site. These volume values are approximate because it may be convenient to use a circular conduit for mixing and a square cross-section for the working section. A circular conduit reduces the required volume of the mixing section (with some minor reduction in power consumption), and a working test section with flat surfaces facilitates optical measurements during characterization and sample placement in the testing phase.

## 2.1 Mixer and flow conditioner

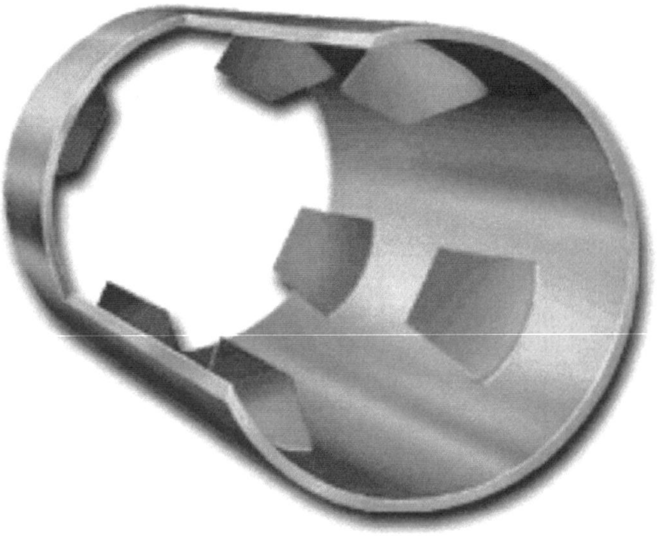

**Figure 2. Typical High Efficiency Vortex (HEV) mixer [12].**

Figure 3 shows a typical High Efficiency Vortex (HEV) mixer which is placed downstream of the fan. This type of mixer was selected because it provides efficient mixing with turbulent flow with minimal pressure drop. For a Kenics™ HEV mixer, Figure 4 shows the relative coefficient of variation, $COV_R$ , defined as the quotient between the outlet and inlet coefficient of variation, as a function of the Reynolds number (defined as $U_\infty\, L/\upsilon$ where $U_\infty$ represent the characteristic velocity in the tunnel, L is the characteristic length and $\upsilon$ is the kinematic viscosity of the gas). The coefficient of variation for the mixing of two fluid represents the quotient of the standard deviation of the concentration and the mean concentration. Therefore, the lower the $COV_R$, the better the mixture quality. The gray zone on Figure 4 represents the range of Reynolds number of this study. Based on Myers et al [13], the pressure drop for our (custom made) apparatus was estimated to be between 1.7 and 17.3 Pa, making it negligible. The HEV mixing process rely on the production of hairpin vortices downstream of the blades of the mixer combined with a pair of counter rotating longitudinal vortices [14]. A mixing section downstream of the device and equal to three times the length of the mixer is recommended to provide for sufficient time for the vortices to complete the mixing process and consequently obtain the predicted $COV_R$. Additional turbulence generated between the mixing section and the test section will only improve the $COV_R$ before the flow enters the test section..

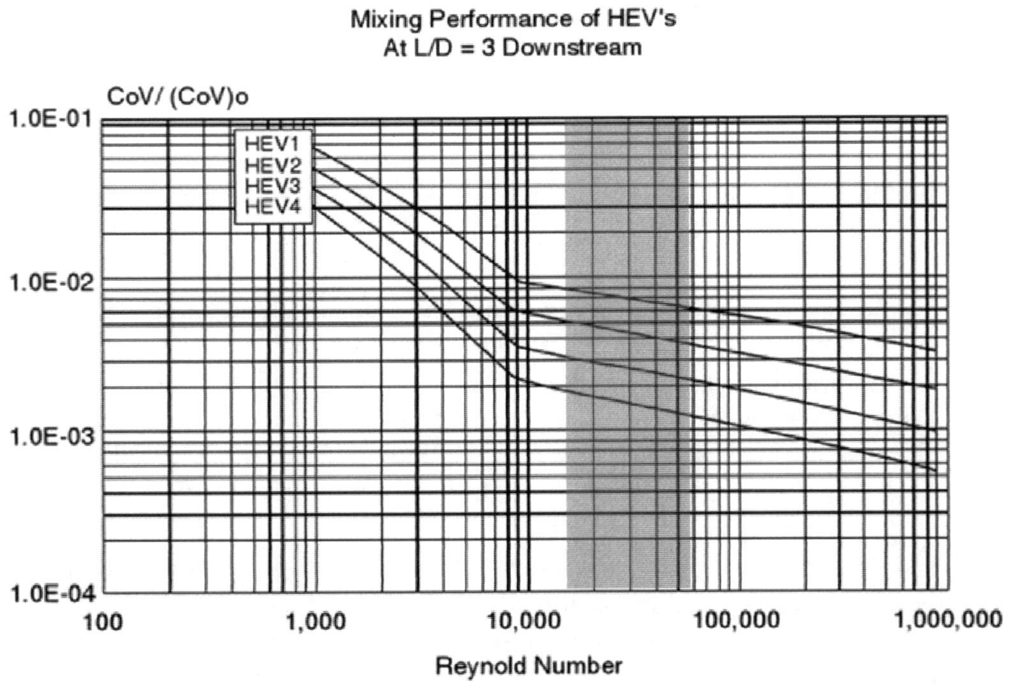


**Figure 3. Relative coefficient of variation ($COV_R$) vs. Reynolds Number and number of tabs row for the Kenics HEV,** [12].

The corners currently consist of two 45° bends with a short duct between them, and the 45° bends are preferred to 90° bends to prevent the accumulation of gases in the dead space. To help maintain good flow distribution in the channel, sets of vanes will be placed in each bend. Operating at low wind speed allows us to consider simple circular cross sectional shapes for the vanes [15].

After the two first corners, the flow conditioner section is planned to eliminate large eddies that might be present in the flow ensuring uniformity of the flow and simultaneously raising the turbulence intensity up to 10%. To achieve this, we plan to use a combination of screens or perforated plates. Samples will be placed in the test section through a removable side with provision for a tray in the floor of the test section to accommodate soil and/or plant roots. After experiments, ventilation of the apparatus will be achieved through two hatches on each bend between the mixing and flow conditioner sections.

## 2.2 The fan

On Figure 2, the motor of the fan is represented outside of the apparatus because chlorine would be expected to react and corrode any of its exposed surfaces. The motor shaft will be supported by Nylon or Teflon bearings located in the side of the apparatus to isolate the motor from chlorine exposure. A centrifugal (squirrel cage) fan is being considered to allow placement of the fan above the apparatus. The fan blades and shaft exposed to chlorine will be treated to limit damage due to chlorine exposure as discussed below.

## 2.3 Building materials

During the first phase of the experimentation, the apparatus will only use a mixture of air and an inert gases will be used for optic velocity and turbulence measurements and concentration measurements (including a propylene tracer). During this preliminary phase, the structure of the test section will be made with plywood, but two of its walls will be made with polycarbonate to allow visualization and optical measurements of the velocity in the section.

In the second phase of the experimentation involving chlorine, all elements in contact with the test mixture must be made with a material or coated with a material capable of resisting high chlorine concentrations. In addition to extending the workable lifetime of the apparatus, reducing the potential for chlorine to react with materials in the test chamber is also desired so that the reduction in chlorine gas concentration can be attributed to reaction with the sample tested as opposed to the chamber itself. In the case of the fan blades and shaft, a commercial coating is planned (such as Cerakote or titanium). Titanium is known to be highly reactive with dry chlorine, however 1% of water may be enough for passivation after mechanical damage to titanium under normal conditions [16]. For the mixing section, a metal or plastic (PVC) duct will be used since no optical access is required. The structure of the corners and the test section will be made with plywood. In the second phase, polycarbonate may be replaced by plywood. The flow straighteners will likely be made of a suitable metal. All surfaces except the fan blades and shaft are planned to be coated with Viton™, a coating material widely used in applications with chlorine in the chemical industry. We are currently evaluating whether the Viton™ coating under consideration will adhere sufficiently to plywood, and if it does not, then the entire apparatus may need to be fabricated from sheet metal with all internal surfaces painted with Viton™.

# 3 Protocol for future experimentations

## 3.1 Experimental resources and characterization of the mixing chamber

Several instruments will be used during the characterization of the chamber and concentration measurement. During both phases of the experimental measurements, the temperature, pressure and humidity of the ambient air will be monitored.

For the characterization of the chamber, the first step will be verifying that the flow entering the test section is uniform and fully mixed through visualization. The working section will be equipped with LED lights and a ROSCO 1500 fogger machine will provide the fog. The fog will be introduce downstream of the fan, at the same point planned for injection of the gases. Video recordings of the fog in the working section will be made with several velocities. After visualization experiments, velocity measurements will be made with a 2D-LDV (Laser Doppler Velocimetry) system using an Argon Ion 300 mW air-cooled laser. This single point technique is non-intrusive and allows simultaneous measurement of two components of the velocity with a good spatial and temporal resolution. Finally, the concentration measurements will be made using an Aurora Scientific mini-PID (photo ionization detectors) The PID measurement method relies on photoionization of a volatile organic compound (in this case a propylene tracer). Small access points will have to be made in the test section for the PID. In the characterization phase, a simulant gas mixture (using sulfur hexafluoride) with be used to reproduce the molecular weight of chlorine. Sulfur hexafluoride is non-toxic, non-flammable, non-reactive and can be used in the PID system.

In conclusion, the characterization of the mixing chamber will assess the quality of the mixing and the flow in the working section, the steps are listed below :

1. Visualization of the flow going in the test section.
2. Velocity profiles taken in the test section will ensure the uniformity of the flow and that the turbulence intensity level is respected.
3. Concentration profiles will provide information on the uniformity of the concentration of the simulant gases in the chamber.

## 3.2 Concentration measurements and evaluation of the reactivity of chlorine with different samples

After the characterization of the working section, the apparatus will be modified to undergo tests with high concentration of chlorine. The changes have been described in section 2.3. Furthermore, the tunnel will have to allow the access of the JAZ sensors in order to measure the concentration of chlorine. As with characterization of the chamber, the evaluation of the reactivity of chlorine with different samples will request several steps listed hereunder.

1. Characterization of the response of the apparatus to chlorine : Tests will be done without samples, so that the chlorine concentration decay due to the apparatus itself can be measured. A correction factor will be calculated and implemented in the following concentration measurements. Multiple tests will be done, in order to ensure of the repeatability of the measurements.
2. Removal rate estimation : Samples will be placed in the tray, allowing the removal rate of these samples to be estimated through concentration measurements.
3. Analysis of the results : The measurement of the concentration decay in the chamber and the study of the samples put in contact with chlorine will provide valid information in order to calculate the removal rate of chlorine for these different samples. Analytical techniques for determining the amount of chlorine uptake by the test material will depend on the material.

# 4 Conclusion

In this article, the design of an innovative apparatus or mixing chamber and its main elements have been presented in details. One of the main difficulties encountered was to scale the device to fit under a six feet long fume hood to limit the hazards involved with the use of chlorine at high concentration level in the laboratory. One other obstacle was to find materials and coating suitable and able to resist chlorine in the long run. The future means and procedures of characterization and measurement have been introduced as well. Once build and characterized, this apparatus will provide new data sets for chlorine deposition on vegetal samples and small objects in conditions similar to the atmospheric environment. In the end, this study aim at the improvement of the understanding and computational modeling of toxic inhalation hazards and the consequences to affected population and infrastructure.

## 5 Aknowledgment

This project is supported by the Chemical Security Analysis Center (CSAC) of the U.S. Department of Homeland Security and Transport Canada. The authors wishes to thank Katherine Elmore for her assistance with this project.

## 6 References

[1] M. B. Dillon, "The role of deposition in limiting the hazard extent of dense-gas plumes," *J. Hazard. Mater.*, vol. 164, no. 2–3, pp. 1293–1303, 2009.

[2] S. Hanna, S. Dharmavaram, I. Zhang, J., Sykes, H. Witlox, S. Khajehnajafi, and K. Koslan, "Comparison of Six Widely-Used Dense Gas Dispersion Models for Three Actual Railcar Accident," in *Process Safety Progress*, 2008, vol. 27, no. 3, p. 248–259.

[3] S. P. Arya, *Air pollution meteorology and dispersion*, 6th ed. 1999.

[4] J. P. Dimbour, A. Dandrieux, and G. Dusserre, "Reduction of chlorine concentrations by using a greenbelt," *J. Loss Prev. Process Ind.*, vol. 15, no. 5, pp. 329–334, 2002.

[5] G. Dusserre and J. Ollivier, "Small scale field experiments of chlorine dispersion," *J. Loss Prev. Process Ind.*, vol. 15, pp. 5–10, 2002.

[6] J. D. Hearn, R. Weber, R. Nichols, M. V. Henley, and S. Fox, "Deposition of $Cl_2$ on soils during outdoor releases," *J. Hazard. Mater.*, vol. 252, pp. 107–114, 2013.

[7] A. C. Hill, "Vegetation: A Sink for Atmospheric Pollutants," *J. Air Pollut. Control Assoc.*, vol. 21, no. 6, pp. 341–346, Jun. 1971.

[8] J. Hearn, J. Eichler, C. Hare, and M. Henley, "$Cl_2$ deposition on soil matrices," *J. Hazard. Mater.*, vol. 237–238, pp. 307–314, 2012.

[9] J. Hearn, J. Eichler, C. Hare, and M. Henley, "Effect of soil moisture on chlorine deposition," *J. Hazard. Mater.*, vol. 267, pp. 81–87, 2014.

[10] D. Brown, "The removal of toxic gases from the air by water, plants and soils.," 2006.

[11] M. L. Wesely and B. B. Hicks, "A review of the current status of knowledge on dry deposition," *Atmos. Environ.*, vol. 34, pp. 2261–2282, 2000.

[12] J. R. Couper, W. R. Penney, and J. R. Fair, *Chemical process equipment revised 2E: selection and design.* Gulf Professional Publishing, 2009.

[13] K. J. Myers, A. Bakker, and D. Ryan, "Avoid agitation by selecting static mixers," *Chem. Eng. Prog.*, vol. 93, no. 6, pp. 28–38, 1997.

[14] W. J. Gretta and C. R. Smith, "The Flow Structure and Statistics of a Passive Mixing Tab," *J. Fluids Eng.*, vol. 115, no. 2, p. 255, 1993.

[15] R. H. Perry, D. W. Green, and J. O. Maloney, *Perry's chemical engineers' handbook.* McGraw-Hill, 2008.

[16] Timet corporation, "Corrosion resistance of titanium," 1997.

# Fire Proofing Cost Reduction Measures for LNG Plant

**Raminaidu Girada**
**Senior Process Safety Engineer, Bechtel Oil, Gas & Chemicals, Inc.**
**3000 Post Oak Boulevard, Houston, TX 77056**
**rgirada@bechtel.com**

**Jegan Babu Arumugakan Thuraiswamy**
**Senior Process Safety Engineer, Bechtel Oil, Gas & Chemicals, Inc.**
**3000 Post Oak Boulevard, Houston, TX 77056**
**jbarumug@bechtel.com**

**Nino Nicotra**
**Process Safety Engineering Chief, Bechtel Oil, Gas & Chemicals, Inc.**
**3000 Post Oak Boulevard, Houston, TX 77056**
**anicotra@bechtel.com**

Prepared for Presentation at
American Institute of Chemical Engineers
2017 Spring Meeting and 13th Global Congress on Process Safety
San Antonio, TX
March 26 – 29, 2017

# Fire Proofing Cost Reduction Measures for LNG Plant

**Raminaidu Girada**
**Senior Process Safety Engineer, Bechtel Oil, Gas & Chemicals, Inc.**
**3000 Post Oak Boulevard, Houston, TX 77056**
**rgirada@bechtel.com**

**Jegan Babu Arumugakan Thuraiswamy**
**Bechtel Oil, Gas & Chemicals, Inc.**
**3000 Post Oak Boulevard, Houston, TX 77056**
**jbarumug@bechtel.com**

**Nino Nicotra**
**Process Safety Engineering Chief, Bechtel Oil, Gas & Chemicals, Inc.**
**3000 Post Oak Boulevard, Houston, TX 77056**
**anicotra@bechtel.com**

**Keywords:** Consequence Modeling, Liquefied Natural Gas (LNG) & Liquefied Petroleum Gas (LPG) and Safe Design

## Abstract

The current industry practice in designing fireproofing requirements for LNG facilities follows the traditional approach of "Liquid Pool Fire". The common industry practice for fire proofing design criteria defines a fire potential equipment (FPE) as the equipment that handles flammable liquid above a specific threshold volume. The FPE generates fire scenario envelope whose dimension ranges are recommended by API 2218. Equipment and structural supports that falls inside the fire envelope are considered for fire proofing.

LNG plants are handling products with very low boiling point, and any leak to atmosphere results in an instant flashing into vapor phase or aerosols. The results from our recent works on pool fire consequence modelling has proved that most of the currently considered fire potential equipment seldom create pool fire. This consequence modeling results clearly demonstrates that the current fire proofing design approach is overdesign, that is not effective. The apparent saving in man-hours for not modeling the liquid release will result in a considerable amount of engineering efforts and material cost being spent on the design and installation of fire proofing for equipment and structure with potentially no real benefits to the inherent safety of the process plant. Considering the current economic trend, a change in approach using consequence modelling will effectively result in considerable cost savings in major LNG projects.

# 1 Introduction

Currently in LNG projects, fireproofing is considered for equipment supports, pipe rack support, structural supports and other auxiliary critical hydrocarbon carrying pipe supports within a fire proofing zone to improve its structural integrity during a fire. The principal value of fire proofing is realized during the early stages of a fire by providing additional time to fire fighters to control or extinguish a fire thus by delaying additional loss of containment which may escalate the incident. The fireproofing requirement is determined by identifying the fire potential equipment and the associated fireproofing zones. Fire potential equipment is defined as an equipment holding flammable liquid above a specific threshold volume, which will have the potential to create a liquid pool and subsequent pool fire if ignited. Per API 2218, the fireproofing zone is defined for every fire potential equipment, that extends 20 ft. to 40 ft. (6 m to 12 m) horizontally and 20 ft. to 40 ft. (6 m to 12 m) vertically from the source of liquid fuel. As a typical approach most projects consider fireproofing zones that extend 30 ft. (9m) horizontally and 30 ft. (9m) vertically from the source.

The above mentioned qualitative approach does not reflect the realistic possibility of pool fire scenario. Therefore, a quantitative analysis may be performed to analyze whether a pool fire is realistic for equipment handling multiple refrigerants and LNG. This analysis has been performed on some process equipment handling multiple refrigerants and LNG using Canary software by Quest Consultants.

# 2 General Technical Guidance

## *2.1 Scope of Fire Proofing*

Fire scenario envelopes for an LNG plant are based on pool fire scenarios. Jet fires are typically not considered as credible for defining fire scenario envelops for fireproofing as described below:

- Jet fires are handled by Prevention, Intervention and Protection. Prevention is by designing the facility in compliance with general fire prevention activities such as those included in NFPA 59A and in the broad application of Process Safety Management(PSM). Once a pressurized jet fire occurs, the intervention and protection can be achieved by a high flow rate firewater stream directed on the point of impingement. This has been proved as an effective mitigation measure in preventing escalation with Jet fires. This will be achieved by strategically locating the fire water monitors around high risk process area, where jet fire is foreseen as a credible scenario.

- Emergency isolation and depressurization system will be provided to lower the pressure of specific isolated sections of the plant that may be involved in a localized jet fire scenario and to de-inventory of gas to flare.

- Efforts on similar LNG projects to extend the fire proofing systems for jet fire scenarios led to enormous impractical level of fireproofing requirement. This could also introduce other hazards such as corrosion.

- Jet fire represents a significant risk or concern for personnel life, working on offshore installations whereas property protection is the basis of pool fire for on-shore LNG facility or petroleum refinery or petrochemical facilities.

### *2.2 Pool Fire Formations*

A pool fire may result via a number of scenarios and it begins typically with the release of flammable material from process equipment. If the material is liquid, stored at a temperature below its normal boiling point, the liquid will collect in a pool. The geometry of the pool is dictated by the surroundings (i.e., diking), but an unconstrained pool in an open, flat area is possible, particularly if the liquid quantity spilled is inadequate to completely fill the diked area. If the liquid is stored under pressure above its normal boiling point, then a fraction of the liquid will flash into vapor, with un-flashed liquid remaining to form a pool in the vicinity of the release.

The pool of flammable liquid may be ignited creating a pool fire. If the release of flammable material from a pressurized process equipment continues, then a jet fire is also likely. If the ignition source occurs at the very beginning of the release, then inadequate time is available for the liquid to form a pool and only a jet fire will result.

Pressurized liquefied gases such as LNG, methane, ethylene and propane will in general disperse as mists or vapor, but may possibly form a small pool if the source operates at low pressure conditions. An exception has to be made for butanes, pentanes and heavier hydrocarbons, which always create pool.

Refer to the following **Figure 1** for pictorial view of the pool fire formation.

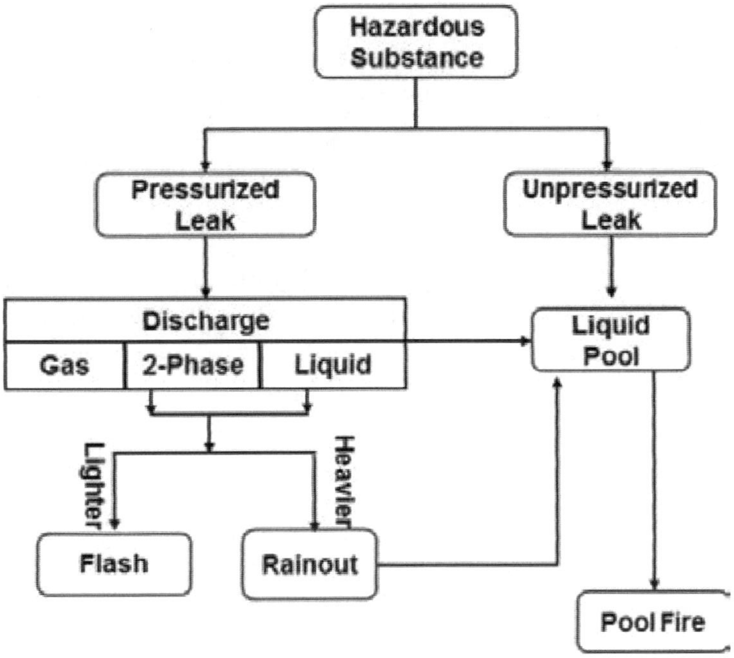


**Figure 1: Pool Fire Formation**

### *2.3 Determination of Pool Fire scenario*

This section provides an overview of the methodology for determining pool fire existence.

In case of accidental release of flammable materials such as Pentanes and heavier hydrocarbons create liquid pool irrespective of their operating conditions. However, the lighter hydrocarbons such as butane and propane may create liquid pool only under specific operating conditions. The potential for creating liquid pool is based on certain parameters as described below:

- **Butanes**

  Butane releases potentially create liquid pool except for high pressure releases and slight sub cooling of release. Further consequence analysis is required to check if a liquid pool is created from the release.

- **Propane and Ethylene**

  Ethylene and propane liquids may or may not pool, depending on conditions. It has been found that a pool formation can be correlated on several parameters: (1) vessel operating pressure, (2) size of rupture hole for self-leakage vessel, (3) the amount of subcooling, which is also important but the assumption is the liquid will be slightly subcooled, and (4) confinement of *fire hazard area*. Together, these variables establish the size of the pool.

  Higher pressure release are more likely to vaporize and mist, and low pressure releases are more likely to pool. Based on a slightly subcooled liquid at pressure, the analysis is divided into three regimes (1) no liquid pools, (2) highly likely to form a significant pool fire, and (3) possible pool fire. Depending on the size of the release, slope, temperature and composition, a consequence model may be necessary to determine if a pool fire is warranted.

- **LNG**

  Similar to propane and ethylene liquids, LNG may or may not pool depending on several parameters: (1) vessel operating pressure, (2) size of rupture hole, (3) the amount of subcooling, and (4) confinement.

Therefore, a quantitative approach was performed in the following section.

## 3 Pool Fire Case Study

In this case study, the following light flammable fluids are modelled: methane, propane, ethylene and LNG. Based on the current fire proofing design approach a total of 28 equipment of the LNG plant that handles the above fluids are qualified as Fire Potential Equipment. All those leakage scenarios are modelled using Canary modelling tool, based on its operating condition.

### *3.1 Leak Scenarios*

To provide a spill rate using CANARY, one has to postulate a leak scenario.

Typical hole size choices range from small (3-10mm) to full bore ruptures (diameter of pipe). Hole size selection can define the general level of risk (larger, high consequence releases are significantly less frequent than small, low consequence release).

CCPS book of "Guideline for Fire Protection in Chemical, Petrochemical, and Hydrocarbon Processing Facilities" defined hole sizes as small (3-10 mm), medium (10-50 mm), large (50-150 mm), and full-bore ruptures. Discussion around these hole sizes also points out that the large and rupture releases are not always the worst-case events due to release duration and dosage effects that should be discounted for.

The Canadian Standard Association (CSA) standards Z276 specifies design spills which define the potential extent of hazard zones that are not allowed to impact the public. Z276 defines, among other (smaller) release scenarios, that the design spill should be "The maximum flow of LNG that could be spilled, from the largest branch connection, 60 mm (2-inch) or smaller..."

Based on the above references, 2-inch is considered as an appropriate hole size for consequence modeling. Therefore, the pool fire modeling has been performed for 2-inch hole size. However, sensitivity analysis has been done considering 6-inch hole to check if larger hole size can result in pool formation. The modeling results for both 2-inch and 6-inch are presented below.

The release elevation was considered to be constant at 1 m height (the lowest height is deemed to result in higher consequences). The release directions in all scenarios were assumed to be horizontal. The consequence modeling input data shown below were used for the calculations. All scenarios were modeled using CANARY software developed by Quest. The software is widely used in industrial consequence modelling studies.

### *3.2 Atmospheric Conditions*

The following more stringent atmospheric conditions that favors/sustains the pool scenario are considered:

Wind Speed: 2 m/s

Stability Class: F

Ambient temperature: 80 °F

### *3.3 Liquid Methane Release at High Operating Pressure*

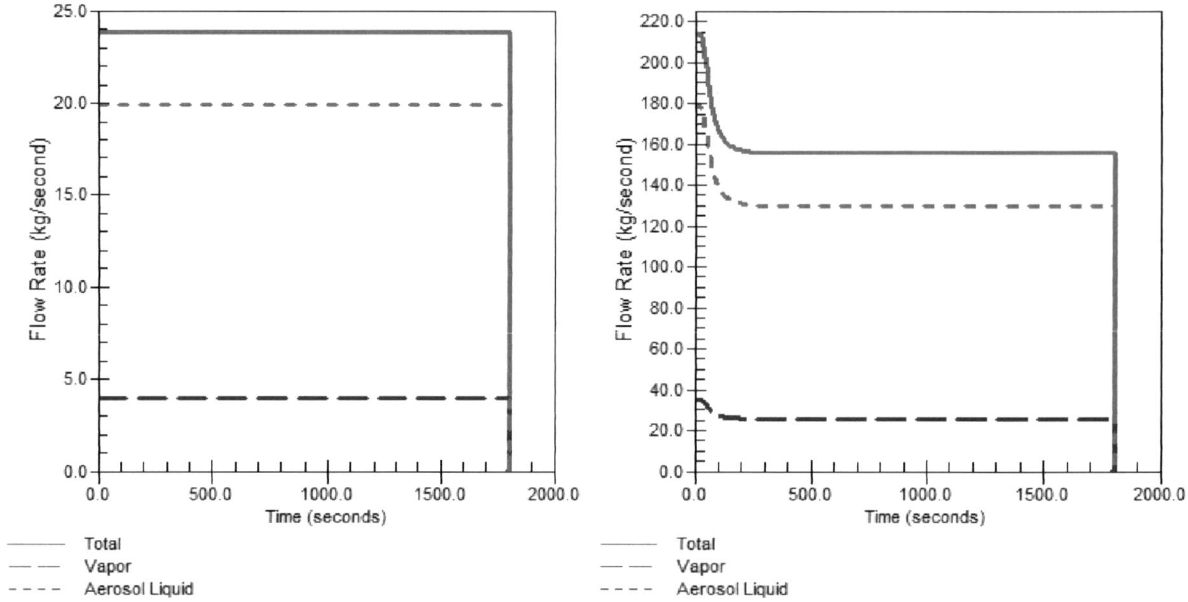


**Figure 2: Liquid Methane Release – 2-inch (left figure) & 6-inch (right figure)**

Consequence modeling for a liquid methane release was performed at specific process conditions for 2-inch and 6-inch hole size. As shown in the above figures, most of the release is flashing into aerosol liquid and remaining release is in vapor phase. No liquid pool is generated from the liquid methane release.

### *3.4 Liquid Propane Release*

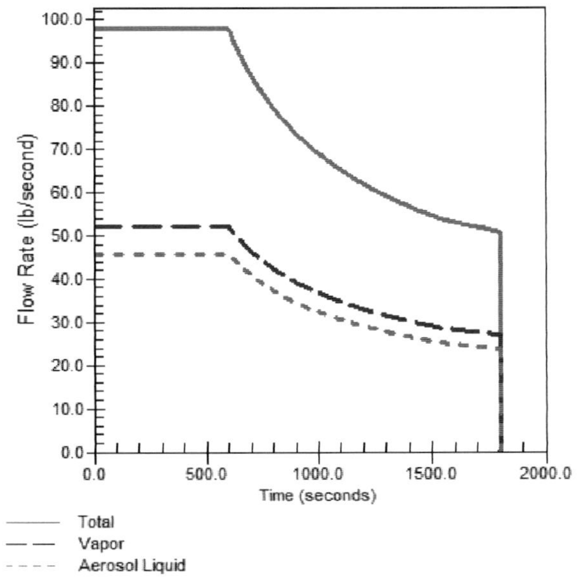


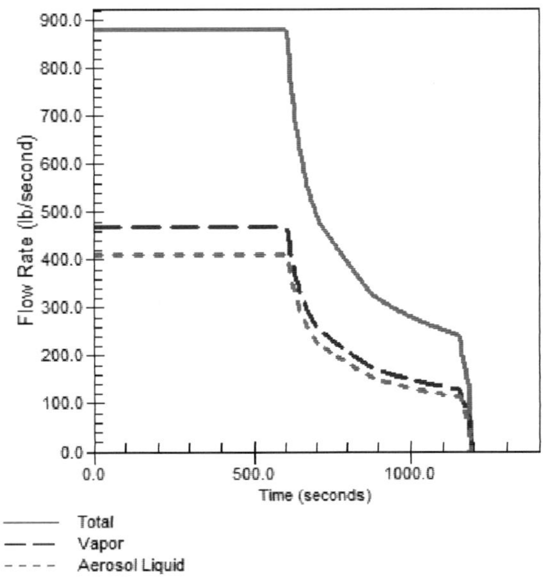


**Figure 3: Propane Release – 2-inch (left figure) & 6-inch (right figure)**

Consequence modeling for a propane release was performed at specific process conditions for 2-inch and 6-inch hole size. As shown in the above figures, the release is flashing into aerosol liquid and vapor phase. No liquid pool is generated from the propane release.

### *3.5 Liquid Ethylene Release*

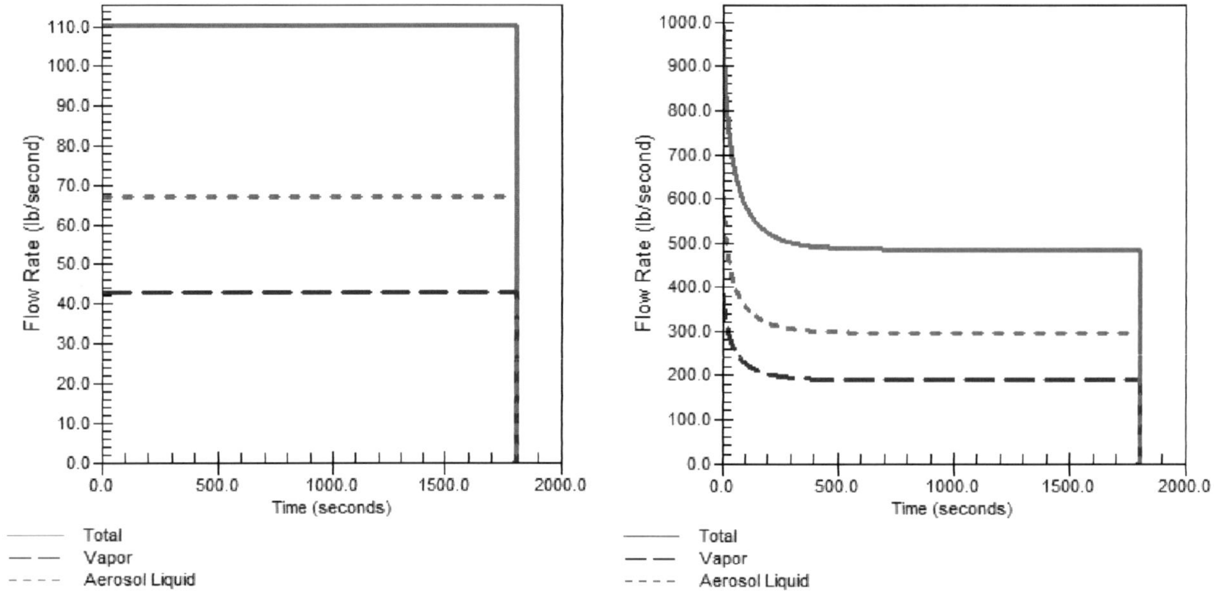


**Figure 4: Ethylene Release – 2-inch (left figure) & 6-inch (right figure)**

Consequence modeling for ethylene release was performed at specific operating process conditions for 2-inch and 6-inch hole size. As shown in the above figures, the release is flashing into aerosol liquid and vapor phase. No liquid pool is generated from the refrigerant release i.e. Ethylene release.

### 3.6 LNG Release at Low Operating Pressure condition

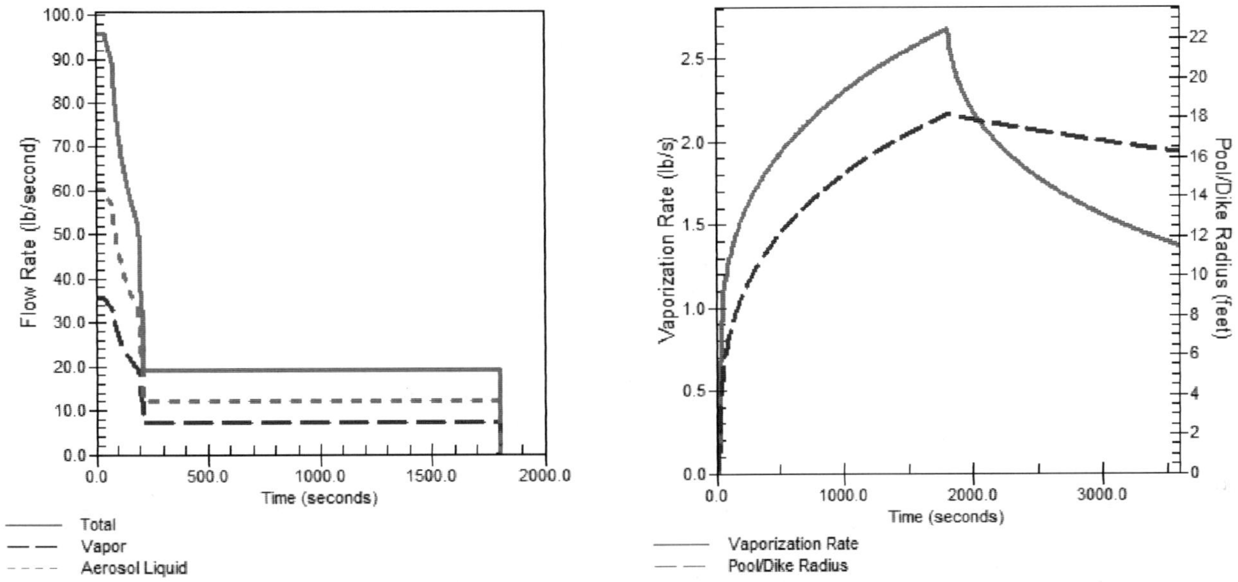


**Figure 5: Liquefied Natural Gas Release – 2-inch (left figure) & 6-inch (right figure)**

Consequence modeling for a liquefied methane was performed at the process conditions of an equipment for 2-inch. and 6-inch hole size. As shown in the above left side figure, the release for 2-inch hole size is flashing into aerosol liquid and vapor phase, without creating any pool. However, as shown in the right side figure, a release from 6in. hole size from the same equipment is creating a liquid pool of 16 ft. radius.

# 4 Conclusions

Consequence modeling results showed that out of the total twenty-eight (28) equipment of LNG liquefaction plant that were normally considered as Fire Potential Equipment (FPE) based on specific threshold volume, thirteen (13) equipment that handles methane, propane, ethylene at some specific operating conditions (e.g. low or medium or high pressure) did not create any liquid pool. Instead they flash along forming a large fraction of aerosol liquids, which never creates liquid pool.

The pool fire modeling was performed for a 2-inch release and a sensitivity analysis was carried out for 6-inch release to check the possibility of liquid pool formation for larger size releases for all the 28 equipment. The 6-inch release did not make a major difference in pool fire modeling results except for one equipment which contain LNG at low operating pressure condition. As stated earlier, thirteen (13) out of twenty-eight (28) equipment had not produced a liquid pool irrespective of the release size.

One (1) out of the remaining fifteen (15) equipment, that handles LNG at lower pressure condition did not generate liquid pool at 2-inch release, however it has generated liquid pool for a larger size (6-inch) release. Therefore, it is recommended to perform the consequence modeling for larger hole size to check the liquid pool formation.

The consequence modeling proved that 46% of the total equipment containing liquid such as methane or propane or ethylene or LNG had not produced a liquid pool. Therefore, considering the current economic trend, consequence modeling based approach for determining fireproofing requirements in conjunction with API 2218 requirement can be used in major LNG projects to justify considerable cost savings.

# 5 References


[1] J.D. Marx and A. Nicotra, "Is a Two-Inch Hole Adequate for Siting Study?", 12th Global Congress on Process Safety, American Institute of Chemical Engineers, Houston, Texas, 2016.

[2] API. "Fireproofing Practices in Petroleum and Petrochemical Processing Plants", American Petroleum Institute, 2013.

[3] CCPS (2003), "Guideline for Fire Protection in Chemical, Petrochemical, and Hydrocarbon Processing Facilities", Center for Chemical Process Safety of the American Institute of Chemical Engineers, 3 Park Avenue, New York, 10016-5991. (ISBN 0-8169-0898-2).

[4] CAN/CSA Z276, "Liquefied natural gas (LNG) - Production, Storage, and Handling", National Standard of Canada, 2015.

# 481560 Remember or Repeat

**Abstract:**

The title of my presentation is "Remember or Repeat" and it is intended to provide a brief historical perspective regarding a number of significant industry process safety events that have occurred over the past several decades, with an admonition (based on Trevor Kletz's famous quote that "organizations have no memory, only people have memories") that if we fail to remember these past events (and their key learnings), we may be setting ourselves up to repeat them. I've also incorporated brief references to three significant NASA events, the Apollo 1 fire in 1967, the Challenger Shuttle explosion in 1986, and the Columbia Shuttle incident in 2003, and plan to highlight the similarities in learnings (e.g., "normalization of deviance") between those events and our own industry experiences. Finally, although I've never personally experienced an event of that magnitude, I have, in one way or another, been "once-removed" from several of them, and plan to highlight how those experiences shaped my views and reinforced my passion for process safety, including the critical importance of never losing a sense of one's own vulnerability.

# 481561 In Harm's Way - After Which Nothing was the Same Again


**Abstract:**

Most incident case studies are presented by a third party who was not directly involved in the incident. As such, the presentations may lack the human element and other insights that can only be properly appreciated and articulated by someone directly involved in the incident. In his early career the author was involved in two major incidents while managing the day-to-day operation of refinery process units. The presentation describes both incidents; the first involving two major fires, and the second where several employees collapsed from exposure to sour gas.

In both incidents, the author found himself too close for comfort and could easily have become a casualty himself. There is nothing quite like raised adrenalin levels and the fear of death to get one's attention! The lessons learned changed a young engineer's attitude to process operations and led to a pursuit of process safety excellence. It has been a long journey, which continues to this day - indeed, a commitment to continuous improvement that never really ends.

# 481562 How a 5 Year Old Learned About Process Safety Management

**Abstract:**

My mother told me that both sides of my family trace back to the early wave of Scotch-Irish that settled the rural areas of New Hampshire in the 1600's. A number of Willey's served the country in the Revolutionary War, the Civil War, WW1, and WW2. They were known for toughness coming from the rugged backwoods of New Hampshire. Such was my father, US Marine, WW2. He worked 24/7 before 24/7 was a term, never missing a day of work until the accident. The accident, in the Spring of 1955 or 1956, occurred when 5 to 7 pine logs fell onto my father while he was loading his logging truck. He barely survived, and was on disability for nearly a year, having fractured his skill, blinded in one eye, and recovering from several broken several back bones. However, he bounced back, more determined than ever to return to logging, and performing the job in a safe manner. He'd often bring me along, and I'd watch his approach to safe operation, always checking things twice, as he chained up his loads. It was a tough way to make a living; however, he never suffered another injury until lung cancer from heavy smoking took him at 43. He always emphasized that safety came first. Do the job right and safe. I was around big equipment and heavy tools frequently, and the surprising thing I remember is my father's attitude towards safe work practice. I received my lessons in process safety management early. Didn't always listen; however, it has led to few incidents in my journey through life, and some of these words of safety first have been passed onto my children and my students.

# 481563 Please, Not Again! - The Second Time Process Safety Hit Home

**Abstract:**

Many Process Industry employees can recall an event or moment when the need for Process Safety became starkly apparent. Perhaps it was something stumbled across in the plant. Maybe it was an incident that came within a hair's breadth of causing a catastrophic incident. These personal watershed moments often create a very real sense of vulnerability to a process incident, and can impact or even completely alter the path of one's career.

Sixteen years after the first, the author experienced another Day that Process Safety Hit Home. This one hit harder than the first. This time, it was personal. It involved an explosion in a new process for which the author had personally led Hazard Studies, Layer of Protection Analyses (LOPAs), Consequence Analyses and Risk Assessment. It was an incident that was not foreseen, despite our best efforts. It caused a self-evaluation of the author's personal competence, and the ability of the tools and methods we used to detect hazards. We speak of having a sense of vulnerability to process incidents, but there can also be a profound sense of vulnerability to our own shortcomings and how they may lead to a process incident.

## 481565 Lessons from the Past

**Abstract:**

This presentation will discuss an incident from the late 1980s which affected the author's choice of career path and learnings from the incident which align with current PSM guidelines. Topics covered will include:

- MoC and PHA
- Emergency Shutdown and Emergency Operations
- Control of Work.

# Help Assure Good PHAs By Enhancing the Sense of Vulnerability

**Scott Berger, CCPSC, FCCPS, FAIChE**
**President**
**Scott Berger and Associates, LLC**

Prepared for Presentation at
American Institute of Chemical Engineers
2017 Spring Meeting and 13th Global Congress on Process Safety
San Antonio, TX
March 26 – 29, 2017

# The Hits Just Keep on Coming

Presented at the 13th Global Congress on Process Safety

by

Scott Berger, CCPSC, FCCPS, FAIChE
President
Scott Berger and Associated, LLC
scottbergerllc@gmail.ciom
+1.609.462.5057

## Abstract

The author describes incidents which helped shape his passion for process safety

## Introduction

This session, "The Day Process Safety Hit Home" is based on the premise that engineers, scientists, operators, and managers in high hazard industries will have, some time in their careers, one experience which positively and permanently shapes their view of process safety. It may be a serious fire, explosion, environmental release, or toxic gas release, or may be a near miss which "scared them straight." The session organizers perhaps hoped that by sharing these case histories with people new to process safety, these individuals can also be scared straight without having to personally have the same experiences as the speakers.

This may or may not be a reasonable assumption. For one, while I accept that some people might have their view of process safety shaped by one very significant incident, I could not think of any one incident that had that impact on me. In fact, my process safety awakening was more a journey than a destination. I expect this is true for most people. For another, I accept that hearing incident case histories and viewing photographs and videos produce some visceral reaction, but I can't believe that this reaction can measure up to being up close and personal.

Therefore, in this paper I am discussing how several incidents which impacted me personally shaped my view of process safety over time. Although I don't expect that hearing or reading about these experience will by itself transform attendees of this session or readers of this paper, I do hope that my experiences can help in that transformation.

**Discussion**

Lesson Learned – Emergency Response Planning

On March 9, 1982 (1), I was working in the process lab at the Rohm and Haas Bridesburg plant, northeast of downtown Philadelphia. At around 3 PM I needed a stretch break, so I stood up and walked to the window, which looked out diagonally across the street to the Allied Signal phenol plant. As I approached the window, I accidentally kicked the trash can. The noise attracted the attention of my colleagues, who started teasing me about the racket. I pointed out the window to where a growing fire was moving through the Allied Plant and said, "Um, guys, that noise wasn't me."

A fireball burst in slow motion across the street, hitting the stone wall bordering the Frankford Arsenal like a wave against a sea wall. The Arsenal had shut down a year earlier, but I didn't know that at the time. So I braced for a second explosion, which fortunately didn't come. We watched as the column of smoke arced over our heads, heading over the river to New Jersey.

At 3:30, there was a plant-wide announcement notifying all non-operating personnel to evacuate. Many plant employees lived in the nearby Bridesburg and Frankford neighborhoods and plant management were trying to be considerate of the families at home. Unfortunately, the incident blocked all but one road out of the plant neighborhood, and traffic was severely gridlocked for hours. A service onramp to the nearby Betsy Ross Bridge could have been opened to allow evacuation across the river to New Jersey, but a police car was parked across the ramp, perhaps as a precaution in case the wind shifted and carried smoke and fumes from the fire over the bridge.

From this incident, I learned about the importance of what became the RBPS element Stakeholder Outreach. Little thought was given in the community and among neighboring plants to coordinate emergency plans. Sheltering in place would have been much more appropriate for both plant workers and people in the community, and greater thought needed to be given to evacuation.

Lesson Learned – Understanding All Process Hazards

In the summer of 1983, I was working at the Rohm and Haas Bayport Plant, starting up a new agricultural chemical plant. Maintenance contractors in the adjacent monomer unit were in the process of cleaning a still. Unknown to them, the residue in the still bottom had been transformed over time to create a substance which could spontaneously ignite in air. As they opened the manway cover, the residue ignited violently, sending a short duration jet flame and blast wave towards the back of the plant. I was walking back from my unit to the control room at the time, and saw the release knock an operator off her bicycle. Miraculously, she was unhurt, and the samples in her basket were undamaged. The deluge system extinguished a small fire near the still base and the maintenance workers got wet from the deluge but escaped unharmed.

This up-close experience involving people I worked with every day taught me the importance of understanding the hazards of all the materials and all the steps in your process, not just the reagents, products, and main reactions.

### Lesson Learned – The Importance of Performing PHAs Like Your Life Depends on it

Shortly after my return from my plant start-up assignment in early 1984, I was assigned to work on improving the safety and environmental impact of a process with a long history of serious health effects. Work began with performing a revalidation of a previous HAZOP. As the youngest member of the process team, I became the scribe. I learned quickly that the person who controls the "pen" controls the meeting (Actually, I used a primitive portable computer and spreadsheet software. It may not have been the first instance of using software to record a PHA, but certainly among the first). I used the power of the pen to ensure that deviations were thoroughly explored, and in the process uncovered quite a few previously unrecognized deviations with significant consequences that had a reasonable chance of occurring, as well as identifying an emission treatment system which was all but ineffective.

When the Bhopal incident occurred late that year, it became clear to all that this plant, sited in a city similar in size to Bhopal, had several potential release scenarios which could equal Bhopal in impact. I spent the next few years developing risk reduction measures for this and similar plants. In the early 1990's, a sister plant had a serious release which was contained and controlled by a mitigation system I developed, preventing a toxic release and allowing no onsite or offsite exposure.

From this experience, I learned the importance of treating PHAs as a matter of life or death, and to work with fellow PHA team members reveling in "the thrill of the hunt." This later became incorporated in my philosophy of Performing PHAs with a Sense of Vulnerability – you can see this paper in the PSM2 PPSS I session.

### Lesson Learned – Competence is Key

I didn't learn about the incident at Hickson Welsh (2) in Castleford, UK until several months after it happened on September 21, 1992. This incident mirrored my experience in Bayport, but escalated by several orders of magnitude. Workers removed the manway from the bottom of a still. The contents ignited violently and a jet flame over 165 meters long cut through the control building and a brick office building behind it, even though it lasted only about a minute. Five workers were killed, including one person in the office building who was in the path of the jet flame.

Perhaps because of the similarity to the Bayport incident, or perhaps simply because of the laser-like focus of the jet flame on both plant and office workers, this incident influenced me deeply. This was much more than weak culture, poor operational discipline or the failure of a PHA, although these factors surely were at play at Hickson Welsh as they were in most incidents. This was a company effectively aiming a cannon at the its own workers. No one would intentionally do this. Such things happen only in the absence of competence.

Lesson Learned – Don't Use Regulations as an Excuse for not Doing the Right Thing

In 2000 I was tasked to perform EHS audits at several of a company's plant sites that were not covered by PSM. One such site, which had been recently acquired, recycled aluminum cans and other scrap into aluminum sheet used for siding and other structural aluminum cladding. Vapors from the aluminum melter contained aluminum as well as soot, which condensed into small particles that were removed in a bag house. The bag house was emptied periodically into a dumpster. I found this dumpster near the bag house, behind the smelter building, near a loading dock. The ground around the dumpster and loading dock was littered with two things – shiny silver aluminum dust that missed the dumpster, and cigarette butts. I learned that the loading dock was the designated smoking area of the plant. Of course, the dust on the ground missed the dumpster because some dust would be suspended and drift away from the dumpster.

The plant EHS manager was quick to point out that the size of the dumpster was small enough such that even if the dust was all aluminum, which he claimed it could not be, it wasn't a covered process. At worst, they had erred in the SARA 313 (annual toxic emission and waste) reporting.

Does the dust explosion pentagon include "covered by regulations" as an enabling condition for a dust explosion? I don't think so! Fortunately, corporate legal recognized the potential lawsuits that could arise in the apparently likely event that the dust ignited. And this incident reinforced to me the importance of addressing hazards, whether or not they are regulated.

Lesson-learned: The Importance of Culture

In late January 2003, CCPS had the opportunity to visit the Kennedy Space Center to watch the final fueling operations of the Space Shuttle Columbia, with a focus on safety. We were very impressed with how the operators carried out their duties. As we posed for a photo in front of Columbia, we talked about how we wished plant personnel in our industry could work as safely with as much operational discipline. Of course, two weeks later, Columbia disintegrated on re-entry. Soon, my phone was ringing off the hook with reporters speculating that some kind of chemical runaway caused the loss of the shuttle, and asking me to confirm (of course, I told them it was too early to say anything). Investigation of the incident revealed a weak safety culture at NASA. As we compared NASA's errors to the process and petroleum industries, the parallels were compelling. CCPS's work influenced the Chemical Safety Board, which in turn recommended that BP empanel a special culture investigation panel. Now, process safety culture is recognized widely as important to a successful process safety effort.

Lesson-learned: Closing out action items

In December, 2004 I had the opportunity to visit the Bhopal Plant on the 20th anniversary of that disaster. As I reported in a paper co-authored with Ron Willey and Dennis Hendershot, we found the site essentially untouched since Union Carbide was expelled from the country in early 1985. Corrugated metal warehouses built to shelter insecticide product from the elements had corroded to the point that rainwater could dissolve insecticide product and carry it away into the groundwater. Vessels held their contents for many years until they ultimately corroded through, leaving only the head still in place. Mercury and asbestos lay on the ground.

Much has been written about the causes of Bhopal, but two additional points were brought to my attention from that visit. First was the scraps of paper lying all over the control room, as you can see in the CSB video on Bhopal. I picked some of these papers up and discovered that they were uncompleted work permits. Hundreds of them. Second, I learned that a year before the disaster, Union Carbide had decided that the plant was too unsafe to continue operation (and the market for the product was not materializing), so they decided to shut down the plant. What both points show clearly is how important – both at the operational level and at the level of corporate leadership – to follow through on your action items. If the maintenance and action items had been resolved promptly, or if the plant was shut down immediately when it was deemed unsafe, Bhopal may not have happened.

Lesson Learned – Process Safety is Personal

The way Bhopal hit home for me has already shown up twice in this paper. I'd like to visit the impact of this tragedy on my view of process safety one last time because of a very important choice it forced me to make. At the time of the incident, I was engaged to a young woman who, I discovered in the days after the tragedy, had a dim view of the chemical industry. She let me know her strong opinion that I should quit the chemical industry and work instead to shut the industry down. Basically, I had a choice – stay in the industry and work to prevent incidents but lose my fiancée, or stay with her and work for an organization like Greenpeace.

You know my choice. So maybe after all, that was one day that process safety hit home for me. But I stand by my assertion that the hits keep on coming. We should all try to internalize the incidents and near misses, whether in our company's facilities or in similar facilities in other companies, and make sure that the process safety message keeps hitting home.

**References**

1. Schwab, R, Explosion & Fire at a Phenol Plant, IChemE Symposium Series No. 10, https://www.icheme.org/communities/subject_groups/safety%20and%20loss%20prevention/resources/hazards%20archive/~/media/Documents/Subject%20Groups/Safety_Loss_Prevention/Hazards%20Archive/S110%20-%20Preventing%20Accidents/S110-16.pdf

2. Patterson, KJ, "Case Study – The Fire at Hickson and Welsh Ltd., Castleford, UK, https://www.icheme.org/communities/subject_groups/safety%20and%20loss%20prevention/resources/hazards%20archive/~/media/Documents/Subject%20Groups/Safety_Loss_Prevention/Hazards%20Archive/S139%20-%20Major%20Hazards%20II/S139-03.pdf

3. Willey, R.J., D. Hendershot, and S. Berger, "The Accident in Bhopal: Observations 20 years later," Process Safety Progress, March 2007

# 481543 Understanding vapor cloud explosions

**Abstract:**

Accidental releases of flammable gases can result in vapor cloud explosions, with potentially severe consequences. This tutorial explains the circumstances that can lead to such an event, how the magnitude of the explosion can be estimated, how severe the consequences could be and potential mitigation strategies.

# 481544 Running the Gamut of Hazard Identification and Risk Analysis Methods

**Abstract:**

Hazard identification and risk analysis (HIRA) methodologies, including those used for process hazard analyses (PHAs), run the gamut from a basic hazard identification exercise all the way to a fully quantitative risk analysis. This tutorial will give a bird's-eye view of these techniques, highlighting similarities, differences, inter-relationships, and discussing the appropriate application of each. The tutorial will show how identification and analysis of incident scenarios is at the heart of all risk analysis techniques, whether the method of choice is a What-If Analysis, a HAZOP Study, a LOPA, or a quantitative risk analysis.

# 481545 Q&A on QRA

**Abstract:**

What exactly is a QRA? How do you do one? What tools are needed? What can a QRA tell you? This tutorial session will help to answer these questions and others, with the goal of providing a basic understanding of quantitative risk analysis (QRA). The nature of consequences and probabilities, and their combination, within the context of chemical process safety, will be explored and explained. The tutorial will provide introductions to consequence modeling tools, probability assessment methods, various risk measures, the computational requirements for developing a QRA, and ways that QRA can be applied to process safety. After the tutorial session, the attendee will be better equipped to know what a QRA is and how it can be used to evaluate the risks inherent to chemical processing.

13TH GLOBAL CONGRESS ON PROCESS SAFETY

# Are your safeguards fast enough to be effective?

**Eloise Roche**
**SIS-TECH Solutions, LP**
**Houston, TX**
**eroche@sis-tech.com**

**Angela Summers**
**SIS-TECH Solutions, LP**
**Houston, TX**
**asummers@sis-tech.com**



Prepared for Presentation at
American Institute of Chemical Engineers
2017 Spring Meeting and 13th Global Congress on Process Safety
San Antonio, TX
March 26 – 29, 2017

# Are your safeguards fast enough to be effective?

**Eloise Roche**
**SIS-TECH Solutions, LP**
**Houston, TX**
**eroche@sis-tech.coms**

**Angela Summers**
**SIS-TECH Solutions, LP**

**Keywords:** Safety Management System, Safety Interlocks, Safety Alarms, Safety Instrumentation Systems

## Abstract

A core attribute of all safeguards is that they must be functional against the loss event in question and one of the essential elements of functionality is speed. How to determine the response time available for a given safeguard to complete its action is a common project question. Several instrumentation and controls design pitfalls should be avoided when specifying the response time. A variety of engineering practices, from basic to advanced, may be used to justify the selected value. Among these are:

- Expert judgment
- Extrapolation
- Mass and energy balance calculations
- First principles process modeling

Key technical issues include hardware response lags, process lags, application program delays, measurement error, and the design and safety margins. Careful attention to this specification is needed from the instrumentation and controls design team, for if an instrumented safeguard performs too slowly compared to the hazardous event, it provides no protection at all.

## 1 Introduction

Functional safety is achieved through implementation of effective process control systems that limit the likelihood of initiating a hazardous event and safeguards that manage the process safety risk remaining after inherently safer design principles have been applied to the process design. Functional safety projects can implement a wide variety of safeguards, including safety controls, alarms, and interlocks (SCAI) [1]. Successful execution of SCAI projects requires significant resource investment. A SCAI project typically involves hundreds of engineering hours to brainstorm hazard scenarios, identify functional safety gaps, and determine the best means to reduce risk. The safeguards may include the use of independent protection layers (IPL) such as a Safety Instrumented Function (SIF) executed within a Safety Instrumented System (SIS). While not all safeguards will meet the criteria of an IPL, an engineering task common to all instrumented safeguards is the specification of the safeguard setpoint and corresponding response time.

The term process safety time has been used in many publications to describe the time interval both for the escalation of the process deviation to the hazardous event and for the safeguard response to prevent the hazardous event. This paper uses terms recently adopted in industry practices:

> Process Safety Time (PST) [2]: Time period between a failure occurring in the process or the basic process control system (with the potential to give rise to a hazardous event) and the occurrence of the hazardous event if the SIF is not performed.
>
> IPL Response Time (IRT) [3, 4, 5]: The time available for the IPL to act given the specified setpoint.

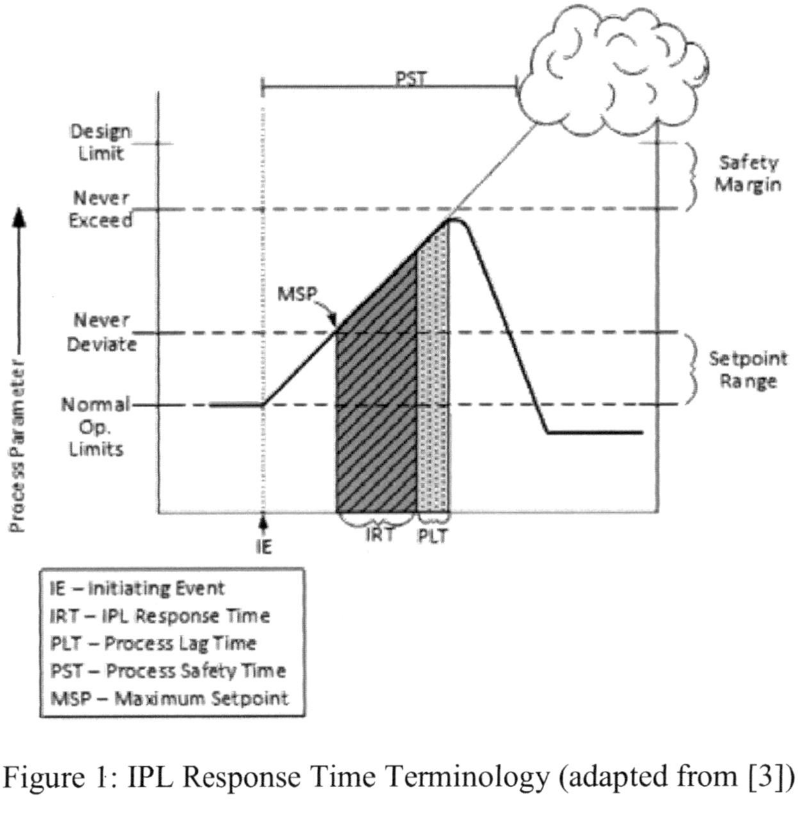


Figure 1: IPL Response Time Terminology (adapted from [3])

The most important distinction is that PST is a property of the process only. In contrast, the IRT (historically referred to as the allocated process safety time) is a property of both the process and the safeguard setpoint. The safeguard must detect the process deviation and complete its action within the IRT to prevent the hazardous event. The determination of the IRT is intrinsically related to the process and its response to the safeguard action, as this determination must account for the process escalation prior to action being taken and any process lag that exists due to retained mass and energy within the process (e.g. cooling of a vessel). Figure 1 depicts the difference between PST and IRT.

For efficient project execution, it is important that the setpoint and IRT specifications are made during front-end-loading. The response time of the final safeguard design should be estimated and compared to the IRT specification during detailed design reviews. Additionally, the safeguard response time measured during pre-startup safety tests is validated against the specified IRT when evaluating whether the functionality of the installed system is acceptable.

# 2 Understanding Process Response to Deviation

To determine the IRT, the engineering team first understands how the process reacts to the process deviation caused by the initiating source of the hazardous event. The PST may be only seconds, which limits the type of safeguards that can be effective. On the other hand, the process may take days to transition from the initiating source to the ultimate consequence, which could allow for a sequence of safeguards. A variety of engineering practices from basic to advanced may be used to better understand how the process responds to the process deviation and to the safeguard action.

## *2.1 Expert Judgment*

Experience is a valuable teacher. Expert judgment involves specifying an IRT based on experience, whether personal or through industry incident reports, and technical analysis. Industry experience may be codified in consensus standards for a particular type of safeguard (e.g., a SIS) or process application (e.g., burner management standards). Expert judgment within a facility may also come from senior operators who have personal experience from historical events.

Expert judgment can be a very useful starting point for a project and may be the only method available for the preliminary IRT specification in early project phases. Expert judgment based on prior observation is often the IRT basis for very short PST events, such as the compressor overspeed reaction shown by process response curve A in Figure 2. However, relying on expert judgment alone may leave the operating facility with a relatively weak justification if a proof test or an incident calls the specified IRT in question. Consider a situation where a SIS has failed a proof test due to the failure mode - too slow of response. When the failure is investigated, the justification for the specified IRT is likely to be requested. Expert judgment is only as good as the individual's expertise.

## *2.2 Extrapolation*

Extrapolation relies upon factored estimates based on equipment differences, such as vessel volume or pipe cross-sectional area, to estimate the IRT for one facility based on a known event in a very similar facility. Such scaling can be used to develop a preliminary IRT specification quickly and has a more clearly documented technical basis than expert judgment alone.

However, not all process responses scale simply. Events that involve multiple phases, energy transfer, or reactive chemistry often do not scale linearly. Furthermore, even minor changes in process equipment layout between the facility that experienced a previous event and the one that is under evaluation can result in potentially substantial differences in response to the same process deviation.

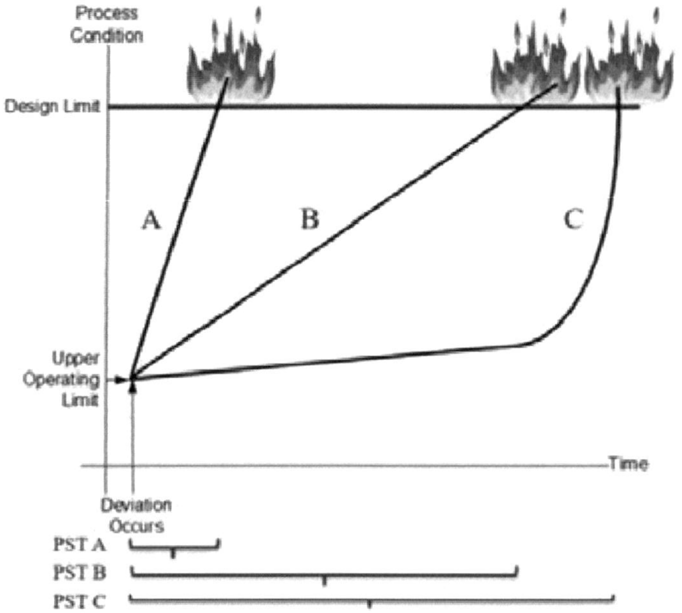


Figure 2: PST is based on process response to deviation

## 2.3 *Mass and Energy Balance Calculation*

The mass and energy balance tools typically used by process engineers during process equipment design can be helpful in the specification of IRT. In particular, iterative runs of a mass and energy balance might be used to predict the process response for relatively straightforward process deviations. For example, the time it takes to overfill a vertical vessel (which might have a response curve like the one depicted by process response curve B in Figure 2) or to pull in a dangerous level of oxygen into a tank after loss of a nitrogen pad can often be estimated using a simple spreadsheet. More complicated design simulators may be useful when the analysis must involve multiple pieces of equipment. For complex process deviations, however, the number of iterative runs of a static mass and energy balance can become burdensome. Due to the nature of the tools, this method is also unlikely to reveal short-term dynamic responses that occur under transitory operating conditions.

## 2.4 *First Principles Process Modeling*

First principles process modeling, sometimes referred to as dynamic simulation, models the response of multiple process parameters to one or more process perturbations as a function of time. In first principles process modeling, the actual process equipment dimensions, including piping configurations, and process chemistry are incorporated into the model. This increased degree of complexity can be used to model complex reactive or multi-phase process reactions and is more likely to reveal short duration dynamic responses that exceed the steady-state predicted values. Often these responses are non-linear and may have one or more points of inflection. For example, process response curve C in Figure 2 might reflect the complex temperature response resulting from the initiation of an auto-catalyzing degradation reaction. The process equipment and instrumentation information needed to build a useful model likely limits the use of this technique to the latter portion of detailed design. Verifying the process simulation against data collected during startup and normal operation helps increase confidence in the model accuracy.

## 3 Establishing Safety Margins

The design limit is a property of the equipment being challenged by the process deviation. Examples of design limits are the maximum allowable working pressure (MAWP) of a pressure vessel [6] or the critical high level [7] in a vessel. However, it is not an inherently safer practice to operate equipment all the way to the design limit. A safety margin, as shown in Figure 1, is usually established by the facility owner and provided to the project team during front-end loading. The minimum safety margin between the design limit and the never exceed limit is often expressed as a percentage in engineering units. Facilities with older or well-used equipment may establish a larger safety margin to account for uncertainty in the equipment's mechanical integrity. The never exceed limit is the maximum (or minimum) value the process parameter can achieve and the process still be considered safe for continued operation. The safeguard action must be successfully completed and successfully affect the process variable before the never exceed limit is reached. Designs which encroach upon this margin should be subject to review by process safety personnel to ensure the results can still satisfy the company risk criteria.

## 4 Understanding Process Lag

The same method(s) used to determine the process response to the deviation can usually be used to determine whether or not there will be a process lag after the proposed safeguard action is complete. In many cases, the process variable of concern will not stop moving in the dangerous direction the instant the final element of the safeguard (e.g., a valve or a motor) goes to the safe state. The level in a tank may continue to rise due to liquid rundown from material in an elevated pipe between the isolation valve and the tank inlet. The temperature and pressure in a distillation column may continue to rise due to the thermal energy stored in the metal of a reboiler, even though the reboiler steam has been shut off.

As shown in Figure 1, the existence of a process lag will require the safeguard action to be completed earlier in the process response curve to ensure the process variable does not violate the never exceed limit established by the design limit and the safety margin. Some scenarios are addressed by multiple independent safeguards that are designed to operate in a sequence. For these scenarios, the setpoints and IRTs should be specified so that the process response to one safeguard is complete before the next safeguard in the sequence is activated.

# 5 Estimating Safeguard Lag

Once the process response curve and process lag are understood, the time available for safeguard action after the process reaches a given setpoint can be determined. This is essential information for specifying the setpoint and IRT. The next step in the process is to determine the least conservative acceptable setpoint. For this task, the estimated response time of the safeguard must be determined. The estimation work can be divided into three tasks:

- Estimating the safeguard typical response time based on components
- Adding delays from the application program(s)
- Adjusting the setpoint for the additional lag caused by measurement error

## *5.1 Safeguard Device Response Time*

The safeguard response time is determined by adding up the individual response times for the equipment needed to execute the required function, along with any time delays introduced within the application program (e.g., filters or delay timers). Detailed instrument and application program design information is often not available at the time of the preliminary hazards and risk analysis (H&RA), so the initial estimates are often made using conservative device response values and assuming no application program delays, such as those depicted in Figure 3. Once device data sheets and detailed application program design documents are available, the initial estimates can be updated.

| Device Type | Typical Device Detection/Action Lag |
|---|---|
| Most analyzer installations (including delays from sampling system) | Tens of minutes |
| Typical temperature installations | 2 to 10 minutes |
| Most other sensors | 5 seconds or less |
| Large valves and variable frequency drive motors | Tens of seconds up to multiple minutes |
| Small valves and other motors | 5 seconds or less |
| Programmable electronic logic solvers | 3 seconds or less |
| Human response | 10 to 40 minutes PLUS |

Figure 3: Typical instrumentation response time [4]

Human response as a safeguard (e.g., an operator response to an alarm) introduces a large and potentially complex response time contribution. There are delays involved in recognizing the presence of an alarm, mentally relating that alarm to the response that must be taken, and overcoming any cognitive biases to taking the safe response action. To have a 90% probability of correct action in an alarm response for which the action is simple and involves no troubleshooting, at least 10 minutes of time for cognitive delays should be added to the estimated safeguard response time. If the response procedure does involve troubleshooting, the cognitive time should be increased to 40 minutes [8]. If the response action requires travelling out into the process facility, the time to get to the destination and take the final action should also be estimated and added to the estimated safeguard response time. The industry consensus values for operator response lags assume that performance shaping factors like training, work/rest schedule, and leadership culture regarding process safety activities are nominal. Deficiencies in any of the operations performance shaping factors can result in an increased human response lag.

## *5.2 Application Program Delays*

In addition to the relatively fixed delays inherent to the physical devices that make up the safeguard, delay can also be introduced through the application program. Delays may be introduced on the safeguard activation, through enabling logic, and through filtering of the signal. Signal filtering for safety applications can be particularly concerning, as a first order filter introduces a delay equal to 2.3 times the time constant just to receive 90% of a signal change. Unless the IRT is particularly generous, it would be inherently safer to address erratic sensor signal problems through correction of the underlying electric system issue or through addition of an independent installation of redundant instrumentation with an NooN voting architecture rather than through the introduction of signal filtering.

The use of intelligent devices and wireless technologies in safeguards create an additional location to inspect for adjustable software delays. Modern sensor transmitters and final element actuators may include customizable filters or even contain activation logic for the safeguard loop, eliminating the need for a separate logic solver. Transmitted signals may also be slowed down through configurable I/O network buffers. As these delays are configurable, they will usually not be included in the base response time documented in the instrumentation datasheet. Ultimately, the design team must identify all sources of software delays and ensure they are accounted for when estimating the response time achieved by a given design.

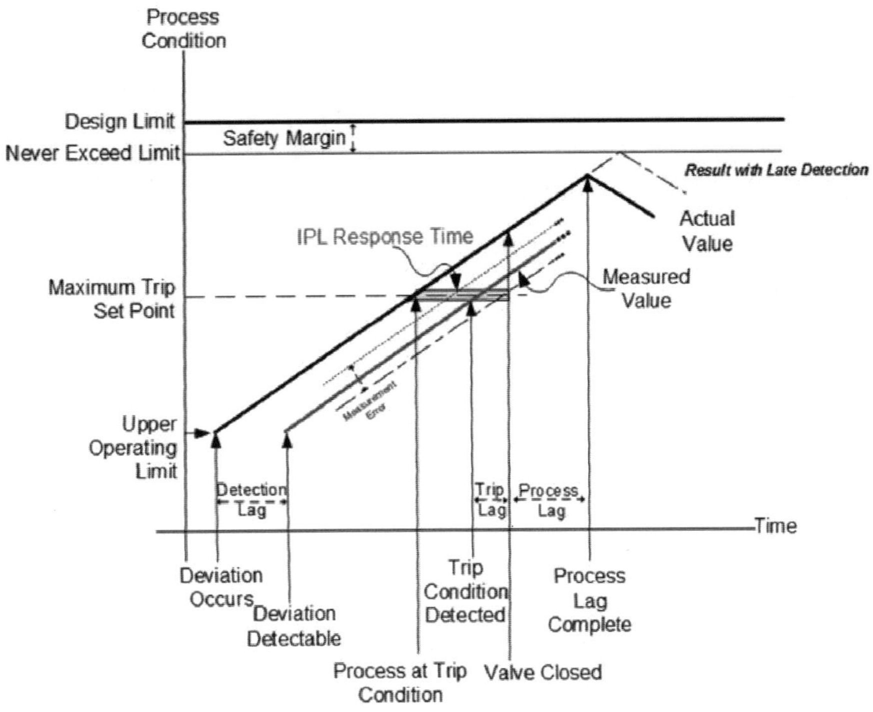


Figure 4: Adjusting setpoint for measurement error (adapted from [4])

### 5.3 *Impact of Measurement Error on Setpoint Selection*

Even for a new sensor that is correctly calibrated and configured, error is expected in the measurement. This normal amount of error is related to the specified accuracy of the device. In determining the final setpoint specification, it should be assumed that the sensor will be reading at the specified accuracy limit away from the setpoint. This would result in late safeguard actuation and the potential occurrence of the dangerous consequence. The setpoint should be adjusted to account for this error. As shown in Figure 4, this adjustment ensures that the process does not violate the never exceed limit even when the measurement is at the less safe extreme of its accuracy at the time of the demand.

## 6 Design Margins and Making the Final Setpoint/IRT Specification

Completion of the above tasks determines the least conservative setpoint value. That is, the highest setpoint value the safeguard could have and still be effective. Moving the setpoint close to the never exceed limit maximizes the operating margin and allows the maximum time for the operator to recover the process and prevent safeguard action, but allows the shortest time possible for the safeguard to act. However, it is rarely a good engineering practice to design a safeguard up to the least conservative (i.e., least safe) limitation.

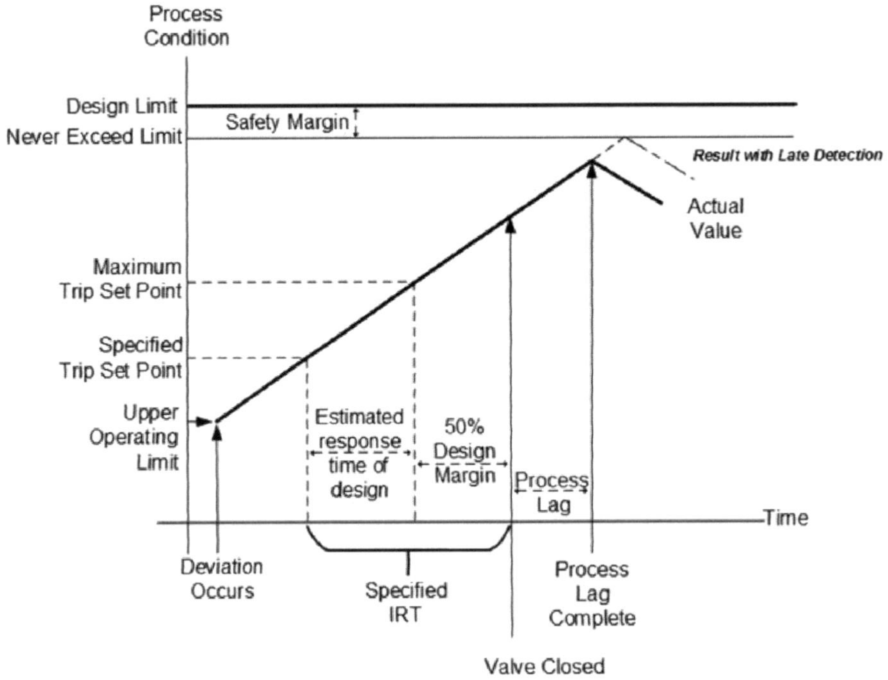


Figure 5: Estimated response time and design margin

Alternatively, the setpoint could be set very close to the normal operating limit, allowing for more response time. The generally accepted margin is to design the safeguard to work in half the time allowed in the specification (depicted in Figure 5). Using such a safety margin helps increase the likelihood that the safeguard is effective even when it does not operate quite as fast as expected once installed in the operating environment. However, having a safeguard setpoint too close to the operating limit can put a facility at risk of more frequent spurious trip events. Spurious trips will cause lost productivity and might come with their own inherent hazards, such as potential equipment damage or industrial hygiene hazards related to the clearing of equipment in preparation for startup.

Usually the optimal setpoint is somewhere between these two extremes. Balancing these two antagonistic design criteria and specifying the setpoint, as shown in Figure 6, often involves a team negotiation that includes members of operations, process safety, and engineering. Where the final selection does not allow a 50% design margin, the justification for the smaller design margin should be clearly documented. Based on the process response to the deviation, the final setpoint specification will determine the IRT specification.

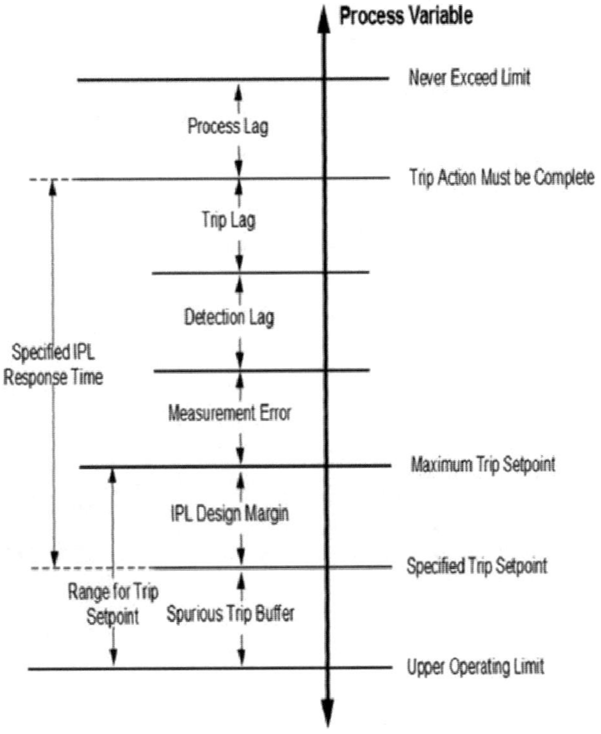


Figure 6: Balancing design margin and spurious trip buffer (adapted from [5] Annex Q)

## 7 Incorporation of IRT into Project Timeline

Due to the varying amount of engineering information required, the different techniques for determining the process response to deviation and specifying the response time of the designed safeguard have an optimum stage for their use during project front-end-loading and detailed design (shown in Figure 7). Just as an iterative process is the most effective for performing the H&RA activities, an iterative approach to the specification of setpoint and IRT can enable the project team to make more informed decisions during front-end-loading. Once engineering details for equipment, piping, application program and instrumentation are available, the design team can take advantage of more advanced techniques to verify and strengthen the documented technical justification for the specified values.

Incorporating this iterative approach into the project schedule is something that should be done during project scoping with the involvement of project management and facility leadership. Project milestone meetings, such as the conceptual design review and detailed design review, should confirm that the H&RA activities and safeguard specification details are at an appropriate level of development. Sufficient time and resources should be scheduled to ensure there will be clearly documented justification for the specified values, as seemingly minor changes in process conditions or equipment dimensions can result in a substantially different process response to deviation. This documentation will be necessary to support the management of change of future proposals for safeguard modification.

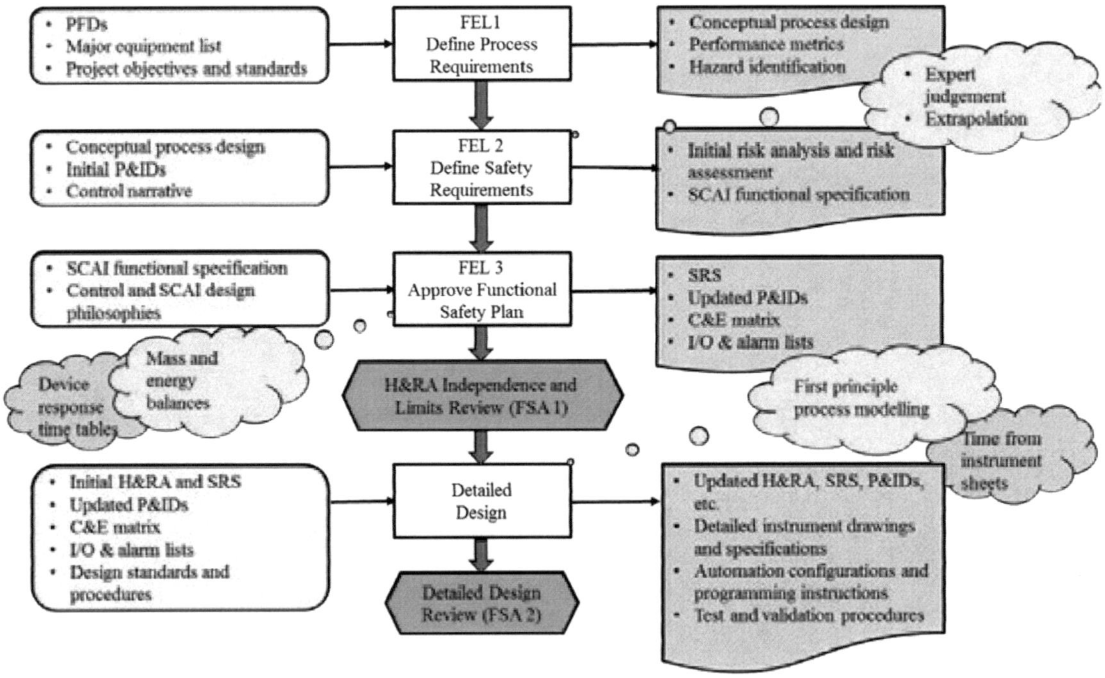


Figure 7: Incorporating response time techniques into project front-end-loading (FEL) and design

## 8 Conclusion

Understanding the process safety time for the process response to a hazardous event is useful when designing multiple safeguards against the same hazard, but the response time specified for a safeguard (the IRT) is a function of the specified setpoint and the process response curve. Multiple engineering techniques exist to analyze the process safety time for a given deviation event and any process lag that might exist after the safeguard action is complete. For best efficiency, these should be applied in an iterative process over the project front-end-loading and detailed design phases. The least conservative setpoint at which the safeguard will be effective is then determined, based on device lags, application programming delays, and adjustments for the expected amount of measurement error. At this point, the competing targets of the desired design margin and spurious trip buffer are used to select the final setpoint specification. Based on the process response curve and process lag, the specification of the setpoint established the IRT specification. The justification for these specifications, including the detailed process and equipment information used to derive them, should be clearly documented to enable future change proposals to be managed effectively.

# 9 References (examples given to show format).

[1] ANSI/ISA. 2012. Identification and Mechanical Integrity of Safety Controls, Alarms and Interlocks in the Process Industry, ANSI/ISA-84.91.01-2012. Research Triangle Park: ISA.

[2] IEC. 2016. Functional safety: Safety instrumented systems for the process industry sector - Part 1-3. IEC 61511. Geneva: IEC.

[3] CCPS. 2014. Guidelines for Initiating Events and Independent Protection Layers in Layers of Protection Analysis. New York: AIChE.

[4] CCPS. 2016. Guidelines for Safe Automation of Chemical Processes. New York, NY: Center for Chemical Process Safety.

[5] ISA. 2015. Guidelines for the Implementation of ANSI/ISA 84.00.01- Part 1, TR84.00.04-2015. Research Triangle Park: ISA.

[6] ASME. 2015. Boiler and Pressure Vessel Code. New York: ASME.

[7] API. 2012. Overfill Protection for Storage Tanks in Petroleum Facilities, 3rd Edition, Standard 2350. Washington, D.C.: API.

[8] CCPS. 2007. Guidelines for Safe and Reliable Instrumented Protective Systems. New York: AIChE.

# Temporary Management of Change: A Problem or a Necessary Part of Managing Process Safety

**Jody Olsen**
**JE Olsen Consulting LLC**
**Anchorage, AK**
**jodyeolsen@gmail.com**

Prepared for Presentation at
American Institute of Chemical Engineers
2017 Spring Meeting and 13th Global Congress on Process Safety
San Antonio, TX
March 26 – 29, 2017

# Temporary Management of Change: A Problem or a Necessary Part of Managing Process Safety

**Jody Olsen**
**JE Olsen Consulting LLC**
**Anchorage, AK**
**jodyeolsen@gmail.com**

**Keywords:** Management of Change, MoC, temporary, risk, OSHA

## Abstract

Temporary Management of Change (MoC) projects seem to be the boogeyman of Process Safety Management (PSM). These temporary changes are frequently needed, but feared by operating company management, compliance groups, and lawyers. Some concerns regarding temporary MoCs are valid while other concerns may be based on false assumptions or incorrect interpretations of regulation. Most alarming, many of the concerns regarding temporary MoCs may lead to worse practices and the potential to introduce more risk into the business. This paper explores the reasons for this aversion to temporary MoCs and proposes a paradigm shift. Instead of considering permanence as the norm, consider that everything is actually temporary. The key is designing, operating, and maintaining for the appropriate window of time. The change will only be approved if the risk falls within accepted company protocol for the required duration of the change. The author will define five categories of temporary MoCs for the purpose of clarifying that one-size-fits-all duration limits and protocols are not appropriate. The author will address three myths regarding practices for managing temporary MoCs that may lead to unintended negative consequences. And finally, the author will propose preferred practices that avoid these unintended outcomes leading to more efficient work processes and safer operations. The need for temporary MoCs and the risks of mismanagement of those temporary MoCs must be well understood by both regulatory agencies and operating companies. Policies and procedures must drive practices that assure safe operation for any time horizon given real world impacts and constraints.

## 1 Introduction

Since the inception of process safety management (PSM) regulation [1], management of change (MoC) has been recognized as one of the most crucial elements of the regulation. Management of temporary changes is often highlighted as a critical and potentially more problematic type of change to be managed.

The incident at Flixborough, 1974 [2][3][4], is frequently cited as an early example of not only inappropriate management of change but also inappropriate management of temporary change. This event involved a temporary bypass that had been installed two months before the incident in order to maintain production while a failed cyclohexane oxidation reactor was repaired. A temporary piping modification – including the pipe, bellows, and pipe support - was designed, installed, and operated without appropriate mechanical review by qualified personnel. The temporary bypass piping failed during a plant start-up that followed unrelated repairs. A vapor cloud explosion followed the release of a large amount of cyclohexane. The explosion killed 28 employees, injured 89 employees and members of the surrounding community, destroyed the plant, and damaged many nearby homes and businesses.

In the intervening years, and especially within the last 25 years since the enactment of the Occupational Safety and Health Administration (OSHA) PSM regulation [1] and similar regulation in other jurisdictions, better PSM practices have become more widely understood and implementation has improved. Most operating companies now routinely apply management of change practices that are intended to address the causes that contributed to the Flixborough incident. Management of Change requirements are widely understood to include appropriate technical and process safety review of the design, drawing and other process safety information (PSI) updates, procedure updates, completion of necessary training and awareness communications, and pre-start-up review to name a few.

Those better practices include demanding equivalent review and approval requirements for all changes – including temporary changes. Where the change is temporary, certain other requirements are stipulated and must be addressed. These additional requirements relate to specifying a duration limit for the temporary change. The temporary change may also include additional safeguards and enhanced monitoring, inspection, and/or maintenance procedures that are specifically applicable for the duration of the temporary change.

This paper examines several myths related to management of temporary changes. While having a limit on the duration for a temporary change is appropriately understood to be an essential requirement for managing any temporary change, this paper addresses several pitfalls where specific interpretations of and practices related to that requirement may actually introduce unintended consequences and risk into the operation. Likewise, the allowance for conversion of temporary MoCs to permanent will be addressed. This practice is often allowed per procedure but is also not advisable or recommended. These practices coupled together can and will often lead to unintended negative consequences. The author offers a preferred approach for defining requirements specifically related to temporary MoCs and outlines a work flow process that places design requirements for the change – including duration and risk assessment - at the front end of the design and approval process.

# 2 Background

Below is a brief recap of a number of background assumptions and some history regarding management of change and management of temporary changes.

## *2.1 Knowns*

- There is a known history of major, minor, and near-miss incidents and failures associated with improper management of change.
- Temporary changes have been part of that failure history.
    - As per OSHA Appendix C, "Temporary changes have caused a number of catastrophes over the years" [5].

- Temporary changes often do not meet company specifications, standards, and/or normal operating limits and, therefore, are expected to have shorter service lives than permanent changes and may also require more rigorous monitoring, inspection, and/or maintenance and may include additional safeguards or layers of protection.

Two examples of familiar temporary changes are shown below:

Figure 1. Blue tarps on roofs after hurricane [6].

Figure 2. Piping repair using pipe clamps [7].

- Due to the above as well as regulatory requirements, approval of temporary MoCs must include a duration limit. Without a duration limit, the operator may lose track of the temporary change and it may effectively become permanent even if not originally intended to be permanent [5].

***2.2 Debunked Untruths*** Some early presumptions are now widely understood to *not* be true.

- It is not true that temporary changes are of such a short duration that review requirements are lesser than standard changes.

   - Review requirements for temporary and permanent MoCs are the same. Procedures must stipulate equivalent review requirements.
      - OSHA 1910.119 (l) provides no allowance for lesser or weaker review requirements for temporary MoCs. No distinctions between temporary and permanent MoCs are mentioned in the PSM regulation [1].
      - As per OSHA Appendix C, "Temporary changes are subject to the management of change provisions [5]."
- It is not true that Process Safety Information updates are waived or are inapplicable for temporary changes [1].
   - Current Process Safety Information (PSI) must be available to personnel involved in the operation regardless of the nature or duration of the change.
      - As per OSHA Appendix C, "Copies of process changes need to be kept in an accessible location to ensure that design changes are available to operating personnel as well as to PHA team members when a PHA is being done or one is being updated [5]."
   - How a company addresses this requirement will vary based on individual company protocols and document management procedures.
      - The regulation does not stipulate the format, such as electronic updates versus hand mark-ups.
      - The regulation does not stipulate the timing for formal updates. Updates for larger studies may require special protocol. Depending on type of study and length of temporary MoC, certain updates may be deferred. Company procedures should ensure that any unincorporated changes are communicated and that a reference to the MoC is made available.

*2.3* ***Myths*** This paper will address the following myths and propose alternative recommendations.

- Myth 1: Most temporary changes are terminated by returning the system to the prior condition, also known as reversing the change.
- Myth 2: The best way to ensure limits on durations for temporary changes is to have a universal duration limit for temporary MoCs prescribed within the company Management of Change procedure.
- Myth 3: Converting a temporary MoC to a permanent MoC is an acceptable practice especially for longer duration temporary MoCs.

# 3 Definitions

At this time, it would be useful to ensure that we have consistent definitions for several terms. Typically, changes are defined within Management of Change procedures as: Emergency, Temporary, or Permanent.

> Note: Emergency changes entail some similar concerns as temporary changes. Due to the highly reactive nature of an emergency change, special provisions are frequently considered allowable for emergency changes that are not allowable for any other type of change. These changes require special considerations and management requirements, but detailed discussion of emergency changes is not within the scope of this paper.

### *3.1 Temporary*

Consider the definition of the word *temporary*:

> Merriam-Webster [10] defines temporary as:
> lasting for a limited time

Below are several definitions provided within reference documentation for the term *temporary change:*

> Temporary Change:
> 1) "implemented for a short, predetermined, finite period [2]"
> 2) "a clear time limit for the validity of the change is defined [4]"
> 3) "planned for limited duration [8]"
> 4) "provided for a defined time period [9]"

The theme is very clear. While the term "short" is a relative term that may have varying interpretations, the reference to a finite, limited, or prescribed time period clearly conveys that the length of time that the change is acceptable to be in service must be stipulated for each temporary change.

### *3.2 Permanent*

Conversely, the antonym *permanent* may be more difficult to define. These references provide the following definitions:

> Permanent:
> Merriam-Webster [10] definition:
> continuing or enduring without fundamental or marked change
>
> Cambridge Dictionary [11] definition:
> Lasting for a long time or forever
>
> Dictionary.com [12] definition:
> existing perpetually; everlasting, especially without significant change.

The definition of *permanent* is less clear, because depending on ones' field of view, all things could be considered to be temporary given that equipment, human life, and even our planet have a lifespan, and therefore, do not last forever or perpetually. In the case of equipment, lifespan is affected by rate of wear, availability of replacement components, potential for obsolescence of mechanical or electronic components, maintenance and operational practices, and system changes affecting design requirements.

For our purposes when considering processing equipment and plants, a good working definition of *permanent change* might be:

Permanent Change:
Expected to remain in place indefinitely. Effective lifespan will frequently be in the range of 30 to 50 years assuming normal operating and maintenance practices. Lifespan typically corresponds to life of the field or the facility.

For the purposes of this paper, *permanent change* will reflect the definition given above.

### *3.3 MoC*
The acronym MoC, Management of Change, refers broadly to requirements associated with implementing changes. When used in the context "the MoC," the acronym is sometimes used as a noun meaning: "the change"; "the project"; "the project ID or record from the MoC application representing a given change or project." The acronym MoC may be used in this context in this paper.

Note: Although temporary changes may include temporary organizational changes, organizational changes are not included within the scope of this paper.

## 4 Regulation and Guidance

### *4.1 Regulatory requirements (United States)*
It would be useful at this time to ensure that we are clear on the language and intent of regulation regarding temporary management of change.

The OSHA PSM regulation, management of change section, reads as follows.

As per OSHA 1910.119 [1]:
1910.119 OSHA Process Safety Management of highly hazardous chemicals
1910.119 (l) *Management of Change*
1910.119 (l)(1) The employer shall establish and implement written procedures to manage changes (except for "replacements in kind") to process chemicals, technology, equipment, and procedures; and, changes to facilities that affect a covered process.
1910.119 (l)(2) *The procedures shall assure that the following considerations are addressed prior to any change:* . . . . (i)technical basis. . . ;(ii) impact . . . on health and safety; (iii) modifications to procedures;
1910.119(l)(2)(iv) *Necessary time period for the change . . .*
(v) authorization requirements . . .
1910.119(l)(3) [Operators, maintenance personnel, and contractors] ". . . shall be informed of, and trained in, the change prior to start-up . . ."
1910.119(l)(4) "process safety information . . . shall be updated accordingly. . ."
1910.119(l)(5) "procedures or practices shall be updated accordingly."

The regulation itself makes no mention of temporary versus permanent, but does indicate that the time period for which the change is approved, installed, and operated must be defined, meaning that the time period must be defined *for all changes*. Given that the period of the change is normally not explicitly defined for each permanent change, companies should formally define the term *permanent* globally within their management of change procedure. This global definition would effectively satisfy the requirement for defining the period of the change for any and all permanent changes as these changes will generally have a consistently understood duration. Given this stipulation in the regulation, it is clear that any change classified as temporary (or emergency) must define the duration that the specific change is allowed to remain in place.

### *4.2 Guidance*

Several clarification and guidance publications support and emphasize the requirement for defining the duration of temporary changes.

> As per OSHA Appendix C:
> It is important that a time limit for temporary changes be established and monitored since otherwise, without control, these changes may tend to become permanent [5].

Likewise, *CCPS Guidelines for the Management of Change for Process Safety* has similar language that refers to the requirement for specifying the allowable duration of the change.

> As per *CCPS Guidelines for the Management of Change for Process Safety*:
> A temporary change is approved with the stipulation that the change last no longer than a specified time and that additional safety features be implemented . . . Temporary changes need prescribed time limits, appropriate authorization, and proper documentation [2].

## 5 Categories of Temporary MoCs: Why are Temporary MoCs needed?

In order to understand issues, constraints, duration limits, and life cycle requirements for temporary MoCs, it is important to recognize the different reasons for these changes. Temporary MoCs will generally fall into one of six categories:

1) Temporary trial use of new equipment, chemicals, process parameters, process configuration, and/or procedures.
2) Temporary modification following a failure but preceding implementation of the permanent, long-term solution. The pending permanent change may be:
    a. A completely new design OR
    b. Return of the system to original specifications.
3) Temporary bypass or failure of safety critical devices or other components. Device or component is out-of-service due to:
    a. Bypass, override, impairment, or removal for proactive maintenance or inspection OR
    b. Failure or discovered defect. Device or component is not functional or not fully functional. Repairs are pending.
4) Temporary connections and/or temporary operations, such as in advance of shut-downs for purging or other occasional operations.
5) "Temporary" deferral of maintenance or inspection procedure due date including turnaround (TAR) deferrals.
6) "Temporary" installation intended to be followed by a future installation having more capacity or capability.

### *5.1 Category 1 - Temporary trial or field test*

Category 1 Temporary MoCs are proactive and planned in nature. These changes would typically be considered trials or pilot studies performed in order to field-test a new idea before major or permanent modifications are made to accommodate the change. The new idea and change could be related to equipment, instrumentation, controls, chemicals, process parameters, process configuration, or procedures.

### *5.2 Category 2 – Temporary modification following a failure*

Temporary MoCs in Category 2 are generally more reactive in nature. The Category 2 MoC is planned, reviewed, and approved but typically on a much shorter timeline in response to an unplanned failure or event. This type of MoC is always considered an interim step until a final planned modification (another MoC project) or repair may be implemented.

This two-step solution may be due to lack of availability of required materials or resources, additional time needed for design and fabrication, or a need for a larger outage in order to implement the permanent change. The long-term solution will typically require a more lengthy and deliberate project management cycle involving detailed review of the causes of the failure, full assessment of the needs, consideration of alternative solutions, detailed design of the new installation, and execution involving construction activities, installation, and start-up.

In some instances (Category 2a), the long-term solution will be a completely new design or modification addressing the cause of the failure. In other instances (Category 2b), the long-term solution may be to return the system to original specifications if the cause of the failure was related to off-specification materials, installation issues, operational or maintenance issues, or an improperly approved prior change. In those cases, other mitigations unrelated to the original design should be implemented to address the cause(s) of those introduced conditions. Although this type of "change" could represent a return to original prescribed specifications and conditions, the long-term solution to restore the original design is now effectively a change. As such, regulatory requirements as well as good project management practices dictate that all required elements of MoC be applied to this change as well. Some straightforward repairs may be excluded and not require an MoC. The permanent change, long-term solution, should be handled as a separate management of change project but should be related to the temporary MoC record addressing the same failure. Start-up of the long-term solution, a permanent modification or repair, will correspond to the closure date of the temporary modification.

Based on the author's field experience, these types of temporary changes represent a significant percentage (estimated at over half) of the MoC projects that are managed within the core MoC application.

### *5.3 Category 3 – Temporary bypass or failure of safety critical devices or other components*

Category 3 Temporary MoCs involve situations where safety critical devices or components are out-of-service or inoperable for any reason including 3a) being bypassed, jumpered, overridden, or removed or 3b) found failed, impaired, not functional, or not fully functional. Category 3 MoCs fit the definition of temporary management of change; however, these types of changes may be managed under a different procedure or protocol [13] due to the nature of the condition, mitigations, and communication requirements, as well as inherent safety concerns.

Category 3a changes are planned, proactive work based on pre-written procedures, such as routine scheduled inspection, testing, and maintenance of relief devices. Category 3b changes are in response to failures and, therefore, represent reactive changes.

Although potentially handled through alternative procedures, databases, and communication tools, Type 3 Management of Change must address all items required under 1910.119 (l) OSHA PSM Management of Change. These requirements include: defining robust authorization requirements for approval of the change; completing an appropriate hazard and risk assessment to determine safety and health impacts, implementation of required mitigations and special procedures while the device is out-of-service; completion of ongoing communication and training requirements; making current PSI available; and defining a duration limit.

### *5.4 Category 4 – Temporary connections and/or temporary operations*

Category 4 Temporary MoCs involve temporary connections to existing process equipment and/or occasional operations. These temporary changes are procedural. The procedures may involve connecting hoses, portable skids, or trucks to process equipment. These procedures may be related to equipment shut-down and isolation activities such as venting and purging of equipment. Or the procedure may represent a routine or occasional operational activity such as off-loading chemicals.

All standard management of change requirements must be addressed for the initial approval of new procedures. If the identical equipment and procedure is used in the future and the procedure addresses all management of change requirements including safety and hazard review and authorization requirements within the procedure steps, the procedure may be reused without creating another MoC record.

### *5.5 Category 5 – "Temporary" deferral of maintenance or inspection procedure due date including turnaround (TAR) deferrals*

Category 5 temporary MoCs involve deferral of planned maintenance or inspection work beyond the scheduled due date. A temporary MoC could be used for a one-time deferral. A permanent MoC would be used for a permanent change to the maintenance or inspection interval prescribed by a particular maintenance plan. Likewise, deferral of turnarounds should be considered in the same category due to the number of planned maintenance and inspection activities that require an outage in order to be executed.

> Note: Companies may chose to consider deferral of an inspection or maintenance workorder due date as a permanent change associated with the single instance of the procedure, inspection, or turnaround activity. If there is sufficient tracking of the incomplete workorder activity through the maintenance management system, that classification would be sufficient. Identifying the change as temporary would add a secondary due date flag providing additional visibility. An MoC is required whether the single instance due date deferral is considered a temporary or a permanent change.

### *5.6 Category 6 – "Temporary*" installation intended to be followed by a future installation having more capacity or capability*

Category 6 temporary MoCs are included for the purpose of completeness. However, it is recommended that this category not be considered temporary. This category represents a two-stage modification – similar to Category 2. However, in this case, both stages of the project are fully planned and proactive. The installation of the final modification will be driven by business rather than technical needs. The interim modification will typically have less capacity or capability than the planned final modification. However, the interim modification is acceptable from a technical and safety standpoint to remain in place indefinitely. In this scenario, it is recommended that the user classify this change as a permanent MoC. Business drivers and budget availability may change. Changes of this type do not require expiration dates or flagging as overdue for closure. Occasional exception may apply.

> * Note: Category 6 changes will not be discussed further, as these types of changes are recommended to be characterized as permanent.

### *5.7 Summary of Temporary MoC Categories*

Table 1 is a summary of the temporary MoC categories discussed above. The administrative management system commonly used to manage the change category is also shown.

**Table 1. Temporary MoC Category Descriptions and MoC Management System**

| | Temporary MoC Categories | | | | | | | |
|---|---|---|---|---|---|---|---|---|
| **Category** | 1 | 2 | | 3 | | 4 | 5 | 6 |
| Sub-Category | | 2a | 2b | 3a | 3b | | | |
| **Temporary MoC Category Description** | Trials, Field Tests | Interim solution following failure | | Bypass or failure of safety device or component | | Temp connections / ops | PM/TAR due date deferral | Staged improvement* |
| **Sub-Category Description** | | Long-term: new design | Long-term: original design | Proactive. Procedural | Reactive: Operate without device | | | |
| **Management (Mgmt) System** | MoC System | MoC System | MoC System or Repair | Bypass Mgmt System; Procedures | Bypass Mgmt System | MoC System; then Procedures | MoC System | MoC System; New Projects |
| **Note:** | | Interim solution is Temp MoC. Long-term solution is Perm MoC. Both MoCs relate to the failure and the other MoC. | | Bypass or failure of safety critical devices could be managed in MoC system, if preferred. | | | Temp or Perm MoC for single instance deferral. | *Call Permanent |

# 6 Dispelling Myths

Section 2.3 summarizes three myths that commonly exist in industry and literature regarding temporary MoCs. Many of these myths, or assumptions, do not accurately reflect the real world usage and service of temporary MoCs. More concerning, many of these assumptions have elicited internal company rules, protocols, and procedures that have a possibility and a tendency to introduce greater risk into the operation.

The author will address each myth with respect to the various categories of temporary MoCs and conclude with preferred recommended practices.

## *6.1 Myth 1: Most temporary changes are terminated by returning the system to the prior condition, also known as reversing the change.*

6.1.1 History/Past Assumptions:
Much literature on the topic speaks to closing temporary changes by returning or reversing the change to its original or prior condition [2][4][5][13][15]. Those references imply a Category 1, 3a, or 4 change is being discussed.

> As per OSHA Appendix C:
> In addition, the management of change procedures are used to [ensure] that the equipment and procedures are returned to their original or designed conditions at the end of the temporary change [5].
>
> As per Marsh:
> For temporary changes (handled under a temporary MoC), one of the key and final actions is the return of the plant to the original design [4].

6.1.2 Alternative Perspective:
Regarding Category 2 or 3b changes - temporary MoCs involving equipment or component failure, reversing the change to the original or prior condition may not be desirable because the original or prior condition may have contributed to the cause of the failure or may not be possible because the failed condition has not yet been corrected.

Consider these examples of temporary changes associated with failures (Category 2):

1. Blue tarps on roofs after a hurricane before roof repair or rebuild.
2. Pipe clamps on pipelines following a leak before permanent pipeline section repair or replacement.
3. De-rating a pipeline to address an inspection finding before permanent section repair or replacement.
4. Removing a pipeline, piece of equipment, or component from service until repair, replacement, or permanent decommissioning.

From the author's experience, Category 2 and 3b, represent the most common types of temporary MoC projects.

**Table 2. Estimated Percentage of Temporary MoCs by Category**

| | Temporary MoC Categories | | | | | | |
|---|---|---|---|---|---|---|---|
| **Category** | **1** | **2** | | **3** | | **4** | **5** |
| Sub-Category | | **2a** | **2b** | **3a** | **3b** | | |
| **Category Description** | Trials, Field Tests | Interim solution following failure | | Bypass or failure of safety device or component | | Temp connec-tions / ops | PM/TAR due date deferral |
| **Sub-Category Description** | | Long-term: new design | Long-term: original design | Pro-active. Procedural | Reactive. Operate without device | | |
| **Management (Mgmt) System** | MoC System | MoC System | MoC System | Bypass Mgmt System; Procedures | Bypass Mgmt System | MoC System; then Procedures | MoC System; New Projects |
| **% of Temp MoCs (MoC System) - Based on Count **** | **5 - 20%** | **50 - 75%** | | | | Initial approval of procedure: **5 - 20%** | **5 - 20%** |
| **% of Temp MoCs (Bypass Mgmt System) - Based on Count**** | | | | **50 - 75%** | **25 - 50%** | | |
| **Based on OOS Period***** | | | | **10 - 50%** | **50 - 90%** | | |
| **Note:** | ** Estimated based on the author's empirical experience. *** Estimated based on period or number of days out-of-service (OOS) rather than number of instances or MoC count, based on author's empirical experience. | | | | | | |

As per Table 2, the majority of temporary MoCs cannot be reversed to the prior condition at an arbitrary date. Many temporary changes are interim responses to failures that require other repairs or projects to be completed before the temporary change can be removed and replaced with another change.

6.1.3 Truth 1: It is important to recognize the differing reasons for utilizing temporary MoCs. Following a failure – which is one of the primary reasons for creating a temporary MoC – the operating company must complete other work before the temporary MoC may be closed. This other work (e.g. repairs, replacements, modifications, etc.) dictates the timeline for the temporary MoC closure and may result in a new configuration.

***6.2 Myth 2: The best way to ensure limits on durations for temporary changes is to have a universal duration limit for temporary MoCs that is not "too long" prescribed within the company Management of Change procedure.***

6.2.1 History:

Operating companies have widely, if not universally, taken the approach that the best way to ensure that limits are established on temporary MoC duration is to prescribe a universal "limit" in the company Management of Change procedure or protocol. Globally established procedural limits for temporary MoC duration typically range from 90 days to one year. Some companies indicate two procedural limits: an initial limit that may be followed by extension up to a maximum limit, usually not exceeding one year. Other companies indicate a "limit" in the procedure, generally not exceeding one year, but allow for extensions beyond that limit, making the initially defined period not a true limit but a target and a checkpoint requiring re-authorization. Some companies establish an ultimate limit by using "shall" or "must" statements that preclude approving extensions. Other companies may not prescribe fixed duration limits, but due to discomfort with longer duration temporary MoCs, simply do not allow or approve lengthy temporary MoCs.

Presumably, this approach is seen as the best way to address the following realities. Temporary MoCs:

- ⇨ involve a change that is approved for only a limited duration
- ⇨ are required by regulation or recommended by guidance literature to have a stipulated duration limit and
- ⇨ are often seen as having a risk or tendency to become effectively permanent if they linger or remain in place "too long."

The consensus solution seems to be: apply a universal limit for temporary MoCs and make it not "too long." Interpretations of not "too long" vary but typically land on one year or less. Some companies may interpret the regulation as requiring a single, fixed, defined limit for all temporary MoCs.

6.2.2 Alternative Perspective:

*This approach of indicating a single, universal duration limit for all temporary MoCs – while well-intended - is inherently flawed and inherently unsafe. This approach will inevitably lead to unintended consequences that could introduce more risk into the operation. Here's why, . . .*

*1) First of all, . . .*
*it is important to recognize the differing reasons for utilizing temporary MoCs as the differing categories of temporary MoCs will have very different drivers for both initiating the temporary MoC and very different time horizons and dependencies for closure of the temporary MoC.*

Category 1 changes, trials and science projects, are entirely proactive. Although each temporary change must have a specifically defined period, this category of temporary change could have a wide range of desired and acceptable durations. The change must be designed to safely meet the intended and necessary life of that change. Thus, globally prescribed limits or even targets are not necessary.

The driver for the duration should be defined within the project scope. Similarly to project cost and schedule estimating, duration estimates should incorporate contingency planning so that the design and assessment properly consider the realistic worst case rather than the best case for how long the change could remain in place. The design should be robust so that the risk falls within the company prescribed risk management boundaries for the realistic duration of the temporary change (See Figure 3). The change may be able to be designed so that no incremental risk is incurred during the term of the temporary change as per the process hazard assessment.

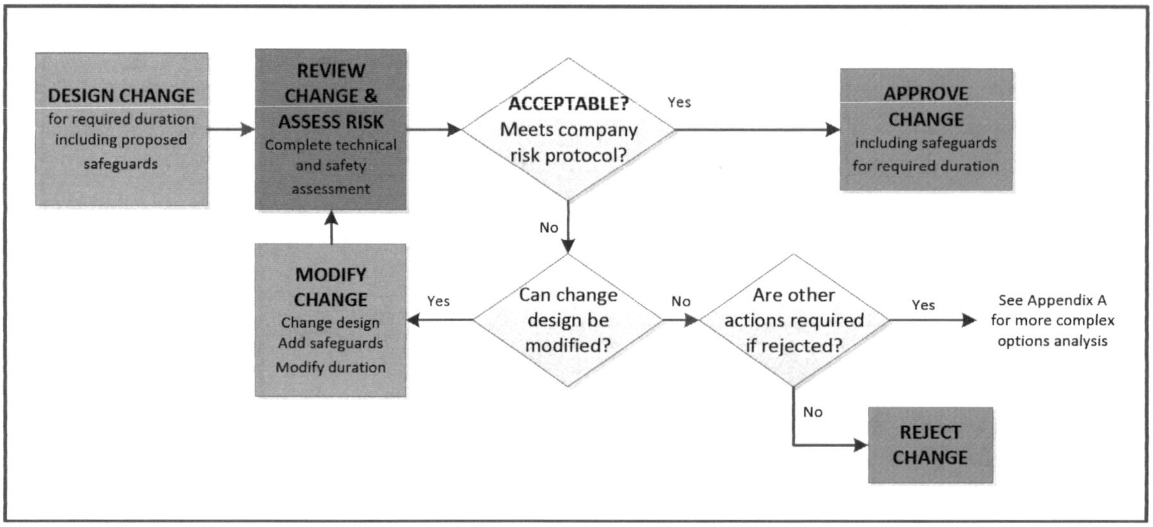


**Figure 3. Simplified Work Flow Process: Temporary MoC Review and Approval**

As with all temporary MoCs, the expiration date prescribed within the specific MoC should be queryable and monitored so that as the deadline approaches, planning will accommodate closure, which means reversal in this case, by the due date. The contingency built-in up front will aid in on-time closure and make it more likely that the original design and initial technical, safety, and operational reviews consider the full-term of the change.

Category 2 temporary changes are interim solutions in response to failures or unplanned events. The duration limits for Category 2 changes are entirely driven by the timeline and critical path for the implementation of the longer-term solution. This long-term solution is directly related to the failure that drove the interim solution. Duration needs for Category 2 MoCs will also be highly variable. A single prescribed limit is not appropriate for this type of MoC.

As with Category 1 MoCs, when defining duration for the Category 2 MoC, it is important to consider the worst-case timeline that incorporates contingency for the long-term project schedule. The long-term solution and all critical path items associated with the long-term solution may not be fully defined at the time that the temporary change will be initially designed and approved. However, some critical path drivers may be known, such as timing for procuring material and future planned shut-down opportunities if a larger unit or plant shut-down is required.

In the case of Category 3a changes - proactive procedures for performing inspection, testing, and maintenance of safety critical devices, these procedures should have very limited duration allowances, typically in the range of 8 to 12 hours (single shift) or less. However, even these procedures should have individually prescribed duration limits based on the specific application and requirements of that procedure. (Refer to Figure 4, Example 1 for an example of a Category 3a change and associated risk assessment.)

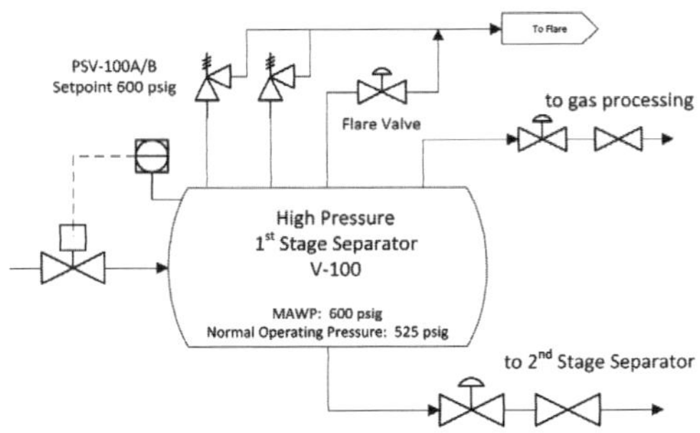


Category 3a: Planned Maintenance Work:
Removing one PSV at a time from service for scheduled inspection, testing, and maintenance is typical Category 3a planned work. Procedures should be prepared in advance which define the hazards (safety risks), required mitigations, and the approved duration for performing this work.

Category 3b: Unplanned Repair Work
In the event that a PSV requires repair or extended removal for any other reasons, the same prepared hazard and risk assessment and mitigation plan may be used as the basis for planning the repair work.

In the case of this PSV removal, the mitigation defined is reducing maximum inflow rate from 200 MMSCFD to 150 MMSCFD. This change will impact production; however, it should be less of an impact than a complete shut-down. The maximum inlet flowrate alarm will be reset to the lower limit. Prescribed operator actions remain the same.

Based on the risk assessment for this PSV setting, the safety risk rank is reduced for the duration of the out-of-service period due to the mitigation of lowering max rate. The flare valve has full relief capacity at the lower rate (could be considered an IPL) and a single relief valve has full relief capacity at the lower rate.

**Risk Rank Comparison - PSV-100A or PSV-100B Out-Of-Service** (a)

| | | | | |
|---|---|---|---|---|
| Relief Capacity per PSV | **150** | MMSCFD | | |
| Relief Capacity - Flare Valve | **175** | MMSCFD | | |
| | **1 Normal Ops** | | **2 1 PSV Out-of-Service** | |
| Normal Max Inflow Rate | **200** | MMSCFD | **200** | MMSCFD |
| Mitigation: Reduce Max Rate by | **N/A** | MMSCFD | **50** | MMSCFD |
| Relief Requirement | **200** | MMSCFD | **150** | MMSCFD |
| Number of PSVs Available | 2 | | 1 | |
| PSV(s) | IPL | Both Rqd | IPL | One Rqd |
| Shut Down Valve | IPL | SIL 1 | IPL | SIL 1 |
| Flare Valve | Not IPL | BPCS | IPL | BPCS |
| Risk Rank | **B2** | | **B1** | |
| Risk Category | **Med - 6** | | **Med - 5** | |

**Risk Rank Matrix** (b)

| CONSEQUENCE | | | LIKELIHOOD (per year) | | | | | |
|---|---|---|---|---|---|---|---|---|
| | | | < 1.E-05 | 1.E-04 | 1.E-03 | 1.E-02 | 1.E-01 | 1.E+00 |
| | SAFETY | | 1 | 2 | 3 | 4 | 5 | 6 |
| | multiple fatality | A | M6 | H7 | H8 | VH9 | VH10 | VH11 |
| | fatality onsite | B | 2 M5 | 1 M6 | H7 | H8 | VH9 | VH10 |
| | perm disability | C | L4 | M5 | M6 | H7 | H8 | VH9 |
| | hospital / rehab | D | L3 | L4 | M5 | M6 | H7 | H8 |
| | medical treatment | E | L2 | L3 | L4 | M5 | M6 | H7 |
| | first aid | F | L1 | L2 | L3 | L4 | M5 | M6 |

Given the safety risk assessment, the duration limit for this Category 3 temporary change will be driven by commercial rather than safety related concerns.

**Figure 4. Example 1: Category 3 Temporary MoC - Separator PSV Out-Of-Service.** Risk comparison is shown as (a) with data and independent protection layers (IPL)s. Risk Rank Matrix is shown as (b).

Category 3b situations - out-of-service or failed safety critical devices or other components - will have initial authorization requirements. Then, these instances will typically have reauthorization and communication requirements prescribed on daily, weekly, and monthly intervals, which are similar to duration limits. Re-authorization requirements typically escalate to higher levels of management as the out-of-service duration extends. However, as with all temporary MoCs, authorization or reauthorization of the 3b temporary MoC must consider all elements of the change including current condition, safeguards, mitigations, and risk. Example 1 in Figure 4 also discusses the PSV removal in the context of a Category 3b reactive change.

In certain cases, appropriately addressing the cause of the Category 3b failure could require extended efforts. If modifications are required, a separate MoC is typically written for the long-term solution. In these cases, the resolution timing for the Category 3b failure would resemble the timelines associated with Category 2 temporary changes.

> Note: Category 3 MoCs are sometimes managed through alternative protocols and procedures. If a modification is required to address the cause of a Category 3b failure, the temporary change and the long-term change may now be managed in two systems.

Regardless of the system in which the change is managed and the semantics of duration limits versus reauthorization or extension, the review of the change for authorization, re-authorization or extension must receive detailed technical and safety review, assurance of adequate safeguards and procedures, evaluation of risk, and consideration of alternative options including process shut-down, as needed. Shut-down considerations must weigh risks associated with shut-down and start-up activities against the risk of continued operation in the current state. The realistic timing for the resolution should be understood and be built into subsequent reauthorization or extension requests. Refer to Appendix A for a more detailed version of Figure 3 work flow where more complex options analysis is required. Although it is desirable for Category 3 temporary changes to have very short durations, a failed device or device prone to failure cannot be returned to service.

Category 4 changes – temporary connections and/or temporary operations - will have duration limits defined by the procedural activity to which these changes relate. These limits may be in the range of one shift to 48 hours. This timing will be specifically dependent on and defined by the activity to be completed as well as the risk assessment. These activities might be handled within the MoC system for only the first instance of approving and using a new procedure.

The Table 3 summarizes the typical range of temporary MoC duration by category of temporary MoC. Defining and understanding the change category or type should aid in understanding the duration needs for a given change, which in turn leads to appropriate design and more robust approval protocol.

**Table 3. Temporary MoC Category Descriptions and Typical Duration Range**

| | Temporary MoC Categories | | | | | | |
|---|---|---|---|---|---|---|---|
| **Category** | 1 | 2 | | 3 | | 4 | 5 |
| Sub-Category | | 2a | 2b | 3a | 3b | | |
| **Category Description** | Trials, Field Tests | Interim solution following failure | | Bypass or failure of safety device or component | | Temp connec-tions / ops | PM/TAR due date deferral |
| **Sub-Category Description** | | Long-term: new design | Long-term: original design | Pro-active: Procedure | Reactive: Operate without device | | |
| **Typical Duration Range** | Variable (Months to Several Years) | Variable (Months to Several Years) | Variable (Months to Several Years) | 8 to 12 hours (single shift) | Days to Months (or longer); intervals prescribed to reauthorize | Hours to Days | Variable (PM: Days to Months; Defer TAR: 1 -2 years) |
| **Duration Tied to** | Temp MoC design intent | Perm MoC timeline | Perm MoC timeline | Procedure guidance | Repair or Perm MoC timing (like Cat 2 Temp MoC) | Procedure guidance | Various: outages, resources |
| **Note:** | Process hazard risk assessment is built into all temporary MoC project or procedure design including prescribing safeguards and mitigations required to bring risk to an acceptable level for the required project duration. (See Figure 3 and Appendix A.) | | | | | | |

*2) Secondly, . . .*
*it is important to recognize that the language in these global procedures – specifically the universal duration limits - does not represent truth or reality. Some temporary MoCs will inevitably need to go longer than one year . . . and some should.*

By recognizing the driver(s) for the duration and defining a plausible timeframe for the temporary change, the technical, safety, environmental, operational, and process safety reviewers will assess the change with a more critical and realistic eye for actual design needs for the project. Placing arbitrary or unrealistic limits on the duration of the temporary MoC will improperly communicate the true design needs to the technical and operational professionals evaluating the change. If the change needs to remain in place over four years to reach the next plant outage opportunity, the design, operational, and maintenance requirements must reflect the needs to safely operate for that period of time. If the change is unacceptable for that duration, a design change or another alternative option must be considered. (Refer to Figure 3 and Appendix A.)

By placing arbitrary limits taken from global standards on the maximum duration for temporary MoCs and then allowing for extension, the true design requirements of the change are not accurately reflected when the design is reviewed and approved. Documentation and approvals must reflect the true anticipated life. If the documented duration of the change does not reflect the intent or need, the MoC could be subject to audit findings.

Even with specificity and contingency built-in, changes to the long-term project scope or definition may affect the implementation timeline and, therefore, require review and re-evaluation of the temporary MoC to consider extension of that change. However, the number and length of extensions can and should be minimized by avoiding use of arbitrary, globally applied duration limits for temporary changes.

Of note, when extension requests are reviewed and approved, the phenomenon of "normalization of deviation" can lead to weaker scrutiny when perpetuating a temporary change that is already in place. Therefore, limiting the number of extensions is desirable.

It is also important to keep in mind that defining realistic and reasonable schedules rather than forced, aspirational but unrealistic schedules will lead to more robust ultimate solutions. Recall the good-fast-cheap triangle [14] and the truth that it teaches us regarding project management.

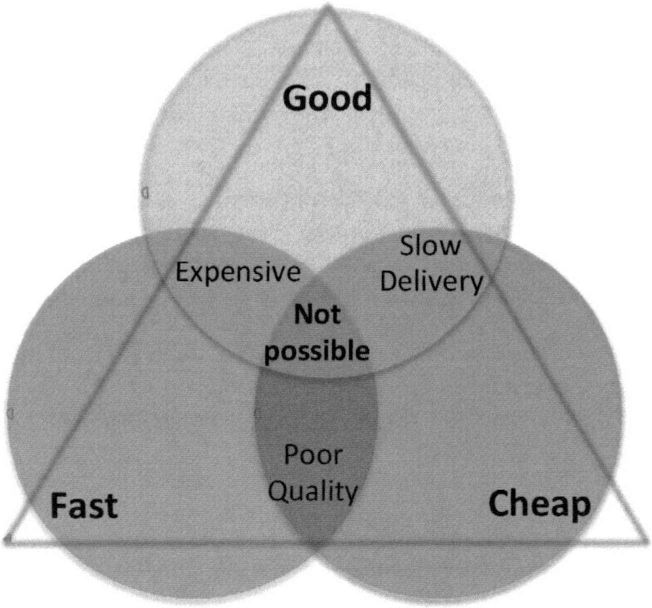


**Figure 5. Good-Fast-Cheap project delivery triangle**

In project management and delivery, you can only pick two of the three of the corners of the good-fast-cheap triangle. And, if you pull too hard on any one corner, you may only get one of the three.

The OSHA PSM regulation requires duration limits for *any* MoC project. Ironically, it is more appropriate to apply a universal definition to "permanent" than "temporary" when defining change duration, although the reverse is generally done.

*3rd, finally and importantly . . .*
*Many companies may misread or misinterpret the language and intent of regulation and guidance.*

The OSHA PSM regulation [1] and well established industry guidance, such as *CCPS Guidelines for the Management of Change for Process Safety* [2] never state that *universal* limits are required for temporary MoCs. The OSHA PSM regulation does *not* define a maximum length for the duration of a temporary change. Likewise, *CCPS Guidelines for the Management of Change for Process Safety* [2] does *not* define a maximum limit. The regulation states and intends that *each* temporary MoC have a duration period stipulated and that the change is appropriately designed, reviewed, and approved considering that duration.

6.2.3 Truth 2: Each Temporary MoC must be designed to meet a specifically stipulated duration limit. Universal caps on duration for temporary MoCs are neither appropriate nor effective.

### *6.3 Myth 3: Converting a temporary MoC to a permanent MoC is an acceptable practice especially for longer duration temporary MoCs.*

6.3.1 History:
Allowance for conversion of temporary MoCs to permanent MoCs is frequently stipulated within company procedures as one of several ways to disposition a temporary MoC that is approaching its deadline.

The following choices are often provided in company MoC procedures as options for addressing the temporary change before the expiration date [4][8][9]:

- At the end of the duration limit, the change shall or must be:
    - Removed or reversed and the plant restored to original condition OR
    - Converted to a permanent change OR
    - Granted an extension

Where extensions are not allowed, the extension option is removed leaving two choices, neither of which may be technically acceptable or feasible.

> Note: Shutting-down the equipment, process, or facility should always be considered as an option, but shut-downs are rarely given as one of the options to consider upon approaching the due date. Of course, shut-downs and start-ups present inherent safety risks as well. But a shut-down may be the only technically acceptable option depending on the options made available.

If a temporary change (of defined limited duration) is converted to a permanent change (of indefinite duration), an essential basis for the prior design and approval has been changed.

Therefore, the original approval is fully invalidated. Any conversion necessarily requires review by any and all prior reviewers and approvers of the temporary MoC (or equivalent discipline representatives).

Regarding conversion administration, in some companies the original MoC record may be redefined as permanent. In other companies, conversions are accomplished by opening and approving a new MoC record describing the same change****. This method of handling the conversion tends to be cleaner administratively allowing for accurate reflection of history, clarity of the scope, insertion of new technical, safety, environmental, operational, and managerial reviews, and documentation of new authorizations. Also, the new permanent MoC will require final updates of all applicable PSI, whereas updates for PSI associated with temporary MoCs may have differing protocol and requirements regarding format and timing. The two MoCs should be related or cross-referenced to each other in the administrative system.

> **** Note: If a new MoC record is opened that describes a new change or condition, then the new MoC is a new permanent change rather than a conversion, although still related to the temporary MoC.

The three disposition options described above, or two where extensions are not allowed, are broadly applied throughout industry. The same choices are also reflected in technical articles [4] [15] on the topic.

6.3.2 Alternative Perspective:

*This approach of offering conversion to permanent as a method for managing temporary MoCs approaching their due date – while well-intended - is inherently flawed and inherently unsafe. This approach will inevitably lead to unintended consequences that could introduce more risk into the operation. Here's why, . . .*

Conversion of a temporary MoC to a permanent MoC should not be allowed unless the conversion is based on one of a set of defined and defensible reasons. Possible legitimate reasons for conversion of a temporary MoC to permanent could include: 1) there was a technical error in the initial assessment, 2) a global company standard or specification has been relaxed specifically impacting this change, 3) another system change has occurred impacting this change (such as safe operating limits have shifted and those new limits have been separately technically approved), 4) the change was a Category 6 change called temporary for business reasons only, not technical or safety reasons, or 5) the change fell into another category due to original duration intent only but meets all standard technical, safety, and operational protocol for permanent operation and is now intended for indefinite operation in the current configuration. **Bumping up against an arbitrary temporary MoC duration limit is *not* one of those reasons.** If criteria are met that allow for conversion, all process safety information updates must occur as per protocol for permanent changes.

Temporary MoCs are typically defined as temporary because the change is unacceptable technically to remain in place for an indefinite timeframe as a permanent installation. Conversion of a temporary MoC to permanent is, therefore, inherently problematic.

Note: Closure of temporary MoCs via conversion to permanent MoCs, if allowed by the company, should be a red flag to auditors. If not based one of the reasons given above or other legitimate reasons defined by the company, the approval of the conversion to permanent may not be justifiable or technically sound.

Temporary MoCs that are inappropriately converted to permanent MoCs place the operating company at higher risk, as the flagging system based on the temporary MoC expiration date will be removed. The change will no longer be flagged in the system and will, therefore, be less visible for further review, monitoring, and closure. Additionally, drivers for outages will be masked when the temporary designation and associated due date is removed.

6.3.3 Truth 3: Converting a temporary MoC to permanent MoC is rarely an acceptable practice. Call a spade a spade, or we induce bad behavior and bad outcomes!

# 7 Two Wrongs Make a Wrong!

Much literature has been written regarding common problems associated with managing temporary MoCs [15]. There seems to be little recognition that the problems may in fact be self-fulfilling prophesies when our management systems steer personnel down paths toward failure. For example, Ozog and Stickles [15], describe common quality "problems" seen with regard to managing temporary MoCs.

> Temporary changes
> The documentation, closure, and communication of temporary changes are often cited as major quality issues in MOC programs. Common issues include:
> a. *The authorized time period expires without removing or making the change permanent.*
> b. *The change is made permanent*, but is not reflected in the piping and instrumentation diagrams (P&IDs) and other process safety information (PSI).
> c. *The change is improperly or incompletely removed* [15].

Causes for these problems are embedded in the problem statements. Conversions to "permanent" are not technically acceptable for a true temporary change, but are frequently inappropriately applied in order to manage deadlines that are neither realistic nor appropriate.

Many of these MoC conversions are being driven by the combination of these two problematic requirements being coupled onto temporary MoCs: arbitrary universal duration limits that are too short *and* allowance for conversion of temporary MoCs to permanent MoCs as a standard, sanctioned method of resolution when MoCs approach these arbitrary deadlines.

*In other words, we are fabricating a problem that should not exist by applying an unnecessary constraint, . . . and we are giving people an inappropriate tool to deal with that "problem."*

Humans that are forced to make a decision between several untenable choices will inevitably make a poor decision. These types of errors in human decision-making that are propagated by management system flaws are routinely found to be causal factors for major and minor industry incidents [16]. As stated in CCPS, *Recognizing Catastrophic Incident Warning Signs in Process Industries*: "Management system deficiencies are a root cause in all effective incident investigation methodologies [16]."

Insidiously, these MoC procedural constraints and requirements may actually build defects into the operation. Auditing to those procedures can introduce even more defects into an operation, as well.

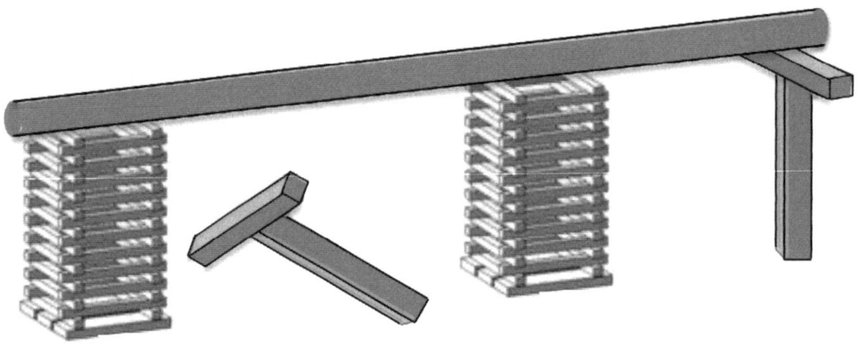

Category 2 Temporary MoC:
Due to soil instability issues, a series of pipe supports servicing a high pressure gas pipeline began falling over.

The timing for implementation of the long-term solution was constrained by seasonal weather windows and a need to consolidate the shut-down with a wider plant outage.

As an interim solution, crib stacks (10' – 15' tall) were specifically designed and engineered by a civil/structural engineer as temporary pipe supports until the new replacement pipe supports could be designed and installed. The crib stacks were reviewed and signed off during the temporary MoC review by all relevant disciplines including mechanical, operations, environmental, and safety. A hazard assessment was conducted.

When the wider plant outage was delayed for one year, the civil/structural and mechanical engineers approved allowing the crib stacks to remain in place for three years with required inspections continuing throughout the period.

However, because of a universal temporary MoC duration limit of one year imposed by the company MoC procedure, the temporary MoC could not be extended. Operations and Engineering Management preference was to convert the MoC to a permanent MoC rationalizing that it would be a "3-year permanent" MoC. This definition of permanent would be in conflict with standard definitions and the intent of permanent installations.

**Figure 6. Example 2: Category 2 Temporary MoC – Interim Pipe Supports, Tall Crib Stacks**

# 8 Conclusions/Recommendations

The following recommendations are offered as preferred practices for managing temporary changes:

1. Recognize that there are distinct categories of and reasons for implementing temporary MoCs.
    - Consider defining the temporary change category in the header information in order to better define likely lifecycle range for that change.
    - Recognize that many temporary MoCs are interim solutions associated with failures (Category 2 or 3b). Those MoCs require implementation of another project, the long-term solution, as a prerequisite for closure of the temporary MoC.

  - Cross-reference temporary MoCs that are interim solutions for failures (Category 2 or 3b) with permanent MoCs that are the long-term solutions for the same failure.
  - Define and monitor critical path item(s) that will define closure timing.
2. Do not dictate universal fixed duration limits for temporary MoCs in company management of change procedures. Design the change considering safeguards and including inspection and maintenance requirements in order to meet company risk protocol and safely operate for the intended duration of the change.
  - Ensure that every temporary MoC has a specified duration limit defined for the specific change as a requirement for approval. In electronic systems, expiration date should be a required field.
    - Note: The OSHA PSM regulation does not require a universal fixed limit on temporary MoC durations.
  - Consider real world constraints when setting the temporary MoC duration such as: materials deliveries, resource availability, shut-down opportunities, etc. Incorporate contingency into setting expiration dates.
  - Refer to Figure 3 and Appendix A for temporary MoC review and approval work flow process that considers duration requirements as an input to the design and review process.
  - Develop internal company protocol that addresses PSI management needs for MoCs of varying temporary lengths.
  - Define the meaning and duration of "permanent change" globally within the company management of change procedures.
3. Conversion of temporary MoCs to permanent should rarely be allowed. An inappropriate conversion could create a hidden or latent process safety hazard.
  - Define a listing of acceptable reasons that temporary MoCs may legitimately be converted to permanent. Allow conversion only if one or more of those conditions is met.
    - Note: A temporary MoC that is approaching an expiration date is not an acceptable reason to convert the MoC to permanent.
  - Clearly define what conditions changed from the prior temporary MoC making the conversion allowable for each temporary MoC that is converted.
  - For the rare instances where conversions are allowable, initiate a new MoC and relate the new permanent MoC to the prior temporary MoC.

Many of the errors that have historically been associated with managing temporary MoCs may be directly attributable to constraints and flaws built into the management systems. Addressing these underlying issues will eliminate some of the causal factors for human error when using these procedures. Allowing personnel the flexibility to design, approve, and manage temporary changes based on the system need as well as the assessed risk will ultimately lead to more efficient and resilient [17] work processes and safer, more reliable operations.

# 9 Acknowledgements

The author wishes to thank Mike Marshall, OSHA; John Wincek, Croda, Inc.; Jan Hayes, RMIT University; Robb Moss, Cal Poly San Luis Obispo; Lance Muniz, Vetergy Group; and David Wilbur, Vetergy Group for their review and comments on this paper.

# 10 References / Resources

## *10.1 Cited References*

1) U.S. Department of Labor Occupational Safety and Health Administration. Process safety management of highly hazardous chemicals, 29 CFR. 1910.119, Jun 1, 1992, as amended Feb 8, 2013.
2) CCPS. *Guidelines for Management of Change for Process Safety*. Center for Chemical Process Safety. American Institute of Chemical Engineers, New York, NY, 2008.
3) T. Takegawa and H. Kobayashi, Hideo, "Disaster of Chemical Plant at Flixborough," Failure Knowledge Database / 100 Selected Cases, Available at http://www.sozogaku.com/fkd/en/hfen/HB1058048.pdf, Accessed on Jan 15, 2017.
4) Marsh. "Management of Change, Risk Engineering Position Paper – 05," 2015, Available at https://www.marsh.com/uk/insights/research/risk-engineering-position-paper-management-of-change-moc.html, Accessed on Jan 15, 2017.
5) U.S. Department of Labor Occupational Safety and Health Administration. Appendix C to 1910.119 – Compliance Guidelines and Recommendations for Process Safety Management (Nonmandatory), Feb 8, 2013.
6) J. Amy, "After Katrina: Roofs remain source of concern," AL.com, Available at http://blog.al.com/live/2010/08/after_katrina_roofs_remain_sou.html, Accessed on Jan 15, 2017.
7) Dalian ZhuHong Mechanical Co., Ltd, Various piping repair pictures, Available at http://www.dlzhjx.com/zz31.asp, http://www.dlzhjx.com/images/7/17.jpg, Accessed on Jan 15, 2017.
8) BASF, "Management of Change Procedure – Wyandotte," Available at http://www2.basf.us/DocSearchWeb/displaydoc?docbase=Wyandotte_MI_US&cabinet=EHS&folder=Publish/WSTD&docname=WYN032.075&rend=pdf, Accessed on Jan 15, 2017.
9) INEOS Nitriles, Green Lake "SOSO Management of Change (MoC) [Procedure]," Available at http://greenlake.biz/soso/soso/IA0900.DOC, Accessed on Jan 15, 2017.
10) Merriam-Websters Dictionary [online], Available at https://www.merriam-webster.com/dictionary/, Accessed on Jan 15, 2017.
11) Cambridge Dictionary [online], Available at http://dictionary.cambridge.org/us/dictionary/english/, Accessed on Jan 15, 2017.
12) Dictionary.com [online], Available at http://www.dictionary.com/browse/, Accessed on Jan 15, 2017.
13) WorkSafe Victoria, "Management of change at a major hazard facility," Available at http://www.worksafe.vic.gov.au/__data/assets/pdf_file/0018/12393/Management_change.pdf, Accessed on Jan 15, 2017.

14) Wikipedia. "Project management triangle," Available at https://en.wikipedia.org/wiki/Project_management_triangle, Accessed on Jan 15, 2017.
15) Ozog and R.P. Stickles, ioMosaic, "Management of Change or Change of Management?" Available at https://www.chemicalprocessing.com/assets/Media/MediaManager/ManagementOfChange.pdf, Accessed on Jan 15, 2017.
16) CCPS. *Recognizing Catastrophic Incident Warning Signs in Process Industries*. Center for Chemical Process Safety. American Institute of Chemical Engineers, New York, NY, 2011.
17) D. Wilbur, Vetergy. "Admitting Human Error in the Workplace," Jan 2016, Available at http://vetergy.com/, Accessed on Jan 25, 2017.

### *10.2 Uncited References*

1) U.S. Department of Labor Occupational Safety and Health Administration. OSHA 3133 Process Safety Management Guidelines for Compliance, 1994 (Reprinted).
2) U.S. Department of Labor Occupational Safety and Health Administration. OSHA 3132 Process Safety Management, 2000 (Reprinted).
3) Y. Malmen, et al. "Shutdowns and Management of Change." *Hazards XXII* Symposium Series No. 156, iChemE, 2011, Available at https://www.icheme.org/~/media/Documents/Subject%20Groups/Safety_Loss_Prevention/Hazards%20Archive/XXII/XXII-Paper-30.pdf, Accessed on Jan 15, 2017.
4) C. Harrison, "Managing Change in Manufacturing & Production Facilities," ByDesign, 2012, Available at http://www.asse.org/assets/1/7/carl_harrison_article.pdf, Accessed on Jan 15, 2017.
5) Safe Work Australia, Model Work Health and Safety Regulations (28 November 2016). Available at http://www.safeworkaustralia.gov.au/sites/SWA/about/Publications/Documents/616/Model-WHS-Regulations-28Nov2016.pdf, Accessed on Jan 15, 2017.
6) National Offshore Petroleum Safety and Environmental Management Authority (NOPSEMA), Available at https://www.nopsema.gov.au/safety/safety-case/safety-case-guidance-notes/, Accessed on Jan 15, 2017.
7) UK Heath and Safety Executive. "Plant modification / Change procedures," Available at http://www.hse.gov.uk/comah/sragtech/techmeasplantmod.htm, Accessed on Jan 15, 2017.

## APPENDIX A

**Work Flow Process: Temporary MoC Review and Approval with more complex alternatives analysis**

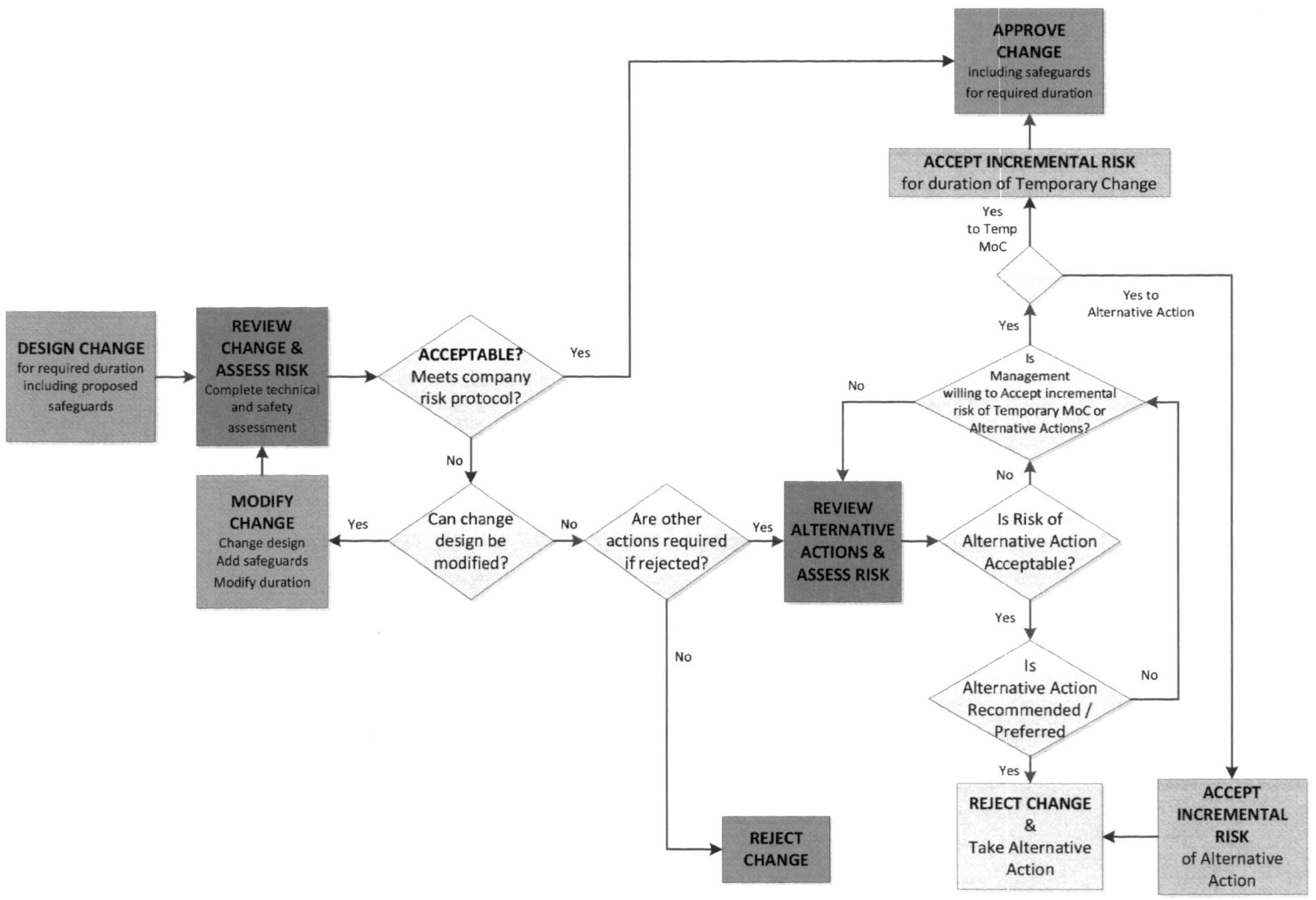

# Don't Do What We Did – Stupidity, Luck, and Making Good Choices

**Mike Moosemiller**
**Baker Engineering and Risk Consultants, Inc.**
**8430 West Bryn Mawr Ave, Suite 720, Chicago, IL 60631**
**mmoosemiller@bakerrisk.com**

**Kathleen A. Kas, PE**
**The Dow Chemical Company**
**4460 State Hwy 225**
**Deer Park, TX 77536**
**kkas@dow.com**



Prepared for Presentation at
American Institute of Chemical Engineers
2017 Spring Meeting and 13th Global Congress on Process Safety
San Antonio, TX
March 26 – 29, 2017

# Don't Do What We Did – Stupidity, Luck, and Making Good Choices

**Mike Moosemiller**
**Baker Engineering and Risk Consultants, Inc.**
**8430 West Bryn Mawr Ave, Suite 720, Chicago, IL 60631**
**mmoosemiller@bakerrisk.com**

**Kathleen A. Kas, PE**
**The Dow Chemical Company**
**4460 State Hwy 225**
**Deer Park, TX 77536**
**kkas@dow.com**

**Keywords:** Human Error, Hazard Awareness, Commissioning, Unsafe Act

## Abstract

Fortunately, the focus on safe behaviors, hazard recognition, and risk management has greatly improved over the past few decades. However, in the early days of the authors' careers, there wasn't the same level of safety awareness - and much more risk-taking behavior. In this paper, the authors recount events in which they (and others) were ignorant, careless, or intentionally put themselves in dangerous situations early in their careers. These poor choices could have, or did actually result in, chemical exposures, fall hazards, overpressure events, and process upsets. The authors describe the circumstances leading to the decisions to take personal risks, what happened (and what could have happened), and the consequences resulting from those poor choices. As George Santayana said in *The Life of Reason* (1905), "Those who do not remember the past are condemned to repeat it." These same events could happen today to inexperienced (or even experienced) engineers, and it is hoped that the lessons learned by the authors will prevent others from putting themselves in risky situations.

## 1 Introduction

The early days of chemical processing were fraught with more hazards to workers than we have today. A classic example was the production of nitroglycerin in the 1870s [1]. Control of the reactor was done manually, with frequent adjustment needed by the operator to keep the equipment within safe operating limits so that an explosion would not occur. (Operators were given one-legged stools so that they would not fall asleep and allow the equipment to migrate outside of safe operating limits – with potentially deadly consequences.)

Fortunately, safety performance in the chemical industry has improved dramatically over the past several decades (Figure 1).

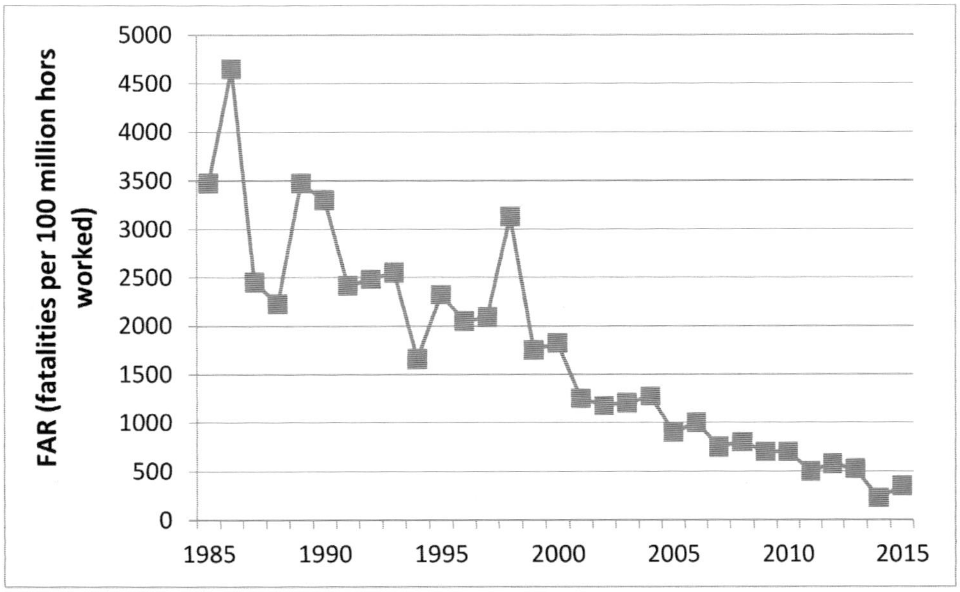


Figure 1: Reported Work Hours and Fatal Accident Rate (1985-2015) [2]

The commitment of employers, as well as government regulations and standards such as the Occupational Safety and Health Administration (OSHA) Process Safety Management (PSM) standard [3], the US Environmental Protection Agency (EPA) Risk Management Program (RMP) rule [4], Directive 82/501/EC (Seveso) [5], the Control of Major Hazards (COMAH) regulation [6], and others have helped to drive improved process safety performance in industry. Certainly, through the learnings from significant industry incidents, chemical companies have recognized the business case – and moral imperative – for eliminating process safety events. Responsible companies no longer tolerate types of behaviors (such as those described in this paper) that may have been practiced 30 or more years ago, and the authors wish to stress that their current and prior employers did not encourage, authorize, or condone the specific activities described in this paper. Although such behaviors would no longer be considered acceptable, the lessons, learned the hard way, should not be lost.

As young engineers, the authors made some classic rookie mistakes. They certainly got themselves into trouble when they did not maintain a keen awareness of their surroundings and the hazards that were present. In some instances, the authors did not recognize the hazards associated with the operation; in other cases, they took it upon themselves to take actions that were either inherently unsafe or for which they had not been properly trained. Poor management of change led to other undesired events, and the authors were also caught in a few traps created by a lack of attention to human factor considerations in equipment design.

The authors (and so presumably, the readers of this paper) are often impacted most strongly by learnings that come through the stories told by those who were personally involved in the incidents. The consequences of the poor choices become so much more "real." From that perspective, the authors are sharing their personal stories regarding mistakes they or others made in their early days and the impact of those poor decisions. It is hoped that the learnings from these mistakes will help newer engineers to have a higher level of awareness of the potential consequences of making dumb decisions as well as a better appreciation for the process safety management systems we have today that help prevent these situations.

# 2 Learning Opportunities: What Could Go Wrong?

## *2.1 But we vented the pressure – didn't we?*

### 2.1.1 The Situation

A new emulsion product was being scaled up in a pilot plant facility. One of the authors was working night shift in support of the new product development, wearing normal street clothes. The emulsion was being cooled by passing it through a heat exchanger using an air-operated diaphragm pump. At the end of the transfer, the author turned off the air to the pump and opened a bleed valve on the product discharge hose to remove pressure from the line prior to disconnection, as illustrated in Figure 2. No material came out of the bleed.

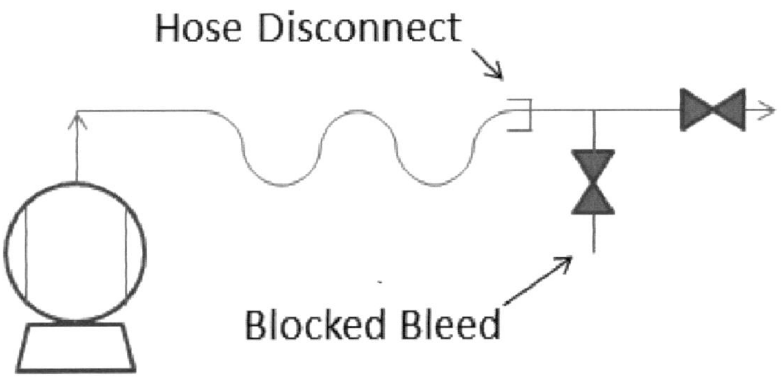


Figure 2: Hose and Bleed Assembly

### 2.1.2 The Poor Choice

The author believed that, because no material came out of the bleed, the hose was vented. Wearing only goggles and gloves, the author disconnected the hose.

### 2.1.3 Potential/Actual Consequences

The hose whipped around and doused the author's legs with warm emulsion. Fortunately, the material was not corrosive and had been sufficiently cooled so as not cause a burn. The author immediately went to the closest bathroom with a shower rather than to the safety shower (another poor choice). Fortunately, there was no injury from the exposure – and someone had left spare work clothes in the shower room, sparing the author the embarrassment of wandering in undergarments in search of substitute clothing.

2.1.4 Key Learnings

The author believed, at that time, that, if no material came out of a bleed valve, the line was cleared. However, materials such as emulsions are prone to fouling due to polymerization. The author should have been suspicious when no material was released; surely, there would have been a few drops of liquid, or a slight sound as residual pressure was relieved.

The author also should have worn heat-resistant work clothes and should have used the closest safety shower rather than a bathroom shower further away.

2.1.5 What Should You Do?

1. Maintain a sense of vulnerability. The author was expecting the hose to be vented, and the lack of material exiting the hose seemed to confirm what the author expected. As a result, the author didn't consider whether the interpretation of the observation data (no flow = no material) was correct

2. Do a Job Safety Analysis. If the engineer had considered what could happen if a vent were blocked, a second vent could have also been opened to verify that the pressure was vented. A whip check could have been used to connect the hose to the piping, preventing the hose from whipping around when disconnected.

3. Follow proper safety practices. Wear the Personal Protective Equipment (PPE) that would be required if an exposure was to occur. Use the local safety shower rather than going to the locker room. Wear clothing that is appropriate for the potential hazards that might be encountered (including thermal burns).

### *2.2 Where common sense did not prevail – or, even if you're the least experienced person, don't keep quiet if you see something that doesn't look right*

2.2.1 The Situation

A stab-in reboiler on a distillation column was being removed for maintenance. The removal required two maintenance people to work on an elevated deck to knock out bolts from the reboiler flange with the column in order to free the reboiler for removal. At some point it was possible that the flange could come free and fall onto the workers, and so its weight (several tons) was being supported by a crane (Figure 3).

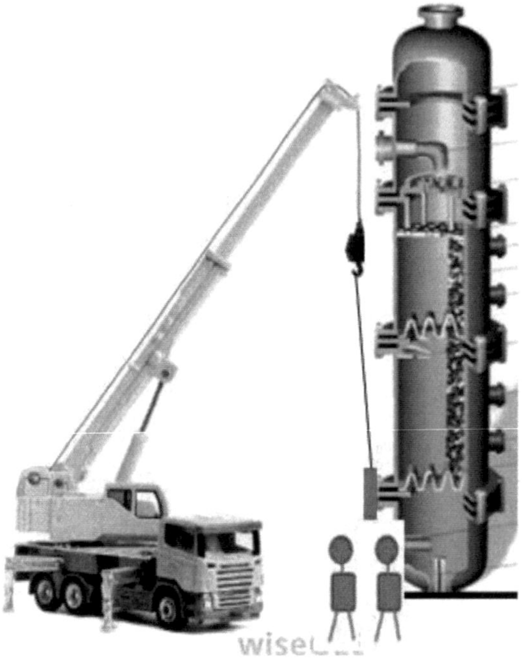

Figure 3: A Crane Under Strain

The author was not associated with this operation, and had only been working in this (the author's first) refinery for two weeks. The author was watching from a distance while several plant supervisors were next to the crane operation. Although the author knew nothing about crane operations, it seemed odd that the crane was pulling so hard on the flange that its rear tires were off the ground. (The nearby plant supervisors were facing away from the wheels and may not have known they were elevated.)

#### 2.2.2 The Poor Choice

Since the author was the least experienced person observing the operation, the author continued to watch but did not engage the plant staff about whether the crane operation was proper.

#### 2.2.3 Potential/Actual Consequences

Once all but two of the bolts holding the flange in place were removed, the final two bolts sheared due to the stress of the over-torqued crane operation. The flange flew up in the air and landed on the backs of the workers who had turned away from it, knocking them to the deck. The flange rebounded and came back down just when one of the workers had staggered back to their feet, hitting him again. The workers were subsequently taken away by ambulance to a hospital, but fortunately did not sustain serious injuries.

#### 2.2.4 Key Learnings

If something doesn't "feel right," then there is a reasonable chance it is not - even if everyone else is comfortable.

### 2.2.5 What Should You Do?

1. Don't let your seniority blind you to what is wrong and don't let lack of experience keep you from speaking out if something doesn't look right.

2. If something isn't right, it isn't right whether you are an employee of the facility or not. Speak up!

## *2.3 Nobody else wore goggles…*

### 2.3.1 The Situation

During a summer internship, the author worked in a quality control laboratory of a cosmetic company, which performed quality tests on incoming raw materials. The author was taught to conduct a number of laboratory tests on these materials. No eye protection was required in the lab. In fact, the author wore contact lenses on a daily basis. One day, the author attempted to inject a solvent into a sodium chloride cell for Infrared (IR) analysis (Figure 4). Unfortunately, the cell port was blocked.

Figure 4: Example of Infrared (IR) Solution Cell [7]

### 2.3.2 The Poor Choice

The author, not wearing eye protection of any kind, applied excessive force to the syringe. During the injection, material squirted back into one eye.

### 2.3.3 Potential/Actual Consequences

The author's eye immediately started to sting and water. The author immediately went to the safety shower to start rinsing the affected eye. However, the contact lens was still in place. The author struggled to remove the contact lens so that the eye could be completely flushed. Fortunately, there was no serious injury.

2.3.4 Key Learnings

1. Eye protection should be worn in any situation where it is possible to be exposed to hazardous liquids. (Many plants and labs require that eye protection be worn at all times, since it is not possible to anticipate all spills, leaks, and drips.)
2. The hazards of a blocked IR cell – and of applying excess force to the syringe – were not identified and documented. The need to apply more than normal force to any task is generally a warning signal.
3. The author was a college student who was not properly trained and supervised. Co-ops, interns, and new hires need to be closely monitored. Set a good example and do not assume that they are fully skilled until they have actually demonstrated competency.

2.3.5 What Should You Do?

1. Wear the required personal protective equipment (PPE) at a minimum but think – should I be wearing more? You are the one at risk! Challenge PPE policies where they don't make sense. A general laboratory safety policy may be inadequate to address the hazards of specific/novel equipment.
2. Train your co-ops and new hires. Watch others' co-ops and new hires carefully. Supervise them closely and do not assume competency. Rely on proper demonstrations of their skills.
3. Conduct Job Safety Analyses (JSAs) to identify potential hazards. Capture the potential hazards and precautions in training materials and operating instructions so that the learnings are not lost.

### *2.4 Use your head – but please, not like that*

2.3.1 The Situation

During new equipment commissioning, or turnaround, there is often the opportunity to view the internals on vessels that is not possible during normal operation. Naturally, one wants to look into vessels – sometimes out of curiosity, sometimes to see if someone you are looking for is inside, etc. (Figure 5).

Figure 5: Looking Inside a Vessel

#### 2.3.2 The Poor Choice

There is often a hesitancy to completely climb into the vessel, either because of safety concerns, time concerns ("it's too much trouble to get a confined vessel entry permit or follow those procedures") or tendencies toward claustrophobia. However, surely just sticking one's head in the manway would be all right …wouldn't it?

#### 2.3.3 Potential/Actual Consequences

The authors are aware of multiple occasions where someone put their head through a vessel manway to 'just have a peek,' and died due to asphyxiation – sometimes with most of their body still outside the vessel. Generally, this is because the vessel was purged with nitrogen.

#### 2.3.4 Key Learnings

Just because an environment *looks* safe doesn't mean that it *is* safe. Beware of the hazards you cannot see as well as the ones that you can see.

#### 2.3.5 What Should You Do?

1. Never allow any part of your body to approach the plane of a vessel (or any confined space) without positive proof that the environment is safe (e.g. testing the atmosphere in the vessel). Verbal assurances are not good enough – the authors know of one case where someone was told that the vessel had been purged with air, when in fact it was purged with nitrogen instead.
2. Be cautious when placing your head around potential leak points on inerted equipment. The authors are aware of a colleague who briefly lost consciousness when looking into the sight glass of a nitrogen-inerted vessel. The sight glass had a leak, and the area around the sight glass was oxygen-deficient.
3. Don't assume that you can just hold your breath and take a quick look – an involuntary reflux that causes you to take a breath could be fatal. Even very short exposures to nitrogen can be fatal.

# 3 Learning Opportunities: Mis-Management of Change

## *3.1 But it worked in the lab!*

#### 3.1.1 The Situation

A new agricultural intermediate process was being operated at a 6,000 gallon scale for the first time. The process included a water wash of the organic phase, followed by a phase separation, to remove salts from the batch. Unfortunately, the batch formed an emulsion and the phases would not separate. This situation had occurred in the lab a few times, and the problem had been remedied by the addition of potassium carbonate salts, which increased the density of the aqueous phase and broke the emulsion. The author took a sample from the commercial vessel, brought it to the plant lab, and added potassium carbonate. The phases then separated.

### 3.1.2 The Poor Choice

The issue occurred on midnight shift. Experienced technical resources were limited. The author decided to try to break the emulsion so the batch could proceed. The author instructed the operator to open the manway of the vessel and add several bags of potassium carbonate to the vessel.

### 3.1.3 Potential/Actual Consequences

Shortly after the bags were added, the batch began to swell and material started "burping" out of the top of the vessel. The operating staff retreated to a safe distance. The material oozed along the floor and immediately solidified due to the colder ambient temperature. The crew later spent hours shoveling waste material off of the floor.

### 3.1.4 Key Learnings

The author did not take into consideration fundamental differences between operating at the lab scale and at the plant scale. In this case, the mixing in the commercial vessel was poor. The bags of salt were dumped and went straight to the bottom of the vessel. Large carbon dioxide bubbles formed at the bottom of the vessel, forcing material from the top. This had not been observed at the lab scale because of better mixing and slower addition of the salt.

### 3.1.5 What Should You Do?

1. Understand the fundamentals of scale-up. Parameters such as mixing and heat transfer change as a process is scaled. It is important to take those changes into consideration when designing different size processes.

2. Conduct a Management of Change Review. The practice of adding salt to break the emulsion had been practiced in the lab but not reviewed at the plant scale. The event occurred on midnight shift, when technical resources were not readily available. As a result, expert input that would likely have identified the potential for this type of event was not obtained. Avoid troubleshooting and creating procedures "on the fly," as there may be unforeseen consequences.

## *3.2 It looked like the right stuff…*

### 3.2.1 The Situation

A new process was being started up at an external manufacturer, and the author was responsible for supporting the start-up. The last reaction step of the synthesis process involved the use of a boiling solvent saturated with a toxic acid gas – a highly corrosive environment [7]. Polytetrafluoroethylene (PTFE) lined piping had been specified as the preferred material of construction. A few hours into the first batch, the recirculation pump was turned on to recirculate the batch for sampling and to provide additional mixing.

### 3.2.2 The Poor Choice

The external manufacturer reused existing pipe. It had a white lining, and it was assumed that the pipe was PTFE-lined. In fact, the pipe was polypropylene-lined, which was not resistant to the process fluid at operating conditions. The Hazard and Operability Study (HAZOP) had not considered the deviation of wrong material of construction.

### 3.2.3 Potential/Actual Consequences

A few hours after the recirculation loop was turned on, local toxic gas detectors alarmed, signaling the presence of acid gas fumes in the process area. The area was evacuated. A gas cloud formed, filled the building, and left the site. The plant was evacuated, the local fire department was called, and a local highway was shut down for several hours. Painful consequences of the event included media coverage, regulatory visits, an incident investigation, clean-up, and replacing all suspect or compromised piping.

### 3.2.4 Key Learnings

Although the author knew that existing vessels would be re-used (and had verified that the vessels were of the appropriate material of construction), the re-use of pipe from a "boneyard" (area for storage of equipment not currently in use) had not been considered or included in the Hazard and Operability Study (HAZOP). Although the proper material of construction to be used for the pipe had been communicated to the external manufacturer, they did not have a robust process to ensure that new pipe of the proper specification was used - or that the material of construction of used pipe was verified.

### 3.2.5 What Should You Do?

1. Determine whether it is appropriate to re-use equipment for a new process. Is it the right material of construction? Has the integrity of the equipment been verified?

2. Consider the use of the wrong material of construction in Process Hazard Reviews (PHAs). Even if the right material of construction is used in initial construction, later maintenance or process changes could result in the inadvertent substitution of the wrong materials of construction. Considering this type of deviation in a PHA would lead to a discussion of the methods to be used to confirm that the right materials are being used and would hopefully lead to the implementation of a work process to validate proper material identification and prevent such incidents. Certainly, positive material identification could be a useful means of verifying the correct material of construction, especially in sensitive applications.

# 4 Learning Opportunities: "I *Can't* Do It Myself!"

## *4.1 It will take forever to get that guy back, so I'll do it myself*

### 4.1.1 The Situation

A level instrument on a vessel that was separating 20 Baume' (14 wt%) caustic from hydrocarbon liquid and vapor phases had just been fixed by an instrument technician. However, the technician had left the area without recommissioning the instrument. (He had left its isolation valves closed.) It had taken a long time to get the technician out to fix the instrument, and calling the technician or an operator to the area to open up the isolation valves would have been a nuisance. (Figure 7)

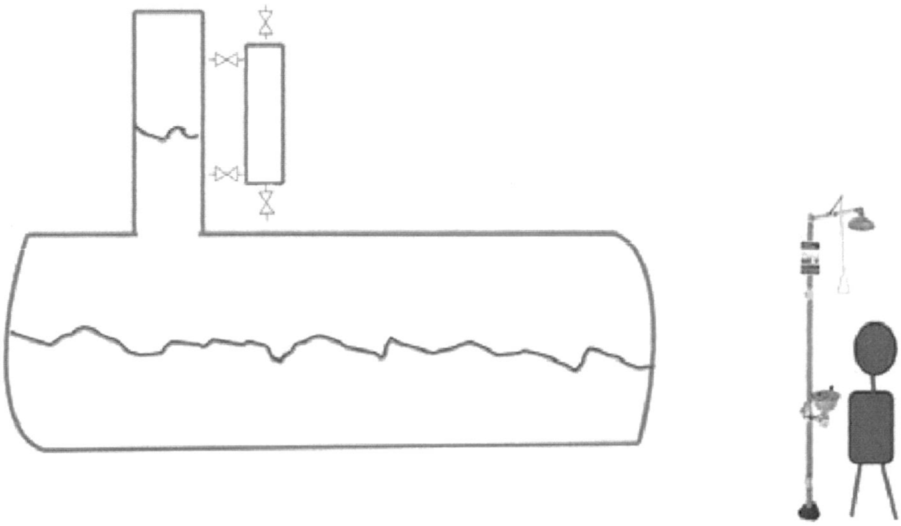

Figure 7: Caustic/Hydrocarbon/Vapor Separator

4.1.2 The Poor Choice

The author believed that, since recommissioning the instrument just involved opening the isolation valves, the author might as well just do it personally rather than try to find someone and persuade them to come back out to the unit. The author checked the instrument drain valve to make sure that it was closed, then opened up the lower valve connecting the process to the instrument.

4.1.3 Potential/Actual Consequences

Unfortunately, the author didn't also check the position of the vent valve from the instrument. About one second after opening the valve, a torrent of caustic solution started raining down on the author's head (covered with a hardhat, at least). The author recognized it as being the caustic since it was slippery. For the same reason, the author decided it was more prudent to jump the six feet from the separator deck to the ground rather than risk grabbing the ladder and slipping off, potentially causing further injury. Soaked to the skin in caustic, the author ran to the nearby safety shower/eyewash station, stripped, and spent the next 15 minutes alternating between the shower and the eyewash. While the author's hardhat changed color from blue to white, the author received only a small minor burn on one arm.

### 4.1.4 Key Learnings

The author believed that it was appropriate to do another person's work for the sake of expediency. The author also assumed that if the instrument drain valve was closed, the vent valve must also be closed, which turned out to be wrong. On the plus side, the author did some things correctly, which prevented a severe personal injury or fatality, including the following:

- Before the startup, the author had been asked where the safety shower/eyewash station for the unit should be located. As it turned out, the author had felt that this area had the highest incident potential, and as a result picked the perfect location for the subsequent event. (It was not practical to put a shower on the deck where the incident occurred.)
- During the startup phase, the eyewash station nozzle was repeatedly broke by workers using it to wash their hands. The author had refused to enter the unit until the eyewash nozzle was permanently fixed – and it had been repaired before the event.
- Having realized that the spray was caustic, the author correctly chose to jump from the separator deck and risk spraining an ankle rather than using the ladder and risk an uncontrolled fall that had the potential to render the author unconscious with no one around.
- The author correctly realized the gravity of the situation and stripped completely naked in order to wash off the caustic as thoroughly as possible.

### 4.1.5 What Should You Do?

1. Don't do a job for which you haven't been trained, and do not make assumptions about the status of equipment based on the status of nearby equipment.

2. Think carefully about the siting of safety equipment – the life you save may be your own!

3. Be aware of your surroundings, and know where emergency equipment is located.

4. Wear the Personal Protective Equipment (PPE) that would be required if an exposure was to occur.

5. Use the local safety shower rather than going to the locker room.

6. A chemical exposure event is no time to be modest. If you have to get buck naked to prevent an injury, go ahead! Better for people to have a story to tell about the engineer that was naked than for them to tell the story about the engineer who died or suffered horrible chemical burns.

## *4.2 Why did the pressure go up so fast?*

### 4.2.1 The Situation

An existing pump had been replaced with a lobe gear pump. The author was monitoring the water testing of the equipment. The next step was to redirect the flow path from the pump to a different location. (Figure 8)

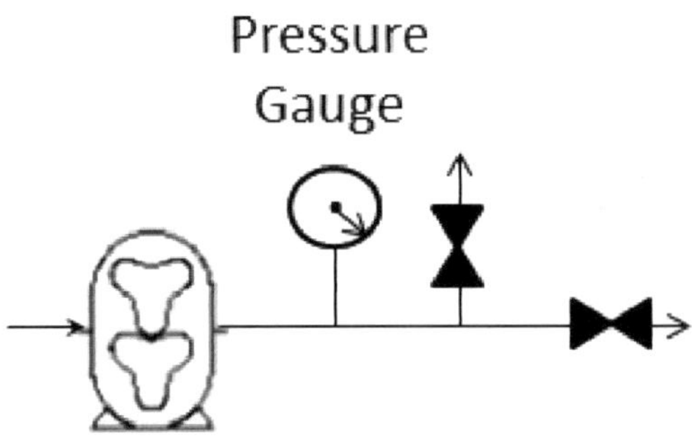


Figure 8: Deadheading a Positive Displacement Pump

4.2.2 The Poor Choice

The author chose not to wait for an operator. Instead of opening the valve to the new flow path first, the operator closed the valve on the discharge of the pump, resulting in "deadheading" the pump.

4.2.3 Potential/Actual Consequences

In an instant, the pressure in the 150 psig line rapidly rose and pegged the pressure gauge. Fortunately, the pressure gauge was clearly visible and author's hand was still on the valve handle. The valve was opened immediately and the line did not fail from the overpressure. (The presence of an internal relief in the pump likely significantly contributed to the successful outcome.)

4.2.4 Key Learnings

The author had worked with operators in the past to troubleshoot centrifugal pumps. Sometimes, when a pump was "air bound," operators would deliberately throttle or briefly close the discharge valve of the pump to remedy the problem. The author did not see the act of closing a pump discharge valve for a few seconds as risky. However, the author did not appreciate that this was a positive displacement pump, which had the capability to quickly generate a pressure sufficient to rupture the pump discharge pipe.

4.2.5 What Should You Do?

1. Understand equipment before you operate it.

2. Be cautious about extrapolating "rules of thumb" and other troubleshooting/operating practices between equipment without first understanding when they are/are not appropriate. Not all equipment is created equal!

### 4.3 *Process control mishap*

#### 4.3.1 The Situation

The author was working in a batch processing plant when an operator asked for assistance in troubleshooting a problem on another process. The Programmable Logic Controller (PLC) would not progress to the next step because it was not receiving a signal from a switch in the field. An instrument technician was not available to fix the issue and product was needed for a customer order.

#### 4.3.2 The Poor Choice

The author had recently received training on PLCs and wanted to use that knowledge. After determining that the field switch was faulty, the author (with limited knowledge of the workings of the PLC and its programming), forced the process control sequence into the next step.

#### 4.3.3 Potential/Actual Consequences

The control sequence did proceed to the next step and started to charge material to the reactor; however, the bottom valve of the reactor was still open! Fortunately, a manual valve was closed, preventing the release of material. The operation was stopped until the valves were aligned in the correct position and a release was averted.

#### 4.3.4 Key Learnings

The author did not recognize that, by advancing the sequence, the ladder logic that would have sent a signal to close the reactor bottom valve was bypassed. Clearly, the author chose to take action without possessing the necessary process control expertise.

#### 4.3.5 What Should You Do?

1. Follow established review and approval protocols when bypassing control systems, advancing sequences, operating manually, or downloading code while equipment is in operation. Equipment might be placed in an improper configuration or safety systems may be bypassed. The author is aware of an incident in another plant, where an engineer attempted to download a new sequence while the equipment was in operation. This resulted in the loss of control of the charging sequence and a deactivation of the high level alarm. A flammable spill resulted; fortunately, it did not ignite.

2. Ensure that you are properly trained on the control system platform and logic, and have a good understanding of the chemical process, before making any changes to the operation of a control system. There may be unintended consequences!

# 5 Learning Opportunities: Human Factors Traps

## *5.1 The pros and cons of ball valves*

### 5.1.1 The Situation

Ball valves are often selected for use in chemical facilities. Ball valves can be very convenient – a simple quarter-turn and they open or close; it doesn't get much easier than that (Figure 9). However, the ease of opening or closing ball valves intentionally also implies that they can be opened or closed not only easily, but *accidentally*.

Figure 9: A Typical Ball Valve

### 5.1.2 The Poor Choice

In one unit, the engineers specified the use of ball valves in hazardous service. In one recent incident, a newly-hired field operator was going through a unit tour when the sleeve of his Nomex suit caught on the handle of a ball valve, causing the valve to open.

### 5.1.3 Potential/Actual Consequences

The valve was on a high pressure hydrogen line, which then discharged up the sleeve of the new-hire. The valve was quickly closed, but he had quite a shocking experience and his arm was badly bruised. He was taken to a local hospital as a precaution; fortunately, there were no other injuries.

### 5.1.4 Key Learnings

The event was a wake-up call that the ease of use of ball valves has a down side, and staff and especially new-hires should be on the lookout for this hazard and consider the potential hazards of specifying ball valves in piping design.

### 5.1.5 What Should You Do?

1. When determining whether to install a ball valve or other type of valve in a given application, consider the implications of both the ease of use and ease of misuse of the valve. In some cases, having a ball valve that can be easily turned could result in unsafe consequences. Some companies restrict the use of ¼ turn ball valves in hazardous service, for example.

2. Orient operators and visitors to a unit that has ball valves to the potential hazards of accidental operation.

3. Be mindful that loose clothing or jewelry can catch on ball valve handles. Not only could this inadvertently operate the valve, but it could also cause an injury. An author observed an operator wearing large hoop earrings bend down and catch one earring on a ball valve handle. Fortunately, the operator recognized the issue and detached the earring before standing up, preventing an injury to the ear or turning the ball valve.

4. The risk of ball valves can be mitigated by using oval handles, spring-loaded handles and plugs/caps in terminal connections Some ball valves also have locking handles, which keep the valve from being inadvertently opened.

## *5.2 The ergonomics of an emergency shutdown (ESD) button*

### 5.2.1 The Situation

A new unit was built, along with a new control room. The control room had a typical layout for its day, with computer control panels and instrumentation along a long wall that had largely open walk-around space in front of it. The unit had been operational for a week or so.

### 5.2.2 The Poor Choice

A colleague was chatting with one of the regular plant staff one shift and casually leaned against the control panel wall. Unfortunately, she happened to place her hand against the wall precisely where the unit emergency shutdown button was located. (Figure 10)

Figure 10: Sometimes, Talk Isn't Cheap

5.2.3 Potential/Actual Consequences

The button depressed, resulting in an unplanned unit shutdown. The outage caused approximately $250,000 in losses.

5.2.4 Key Learnings

1. During the design phase, it is important to give sufficient consideration to the potential for inadvertent activation of safety devices.

5.2.5 What Should You Do?

Design for ergonomics and the potential for human error. After the event occurred, the plant purchased a plastic cover for about $1 and placed it over the ESD button. A better design that would have prevented the event entirely is if the ESD was operated as a pull button rather than a push button, or locating the ESD at an easily reached position not at convenient "accidental" activation locations.

# 6 Learning Opportunities: Situational Unawareness

## *6.1 Watch your step, even if you're tired*

6.1.1 The Situation

A new distillation tower had been installed but not yet commissioned. Part of pre-commissioning activities included climbing the tower to make sure that all equipment was in place. The tower was approximately 230 feet tall. Reaching the top of the tower involved climbing a series of caged ladders that went from one deck to another in about 20 foot intervals.

### 6.1.2 The Poor Choice

The author had many activities to check on, so (being in much better shape in those days) the author climbed the tower fairly fast, to the point where the author was winded after reaching the deck at the 175-foot level.

Upon stepping off the top of one ladder onto the deck, the author almost placed a foot through a section of missing grating adjacent to the ladder, which would have resulted in a free-fall of 175 feet. (Figure 11)

Figure 11: Getting Fresh Air Without Benefit of a Parachute

### 6.2.3 Potential/Actual Consequences

Fortunately the author managed to stop just in time, so nothing bad happened (except for a rapid acceleration of his heartbeat for a minute or two, and experiencing the sensation of having his life pass before his eyes). The author provided an emphatic description of the safety issue and pushed for prompt resolution.

### 6.2.4 Key Learnings

The author allowed himself to become winded, increasing the chance of falling through the hole.

### 6.2.5 What Should You Do?

1. Do not allow yourself to become winded – it leads to inattention, and inattention can lead to a bad outcome in a hazardous situation.

2. If you see an unsafe situation, do what you can to mitigate the problem. In this case, the author found some red plastic ribbon trash that was stuck at that deck, and wrapped the ribbon around the grating at the gap to try to warn someone else of the situation if they happened to enter the area before the problem was fixed.

3. If you see an unsafe situation, immediately notify the appropriate people in the unit. Continue to monitor the situation occasionally, and remind people of the issue until it is fixed.

### *6.3 The hazards of following piping in a pipe rack*

#### 6.3.1 The Situation

It is common for anyone who is trying to learn a new process unit to walk down the equipment in the field. This includes following runs of piping in pipe racks that may swerve back and forth in the rack, or at times might be partially obscured by other piping in the pipe rack. (Figure 12) Sometimes it requires a great deal of focusing on the line to not lose track of its route.

Figure 12: A Typical Pipe Rack

#### 6.3.2 The Poor Choice

A co-worker of the author was undertaking such an activity and walked for an extended distance while continually looking up in the pipe rack.

#### 6.3.3 Potential/Actual Consequences

Unbeknownst to this co-worker, some distance from the start of this pipe run was an unmarked open manhole below the pipe run. The co-worked walked into the manhole and fell into the sewer. Fortunately, there was no injury from the fall. However the person's smell was so offensive that his carpool colleague refused to allow him in the car at the end of his shift – resulting in a 10-mile walk home for the unfortunate sewage-drenched co-worker after the shift.

6.3.4 Key Learnings

The person incorrectly assumed that there would be no obstructions or dangers along their path.

6.3.5 What Should You Do?

Before tracing long runs of pipe, view the entire path to make sure there are no hazards along the path. One can also stop periodically and check one's surroundings.

# 7 Conclusion

The authors and others clearly made poor choices as a result of poor situational awareness, lack of hazard recognition, inadequate management of change, inadequate human factors design, and taking it upon themselves to do tasks which were inherently unsafe or for which they had been inadequately trained. As the authors reflected on this collection of cautionary examples, they feel lucky to have survived the stupidity and poor choices of their formative engineering years largely intact (except perhaps for their collective pride)! Fortunately, the authors are somewhat older and wiser from these experiences. More importantly, the authors are grateful that process safety behaviors, management systems, and equipment design have evolved to the point where these activities would not be necessary nor tolerated today.

However, it should be noted that, although processes have become better designed and automated, people have become less hands-on with the process and equipment. There is a risk that operations personnel (including engineers) will lose the deep level of process knowledge that hands-on operation once enabled. Although it is quite fortunate that today's engineers are not placed in a position to take the risks that were taken several decades ago, it is important that they do not lose their appreciation for what can happen when these types of poor decisions are made. As George Santayana said, "Those who do not remember the past are condemned to repeat it."[9] It is hoped that the lessons learned by the authors will heighten the awareness of newer engineers regarding the potential consequences of putting themselves into risky situations such as these, and perhaps inspire the next generation of engineers to make better, smarter choices than the authors did.

# 8 References

[1] "Alfred Nobel in Scotland," The Official Website of the Nobel Prize, Available at: https://www.nobelprize.org/alfred_nobel/biographical/articles/dolan/. Accessed on Dec. 22, 2016.

[2] "Safety performance indicators - 2015 data," International Association of Oil & Gas Producers, June 2016.

[3] US Department of Labor Occupational Safety and Health Administration (OSHA) Standard 29 CFR 1910.119, "Process Safety Management of Highly Hazardous Chemicals," 1991.

[4] US Environmental Protection Agency (EPA) Rule 40 CFR Part 68, "Risk Management Program," first promulgated in 1991.

[5] European Union Directive 82/501/EC, "Directive on the control of major-accident hazards involving dangerous substances," first promulgated in 1982.

[6] UK Health and Safety Executive (HSE), "Control of Major Accident Hazards," 1999.

[7] Infrared Spectroscopy. (n.d.) In Wikipedia. Photo retrieved from https://commons.wikimedia.org/wiki/File:SolnIRcell.jpg#filelinks and modified on January 15, 2017.

[8] Hendershot, D., Johnson, S., and Kas, K., "What You Ask For Isn't Always What You Get," presented at the Mary Kay O'Connor Process Safety Center Symposium at Texas A&M University, 2000.

[9] Santayana, George, *Reason in Common Sense*, P. 284, volume 1, of *The Life of Reason*, 1905.

# Dust-Explosion Hazards: Housekeeping Quantified

**Richard W. Prugh, PE, CSP**
**Principal Process Safety Engineer**
**Chilworth Technology, Inc., a Dekra Insight Company**
**609-799-4449; richard.prugh@dekra.com**

Prepared for Presentation at
American Institute of Chemical Engineers
2017 Spring Meeting and 13th Global Congress on Process Safety
San Antonio, Texas
March 26 to 29, 2017

# Dust-Explosion Hazards: Housekeeping Quantified

**Richard W. Prugh, PE, CSP**
**Principal Process Safety Engineer**
**Chilworth Technology, Inc., a Dekra Insight Company**
**609-799-4449; richard.prugh@dekra.com**

**Keywords: housekeeping; combustible dust; secondary explosion; dust accumulation; electrical classification; personal protection**

## Abstract

Limiting the accumulation of combustible dusts in the workplace is critically important in the prevention of primary and secondary dust explosions. Guidance on accumulation limits began in 1997 by the NFPA in the form of a layer-thickness criterion. As stated in NFPA 654, "the hazard area shall include areas where dust accumulations exceed 1/32 inch (0.8 mm)." Since that date, there have been several modifications of dust-layer guidance, and the relationship of dust accumulations to electrical classification was introduced in 2008. The 2013 issue of NFPA 654 introduced four additional criteria, for maximum weights of dust accumulations. These criteria can be expressed in terms of grams per square foot, and the results of calculations lead to very low "allowable" values. Two of these "mass" methods require an assumption of the extent of entrainment of dust from accumulations, as caused by a primary explosion; the available data indicate that the suggested entrainment fraction is not "conservative".

This paper compiles the current methods for determining the amounts of dust that would be hazardous as fuel for flash fire or secondary explosion. Also presented is guidance for housekeeping to minimize these hazards while providing an environment where general-purpose electrical equipment could be used, in an "unclassified" area. Further, guidance is given for the personal protective equipment that should be worn by persons who perform housekeeping, particularly where large spills are involved.

# 1 Introduction

The purpose of this presentation is to provide a review of previous and current standards and the guidance that these standards provide for "housekeeping". For combustible dusts, the term "housekeeping" is applied to the removal of accumulations in the workplace before they become hazardous [Reference 1]. Thus, the hazards of fires, flash-fires, and "secondary" explosions would be minimized [References 2 and 3]. An additional objective of this document is to compile the references to similar previous presentations, at Loss Prevention Symposia, and at other conferences. Thus, the overall objective is to assist persons who are tasked with establishing housekeeping policies and procedures, and for auditing the conformance with in-house and regulatory housekeeping standards.

Determining the magnitude of hazardous accumulations of dust has been mentioned previously, with descriptions of the available methods for [References 4, 5, and 6]. Also, the entrainment of such deposits – as the result of a "primary" event – to provide the fuel for a "secondary" explosion or flash-fire has been extensively investigated [References 7, 8, and 9].

# 2 Combustible-Dust Accumulation Limits, for "Organic" Dusts

## *2.1 The "Layer-Depth" Criterion*

The National Fire Protection Association [NFPA] has included a dust-accumulation limit of 1/32 inch since 1994, in NFPA 654, with the following statement: "the hazard area shall include areas where dust accumulations exceed 1/32 inch [0.8 mm]." [Reference 10]. This requirement for prevention of secondary dust explosions was repeated in the 1997 issue [Reference 11], with supporting calculations.

The intention of these calculations was to show how a 1/32-inch layer of a dust having a density of 75 pounds per cubic foot could produce a dust cloud of "optimum" concentration that was 10 feet high over the area that was covered by the dust. A conclusion was that a 1/32-inch layer of combustible dust over 5% of the floor area "is capable of creating a hazardous condition". A companion observation was that the available surface area of overhead beams and ledges would be about 5% of the floor area. Thus, overhead accumulations would be hazardous even if the floor was kept clean. This issue of the standard [Reference 12] cautioned against the use of "compressed-air cleaning" since the smaller particles of powders and dusts can be lofted onto such overhead surfaces, where cleaning is difficult and hazardous accumulations can then occur.

In the 2000 edition of NFPA 654, the dust-layer criterion was modified to allow consideration of the bulk density of dusts, since the previous density would not apply to most organic dusts [Reference 13]. Thus, the allowable thickness would be given by:

$$t_{ALLOWABLE} = [1/32]\,[75/BD] \quad \text{inch, or} \tag{1}$$

$$t_{ALLOWABLE} = [\,0.8\,]\,[75/BD] \quad \text{millimeters} \tag{2}$$

where BD is the bulk density, in pounds per cubic foot. Thus, the following Table 1 can be constructed:

**Table 1. Allowable Dust-Layer Accumulations**

| BULK DENSITY [pounds per cubic foot] | | | | | | | | | |
|---|---|---|---|---|---|---|---|---|---|
| BD | 5 | 10 | 15 | 20 | 25 | 30 | 45 | 60 | >75 |
| Example | | Flock | Most Organic Dusts | | | | | | Metals |
| $t_{ALLOWABLE}$ | 0.5" | 0.25" | 1/6" | 1/8" | 0.1" | 1/12" | 1/20" | 1/25" | 1/32" |
| $t_{ALLOWABLE}$ | 12 mm | 6.0 mm | 4.0 mm | 3.0 mm | 2.5 mm | 2.0 mm | 1.3 mm | 1.0 mm | 0.8 mm |

Bulk densities can cover a very wide range [Reference 14] and can also be a function of particle-size distribution.

The Occupational Safety and Health Administration [OSHA] has incorporated the dust-layer criterion in its guidance to compliance officers as: "Annex D of NFPA 654 contains guidance on dust layer characterization and precautions". It indicates that immediate cleaning is warranted whenever a dust layer of 1/32- inch thickness accumulates over a surface area of at least 5% of the floor area of the facility or any given room [Reference 15]. This criterion was based on a bulk density of 75 pounds per cubic foot, but the bulk-density modification was introduced in a later memorandum [Reference 16].

FM Global provides similar guidance, with a dust-accumulation limit of 1/16" inch for a dust having a bulk density of 36 pounds per cubic foot, and with this accumulation being limited to 5% of the floor area [Reference 17].

A measure of the combustible-dust "density" – in terms of grams per square foot – can be obtained by multiplying the layer depth by the bulk density:

$$D_{ACCUMULATION} = [453.5/(25.4*12)]\; t_{ALLOWABLE}\, BD = 1.49\; t\; BD \tag{3}$$

$$D_{ACCUMULATION} = 1.49\,[\,0.8\,(\,75/BD\,)\,]\,BD = 90 \text{ grams per square foot of floor area.} \tag{4}$$

Thus, based on the layer-depth criterion alone, the maximum allowable "density" – or the "threshold dust mass/accumulation" – of accumulated dust would be about 90 grams per square foot, as shown in Table 2.

**Table 2. Allowable Dust-Layer Densities**

| BULK DENSITY [pounds per cubic foot] | | | | | | | | | |
|---|---|---|---|---|---|---|---|---|---|
| BD | 5 | 10 | 15 | 20 | 25 | 30 | 45 | 60 | >75 |
| Example | | Flock | Most Organic Dusts | | | | | | Metals |
| $t_{ALLOWABLE}$ | 0.5" | 0.25" | 1/6" | 1/8" | 0.1" | 1/12" | 1/20" | 1/25" | 1/32" |
| $t_{ALLOWABLE}$ | 12 mm | 6.0 mm | 4.0 mm | 3.0 mm | 2.5 mm | 2.0 mm | 1.3 mm | 1.0 mm | 0.8 mm |
| g./sq.ft. | 90 | 90 | 90 | 90 | 90 | 90 | 90 | 90 | 90 |

However, this high density of combustible dust would only be "allowable" in an area that is 5% of the total floor area. Thus, the averaged dust density over the whole floor area would be 5% of 90 grams per square foot or about 4.5 grams per square foot of floor area [about 10 pounds over 1,000 square feet], for all bulk densities. This corresponds to the guidance provided in an earlier presentation [Reference 18].

As will be shown later, this threshold dust mass/accumulation of 4.5 g./sq.ft. may not be sufficiently "conservative", particularly for prevention of flash-fire hazards.

### *2.2 Mass Methods*

The following statements are from the "Origin and Development of NFPA 654", in the 2013 issue of NFPA 654 [Reference 19]:

*The most important change is the inclusion of four methods to determine whether a dust fire or explosion hazardous condition exists in a facility. The methods include those based on mass accumulation and risk evaluation as well as the layer depth criterion, which was described in the 2006 edition. Each of the methods is viewed as equivalent in establishing that a hazardous condition exists for either fires or explosions, so the change offers options for users of the standard when determining this fundamental condition. [This] standard includes changes to the housekeeping requirements, [involving] the determination of a dust hazardous condition and prompts a cleaning frequency based on the nature of the dust layer or dust mass.*

The four methods are [Reference 20]:

(1) Layer depth criterion method
(2) Mass method A
(3) Mass method B

(4) Risk evaluation method

The equivalency of the above four methods is stated as: "Each of the methods shall be deemed to provide equivalent levels of safety."

The following paragraphs present Mass Methods A and B.

*2.2.1 Mass Method A, for Protection Against Flash-Fire Hazards*

According to NFPA 654: "The threshold dust mass[/accumulation] establishing a building or room as a **dust flash-fire hazard** area, $M_{BASIC\text{-}FIRE}$, shall be determined by the following equation:

$$M_{BASIC\text{-}FLASHFIRE} = 0.02 \cdot A_{FLOOR} \qquad \text{kilograms"} \qquad (5)$$

where:
$M_{BASIC\text{-}FLASHFIRE}$ = threshold dust mass (kg.) based on personnel fire-exposure criterion; and
$A_{FLOOR}$ = lesser of enclosure floor area (sq.m.) or 2000 sq.m.

The above relationship can be rearranged and converted to useful units:

$$M_{BASIC\text{-}FLASHFIRE} / A_{FLOOR} = 1.9 \text{ g./sq.ft., or about 4 lbs. per 1,000 sq. ft. of floor area} \qquad (6)$$

Note that this criterion is independent of bulk density – or other properties – of the dust, and is independent of the size or pressure-resistance of the room or building. However, the total allowable weight of dust in a room or building is directly proportional to the floor area.

*2.2.2 Mass Method A, for Protection Against Explosion Hazards*

According to NFPA 654: "The threshold dust mass[/accumulation] establishing a building or room as a **dust explosion hazard** area, $M_{BASIC\text{-}EXPLOSION}$, shall be determined by the following equation:

$$M_{BASIC\text{-}EXPLOSION} = 0.004 \cdot A_{FLOOR} \cdot H \qquad \text{kilograms"} \qquad (7)$$

where:
$M_{BASIC\text{-}EXPLOSION}$ = threshold dust mass (kg) based on building damage criterion;
$A_{FLOOR}$ = lesser of enclosure floor area (sq.m.) or 2000 sq.m.; and
$H$ = lesser of enclosure ceiling height (m.) or 12 m.

The above relationship can be rearranged and converted to useful units:

$$M_{BASIC\text{-}EXPLOSION} / A_{FLOOR} = 0.00025\, H = 0.11\, H \text{ g./sq.ft.,} \tag{8}$$

or 0.24 *H* lbs./1000 sq.ft. of floor area

where:
$H$ = lesser of enclosure ceiling height (feet) or 39 ft.

Note that this criterion also is independent of bulk density – or other properties – of the dust, and is independent of the floor area or the pressure resistance of the room or building. For a ceiling height of 12 feet, the above relationship would limit the dust accumulation to 1.3 g./sq.ft., or about 3 pounds over a 1,000-square-foot area.

The limiting dust "densities" for flash-fire and explosion hazards would be equal – at about 1.9 g./sq.ft., or about 4 pounds per 1,000 square feet – for a room height of about 17 feet, or about 5.3 meters.

### *2.2.3 Mass Method B, for Protection Against Flash-Fire Hazards*

To provide an estimate of the effects of flash fire on an exposed person, the NFPA standard apparently visualizes that dust accumulations having a mass $W_{SURFACE}$ on elevated surfaces are dislodged by a primary event, such as an explosion. The dust particles then descend essentially to floor level as a dense "shower" that has a "worst-case" concentration, $C_{WORST}$, at which the Deflagration Index, $K_{ST}$, and/or $P_{MAX}$ values are determined. Thus, the initial volume of the dust cloud, before combustion occurs, would be given by:

$$V_{INITIAL} = W_{DISPERSED} / C_{WORST} \qquad \text{cubic meters} \tag{9}$$

where $W_{SURFACE}$ is the weight of dust that is dislodged from elevated surfaces, in grams;
$C_{WORST}$ is the worst-case dust concentration, in grams per cubic meter.

The weight of the accumulated dust that is dispersed is related to the weight of accumulated dust by:

$$W_{DISPERSED} = W_{DUST}\, \eta_{DISPERSION} \qquad \text{grams} \tag{10}$$

where $\eta_{DISPERSION}$ is the fraction of the accumulated dust that is dispersed by the initial event.

Thus, the "initial" volume of the layer of dust/air mixture above the floor would be given by:

$$V_{INITIAL} = [\, W_{DUST}\, \eta_{DISPERSION} \,] / C_{WORST} \quad \text{cubic meters} \tag{11}$$

To evaluate the flash-fire hazard to personnel in the enclosure, it would be appropriate to determine how a burning dust cloud could envelop those personnel. If ignition of the initial layer of dense dust/air mixture were to occur, this cloud would expand upward as a result of combustion. The volume of a cloud that would have a height, D, equal to that of an exposed person can be calculated. The cloud expansion is directly related to the increase in temperature, and the increase in temperature is closely related to the increase in the absolute pressure that results from confined combustion or explosion; that is:

$$V_{FINAL} / V_{INITIAL} = [\, P_{MAX} + P_{INITIAL} \,] / P_{INITIAL} = [\, A_{FLOOR}\, D \,] / V_{INITIAL} \tag{12}$$

or:

$$V_{INITIAL} = [\, A_{FLOOR}\, D \,]\, [\, P_{INITIAL} / [\, P_{MAX} + P_{INITIAL} \,] \,]$$
$$= [\, W_{DUST}\, \eta_{DISPERSION} \,] / C_{WORST} \quad \text{cubic meters} \tag{13}$$

The above relationships can be combined to give:

$$W_{DUST} = C_{WORST}\, [A_{FLOOR}\, D]\, [\, P_{INITIAL} / [P_{MAX} + P_{INITIAL}] / \eta_{DISPERSION} \quad \text{grams} \tag{14}$$

There is an assumption that the probability of one or more persons avoiding exposure to the flash fire is very low, and the standard suggested a probability of 5% or less [a]. Thus, the "allowable" weight of dust accumulations should be reduced by this factor. Thus, for a 5% probability of avoiding flash-fire exposure of 5%, the above relationship would become:

$$W_{DUST} = 0.05\, C_{WORST}[A_{FLOOR}\, D][P_{INITIAL} /[P_{MAX} + P_{INITIAL}] / \eta_{DISPERSION} \quad \text{grams} \tag{15}$$

The above derivation yields the statement in NFPA 654: "The threshold dust mass[/accumulation] establishing a building or room as a **dust flash-fire hazard** area, $M_{FLASHFIRE}$ shall be determined by the following equation:

$$M_{FLASHFIRE} = P\, C_{WORST}\, [P_{INITIAL}/(P_{INITIAL}+P_{MAX})][A_{FLOOR}\, D / \eta_{DISPERSION}] \quad \text{kg."} \tag{16}$$

where:

$M_{FIRE}$ = threshold dust mass (kg) based on personnel fire exposure criterion;

P = probability of [a person avoiding] flame impingement ~~on a person~~, not to exceed 0.05 (5 percent probability) [a];

[a] This definition departs significantly from the definition in NFPA 654, paragraph 6.1.5.2; however, this definition meets the intention of the evaluation of flash-fire exposure.

$C_{WORST}$ = worst-case dust concentration (kg./cu.m.) for the maximum rate-of-pressure-rise;
$P_{INITIAL}$ = 1 bar absolute;
$P_{MAX}$ = maximum pressure (bar g) developed in tests on the accumulated dust;
$A_{FLOOR}$ = enclosure floor area (sq.m.);
D = nominal height of a person (2 meters); and
$\eta_{DUST}$ = entrainment fraction = 0.25 [non-conservative[b]]

The above relationship can be rearranged, with the above "default" values accepted (with a more-realistic nominal height of 1.75 meters or 5'-10"), and converted to useful units:

$$M_{FLASHFIRE} / A_{FLOOR} = [0.05][C_{WORST}][1/(1+P_{MAX})][(5.75)/35.3][1/0.25][2.205][453.5] \text{ g./sq.ft.} \quad (17)$$

$$M_{FLASHFIRE} / A_{FLOOR} = 32.5 \, [C_{WORST}] \, [1/(1+P_{MAX})] \text{ g./square foot of floor area.} \quad (18)$$

For typical organic dusts, the concentration $C_{WORST}$ at which the maximum burning rate occurs – and the concentration at which the value of the Deflagration Index $K_{ST}$ is obtained – is about 0.65 kilogram per cubic meter [References 21 and 22]. The maximum explosion pressure attained by organic dusts is about 9.0 barg [Reference 23]. Thus, for many organic dusts, the preceding equation becomes:

$$M_{FLASHFIRE} / A_{FLOOR} = 32.5 \, [0.65] \, [1/(1+9)] = 2.1 \text{ g./sq.ft., or about 4.5 lbs./1000 sq.ft.} \quad (19)$$

*2.2.4 Mass Method B, for Protection Against Explosion Hazards*

The NFPA standard for explosion venting [Reference 24] contains a relationship for "partial-volume" combustion in an enclosing structure, where the combustion would not rupture the confining walls and roof:

$$V_{CLOUD} / V_{TOTAL} = P_{REDUCED} / P_{MAX} \quad (20)$$

where $V_{CLOUD}$ is the volume of the dust/air cloud at an explosible mixture, in cu.ft. (or cu.m.);
$V_{TOTAL}$ is the total volume of the enclosure, in cubic feet (or cubic meters);
$P_{REDUCED}$ is the pressure that would result from the combustion, in psig (or barg); and
$P_{MAX}$ is the explosion pressure when the mixture fills the volume, in psig (or barg).

[b] "Non-conservative", as explained later in this paper.

This relationship also can be shown as:

$$[\, W_{DISPERSED} / C_{WORST} \,] / V_{TOTAL} = P_{REDUCED} / P_{MAX} \qquad (21)$$

where $W_{DISPERSED}$ is the weight of dust in the suspended mixture of dust and air, in grams;
$C_{WORST}$ is the concentration at which the Deflagration Index, $K_{ST}$, and/or $P_{MAX}$ values are determined, in grams per cubic foot (or grams per cubic meter);
$V_{TOTAL}$ is the volume of the enclosure, in cubic feet (or cubic meters).

The weight of dust in the suspended mixture of dust and air is related to the fraction of the accumulated dust that is raised and dispersed from horizontal surfaces (or from the walls of the enclosure): the "entrainment factor", $\eta_{DISPERSION}$; thus, $W_{DISPERSED} = W_{SURFACE}\, \eta_{DISPERSION}$. The total volume of the enclosure is the product of the floor area and the height of the enclosure: thus, $V_{TOTAL} = A_{FLOOR}\, H$. The value of $P_{REDUCED}$ is to be increased by a "Dynamic Load Factor", DLF, to obtain the strength of the enclosure walls and ceiling/roof, $P_{ES}$, in response to a brief-overpressure event: thus, $P_{ES} = P_{RED}\, DLF$, or $P_{RED} = P_{ES} / DLF$.

The preceding relationship would then become:

$$W_{SURFACE}/A_{FLOOR} = [\, P_{ES} / DLF \,]\,[\, C_{WORST} / P_{MAX} \,]\,[\, H / \eta_{DISPERSION} \,] \ \text{g./m}^2 \qquad (22)$$

where $P_{ES}$ and $P_{MAX}$ are in barg (or psig);
$C_{WORST}$ is in grams per cubic meter; and
$H$ is in meters.

or, for the enclosure height in feet, and the floor area in square feet:

$$W_{SURFACE} / A_{FLOOR} = 0.028\, [\, P_{ES} / DLF \,]\,[\, C_{WORST} / P_{MAX} \,]\,[\, H / \eta_{DUST} \,] \ \text{g./ft}^2 \qquad (23)$$

The above derivation yields the statement in NFPA 654: "The threshold dust mass[/accumulation] establishing a building or room as a **dust explosion hazard** area, $M_{EXPLOSION}$, shall be determined by the following equation [to contain a combustible dust explosion, and minimize damage to the walls and ceiling]:

$$M_{EXPLOSION} = [\, P_{ES} / DLF \,]\,[\, C_{WORST} / P_{MAX} \,]\,[\, A_{FLOOR}\, H / \eta_{DUST} \,] \ \text{kg.} \qquad (24)$$

where:
$M_{EXPLOSION}$ = threshold dust mass (kg) based on building damage criterion;
$P_{ES}$ = enclosure strength evaluated based on static pressure calculations for the weakest building structural element not intended to vent or fail (bar g);

$DLF$ = dynamic load factor; a [worst-case, conservative] value of 1.5;
$C_{WORST}$ = worst-case dust concentration (kg./cu.m.) for the maximum rate-of-pressure-rise;
$P_{MAX}$ = maximum explosion pressure (bar g) developed in tests on the accumulated dust;
$A_{FLOOR}$ = enclosure floor area (sq.m.);
$H$ = enclosure ceiling height (m); and
$\eta_{DUST}$ = entrainment fraction; a [best-case, non-conservative[b]] value of 0.25.

The product of the above values of DLF and $\eta_{DUST}$ is about 0.4 and represents a combination of "conservative" and "non-conservative" values. The value of DLF is very unlikely to be 1.0 or less, and the value of $\eta_{DUST}$ is unlikely to be much above 0.6, for a product of 0.6. Thus, the "permitted" values of 1.5 and 0.25 appear to be reasonable.

The above relationship can be rearranged, with the above "default" values accepted, and converted to useful units:

$$M_{EXPLOSION} / A_{FLOOR} = [1000/35.3]\,[P_{ES}/P_{MAX}]\,[C_{WORST}/1.5]\,[H/0.25] \quad \text{g./ft}^2 \qquad (25)$$

$$M_{EXPLOSION} / A_{FLOOR} = 76\,[P_{ES}/P_{MAX}][C_{WORST}]\,[H] \quad \text{g./square foot of floor area} \qquad (26)$$

For typical organic dusts, the concentration $C_{WORST}$ at which the maximum burning rate occurs – and the concentration at which the value of the Deflagration Index $K_{ST}$ is obtained – is about 0.65 kilogram per cubic meter [References 21 and 22]. The maximum explosion pressure attained by organic dusts is about 9.0 barg [Reference 23]. Thus, for many organic dusts, the preceding equation becomes:

$$M_{EXPLOSION} / A_{FLOOR} = 76\,[\,P_{ES} / 9\,]\,[\,0.65\,]\,[H]\,[1/14.5] \qquad \text{g./ft}^2 \qquad (27)$$

$$M_{EXPLOSION} / A_{FLOOR} = 0.38\,P_{ES}\,H \qquad \text{grams per square foot of floor area} \qquad (28)$$

where:
$P_{ES}$ = enclosure strength (psig); and
$H$ = enclosure ceiling height (ft.).

[b] "Non-conservative", as explained later in this paper.

Guidance concerning the pressure resistance of wall and ceilings is limited, and there do not appear to be any "rules of thumb" for wall strengths. However, the following "data" might be useful:

1. The maximum wind speeds were probably considered in the design of metal or composite-material siding (on steel frames), and the required wind-pressure resistance would be given by: $P = 0.0025\ V^2$ pounds per square foot, where V is the wind speed in miles per hour [Reference 25]. For example, a wind speed of 120 MPH could produce a pressure of about 40 psf, or about 0.28 psig.

2. A steel-sided room within a building would suffer 50% damage if exposed to a pressure of about 0.55 psig [Reference 26].

3. Generally, buildings cannot withstand more than about 1 to 2 psig [Reference 27].

4. A brick wall would have very little ductility, and pressure resistance would be determined by the mass of the wall. An 18-inch wall could withstand about 10 psig [Reference 28], so a 4-inch wall probably could withstand about 2 psig], and 8 to 12-inch walls could withstand 7 to 8 psig [Reference 29].

5. Unreinforced concrete-masonry-unit [block] walls could withstand 2 to 3 psig [Reference 29]

6. A corrugated sheet-steel wall, 1/4-inch thick, could withstand 8 psig [Reference 30].

7. A concrete-masonry-unit [block] wall could withstand 10 psig if fully grouted and with double reinforcement [Reference 28].

8. Reinforced-concrete walls can withstand pressures of 10 psig or more [Reference 31].

As an example of the use of the preceding equation, the dust-accumulation limit for a room having a pressure resistance of 0.55 psig [about 80 pounds per square foot] and a 16-foot ceiling would be given by:

$$M_{EXPLOSION} / A_{FLOOR} = 0.38\ [0.55]\ [16] = 3.3\ \text{g./ft}^2\text{; about 7 lbs./1,000 ft}^2 \qquad (29)$$

For very large rooms of buildings, and for very strong rooms or buildings, the flash-fire limits on dust-accumulation weights will be smaller than the explosion limits. Thus, the flash-fire limits usually should be used as the housekeeping criteria.

## *2.3 Summary of Combustible-Dust Accumulation Limitations*

The results of the calculations in the preceding sections can be tabulated as follows:

**Table 3. Summary of Calculations**

| Properties | Method for Determining the Threshold Dust Mass/Accumulation [g./sq.ft.; lbs./1000 sq.ft.] | | | | |
|---|---|---|---|---|---|
| | Flash-Fire Hazard | | "Secondary" Explosion Hazard | | |
| | Mass A | Mass B | Layer | Mass A | Mass B |
| Limit | 1.9 g/ft$^2$ | 2.1 g/ft$^2$ | 4.5 g/ft$^2$ | 1.3 g/ft$^2$ | 3.3 g/ft$^2$ |
| | 4 lbs/1,000 ft$^2$ | 4 lbs/1,000 ft$^2$ | 10 lbs/1,000 ft$^2$ | 3 lbs/1,000 ft$^2$ | 7 lbs/1,000 ft$^2$ |
| Variables | None | $C_{WORST}$; $P_{MAX}$ | None | H | $C_{WORST}$; $P_{MAX}$;$P_{ES}$; H |

It may be of interest to note that equal dust-accumulation limits of 1.9 g./sq.ft. [or about 4 pounds per square foot of floor area] – are obtained only if:

a. The allowable area for the Layer method is only about 2% of the floor area, rather than 5%.

b. The room height is 17 feet, for the "Explosion-A" method.

c. The ratio of $C_{WORST}/(1+P_{MAX})$ is about 0.058, for the "Flash Fire-B" method; that is, for a value of $C_{WORST}$ of 0.65 g./cu.m., the value of $P_{MAX}$ would be about 10 barg (typical of a metal dust).

d. The ratio of $C_{WORST}/(1+P_{MAX})$ is about 0.058, and the room height is 17 feet, and the value of $P_{ES}$ – the enclosure strength – is about 0.023 barg, or about 50 pounds per square foot, for the "Explosion-B" method.

## *2.4 Example of Dust Collection*

In the recently-published "NFPA Guide to Combustible Dusts" [Reference 32], an example of hypothetical dust accumulation is provided. The "data" can be shown – with extended calculations that are based on a bulk density of 30 pounds per cubic foot – in the following Table:

**Table 4. Summary of Housekeeping Results for a Segregated Room [Example]**

| Location | Area | Interval | Dust | Rate | Thickness | Density |
|---|---|---|---|---|---|---|
| Floor | 500 sq.ft. | 8 hours | 25 lbs. | 75 lbs./day | 0.02 in.; 0.5 mm. | 23 g./sq.ft. |
| Equipment | 100 sq.ft. | 7 days | 10 lbs. | 1.4 lbs./day | 0.04 in.; 1 mm. | 45 g./sq.ft. |
| Overhead | 500 sq.ft. | 60 days | 30 lbs. | 0.5 lbs./day | 0.024 in.; 0.6 mm. | 27 g./sq.ft. |
| Totals | 1,100 sq.ft. | [22% of area] | 65 lbs. | - | - | 27 g./sq.ft. |
| Summary | 5,000 sq.ft. [4,600 sq.ft. floor] | | 65 lbs. | - | - | 5.9 g./sq.ft. |

As stated by the authors, the layer-depth criterion would be 0.078 inch [2 mm.], but only over 5% of the total horizontal-surface area of 5,000 square feet. Thus, the total "allowable" dust accumulation would be limited to 1.6 cubic feet, or about 50 pounds of dust. Therefore, it is very likely that the layer-depth criterion would be exceeded, particularly prior to the 60-day cleanup.

Further, the authors state: "The total dust accumulation in this example exceeds the allowable for the room footprint area; thus, the current cleaning frequencies result in a dust deflagration hazard in the room."

The situation that is described by the preceding Table 4 is particularly hazardous because "Any dust that is elevated above floor level in joists, beams, equipment tops, etc., should be considered as being capable of being thrown into suspension." [Reference 33]. A 40-pound accumulation of combustible dust over a floor area of 4,600 square feet would represent a "density" of 4 grams per square foot. This exceeds the "Mass A" criteria for a flash-fire hazard and for an explosion hazard unless the ceiling was more than 35 feet above floor level. A density of 4 g./sq.ft. also would exceed the "Mass B" criteria for a flash-fire hazard and for an explosion hazard for most combustible dusts in typical enclosed process buildings.

As also stated by the authors, there are three alternatives:

1. Change the cleaning frequencies/intervals to limit the accumulation to 50 pounds;

2. Provide personnel protection from the potential flash-fire hazard, and protect the structure from overpressure/explosion damage; or

3. Provide engineering controls to reduce and/or capture fugitive dust releases from the equipment.

Facilities that have been experiencing "housekeeping problems" with accumulations of combustible dust should begin weighing the dust that is swept or vacuumed from floors; from elevated surfaces such as roof supports and ductwork; and from the surfaces of equipment, including motors, electrical panels, and process equipment. The purposes are:

1. The accumulated-dust "densities" – in terms of grams per square foot, or pounds per 1,000 square feet of floor area – could be used to establish and/or adjust the frequency of cleaning floors and elevated surfaces.

2. The trend in recovered-dust weight divided by the interval between such housekeeping activities – such as pounds per week – can serve as a "leading indicator" concerning the adequacy [or inadequacy] of the control of flash-fire and secondary-explosion hazards.

3. Guidance for minimizing the flash-fire and secondary-explosion hazards resulting from accumulations of combustible dust are:

    a. "Most-conservative" limitation for flash fire and explosion: 1.9 grams per square foot, or 4 pounds per 1,000 square feet of floor area.

    b. "Least-conservative" limitation for flash fire: 1.9 g./sq.ft., or 4 lbs./1,000 sq.ft., as based on the properties of "typical" organic dusts.

    c. "Least-conservative" limitation for explosion: 3.3 g./sq.ft., or 7 lbs./1,000 sq.ft., as based on (1) the properties of "typical" organic dusts, (2) 180-MPH wind-resistance walls, and (3) a 16-foot ceiling.

    d. The dust accumulations that would be "allowed" by the Layer method [about 4.5 g./sq.ft., or about 10 lbs./1,000 square feet of floor area] appear to be "excessive" when compared to the limitations that would be imposed by the above "Mass Methods".

There is another important aspect in the control of dust accumulations, and that is the relationship between the amount [particularly, the layer depth] of combustible-dust accumulations and the electrical classification that would apply to areas in which combustible-dust accumulations occur. Of concern regarding dust deposits on electrical devices is the hazard associated with the thermal insulating properties of the layer, which can limit the heat dissipation and allow powder temperature rise. If a dust layer gets too hot, it can self-ignite. Such smoldering dust layers present both an immediate fire hazard as well as a high energy ignition source for the expansion of the fire or ignition of dust clouds that could come into contact with the embers.

### *2.5 Housekeeping Guidance for "Chronic" or "Continual" Releases of Combustible Dusts*

Where "episodic" releases of combustible dusts occur – frequently, or infrequently – guidance is provided in NFPA 654, NFPA 652, and NFPA 484 for "unscheduled" housekeeping [References 34, 35, and 36]; that is, removal of dust spillages or other types of "transient" dust releases. In the following Table 5, "TDMA" is the Threshold Dust Mass/Accumulation of about 90 grams per square foot, as determined from the "Layer Method" calculation that is described above. Since this dust "density" would be hazardous if covering a large area, other "more-conservative" values of "TDMA" can be selected, as justified by use of the "Mass A" or "Mass B" methods.

**Table 5. Unscheduled Housekeeping Guidance [NFPA 654]**

| Accumulation Level [NFPA 652] | Accumulation on the Worst Single Square Meter [10 square feet] of Surface | Average Layer Depth at 30 Pounds per Cubic Foot [480 kg./cu.m.] | Max. Spill-Area-to-Room-Area Ratio [%} | Longest Time to Complete Local Cleaning of Accessible Surfaces | Longest Time to Complete Local Cleaning of Remote Surfaces |
|---|---|---|---|---|---|
| 1 | 1X < TDMA < 2X | 5/64" to 5/32"<br>2 to 4 mm. | 2% to 1%,<br>for 1.9 g./cu.m. average | 8 hours | 24 hours |
| 2 | 2X < TDMA < 4X | 5/32" to 5/16"<br>4.2 to 8.3 mm. | 1.0% to 0.5%, for 1.9 g./cu.m. average | 4 hours | 12 hours |
| 3 | >4X TDMA | >5/16"<br>>8 mm. | <0.5%, for 1.9 g./cu.m. average | 1 hour | 3 hours |

The preceding Table 5 could be utilized to establish the maximum area – as a percentage of the total room area – that is "allowed" to be covered with dust at a given depth. Thus, if frequent spillages or deposits occur rather frequently – as at bagging stations – it might be advisable to provide floor markings that would show the maximum area, to prompt cleanup of that area.

# 3 Relationship Between Dust Accumulations and Electrical Classification of Dusty Areas

In the U. S., there are two NFPA publications that are concerned with the electrical classification of areas in which combustible dust might be present in the atmosphere or as accumulations on surfaces in the work area. The following publications provide dust-accumulation limits in the appendices that are "*for informational purposes only*":

1. NFPA 654, Table A.6.5.2 [Reference 37].

2. NFPA 499, Tables A.6.3.2(a) and A.6.7 [Reference 38].

The following Table 6 copies information from the NFPA 654 standard [in red], with the addition of calculated "Dust Densities" and "Maximum Ratio of Spill Area to Room Floor Area" (in percent).

**Table 6. Guidance for Area Electrical Classification [NFPA 654; Table A.6.5.2]**

| Depth of Dust Accumulation | Dust Density [g./sq.ft.] | Max. Spill/Room Area % | Frequency | Housekeeping Requirement | Electrical Classification |
|---|---|---|---|---|---|
| Surface color discernible | Near zero; Negligible | 100% | No spills or releases | Maintain negligible | Unclassified |
| 0 to 1/32" [0 to 0.8 mm.] | 0 to 35 g./sq.ft. | 100% to 5%, for 1.9 g./cu.m. average | 2 to 3 times per year | Clean up during same shift | Unclassified |
| | | | >3 times/year; 24-hr. accumulation | Maintain <1/64"/0.4 mm. | Unclassified, but dust-tight; NEMA 12 |
| 1/32" to 1/8" [0.8 to 3 mm.] | 35 to 140 g./sq.ft. | 5% to 1.5%, for 1.9 g./cu.m. average | 2 to 3 times per year | Clean up during same shift | Unclassified, but dust-tight; NEMA 12 |
| | | | >3 times/year; 24-hr. accumulation | Maintain <1/16"/1.6 mm. | Class II, Div. 2 |
| > 1/8" [>3 mm.] | >140 g./sq.ft. | <1.5%, for 1.9 g./cu.m. average | 2 to 3 times per year | Immediate shutdown; clean | Class II, Div. 2 |
| | | | >3 times/year; 24-hr. accumulation | Minimize accumulation | Class II, Div. 1 |

The following Table 7 copies information from the NFPA 499 standard [in red], with the addition of calculated "Dust Densities" and "Maximum Ratio of Spill Area to Room Floor Area" (in percent). When Tables 6 and 7 are compared, it is clear that there are some significant differences between these two standards that have yet to be resolved by the NFPA.

**Table 7. Guidance for Area Electrical Classification [NFPA 499; Tables A.6.3.2(a); A.6.7]**

| Depth of Dust Accumulation | Dust Density [g./sq.ft.] | Max. Spill/Room Area % | Frequency | Housekeeping Requirement | Electrical Classification |
|---|---|---|---|---|---|
| Surface color discernible | Near zero; Negligible | 100% | No spills or releases | Not Specified | Unclassified |
| 0 to 1/32" [0 to 1 mm.] | 0 to 35 g./sq.ft. | 100% to 5%, for 1.9 g./cu.m. average | Infrequent Release | Clean as appropriate | Unclassified |
| 1/32" to 1/8" [0.8 to 3 mm.] | 35 to 140 g./sq.ft. | 5% to 1.5%, for 1.9 g./cu.m. average | Infrequent Release | Maintain less than 3 mm. [1/8"] | Class II, Div. 2 |
| | | | Occasional Cloud Formation | Minimize additional accumulation or cloud formation | Class II, Div. 1 or Div. 2 |
| >1/8" to 1/2" [>3 mm. to 12.5 mm.] | 140 to 570 g./sq.ft. | 1.5% to 0.3%, for 1.9 g./cu.m. average | Continuous; Frequently | Minimize additional accumulation | Class II, Div. 1 |
| >1/2" [>12.5 mm.] | >570 g./sq.ft. | <0.3%, for 1.9 g./cu.m. average | Infrequent thick dust layer or extensive dust cloud | Immediate shutdown; clean | Class II, Div. 1 |

Despite the differences in the two standards – NFPA 654 and NFPA 499 – the above Tables 6 and 7 provide good guidance concerning minimum-hazard dust-layer thicknesses and maximum thick-dust areas. Thus, the preceding two Tables can be combined to give "minimum essential" guidance for protection against flash-fire and explosion hazards resulting from combustible-dust accumulations, as shown in Table 8.

**Table 8. Suggested Guidance for Area Electrical Classification [Combined Standards]**

| Depth of Dust Accumulation | Dust Density [g./sq.ft.] | Max. Spill/Room Area % | Frequency | Housekeeping Requirement | Electrical Classification |
|---|---|---|---|---|---|
| Surface color discernible | Near zero; Negligible | 100% | No spills or releases | Not Specified | Unclassified |
| 0 to 1/32" [0 to 1 mm.] | 0 to 35 g./sq.ft. | 100% to 5%, for 1.9 g./cu.m. average | 2 to 3 times per year | Remove during same shift | Unclassified |
| | | | >3 times per year; 24-hr. accumulation | Maintain <1/64"/0.4 mm. | Unclassified, but dust-tight; NEMA 12 |
| 1/32" to 1/8" [0.8 to 3 mm.] | 35 to 140 g./sq.ft. | 5% to 1.5%, for 1.9 g./cu.m. average | 2 to 3 times per year | Remove during same shift | Unclassified, but dust-tight; NEMA 12 |
| | | | >3 times per year; 24-hr. accumulation Occasional cloud | Maintain <1/16"/1.6 mm. | Class II, Div. 2 |
| >1/8" to 1/2" [>3 mm. to 12.5 mm.] | 140 to 570 g./sq.ft. | 1.5% to 0.3%, for 1.9 g./cu.m. average | 2 to 3 times per year | Immediate shutdown; clean | Class II, Div. 2 |
| | | | >3 times per year; 24-hr. accumulation; Occasional cloud | Minimize accumulation | Class II, Div. 1 |
| >1/2" [>12.5 mm.] | >570 g./sq.ft. | <0.3%, for 1.9 g./cu.m. average | Infrequent thick dust layer or extensive dust cloud | Immediate shutdown; clean | Class II, Div. 1 |

The above guidance for Class and Division electrical area classifications can also be applied to Zones, where Zone 22 corresponds to Class II, Division 2, and Zone 21 corresponds to Class II, Division 1. Zone 20 is essentially limited to equipment, where a hazardous concentration of dust occurs very frequently or continuously, and the nearest equivalent is Class II, Division 1.

## 4 Consideration of Non-Electrical Ignition Sources

There are several types of "non-electrical" ignition sources that are prevalent in chemical plants, and they also need isolation from combustible dusts, just as combustible dusts are isolated from electrical arcs and sparks. Typical "non-electrical" ignition sources – for "primary" dust fires, flash-fires, and explosions – include the following, in approximate descending order of importance [References 39 and 40]:

**Table 9. Non-Electrical Ignition Sources**

| | |
|---|---|
| Open flames | [boilers, incinerators, air heaters] |
| Welding and cutting | [torches or arcs] |
| Explosions in equipment | [ignition of dust/air, gas/air, vapor/air mixtures] |
| Hot surfaces of equipment | [ovens, driers, piping for heating fluids] |
| Accidental Fire | [combustibles at hot spots] |
| Frictional hot spots | [bearings, metal-to-metal rubbing] |
| Frictional sparks | [grinding, polishing] |
| Vehicles | [exhaust systems] |
| Impact sparks | [thermite, mechanical systems] |
| Unstable materials | [peroxides, pyrophorics, spontaneous combustion] |
| Adiabatic compression | [diesel effects] |
| Electromagnetic radiation | [stray fields] |
| Ionizing radiation | [instrumentation] |
| Lightning | [storms] |

A method for studying and correcting the undesired effects of these "non-electrical" ignition sources on combustible dusts [Reference 41] is the "Mechanical Equipment Ignition Risk Assessment" [MEIRA]. Assessment of mechanical ignition hazards is outside the scope of this presentation. However, the control of dust accumulations near such ignition sources is important in preventing fires, flash-fires, and explosions. Persons performing Dust Hazards Analysis should be aware of these other types of ignition sources and determine whether or not conformance to the area electrical classification adequately isolates these types of ignition sources from combustible dusts.

## 5 "Experimental" Data for "Organic" Dusts

During a recent Dust Hazards Analysis, an operator was observed sweeping-up dust in a powder-storage area, and he was asked to weigh the dust that he had gathered. The weight was about 5 pounds, and the floor area was about 2,500 square feet. Thus, the average dust "density" was about 2 pounds per 1,000 square feet, or about 0.9 gram per square foot. Thus, this amount of dust accumulation was unlikely to be sufficient to "fuel" a secondary dust explosion.

At another location, the site management was asked to collect the dust that was (1) on the ground floor at a bagging station and (2) on equipment and the building structure in an "upstairs" operating area. The weight that was collected at the first location was about 4 pounds, and the floor area that was swept-up was about 240 square feet. Thus, the average dust density was about 17 pounds per 1,000 square feet, or about 7.5 grams per square foot. This is a high dust density. However, the floor and surfaces in the rest of this 1,150-square-foot room did not contain significant dust; that is, the surface colors were not obscured. Thus, the average dust density was about 3.5 pounds per 1,000 square feet, or about 1.6 grams per square foot. Therefore, this amount of dust accumulation could be considered to be marginally acceptable, at the time of this "fortuitous" dust cleanup.

At the second location – which was isolated from the ground floor by solid flooring – about 10.2 pounds of dust were collected, and the floor area was 1,150 square feet. Thus, the average dust density was about 9 pounds per 1,000 square feet, or about 4.0 grams per square foot. This dust density exceeds the NFPA 654 "Mass Method" guidelines and indicated that more-frequent housekeeping – or better dust containment – was needed.

Figure 1, below, shows a 30-gram deposit of "dust" [sodium bicarbonate; baking powder] uniformly sifted onto a one-square-foot area; thus, 30 grams per square foot [g./sq.ft.]. The depth of the deposit is about one millimeter, as indicated by the "standard" paper clip. A wide-spread deposit of combustible dust having a "density" of 30 g./sq.ft. would significantly exceed the NFPA criteria for the prevention of flash fire and explosion for typical chemical-plant constructions.

Figure 2, below, shows a 3-gram deposit of baking powder sifted onto a one square foot area. As indicated by this photograph, a deposit of 3 g./sq.ft. would marginally obscure the color of the surface under the deposit. As indicated by the preceding discussion, a combustible-dust density of 3 g./sq.ft. would exceed several of the NFPA critera for flash-fire and explosion prevention.

## 6 Combustible-Dust Accumulation Limits, for "Metal" Dusts

### *6.1 The "Layer-Depth" Criterion*

In the NFPA 484 standard for metal dusts, the Threshold Limit is defined as: "*An established value of dust accumulation, determined as part of the hazard analysis for a particular enclosure of facility and specific metal, for taking an appropriate action.*"

*There have been several proposed methods for determining an appropriate threshold value for fugitive dust accumulations. It should be noted that some methods might not be appropriate in all situations. Fugitive dust shall not be allowed to accumulate to a level that obscures the color of the surface beneath it. It shall be permissible to establish, in a building or room, an alternate housekeeping dust accumulation threshold based on a documented hazard assessment acceptable to the Authority Having Jurisdiction [AHJ].*

*Unscheduled housekeeping should be performed to limit the time that a local spill or short-term accumulation of dust is allowed to remain before cleaning the local area to less than the threshold dust mass/accumulation. Table A.7.4(b) [of the NFPA standard] shows approximate* *equivalent depths for the accumulation values in Table A.7.4(a) [of the NFPA standard] when the threshold dust mass/accumulation is 1 kg/m$^2$.*

It is likely that a double layer of dust particles would provide significant obscuration of the color of the surface under the layer of dust. For example, a double layer of dust particles having a uniform diameter of 100 microns and a density of 1 g./cu.cm. would have a "dust density" of about 10 grams per square foot, as based on:

$$W / A = 100\ d\ \rho \text{ grams per square foot} \tag{30}$$

where d is the diameter of the particles, in millimeters, and
$\rho$ is the particle density, in grams per cubic centimeter.

The thickness of the dust layer would be about twice the particle diameter, or about 0.2 mm.

Since the above relationship indicates that the "dust density" is directly proportional to particle density, an obscuring layer of 100-micron steel particles – for a particle density of 7.8 g./cu.cm. – would have a "dust density" of about 78 grams per square foot, with the same dust thickness, near 0.2 mm. If the layer-depth criterion of NFPA 654 were to be used to determine a hazardous accumulation for iron dust, the result would be:

$$t_{ALLOWABLE} = [\,0.8\,]\,[75/BD] = [\,0.8\,]\,[75/180] = 0.33 \quad \text{millimeter} \tag{31}$$

and this is close to the "obscuring" depth. The following Table provides the results of similar calculations for other metal dusts:

**Table 10. Dust-Accumulation Thresholds Based on Layer Depth**

| Dust | | Aluminum | Copper | Iron | Magnesium | Titanium | Zirconium |
|---|---|---|---|---|---|---|---|
| Bulk Density | (lbs./cu.ft.) | 44 | 101 | 180 | 46* | 60 | 100* |
| Layer Thickness | (mm.) | 1.4 | 0.6 | 0.3 | 1.3 | 1.0 | 0.6 |

* approximate bulk density [based on particle density]

### 6.2 *"Mass Method" Criteria for Metal Dusts*

The "Mass Method A" and "Mass Method B" calculations that are described above for organic dusts are derived from considerations of heat of combustion and the results of combustion-rate tests on dispersed dusts [and the effects of thermal radiation and blast pressures on humans]. It might be expected that similar "Mass Method A" and "Mass Method B" calculations will be available for metal dusts in the near future, also.

# 7 Estimation of Entrainment Fraction

In the NFPA 654 standard, a procedure for determining the maximum "allowable" dust accumulations is presented as "Mass Method B". In the equations for flash-fire and explosion hazards, an "Entrainment Fraction" defines the fraction of dust that is removed from the [top of] the dust deposits. An "entrainment fraction" of 0.25 is suggested, with the provision that *"an alternative value of* $\eta_D$*, based on a risk assessment that is acceptable to the AHJ is permitted".* However, two sets of tests that were conducted to determine the extent of entrainment of dust deposits – when exposed to blasts of combustion gases [Reference 42] – indicate that a value of 0.25 probably is not "conservative" for thin layers of combustible [organic] dusts.

## *7.1 NGFA Tests*

In the referenced study, the results of two sets of tests were reported. The results of five tests that were supported by the National Grain and Feed Association [NGFA] are presented in Table 11. Cornstarch was spread on the floor of an 8'x8'x80' gallery, and the layers of dust were subjected to an explosion of a cornstarch/air mixture at one end of the gallery. The bulk density of the cornstarch was about 42 pounds per cubic foot [0.67 g./cu.cm.], and the particle density of cornstarch is about 1.5 g./cu.cm. [94 pounds per cubic foot].

**Table 11. Interpreted Results of NGFA Tests on Deposits of Cornstarch Powder**

| Test No. | Initial Dust Density | Initial Layer | End Dust Density | End Layer | Entrainment Fraction |
|---|---|---|---|---|---|
| 50 | 17 g./sq.ft. | 0.27 mm. | 7 g./sq.ft. | 0.10 mm. | 0.62 |
| 56 | 21 g./sq.ft. | 0.33 mm. | 10 g./sq.ft. | 0.16 mm. | 0.53 |
| 53 | 23 g./sq.ft. | 0.37 mm. | 15 g./sq.ft. | 0.25 mm. | 0.34 |
| 51 | 35 g./sq.ft. | 0.55 mm. | 21 g./sq.ft. | 0.33 mm. | 0.40 |
| 52 | 64 g./sq.ft. | 1.02 mm. | 44 g./sq.ft. | 0.71 mm. | 0.30 |

The results of these tests show that the Entrainment Fraction increases as the thickness of the dust layer decreases. An equation that describes this behavior [approximately] is:

$$E = 1 - e^{-12.5 / D^{0.925}} \quad (32)$$

where D is the "dust density" in grams per square foot.

## *7.2 NIOSH Tests*

The results of tests that were supported by the National Institute of Occupational Safety and Health [NIOSH] involved mixtures of coal dust and limestone dust. The mixtures were spread between two 1-inch rails in a 20-foot-wide, 8-foot-high, and 1600-foot long tunnel. The particle densities for coal dust and limestone dust are about 1.33 and 2.75 g./cu.cm., respectively [83 and 172 pounds per cubic foot], and the bulk densities of the mixtures were about 0.85 g./cu.cm. [about 53 pounds per cubic foot].

Ten tests are reported in the cited reference, with the coal-dust concentrations ranging from 20 to 35 weight-percent. The dusts were screened though a 74-micron screen, and the tests used dusts in which 20% to 80% passed through the screen. The dust-layer thicknesses were 1.0 inch [25.4 mm.] in all of the tests. Thus, the "dust density" was about 2,000 grams per square foot. The layers of dust were subjected to explosion of a methane/air mixture at one end of the tunnel.

The amounts of dust that were removed from the deposits ranged from 1.0 to 2.6 millimeters, with an average of 1.70 millimeters. Thus, the entrainment fractions ranged from 0.039 to 0.102. The results of the tests indicated that there were no relationships between the percentage of coal dust and the particle-size distribution – within the above ranges – and the amount of dust that was entrained by the explosion-generated flow of combustion gases.

### *7.3 Conclusions Resulting from the Tests*

The results of these tests indicate that the Entrainment Fraction is a strong function of the amount of dust in the layer: high values of the Entrainment Fraction for thin layers of dust, and low values of the Entrainment Fraction for thick layers of dust. Thus, a conservative approach to the use of the "Mass Method B" procedures for determining the maximum "allowable" dust accumulation – in terms of grams per square foot – would be to calculate with an Entrainment Fraction of 0.25 and then recalculate with an entrainment fraction that would be obtained from the above equation (32). Repetitive calculations might be required to obtain the final result.

## 8 Other Quantifiable Aspects of Housekeeping

It is important that personnel who are expected to clean-up spills or releases of combustible dusts (1) be trained to avoid creating large dense [explosible] clouds of combustible dust and (2) be well-protected against the hazard of flash-fire that might result from ignition of smaller clouds of dust. This would include the members of the spill-response team, contractors, operators, and/or mechanics that would be involved in collection and removal of combustible-dust spills or released. Careful sweeping and cautious handling of the collected dust can minimize the likelihood of large dust clouds and the hazard of explosion. For dusts that could be ignited by electrostatic charges on humans – 25 millijoules or less [Reference 43] – the cleaning implements should be static-dissipative or semi-conductive, with a resistance to ground that is less than 100 megohms [Reference 44]. Such "grounding" could be through the person doing the cleaning, if gloves, footwear, and working surface have similar conductivities.

For protection against flash-fire, persons who clean-up dust deposits should be wearing static-dissipative footwear, to minimize the ignition hazard. Also, they should be wearing flame-resistant outer clothing, with face and head covering, for protection against possible flash fire, since the mortality likelihood is about 25% for ignition of "everyday" clothing [Reference 45]. Further, if the clean-up involves "pyrophoric" metal dusts – such as titanium, zirconium, aluminum, lithium, magnesium, cesium, or sodium and potassium "sand" [Reference 46] – a metallized "flash oversuit" should be worn. The flash-fire temperatures of metal dusts are much higher than those of most organic dusts. The adiabatic flame temperatures of metal dusts is limited only by the boiling points of the oxides: about 3,500°C for aluminum oxide and about 3,600°C for magnesium oxide [Reference 47]. This is in contrast with the adiabatic flame temperatures for organic dusts – about 2,000°C [Reference 48] – as limited by the decomposition of carbon dioxide and water vapor [References 49 and 50].

The preceding discussion concerning personal protective equipment would also apply to persons who are – or could be – exposed to dense clouds of combustible dust, either routinely or during "transient" activities. The former exposure would apply to slitting and dumping bags of small-particle powder into mixers, or for recovery or re-work of product. The latter exposure could involve cleaning of ductwork or emptying dust collectors. Work within dust collectors will also present additional hazards associated with confined space entry.

Cleaning of equipment and surfaces with compressed air should be prohibited, since such dispersion of dusts in air tends to "classify" the dust [References 51 and 52]. The larger particles fall rather rapidly to the floor, but the smaller particles can be lofted by air currents onto elevated surfaces, where they could become fuel for secondary explosions [Reference 17]. The settling velocity for 75-micron [dust] particles is about 0.3 meter per second [Reference 53], and this corresponds to an air velocity of about 0.7 MPH. Thus, a very small upward air current – such as a thermal updraft – could keep such particles suspended indefinitely.

# 9 Path Forward

There are some aspects of housekeeping that should be resolved in the NFPA standards:

1. Provide better guidance concerning the "Entrainment Fraction" that is used in the "Mass Method B" equations for maximum dust accumulations, and

2. Provide guidance for metal dusts concerning the flash-fire and explosion hazards in NFPA 484, similar to that which is given in NFPA 654 for "organic" dusts, and

3. Tests should be conducted to determine the "Entrainment Fraction" for metal dusts, which typically have bulk and particle densities that are significantly greater than those of "organic" dusts.

## 10 Conclusions

The publications of the National Fire Protection Association provide good guidance concerning limitations of combustible-dust accumulations, to avoid flash-fire and explosion hazards. A chemical plant – or other site where combustible dusts are handled – could use the results of the calculations based on the NFPA guidance to establish housekeeping policies and procedures.

Additional guidance is needed from the NFPA concerning the maximum "allowable" accumulations of metal dusts, to minimize the flash-fire and explosion hazards, like the "Mass Method B" procedures for "organic" dusts.

The NFPA publications also show the close relationship between combustible-dust hazard and electrical classification, to minimize the likelihood of electrical arcs, sparks, and high temperatures in electrically-powered equipment from igniting combustible dusts. With sufficiently-frequent cleaning, the powder-handling and dust-handling areas could be considered as "unclassified", and "general-purpose" electrical equipment could be used in these clean areas. Otherwise, the use of electrical equipment should be restricted to those types that meet the electrical classification of the area. Other types of ignition sources also should be isolated from dense clouds and layers of combustible dusts.

It is important that persons who respond to combustible-dust emergencies and persons who could be exposed to dense clouds of combustible dust be protected against flash-fire hazards. Good containment of dusts within equipment minimizes these hazards, and activities that could lead to formation of dust clouds or to the accumulation of dusts on elevated surfaces should be strictly limited.

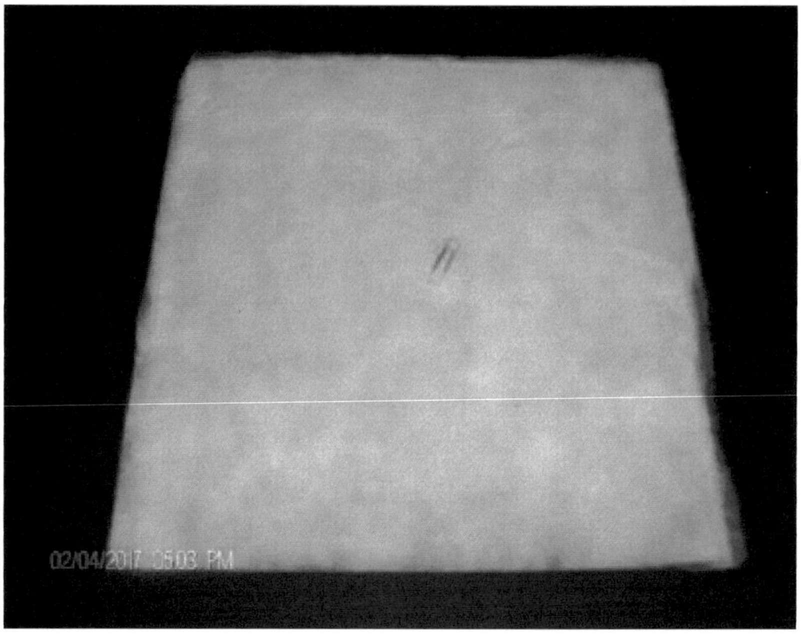


**Figure 1 30 grams of Sodium Bicarbonate Dust on One Square Foot**

**Figure 2 3 grams of Sodium Bicarbonate Dust on One Square Foot**

## 11 References


1. W. L. Frank, S. A. Rodgers, and G. R. Colonna, *NFPA Guide to Combustible Dusts*, NFPA, Quincy MA, Pages 4, 52, 130, 132, 136, 150, 179, and 234 to 246 (2012).

2. W. L. Frank, S. A. Rodgers, and G. R. Colonna, *NFPA Guide to Combustible Dusts*, NFPA, Quincy MA, Pages 5, 15, 20, 144, 234, 242, and 311 (2012).

3. V. Babrauskas, *Ignition Handbook*, Fire Science Publishers, Issaquah WA, Page 144 (1992).

4. H. L. Febo, Combustible Dust Hazard Recognition – An Insurer's View, Process Safety Progress, 30 (1), Page 84 (March 2011).

5. T. J Myers and A. F. Ibarreta, Tutorial on Combustible Dust, Process Safety Progress, 32 (3), Page 305 (September 2013).

6. W. L. Frank G. R. Colonna, On-Going Developments in Addressing Combustible Dust Hazards, Process Safety Progress, 34 (1) Page 29 (March 2015).

7. E. A. Ural, Dust Entrainability and Its Effect on Explosion Propagation in Elongated Structures, Process Operations Progress, 11 (3), Page 176 (July 1992).

8. E. A. Ural, Towards Estimating Entrainment Fraction for Dust Layers, Fire Protection Research Foundation, NFPA, Quincy MA (June 2011).

9. R. K. Eckhoff, *Dust Explosions in the Process Industries*, 2nd Ed., Butterworth Heinemann, Woburn MA, Pages 226 to 234, 246, 490, 493 (Figure 7.10), and 572 (1999).

10. National Fire Protection Association, *Standard for the Prevention of Fire and Dust Explosions from the Manufacturing, Processing, and Handling of Combustible Particulate Solids*, NFPA 654, NFPA, Quincy MA (1994).

11. National Fire Protection Association, *Standard for the Prevention of Fire and Dust Explosions from the Manufacturing, Processing, and Handling of Combustible Particulate Solids*, NFPA 654, NFPA, Quincy MA, Paragraphs 2-2.3.1, 4-2.1, and A-4-2-1 (1997).

12. National Fire Protection Association, *Standard for the Prevention of Fire and Dust Explosions from the Manufacturing, Processing, and Handling of Combustible Particulate Solids*, NFPA 654, NFPA, Quincy MA, Paragraphs 4-2.2 and A-4-2.1 (1997).

13. National Fire Protection Association, *Standard for the Prevention of Fire and Dust Explosions from the Manufacturing, Processing, and Handling of Combustible Particulate Solids*, NFPA 654, NFPA, Quincy MA, Paragraphs 2.2.3.1 and A.2.2.3.1 (2000).

14. Tapco, Inc., Bulk Material Density Table, Available at http://www.tapcoinc.com/content/product_data/tapco_catalog_09_p88-94.pdf. Accessed on January 12, 2017.

15. U. S. Dept. of Labor, Occupational Safety and Health Administration, OSHA Instruction – Combustible Dust National Emphasis Program (Reissued), PL 03-00-008, GPO, Washington DC, Paragraph IX.E.3.c (Oct. 1, 2015).

16. U. S. Dept. of Labor, Occupational Safety and Health Administration, OSHA Memo – Evaluating Hazardous Levels of Accumulation Depth for Combustible Dusts, GPO, Washington DC, (June 15, 2015).

17. FM Global, Property Loss Prevention Data Sheet 7-76 – Prevention and Mitigation of Combustible Dust Explosion and Fire, FM Global, Johnston RI, Paragraph 3.1.1 (2014).

18. S. A. Rodgers, Application of the NFPA 654 Dust Layer Thickness Criteria – Recognizing the Hazard, Process Safety Progress, 31 (1), Page 24 (March 2012).

19. National Fire Protection Association, *Standard for the Prevention of Fire and Dust Explosions from the Manufacturing, Processing, and Handling of Combustible Particulate Solids*, NFPA 654, NFPA, Quincy MA, Page 654-1 (2013).

20. National Fire Protection Association, *Standard for the Prevention of Fire and Dust Explosions from the Manufacturing, Processing, and Handling of Combustible Particulate Solids*, NFPA 654, NFPA, Quincy MA, Paragraphs 6.1.1.3, 6.1.1.4, 6.1.4, 6.1.5, and 6.1.6 (2017).

21. W. L. Frank, S. A. Rodgers, and G. R. Colonna, *NFPA Guide to Combustible Dusts*, NFPA, Quincy MA, Page 66 (2012).

22. R. K. Eckhoff, *Dust Explosions in the Process Industries*, 2nd Ed., Butterworth Heinemann, Woburn MA, Page 40 (1999).

23. W. L. Frank, S. A. Rodgers, and G. R. Colonna, *NFPA Guide to Combustible Dusts*, NFPA, Quincy MA, Pages 66 to 70 (2012).

24. National Fire Protection Association, "Standard on Explosion Protection by Deflagration Venting", NFPA 68, Quincy MA, Paragraphs 7.2.7 and 8.3.2 (2013).

25. E. A. Avallone, T. Baumeister, and A. M. Sadegh, *Marks' Standard Handbook for Mechanical Engineers*, 11th Ed., McGraw-Hill, New York NY, Page 12-20 (2007).

26. Center for Chemical Process Safety, *Guidelines for Evaluating Process Plant Buildings for External Explosions, Fires, and Toxic Releases*, 2nd Ed., AIChE, New York NY, Figure 4.1, Page 48 (2012).

27. S. Mannan [editor], *Lees' Loss Prevention in the Process Industries*, 3rd Ed., Elsevier, Boston MA, Page 17/251 (2005).

28. D. O. Dusenberry [editor], *Handbook for Blast-Resistant Design of Buildings*, John Wiley & Sons, Edison NJ, Pages 285 and 430 (2010).

29. D. A. Crowl and J. F. Louvar, *Chemical Process Safety*, 3rd Ed., Prentice Hall, Upper Saddle River NJ, Page 289 (2011).

30. Safe Havens Enterprises, Inc., Blast-Resistant Modular Buildings. Available at http://SafeHavenEnterprises.com. Accessed on January 12, 2017.

31. C. Naito, Blast-Resistant Design Considerations, Ascent Publishers, Allahabad India, Page 31 (Winter 2011).

32. W. L. Frank, S. A. Rodgers, and G. R. Colonna, *NFPA Guide to Combustible Dusts*, NFPA, Quincy MA, Table 5.13 on Page 241 (2012).

33. E. A. Ural, Towards Estimating Entrainment Fraction For Dust Layers – Final Report, Fire Protection Research Foundation, NFPA, Quincy MA, Pages 34 to 42 (June 2011).

34. National Fire Protection Association, *Standard for the Prevention of Fire and Dust Explosions from the Manufacturing, Processing, and Handling of Combustible Particulate Solids*, NFPA 654, NFPA, Quincy MA, Paragraphs Tables A.8.2.1.3 (a) and (b) (2017).

35. National Fire Protection Association, *Standard on the Fundamentals of Combustible Dust*, NFPA 652, NFPA, Quincy MA, Paragraph 8.4 and Table A.8.4.6.3 (2016).

36. National Fire Protection Association, *Standard for Combustible Metals*, NFPA 484, NFPA, Quincy MA, Paragraph 5.4, Chapter 7, and Paragraph 8.8 (2015).

37. National Fire Protection Association, *Standard for the Prevention of Fire and Dust Explosions from the Manufacturing, Processing, and Handling of Combustible Particulate Solids*, NFPA 654, NFPA, Quincy MA, Table A.6.5.2 (2017).

38. National Fire Protection Association, *Recommended Practice for the Classification of Combustible Dusts and of Hazardous (Classified) Locations for Electrical Installations in Chemical Process Areas*, NFPA 499, NFPA, Quincy MA, Paragraph 6.3.2 and Table A.6.7 (2013).

39. S. Mannan [editor], *Lees' Loss Prevention in the Process Industries*, 3rd Ed., Elsevier, Boston MA, Pages 17/260 and 17/264 (2005).

40. V. Babrauskas, *Ignition Handbook*, Fire Science Publishers, Issaquah WA, Page 736 (1992).

41. British Standards Institute, *Methodology for the Risk Assessment of Non-Electrical Equipment and Components for Intended Use in Potentially Explosive Atmospheres,* BS EN 15198 (2007).

42. E. A. Ural, Towards Estimating Entrainment Fraction For Dust Layers – Final Report, Fire Protection Research Foundation, NFPA, Quincy MA, Pages 34 to 42 (June 2011).

43. V. Babrauskas, *Ignition Handbook*, Fire Science Publishers, Issaquah WA, Page 560 (1992).

44. L.G. Britton, *Avoiding Static Ignition Hazards in Chemical Operations*, AIChE/CCPS, New York NY, Pages 72 and 78 to 81 (1999).

45. S. Mannan [editor], *Lees' Loss Prevention in the Process Industries*, 3rd Ed., Elsevier, Boston MA, Page 16/251 (2005).

46. P. G. Urben [editor], *Bretherick's Handbook of Reactive Chemical Hazards*, 6th Ed., Butterworth Heinemann, Woburn MA, Pages 20, 1492, 1723, 1749, 1758, 1815, 1913, and 1924 (1999).

47. V. Babrauskas, *Ignition Handbook*, Fire Science Publishers, Issaquah WA, Page 356 (1992).\

48. V. Babrauskas, *Ignition Handbook*, Fire Science Publishers, Issaquah WA, Pages 1024 to 1053 (1992).

49. J. H. Perry, *Chemical Engineers' Handbook*, 4th Ed., McGraw-Hill, New York NY, Page 9-40 (1963).

50. E. A. Avallone, T. Baumeister, and A. M. Sadegh, *Marks' Standard Handbook for Mechanical Engineers*, 11th Ed., McGraw-Hill, New York NY, Page 4-30 (2007).

51. National Fire Protection Association, *Standard for Combustible Metals*, NFPA 484, NFPA, Quincy MA, Paragraph 7.6.1 (2015).

52. National Fire Protection Association, *Standard on the Fundamentals of Combustible Dust*, NFPA 652, NFPA, Quincy MA, Paragraph A.8.4.2.6.1 (2016).

53. R. K. Eckhoff, *Dust Explosions in the Process Industries*, 2nd Ed., Butterworth Heinemann, Woburn MA, Page 218 (1999).

Note: The opinions that are expressed in this presentation are those of the author and may not represent the opinions, or risk-avoidance criteria or guidance, of Chilworth Technology, Inc. or Dekra Insight.

# Implementing an Effective Management of Change (MOC) System for E&P Business and Non-PSM Facilities

**Sandipan Laskar**
**Risk Engineering Leader**
**Occidental Oil and Gas Corporation**
**5 Greenway Plaza**
**Houston, TX 77046**
**Sandipan_Laskar@Oxy.com**

Prepared for Presentation at
American Institute of Chemical Engineers
2017 Spring Meeting and 13th Global Congress on Process Safety
San Antonio, TX
March 26 – 29, 2017

# Implementing an Effective Management of Change (MOC) System for E&P Business and Non-PSM Facilities

**Sandipan Laskar**
**Risk Engineering Leader**
**Occidental Oil and Gas Corporation**
**5 Greenway Plaza**
**Houston, TX 77046**
**Sandipan_Laskar@Oxy.com**

**Keywords:** Process Safety Management (PSM), Process Safety Information (PSI), Management of Change (MOC), Safety Management Systems.

## Abstract

The upstream oil and gas industry involves processing high volume of live data during different operations (such as drilling, production, well servicing etc.) and thereby generating high volume of process or procedure changes. It is very important to manage these changes as based on the past incidents in the industry it can be noted that accidents often occurs after changes. Any change that has safety implications should be carefully evaluated, including performing a hazard analysis, before it is allowed. One of the practicable approach in managing these changes is by implementing a Management of Change (MOC) system.

Most companies have policies for management of change, but the implementation and enforcement of these policies can vary greatly. An effective MOC work process includes a review and authorization process for evaluating proposed adjustments to facility design, operations, organization or activities prior to implementation to make certain that no unforeseen new hazards are introduced and that the risk of the existing hazards to employee, the public, or the environment is not unknowingly increased. It also includes steps to help ensure that potentially affected personnel are notified of the change and that pertinent document, such as procedures, process safety knowledge, and so forth, are kept up-to-date.

The main aim of this paper is to address the three basic questions (what? why? and how?) in implementing an effective MOC system. There are more robust MOC program available for the facilities covered under OSHA Process Safety Management program. However the management of change (MOC) work process mentioned in this paper is fit for purpose for upstream oil and gas operation. This paper discusses challenges in implementing MOC system for facilities not covered under OSHA Process Safety Management program and addresses common misconceptions of Management of Change system.

## Introduction

Most oil and gas production, refining and petrochemical organizations have some kind safety management program that stems from the requirements outlined under OSHA's CFR 1910.119. Some have a very extensive program that covers most of the 14 elements of Process Safety Management (PSM) and is managed via an integrated platform while others follow an ad hoc approach that does not span all of the elements or is disjointed.

The OSHA PSM guidelines are enforceable throughout the United States and specific US holdings for facilities having a threshold volume of hazardous chemicals, but are not applicable where another federal agency has occupational safety and health jurisdiction, including in federal waters controlled by Bureau of Ocean Management Regulation and Enforcement (BOEMRE, formerly Minerals Management Service (MMS)). Oil and gas drilling and servicing operations and normally unmanned remote facilities are specifically excluded from the standard, but typical facilities in the upstream industry such as production facilities, gas oil separation plants, and gas plants are covered if the threshold chemical volumes, typically hydrocarbons, are present.

For offshore facilities, a more holistic approach applying to both production and drilling and without a threshold volume was published originally in 1998 by the API, with the most recent update in 2004 (API 2004). This recommended practice 75 (API RP 75) establishes methods to develop a Safety and Environmental Management Program (SEMP) covering 12 target elements, which addresses all the major aspects of an OSHA PSM program. Although the MMS and industry groups ran a number of audit surveys during the initial implementation period of API RP 75 (BOEMRE 2010), recent compliance status data are not publicly available.

The MMS published an Advance Notice of Proposed Rulemaking (ANPR) in the May 2006 US Federal Register related to the creation of an enforceable PSM/SEMP, and provided a report in 2009 on the comments and data gathered during the voluntary PSM/SEMP period (MMS 2009). Data analysis was performed on a number of MMS incident reports, noncompliance notices, and spill reports, showing a preponderance of incidents resulting from failures in one or more of four main program areas.

Historically the MMS has not been regarded as a strong enforcement agency, though with the reorganization announced by the Interior Secretary in May 2010 (Straub 2010) splitting the field development, enforcement, and revenue functions of the former MMS body, in combination with the promulgation of an enforceable PSM/SEMP and the recent major catastrophe in the Gulf of Mexico, one might have expected a more stringent regulatory environment with greater emphasis on design review and prevention as well as increased penalties and fines should incidents occur. Following a flurry of activity in the second half of 2010, the SEMP as described in API RP 75 became mandatory 15 November 2010 (BOEMRE 2010).

## Management of Change – What?

Management of Change is defined as a formal process that systematically manages Changes to the organization, operations, process equipment, technology, procedures, chemicals, or for any temporary modifications that could impact the original design, operating & maintenance criteria, and/or environmental performance. The MOC process is intended to control the potential impact of any proposed Changes (other than for replacement-in-kind) that might impact personal safety, legal & regulatory compliance, safeguard the environment and/or the safety or emergency response of the facility or equipment. This shall also include personnel or organizational Changes which might impact the safety or emergency response provided to the Process. So in an essence management of changes, or MOC is a best practice used to ensure that safety, Health and environmental risks are controlled when a company makes changes in their facilities, documentation, personnel or operations.

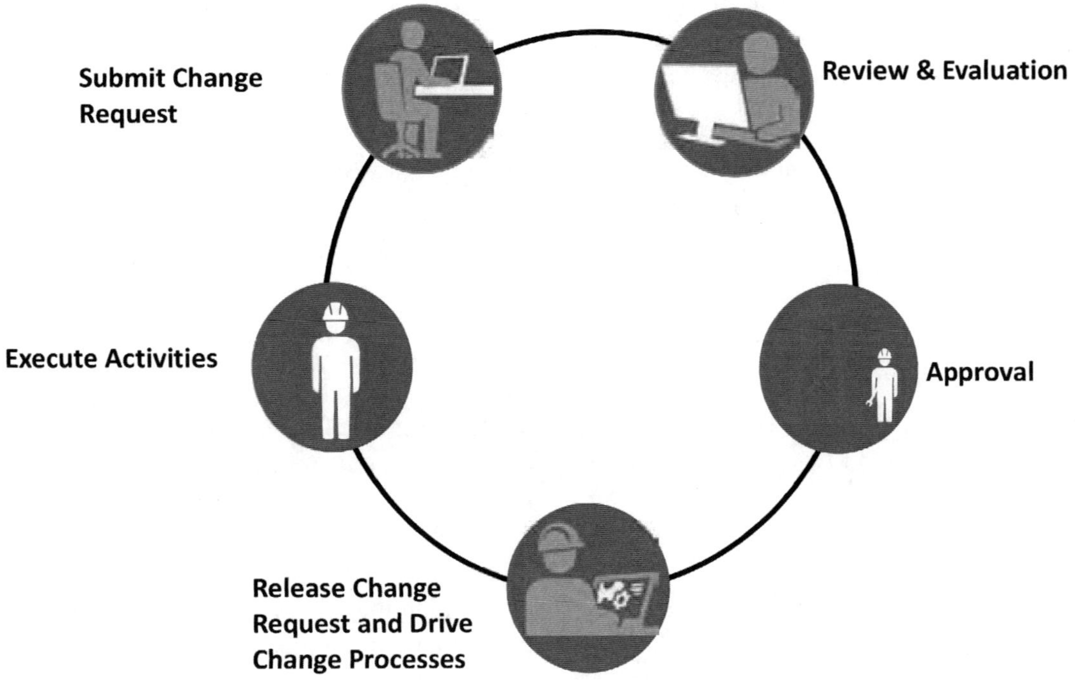


**Figure 1- An Overview of Management of Change**

The above figure illustrates a simple overview of the management of change work process and each of those activities could be further broken down into more detailed workprocess having multiple layers of activities attached to them.

The MOC process must address changes that range from minor changes to major capital projects as well as handling temporary and permanent changes. This can be accomplished in many ways. Some companies have developed "one-size-fits-all" MOC processes that are used for every change. Other companies find it easier to have multiple MOC systems. The multiple systems can be split based on the type of change (equipment vs. procedural), the duration (temporary vs. permanent) or the size of the change (minor vs. major). The decision to use an electronic MOC process or a paper process should also be considered. The important thing to remember is that the MOC process or processes should be flexible, easy to use and fit your facilities culture.

A typical management of change process should address following changes:

- Facility Changes: The scope of the facility changes include any modifications made to the equipment within a facility. For a piping system, this may include replacing a carbon steel pipe replacing 2-in. pipe with 3-in. pipe, replacing 150-lb flanges with 300-lb flanges, adding tracing or insulation, substantially changing the pipe routing, tying the piping into another line, or anything that changes the pipe class on a P&ID. Other facility changes include replacing a centrifugal pump with a diaphragm pump, re-tubing a heat exchanger with a different gauge tubing or material of construction etc.

  Facility changes also includes changes related to the control system. Some example could be changes to the programming or control logic, changes to who has access to the logic, resetting interlock timers, bypassing any circuit element or protective device, or changes to an alarm point that takes it outside of the safety or quality limits specified in the operating procedure.

- Procedural Changes: This category includes any changes to previously established safety, quality, or operating limits in the operating procedure.

- Organizational Changes: The Organizational MOC is intended to ensure that the site operating organization can provide safe operation during both normal and emergency situations. Examples of an Organization MOC include: changes in the number of operating personnel at the site, changes in hours worked (e.g. 5-day maintenance coverage to 4 day maintenance coverage), changes in delegation of authority where safety is impacted, changes in organizational structure (e.g. Elimination of a lead operator or HES position), changes from company to contract operators and/or maintenance personnel. It should be noted that the above examples are not exhaustive list and other Organizational MOC situations may exist.

Changes may be permanent or temporary. If a temporary change is made, it must be tracked and properly reversed. Companies should also have a procedure to handle emergency changes, which are those that must be made quickly during off-hours *(e.g.,* nights. weekends, and holidays) when the regular MOC approvers may not be available. For example, approval could be obtained verbally rather than in writing (as long as it is properly documented) or approval could be obtained from authorized on-call personnel rather than the regular approvers.

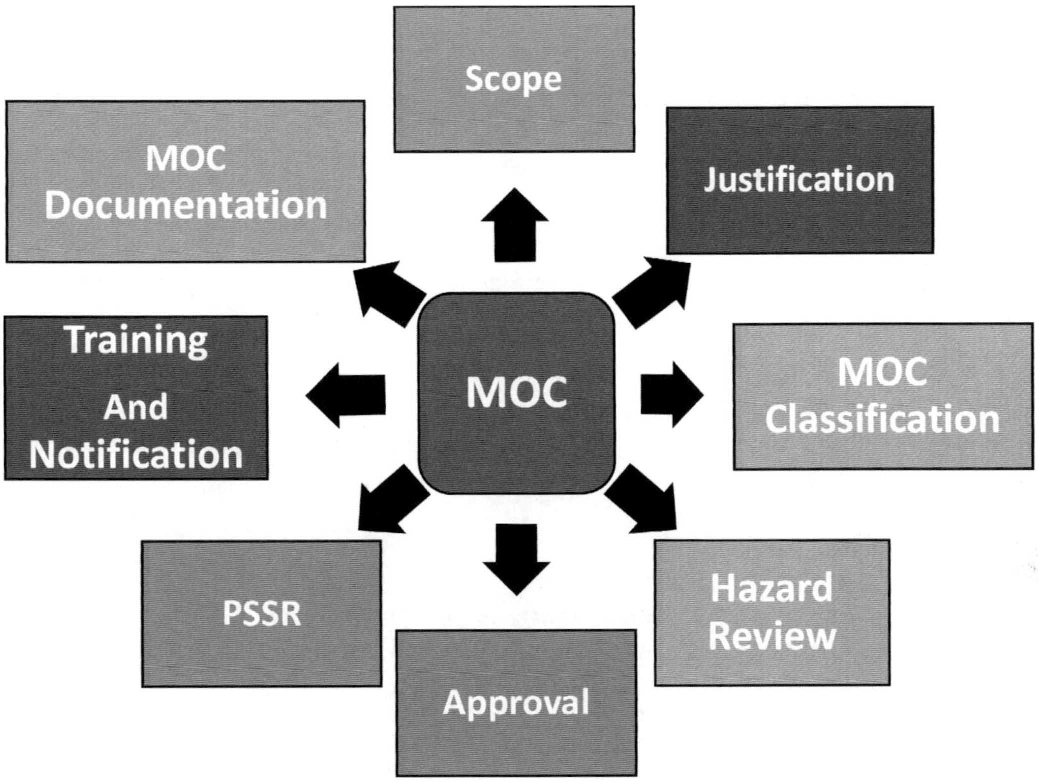


**Figure 2- Fundamental Building Blocks of Management of Change**

The fundamental building blocks of a Management of Change process are shown in the figure above. All the fundamental building blocks will be clearly explained in the management of change implementation section.

## Management of Change - Exception

A management of change is changes related to Facilities, Procedures or Organization except for Replacement in Kind or identical changes. A Replacement in Kind (RIK) change can be an item (equipment, chemical, procedure, etc.) that meets the design specification of the item it is replacing. This can be an identical replacement or any other design alternative specifically provided for in the design specification, as long as the alternative does not in any way adversely affect the use of the item or associated items.

## Management of Change – Why?

Some of the benefits of an effective Management of Change process are:

- It minimizes unplanned adverse impacts on system integrity, security, stability, and reliability for the business process being altered or added.
- It maximizes the productivity and efficiency of staff planning, coordinating, and implementing the changes.
- It provides a stable production environment.
- It ensures the proper level of technical completeness, accuracy of modifications, and testing of systems before implementation.
- It provides an appropriate level of management approval and involvement.

The safety program of a facility typically consists of process hazard analyses (PHAs), a mechanical integrity (Ml) program, personnel training, PSSRs, PSI and MOC. MOC plays a central role in this program because:

- It provides updates to PSI
- It identifies when a PSSR is needed
- It ensures that employees are trained to carry out new procedures
- It adds new equipment to mechanical integrity test and inspection schedules

The bottom line is that Management of Change helps prevent incidents.

## Implementing a Management of Change – How?

A management of change is a procedure for safely implementing a change to a facility and/or procedure by identifying the impact of the change. The figure below illustrates a simple management of change work flow which includes all the different building blocks for an effective and robust management of change program. A more detailed management of change work process with key players are shown in figure 4. The main objective of this simple work flow is to enunciate flow of information between different building blocks of the management of change program.

An effective MOC process consists of following basic steps:

1. Identify the need for change (MOC) - Initiation
2. Review and approval – Initial Review
3. Evaluate the Risk (PHA)
4. Review and Approval Process – Final Review
5. Construct/pre-commissioning
6. Effective Training and Communication
7. Pre-Startup Safety Review (PSSR)
8. MOC Closure

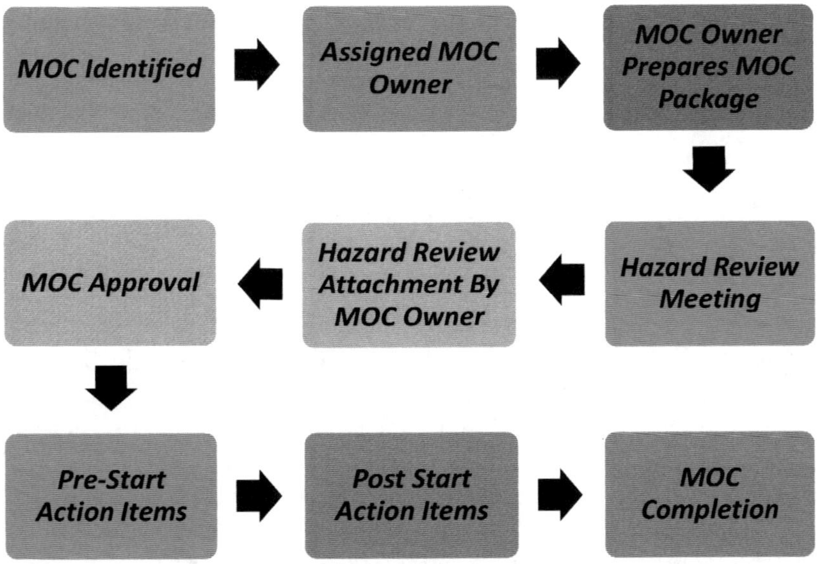


**Figure 3- A Simple Management of Change Workflow**

_Identify the need for change_: In the initiation step, any employee should be able to propose a change to a procedure, equipment or technology of the process/facility. The MOC forms should allow the initiator to document the what, why and when of the proposed change. The initiator should request the change in writing. This should include a description of the change, the technical basis for the change, the potential impact on safety and required PSI updates. The initiator then forwards the information to a member of the facility management team for review and approval.

_Review and Approval – Initial Review:_ The preliminary review consists of a "go/no go" decision based on the information supplied by the initiator. This may include discussions between the initiator and the reviewer to clarify the details of the proposed change and alignment with the facility's needs. If the decision is made to proceed with the change, the preliminary reviewer should, as much as s/he is able, document the technical basis for the change, impact on safety and health and the timeframe to accomplish the change. If the change is temporary, the preliminary reviewer should also ensure that the date that the temporary change will be reverted to normal is included in the MOC documentation.

_Evaluate the Risk (PHA)_: The risk assessment/technical review step is one of the most important activities within the MOC process. It is critical to involve the facility engineer, technical authority or someone with a deep understanding of the process in this step. The risk assessment should match the complexity of the change. For minor changes, it may be okay to ask the question "What if?" More complex changes require tools like process hazard analysis or failure mode-effect analysis. Some companies find it useful to include a risk assessment checklist in their MOC process. The checklist includes questions like "Does this change involve a shutdown or safety system?" and "Does this change increase or decrease the pressure, temperature or flow rates within the process?" The completed checklist can be the risk assessment or it can point the user toward a more sophisticated risk assessment tool or additional review by safety, environmental or technical experts. Regardless of the risk assessment method, it is important to document what could reasonably go wrong and the steps required to prevent it within the MOC prior to obtaining approval to design/construct.

*Review and Approval – Final Review:* The approval to design/construct step is typically granted by the facility superintendent or manager. This person is committing resources toward completing the change and agreeing that the change can safely be made based on the information provided. The final review is based on the availability of all design documentation along with a process hazard analysis to ensure proper safeguards.

*Construct/ Pre-Commissioning:* Most of the "real work" is accomplished in the design/construct/precommission step. In this step, final designs are completed, piping and equipment constructed, procedures updated, redline drawings created and equipment installation verified. If the design changes during this step, another round of risk assessment and approval is critical to ensure that risks associated with the new design are properly managed

*Training and Communication:* This step is part of the pre-startup safely review (PSSR) and involves critical activities that should be completed prior to putting the change into service. Notification involves informing all affected persons and departments that the change is about to occur. Training by a knowledgeable instructor is conducted (as needed) for all personnel who will be operating the process after the change has been made.

*Pre-Startup Safety Review:* Field verification is an inspection to ensure that what was done in the field matches the specifications in the engineering design package. The PSSR step consists of the set of verifications and assurances that each portion of the work associated with the change has been properly completed. The PSSR items are typically assigned to multiple people, including facility engineers, operations leaders and maintenance leaders. A PSSR must include the following items:

- equipment meets the design specifications and has been properly installed;
- adequate safety, operating, maintenance and emergency procedures have been developed or existing procedures modified accordingly;
- process hazard analyses have been completed for new facilities and recommendations have been resolved or implemented before startup;
- employees involved in operating the PSM-covered process have been trained on or notified of the change as appropriate;
- safe work practices have been revised accordingly;
- process safety information used by employees (e.g., piping and instrumentation diagrams, equipment spec sheets) has been updated accordingly;
- Items that will be needed in the future have been assigned for completion. This typically includes items like coordinating spare parts inventory with purchasing.

*MOC Closure:* Once all the post startup action items are completed then the MOC is closed. It is of utmost importance to ensure that all the process safety information is updated before the MOC is closed.

## Conclusion

A simple, yet robust, MOC process can help ensure that risks associated with change are understood and properly managed, employees are trained on how the change affects them, accurate procedures are in place for maintenance and operations, equipment files are up to date and spare parts are available when they are needed. When these elements are in place, a facility can meet their goal of safe, reliable operations. Thus is could be concluded that a well-designed MOC process is an essential loss prevention tool for any business. It is not just for certain hazardous industries. The process applies to any company that wishes to avoid future losses resulting from today's changes. MOC does not have to be overwhelming or so difficult to use that it inhibits change. It cannot effortlessly compensate for past omissions, only reduce your future risk. So the time to start developing and implementing an effective MOC process is today.

## Disclaimer

The information contained in this document represents the current view of the author at the time of publication. Process safety management is complex and this document cannot embody all possible scenarios or solutions related to compliance. This document is for informational purposes only. Oxy makes no warranties, express or implied, in this paper.

## References


[1] "API RP 75, Recommended Practice for Development of a Safety and Environmental Management Program for Offshore Operations and Facilities", third edition. 2004. Washington, DC: API

[2] "Oil and Gas and Sulphur Operations in the Outer Continental Shelf--Safety and Environmental Management Systems", Bureau of Ocean Energy Management, Regulation and Enforcement Implementation (BOEMRE), *Final Rule*, 30 CFR Part 250, Volume 74, Number 199, (October 15, 2010)

[3] "Petroleum Refinery Process Safety Management National Emphasis Program", U.S. Occupational Safety and Health Administration (US OSHA), Directive CPL 03-00-004, (June 7, 2007)

[4] "PSM Covered Chemical Facilities National Emphasis Program", U.S. Occupational Safety and Health Administration (US OSHA), Directive CPL 09-06-002, (July 27, 2009)

[5] "Survey for API RP 75", Bureau of Ocean Energy Management, Regulation and Enforcement Implementation (BOEMRE), (September. 17, 2010)

[6] Occupational Safety and Health Administration. *29 CFR 1910.119, Process Safety Management of Highly Hazardous Chemicals*. 1992.

[7] "Precise Process Safety Information – A Key for Successful Process Safety Management Program", S. Laskar, *Global Process Safety Conference*, (April. 29, 2013)

# Improving Human Factors Review in PHA and LOPA

**Dave Grattan, P.E., CFSE**
**aeSolutions**
**10375 Richmond Ave. Suite 800**
**Houston, TX 77042**
**dave.grattan@aesolns.com**



Prepared for Presentation at
American Institute of Chemical Engineers
2017 Spring Meeting and 13th Global Congress on Process Safety
San Antonio, TX
March 26 – 29, 2017

# Improving Human Factors Review in PHA and LOPA

**Dave Grattan, P.E., CFSE**
**aeSolutions**
**10375 Richmond Ave. Suite 800**
**Houston, TX 77042**
**dave.grattan@aesolns.com**

**Keywords:** Human Factors, Human Error, PHA, LOPA, Facilitator, Task Analysis, Bias, Cognitive Psychology

## Abstract

Human Reliability practitioners utilize a variety of tools in their work that could improve the facilitation of PHA-LOPA related to identifying and evaluating scenarios with a significant human factors component. These tools are derived from human factors engineering and cognitive psychology and include, (1) task analysis, (2) procedures and checklists, (3) human error rates, (4) systematic bias, and (5) Barrier effectiveness using Bow-tie.

Human error is not random, although the absent minded slips we all experience seem to come out of nowhere. Instead, human error is often predictable based on situations created external or internal to the mind. Human error is part of the human condition (part of being a human) and as such cannot be eliminated completely.

For example, a task performed at high frequency (e.g., daily or weekly) develops a highly-skilled operator with an expectation of a low error probability for that task. However, as the operator' skill increases, their reliance on procedures decreases, leaving them open to memory lapses caused by internal or external distractions. The fact that a skilled operator becomes less dependent on procedures is not a conscious decision. It is part of the human condition. Forcing a skilled operator to read the procedure while performing the task they are skilled at, is like asking you to think about what your feet are doing as you walk down a flight of stairs. In both cases a loss of adroitness will occur.

A large portion of this paper will be to describe with practical examples the five tools mentioned above. Task analysis is a talk-through and walk-through exercise of a task (typically focusing on one or two critical steps of a procedure) that is used to identify error likely situations (ELS). Quantitative human error rates can be attached to the ELS depending on if the error is associated with skill, rule, or knowledge (SRK) based performance. Systematic biases produced by Type 1 (fast) thinking cause judgment and diagnosis errors related to response to abnormal situations.

Having a working knowledge of these five tools will improve a PHA-LOPA facilitator's awareness and ability to better evaluate human error related scenarios and Barrier failure. In addition the facilitator will feel confident about recommending the need for a more detailed follow-up study such as an HRA (Human Reliability Analysis).

# 1 Introduction

This paper is written for facilitators of PHA and LOPA and their managers. The message is simple: Improve human factors review in PHA and LOPA by applying the already established tools of human factors and human reliability. How can we improve something we do not practice ourselves as facilitators? There is a caveat however. Much of the heavy lifting has to be done outside of PHA and LOPA studies. The argument for this is presented below in **Section 2.2**.

The LOPA method itself presents challenges and opportunities for using human factors and human reliability. Any LOPA scenario handling human error as initiating cause, human response, or administrative control barriers is a candidate for a safety critical task analysis. There is ample low-hanging fruit with regard to human factor based scenarios already coming out of PHA and LOPA. One of the tenets of this paper is that any credit given for human response to a critical alarm or other human barrier, or any human error initiating a catastrophic scenario, should undergo a safety critical task analysis, regardless if the gap is closed using generic credits (see **Section 2.4.5** for why).

The following tasks have been successfully treated with the qualitative and quantitative methods and concepts discussed in this paper. It is important to note that each of these tasks were well-defined, bounded, primarily skill and procedurally based tasks involving one or more personnel.

- Specific task performed during normal shutdown
- Specific task performed during start-up
- Safing for maintenance work
- Safing for N2 back-up in Instrument Air
- Response to critical alarm (including diagnosis)
- Restoration of critical equipment following ITPM (inspection, test, preventive maintenance)
- Catalyst drum change-out
- Rupture disk Barrier effectiveness
- Tanker Truck Unloading

Organizational factors are recognized as important human factors. Organizational factors such as social and institutional influences are discussed in this paper qualitatively. The quantitative treatment of organizational factors as they pertain to predicting incidents is complex (some would argue currently intractable) with large uncertainty. This paper is primarily written addressing skill or rule based tasks performed by one or more operators.

# 2 Why Human Factors and Human Reliability?

*"Most industrial accidents are caused by human error: estimates range between 75 and 95 percent. How is it that so many people are so incompetent? Answer: They aren't. It's a design problem. If the number of accidents blamed upon human error were 1 to 5 percent, I might believe that people were at fault."* Don Norman [1]

*"Saying most accidents are due to human error is a bit like saying falls are due to gravity."* Trevor Kletz [2]

### *2.1 We are Under-predicting Process Safety Incidents*

There is published evidence [3] stating that current methods (HAZOP, LOPA, QRA) are predicting only about one-third of actual process safety incidents. The under-prediction is coming from the frequency (or likelihood) part of the risk equation. Identifying hazards and estimating consequences has been comparatively more robust.

Could this observed gap be coming from Human Factors impact?

### *2.2 The Case for Why PHA and LOPA are Not Enough*

PHA and LOPA, and their derivatives including Procedural HAZOP, generic Human Factors checklists (see **Section 9** for a comparison with a Task Analysis checklist), etc. are targeted to evaluating states and conditions (e.g., is the alarm configured as priority, or, what-if this step in the procedure is omitted?) as opposed to behavior (e.g., how will the console operator interpret the alarm, or, how likely is the operator to skip that step and why?). States and conditions are important to validate, however behavior is just as important. The existing tools available to evaluate human (and organizational) behavior are task analysis and human reliability analysis. PHA and LOPA methodology cannot get you there.

Another reason is related to Barrier effectiveness. Barriers are listed as providing protection in a LOPA study. The hardware reliability of a Barrier can be and is typically estimated (or assumed to be a generic value). So to can the human impact on a Barrier be estimated, except it's typically not. Sometimes it's assumed the generic data value for hardware failure includes systematic (human) failures. See **Section 2.4.5** for why this is a myth. Human and organizational factors are at least as important as random hardware failures on a barrier, and they are not being accounted for. Methods outside of PHA and LOPA are needed.

### *2.3 Human Factors*

#### 2.3.1 What is it

A human factor is anything that affects human performance, good or bad. The goal of human factors is to match the task environment and system interface(s) to the abilities of the operator. If the physical or mental requirements of a task exceed the operator's ability, errors will occur.

### 2.3.2 Affordances, Signifiers, Forcing Functions, and Mapping

When performing a task analysis the facilitator must be able to spot deficiencies in the task environment. This can only come with practice. The following tools are used by human factors specialists [1]:

**Affordances** are used to afford a location to perform work. A simple example is a valve handle. It provides a place for the operator to place their hands. Affordances can be used improperly, for example, an operator may have to stand on said valve handle (or the piping) to reach a manual valve above him.

**Signifiers** tell us how to interact with the affordance. A round valve handle turns. A levered valve handle rotates. The best signifiers are intuitive (don't require conscious thought, i.e., Type 2 thinking – see **6.2**). Signs require effort and conscious thought (to see, read, and comprehend), and are therefore not the best use for signifiers (this is not to suggest getting rid of signs, but to understand the limitation). Populational stereotypes are learned signifiers, for example, "righty-tighty, lefty-loosey." Designs that violate populational stereotypes are considered error-likely.

**Forcing functions** force desired behavior. For example, car-seals, chain locks, and captive key systems placed on valves are recognized as reducing errors of omission and commission related to valve manipulation.

And, **mapping** is used to spatially connect controls with affordances. For example, procedures that use labels and drawings to map said valve to the actual valve in the field.

These concepts are only useful in the context of the operator performing the task, and so task observation of the operator with the equipment and procedures is necessary to be complete a successful analysis.

### 2.3.3 Human Nature

Slips and lapses (defined in **Section 3.4**) that sometimes occur are part of human nature. This type of human error cannot be trained, punished, rewarded, designed, or otherwise removed out of humans. It is part of being human. Policy, procedures, and good human factors help, but we should still expect the occasional human error. We should design systems to be tolerant of human error or allow for their recovery. If an operator commits an error, we should not look at the operator, but instead look at the task, procedures, environment, equipment interfaces, etc. the operator has to work with, because it is more likely those are the root cause (not the human).

### 2.3.4 Procedures

Different types of procedures are in use, including,

- Routine Operating
- Preventive Maintenance (e.g., for critical instrumentation)
- Response to an abnormal situation (critical alarm based response)

Each type of procedure is amenable to human factors and human reliability analysis via task analysis.

### *2.4 Human Reliability*

Human reliability study is the quantification of human factors. Human reliability puts numbers to (i.e., attempts to quantify) human factors influences on tasks and can be used to help "sharpen the pencil" for error estimates to use in PHA-LOPA or QRA. Human reliability has a long history (see for example reference [4]) and is part of the renaissance of quantitative methods in the process industries [5]. Any number that is produced has an uncertainty associated with it (see **2.4.2** and **Section 8**).

#### 2.4.1 Tractability

How can we predict human error? Humans are unpredictable (i.e., often irrational).

There are several ways to answer this question. The easiest and most relevant is the fact that there is so much low-hanging fruit associated with improving human (and organizational) factors related to hazard scenarios and Barrier effectiveness that it would take a generation worth of work in this field applied to the Process Industries to overtake the capabilities of the current methods.

The current methods to evaluate human behavior that include the SRK (skill, rule, knowledge) model and many human reliability models (e.g., THERP (Technique for Human Error Rate Prediction) – see ref [12]) have been validated as effective at predicting human error. The issue of human irrationality in the context of plant accidents even has a name: local rationality (what is in the head of the operator at the time they made that catastrophic error) [9].

The catastrophic accidents that post to the CSB (Chemical Safety Board) website at a steady rate, do not involve unpredictable human error.

One of the most incredible human factor case studies is the Browns Ferry fire of 1975. Browns Ferry nuclear plant is located in Alabama. The maintenance team had just installed some new cabling into a control room, and were leak testing the cable seals by looking for air flow in the duct, using the naked flame of a candle. The foam insulation caught fire, and burned through safety and control cables (common duct), and the operations team narrowly avoided a complete meltdown. One might ask how something like this could have been predicted a priori. It would be nearly impossible. But this and other extraneous acts often lie in wait for a number of years before causing an incident. The candle flame method to leak test had been in use for several years and was even recognized as a hazard.

#### 2.4.2 Model Error

As mentioned above, the process industry has seen a "renaissance" in use of quantitative methods including SIL (safety integrity level) calculations, Fault Tree Analysis for likelihood of hazard, Event Trees for distribution of outcomes, and Bow-Ties (combined fault tree and event tree) as a means of communicating hazards and identified barriers to Operations staff, and identifying process safety leading indicators.

In any quantitative model of system performance, the following are typically recognized as sources of error:

- Equipment data (a function of equipment type, service, quality of maintenance, etc.)
- Incorrect model (incorrect logic)
- Human reliability impact
- Wrong assumptions (about plugging for example)
- Unforeseen hazards or ways in which hazards can occur (see **Section 8.1**).

The field of human reliability attempts to account for the uncertainty introduced by the impact of human factors on system performance. This may represent the most significant source of model error, next to bad assumptions.

### 2.4.3 HRA Methodologies

There are a myriad of HRA methodologies, some public, some proprietary, as evidenced by the veritable alphabet soup of acronyms associated with the methodologies.

The Health and Safety Executive (HSE) in the UK considered that it would be useful to be up to date with developments in the field of quantitative HRA methods and to have knowledge of the capability of the tools and an understanding of their strengths and weaknesses, to improve consistency, and determine acceptability of their use [6].

To sort this out, in 2009 the HSE published a review of 72 HRA methods that were "potentially relevant to HSE major hazard directorates." Of the 72 potential HRA tools, 17 were considered useful for major hazard directorates [6].

THERP (Technique for Human Error Rate Prediction) [12] is one of the 17 methods considered to be useful by the HSE.

### 2.4.4 THERP

THERP is considered a first generation human reliability model (in the public domain the models only go up to second generation). ATHEANA (A Technique for Human Error Analysis) is the second generation model developed for the USNRC (US Nuclear Regulatory Commission) that is intended to improve THERP, especially related to cognitive behavioral mechanisms, organization, and safety culture, in the context of response to abnormal events. But ATHEANA would benefit from further development and ultimately more widespread industry acceptance [6].

THERP is a perfectly acceptable methodology, especially for the low-hanging fruit coming out of current LOPA's. Some of the benefits of THERP:

- It's easy for someone with a PHA-LOPA-QRA-facilitator background to use
- It includes its own extensive human error data set
- It incorporates Task Analysis to uncover error mechanisms

- It has a very nice dependency model built-in
- It uses "recovery factors" such that unrecovered human error is the metric of interest
- It flanges up nicely with LOPA, QRA, and Bow-tie

THERP does take practice to get good at, and it does require an eye for human factors, and some study of cognitive behavior.

2.4.5 The Myth that Generic Equipment Failure Data includes Human Error

There is a belief that generic failure data (e.g., generic data tables for initiating events and protection layers) already includes data on human error. (It can be argued that *all* failure is ultimately human induced, but this is not the point I'm making.) This myth, however, is partly true. For example, failure associated with the quality of maintenance certainly contributes to the failure data of generic equipment. However, the following types of human errors are not included in generic equipment failure data [4]:

- Insertion of an incorrect component or device
- Wrong specification
- Miscalibration
- Left in wrong mode (manual, test, inhibit, etc.)
- Failure to restore system following maintenance (block valves)
- Inadvertent bypass during normal operation

But even if the above myth were true, it misses the point entirely. Hardware failure is considered random (statistical). Human error is systematic (i.e., deterministic). The two types of error differ only by the amount of information known about each. Behavioral models and task analysis provide information of why human error occurs or might occur. We can use this information to improve work processes, work situations, policy, competency, staffing, communication, contractor roles, etc. so that we might prevent the error from occurring. If we bury human error as a generic random variable, we miss this opportunity. Claiming that a generic value of human error is 1/10 and that it is conservative, is misleading and dangerous, because of the nature of determinism, the error likelihood (for an identified cause) is either 1 or 0. The same point could be made when discussing the numbers in **Section 5.2**. The way out of this problem is to *always* couple the quantitative analysis with the qualitative Task Analysis.

# 3 Tool 1: Task Analysis

## *3.1 Introduction*

Human factors and human reliability share a technique called 'safety critical task analysis' or 'task analysis' for short. As the name implies, task analysis is a systematic analysis of a task (or barrier) that can help identify problems and reduce the likelihood of human error associated with that task (or barrier). Task analysis looks at all the ways that humans touch a system, how failure can occur, and ways to improve it. Task analysis is used to ferret out potential human errors related to a specific task, and make recommendations for improvement as well as estimate likelihood of occurrence.

Task analysis could look at the critical steps of a procedure (usually one or two steps of a large procedure), responding to a priority alarm, or restoring a critical system following maintenance. Task analysis can also be conducted on the escalation factors and escalation factor controls related to a Barrier. Task analysis is conducted as a team exercise, and includes talk-through and walk-through activities. Typically one task (or barrier) takes one day of team time to perform the task analysis, with additional follow-up reporting and some iteration.

You don't have to go far to find a task to analyze. They are right there in your PHA-LOPA study. Any catastrophic hazard scenario that has a significant human factors component including human barriers is a candidate for task analysis.

Task analysis can be picked up quickly by an experienced PHA-LOPA facilitator. The caveat is the facilitator needs to have an eye for human factors issues and an understanding of cognitive bias (see **Section 6.1**). It is recommended that a facilitator experienced in the methodology be engaged as a mentor.

The Functional Safety Assessment required by ANSI/ISA 84 as an audit of the Safety Instrumented System is really just a task analysis.

See **Section 9** for sample Task Analysis checklists.

### *3.2 Error-Likely Situations*

A primary goal of task analysis is to identify error likely situations. This is partly accomplished by talking to operators that perform the task, reviewing procedures, and actually observing (real or a simulated walk-through) the task being performed. Here is where an eye for human factors is necessary. Much of human factors is common sense. If you talk to an operator who has been on the job for 30 years and he tells you this is the first time someone has ever come out to look at how he works, the fruit will be lying on the ground.

A lot of error likely situations are obvious, for example,

- Reading errors caused by missing valve handles or indicators (valve position), or instrumentation that is not back-lit.
- Omission errors caused by equipment that is not accessible.
- Commission errors caused by equipment layout.
- Non-response errors because alarms are buried in a DCS graphic, or the audible alarm volume is turned down too low.
- Interpretation error (of a signal) caused by failure to trend that signal on a DCS graphic (or use of process switches versus transmitters).
- Interpretation error of the state of equipment because there is not enough information for the Operator.
- Not believing an alarm is real because of so many other spurious alarms.
- Not knowing how to respond to an alarm because there is no training or documentation.
- Reliance on short-term memory because of deficient design.
- Spatial disorientation caused by layout (e.g., stairs) of identical equipment trains.

There are other error likely situations related to deficient procedures, work practices, communications, staffing, task load, competency, etc.

To be successful at identifying error likely situations takes practice. But a basic understanding of the types of human error possible is the way to begin.

### *3.3 Human Error is not a 'Cause'*

Human error does not occur randomly (in a statistical sense). Human error is deterministic (i.e., it can be predicted using an appropriate behavioral model and task analysis). Examples of when human error will occur:

- When a person is rushing, distracted, fatigued, or complacent. (These are the Big 4.)
- Spatial disorientation (e.g., working on the wrong but identical equipment)
- Tasks that rely on short term memory (short term memory can hold approximately 5 items at most under ideal conditions)
- Interpreting a signal as false
- Knowledge based tasks (a novel situation) has higher error rates

We can reduce (but not eliminate) the likelihood of human error by evaluating the systems under which it occurs.

### *3.4 Human Error Classification*

The following classification of error types has been established by engineering psychologists (a.k.a. human factors specialists) and has been used for many years [13].

#### 3.4.1 Slips and Lapses

A 'slip' is an error of commission (a task done incorrectly).

A (memory) 'lapse' is an error of omission (forgetting a task).

Slips and memory lapses primarily occur during Type 1 thinking (see **Section 6.2**).

#### 3.4.2 Mistakes

Mistakes are cognitive, that is they occur primarily during Type 2 thinking (see **Section 6.2**).

Examples of mistakes include:

- PHA errors
- Design errors
- Errors made writing procedures (not typos)
- Mis-diagnosing an abnormal situation

Mistakes primarily occur when faced with a new, ad hoc, or novel situation.

### 3.4.3 Craft Errors

Lack of adroitness related to tasks performed by maintenance, operations, pipe-fitters, etc. For a skilled craftsmen, these error likelihoods are lower than slips, lapses, and mistakes.

### 3.4.4 Social and Institutional Influences

These types of errors are the most insidious and can produce the most catastrophic incidents. Coupled with extended time and economic pressures, some hazard scenarios are especially susceptible to extreme outcomes. The added time a scenario may play out (e.g., days) allows multiple people (and therefore personalities) and conflicting objectives (safety v. economic) to creep-in. De facto Company policy, culture, remuneration, and the lines of communication are the organizational factors that are important.

Social and institutional factors can influence specific skill, rule, and knowledge based tasks. Task analysis has the potential to uncover some of these issues (e.g., by asking about non-compliance to a procedure), and can make these issues more discoverable.

### 3.4.5 Non-Compliance

A violation of a procedure or certain steps of a procedure. Not working to the procedure. Often times, the procedure is wrong, or the operator has found a better way of working. If there is a work process and culture to support operators writing their own procedures with appropriate checks of course, these types of non-compliance will not be an issue. On the other hand, ad hoc or unauthorized bypassing can be a serious non-compliance issue.

### 3.4.6 Summary of Error Classification

**Table 1. Human Error Classification**

| Type of Error | Example | Most Likely Cause | Recovery Factors |
|---|---|---|---|
| Slip | Opening the wrong valve | Distraction | Feedback (usually immediate) |
| Memory Lapse | Forgetting a step in a procedure | Distraction | Checklist (see **4.1.5**) |
| Mistakes | Ignoring a real signal | Bias | Mistakes are hard to detect and correct |
| Craft | Loose Cam-lock fitting | Maladroitness | Feedback (usually immediate) |
| Social and Institutional | We've done it this way for many years and never had an accident. | Turkey Trap bias (see **8.1**) | Culture change |
| Non-compliance | Not following procedure | Poor procedures | Administrative |
| Knowledge based | Abnormal situation | Novel situation | Planning, Trial and error |

# 4 Tool 2: Using Procedures and Checklists

## *4.1 How Routine Operating Procedures are Used*

"*We have an opportunity before us, not just in medicine but in virtually any endeavor. Even the most expert among us can gain from searching out the patters of mistakes and failures and putting a few checks in place. But will we do it? Are we ready to grab onto the idea? It is far from clear.*" Atul Gawande [7]

Procedures are needed for safe operations and reliable maintenance. Working without reference to a procedure creates an error likely situation. Procedures are an important human factors tool.

### 4.1.1 Referencing Procedures In PHA-LOPA

In a PHA-LOPA, references to specific procedures where human action is involved in the study should be made. For example, operator error as an initiating cause, or operator response to a critical alarm. A laptop should be open in the study and should be able to pull up and reference the specific procedures as needed. At times, a quick read through of the procedure with the team is warranted. If a human action shows up in a PHA-LOPA without a procedural reference and none exists, consideration should be given for the need of one.

### 4.1.2 Using Procedures to Operate the Plant

Step-by-step routine procedures used by operations to run the plant are a critical administrative control system. Written procedures establish minimum requirements and expectations, which should be communicated via initial operator qualification and periodic refresher training thereafter.

Procedures used in plant practice have been successful at keeping human task error rates lower compared to other fields that do not use any procedures, for example, the error rates of the medical profession as described in reference [7].

In practice, procedures are used or referenced (or not used) in a variety of ways.

- Procedures may not be accessible, or don't exist
- Procedures may be printed and read before a job
- Procedures may be carried in the truck or in the back pocket
- Procedures may be carried by a scribe as part of a checklist
- Procedures may be posted at the location the task is performed
- Procedures may be placed at the location where data readings are taken to record
- Procedures are committed to long term memory
- Operators don't have enough hands to carry and read a procedure while performing the actual task at the same time (walking, bending, carrying tools, turning valves, wrenches, pushing buttons, etc.)

### 4.1.3 Working from Memory versus Procedure

Experienced operators that perform regular tasks are working primarily from long term memory. New operators rely heavily on written procedures. New operators will often read the procedure just prior to the task (for which they are qualified), while experienced operators will refresh their long term memories on a task performed infrequently. Experienced operators may carry the procedure in their back pocket for reference. New operators will shadow senior operators and be mentored in a task (to be qualified). In all cases procedures should be made available. Procedures should be reviewed annually and refresher training provided per regulatory requirements.

An experienced operator will not perform a task by reading a step in a procedure, performing that step, then reading the next step and performing that step, and so on. Experienced operators work from long term memory as mentioned above. Besides, procedures are insufficiently detailed to be used in that manner, nor should they be made to be. An operator will quickly lose interest in a procedure that spells out every mundane step. All procedures could be improved. Some procedures contain mistakes or critical omissions. If operators actually followed these procedures to the letter, errors would be made every time they were used.

As a skill is perfected over time, Human nature directs us to use Type 1 thinking (see **Section 6.2**), and the tasks we perform are largely done on auto-pilot (i.e., without conscious thought). Potential for slips and memory lapses creep in. This is often why newly qualified operators can often be more reliable than experienced ones (new operators may make more mistakes, but experienced operators have more potential to make the critical errors). New operators tend to spend more time in Type 2 thinking during a task.

### 4.1.4 Checklist v. Procedure

A checklist can be an effective forcing function (something that forces human behavior in a desired way). The caveat is, a checklist will not be effective if it is not used properly. A checklist will not be used properly if for example it is poorly crafted (e.g., too long) or not used in the correct situation (e.g., where long term memory is good enough). In these cases, checklists become equivalent to procedures, with no added benefit. Many checklists are "pencil-whipped" because it is difficult to check boxes at the same time as holding tools and turning valves. In these cases where a checklist is deemed absolutely necessary, a scribe should be used to initial the steps as they are performed by a different person. The next section discusses improving the use of checklists.

### 4.1.5 Improving use of a Checklist

Checklists are less useful to experienced operators because much of what an experienced operator can do comes from long term memory, and therefore long checklists are redundant to procedures, tedious, and a waste of good time. Let an experienced operator use their training, experience, intuition, judgment, competence, etc. to perform a task, but supplement this with short (60 to 90 second) "pause points" at which juncture a short check-list is reviewed for the most critical steps.

# 5 Tool 3: Human Error Rates

Quantification of human error rates provides several benefits beyond having a number to use in LOPA or QRA.

- It forces decomposition of the task (which is more accurate than guessing the whole) see **Section 5.2**
- It forces a deeper dive to understand qualitative factors

There are caveats to using numbers. At times too much effort is spent on the numbers. The uncertainty is too large to justify it. Without qualitative understanding, the numbers are meaningless.

## *5.1 Failure Modes Distribution*

The following is an approximate distribution of error likelihoods for different error types (the more likely being listed first).

- Knowledge based errors. Those that require planning and trial and error.
- Selection errors or slips coupled in time and space (e.g., on the flight deck, selection errors on a DCS graphic, or the myriad of small electronic gadgets that pack multiple modes and functions into a small space). In our personal lives these types of errors may occur *daily.*
- Memory Lapses most typically caused by external interruptions (being pulled to another task) or internal interruptions (what we call 'absent-mindedness', our mind is occupied with something other than the task at hand). The Big 4 that can lead to absent-mindedness are: rushing, distraction, fatigue, or complacency.
- Selection errors decoupled in time and space (e.g., two identical valves 10 feet apart operated 60 seconds or more apart).
- Skill-based errors or craft error based on physical skill.

## *5.2 Using Human Reliability Data*

"*If we break problems down into component questions (What is the chance of A happening? What is the chance of B happening? What is the chance that a man will be in the area when both happen? What is the chance of a fatality if he is there?) answering them with facts where possible and with opinions only when there are no facts available we are more likely to get a correct answer than if we try and guess the answer to the whole problem.*" Trevor Kletz [8]

### 5.2.1 Quantifying Human Error

The formula for calculating human error probability (HEP) is simple:

$$\text{HEP} = \frac{\text{Number of errors occurred}}{\text{Number of opportunities for error}}$$

The application of the equation is difficult for several reasons, and is why there exists a sparsity of human performance data.

Firstly, the numerator is hard to get because self-reporting of errors or reporting someone else's error has obvious problematic connotations.

Second, the denominator is hard to get because often we just don't know an accurate estimate of the number of opportunities for a given error.

### 5.2.2 Being better than 1/10

We are conditioned by LOPA that 1/10 is a typical and acceptable human performance value to use as a generic credit. That's fine (assuming we ignore the deterministic nature of human error, see **Section 2.4.5**). But human factors psychologists have recommended for us to consider error rates in our own personal lives that are as good as 1/1,000 and even under certain circumstances as good as 1/10,000. If you do this, you will convince yourself very easily that better than 1/10 is entirely possible and reasonable.

How good can humans be? Recognized and published human performance limiting values can be found in the literature, for example in Kirwan [4] as follows,

- Single operator carrying out task(s) on plant .. 1e-4
- Operators carrying out task on plant.. 1e-4 to 1e-5
- Control room based team.. 1e-5

These are HEP (human error probabilities). These are <u>performance limits</u>. Often human error rates for a specific task are found to be much higher than this.

To put this in perspective, one would need data for three years performing a daily task error free to demonstrate a HEP of 1e-3. One would need 30 years of data performing a daily task error free to demonstrate a HEP of 1e-4. For this reason, as a human reliability practitioner, one would be hesitant claiming as low as 1e-4 for a task performed by the same person even with good (self) recovery factors and good human factors.

For typical slips and memory lapses under Type 1 thinking (see **Section 6.2**) human error probabilities between 1e-3 and 1e-4 (for example, per step of a procedure) are perfectly reasonable, depending of course on the human factors.

### 5.2.3 The Relationship between Task Frequency and Human Error Rate

Operator tasks that are performed frequently (e.g., every shift) are especially problematic to handle quantitatively because the opportunity for error is so great, yet slips and lapses will occur at the same rate (there is no data to say they get lower). To convert the HEP (human error probability) into a frequency, one simply multiplies by the number of opportunities per time. Still, this issue probably has more to do with the fact that the HEP assumed are on the conservative side. Use of recovery factors, human redundancy, and inherent redundancy in the procedure itself are ways to handle this issue.

# 6 Tool 4: Understanding Systematic Bias

## *6.1 Bias and Heuristics*

The following are some ways a bias can be described:

- A trap
- A distortion (the difference between what is seen and what is there)
- A systematic error (not statistical)
- Has the potential to be corrected for if it can be identified

A heuristic is a shortcut that bypasses cognition (i.e., thinking). Different biases provide the short-cut mechanism. We expect cognitive thinking to be less prone to bias (see **Section 6.2**).

Some examples of bias that have relevance to predicting and controlling major accident hazards are:

Confirmation bias and belief perseverance – we naturally look for corroboration for what we already know or believe, and we hold on to it. Rarely do people look to disconfirm. In the context of major accident hazards, what we don't know is far more relevant than what we do know. Why do we struggle with distinguishing between severity x or y when the hazard is already identified? Is the issue of taking multiple BPCS credits really that controversial when every barrier is already coupled by biased thinking (and we don't even realize it)? Are the consequences of summing versus not summing LOPA causes meaningful when accident sequences never play out in a linear fashion? I'm not arguing to ignore these issues, but to understand the context and prioritize. We already know these things, and the uncertainty is small. The uncertainty is very large for what we do not know, and worse if we don't look for it.

Hindsight bias – our inability to correctly remember or interpret past events, and our need to find linear causality (simplification) for explaining past events, produces a powerful distortion. If you've ever watched an accident video and thought, "How could they let that happen," you've fallen victim to hindsight bias. If you've ever read multiple accident reports from different sources on the same accident, and each report states a different cause or causes, this demonstrates hindsight bias at work.

Information bias – more information produces better results. Not always. What more information does do is make us more (falsely) confident in our results. This bias runs rampant in quantitative analysis. We drill down and sharpen certain variables in a calculation only to ignore other variables whose uncertainty far exceeds what we are attempting to calculate.

Tunneling bias – ignoring or missing sources of uncertainty. This bias may be the umbrella for the others. Any number or calculation produced should have an uncertainty associated with it.

Commitment bias and capture error – these manifest themselves in many accident sequences where an operator has gotten on the wrong equipment, or followed the wrong procedure or action.

Independence trap – simply becoming conscious of an IPL (independent protection layer) or barrier results in a loss of independence of that barrier. This is the most insidious of all traps and has been written about by McLeod [9] and Hopkins [10]. How does this trap work? Say I have two IPLs, x and y, and I am aware of both. Any decision made for one will be influenced by the other. For example, if x is bypassed we still have y. If x fails we still have y. If x works we don't need y. And, it would be nearly impossible for x and y to fail at the exact same time. While this reasoning may be technically and logically true, it can produce a gradual and subtle increase in risk and worst-case it can defeat all of the safety barriers for a scenario.

Normalization of deviation – There have been many examples of this since the term was first coined, and is sometimes why accidents take years to manifest themselves.

Group-think – the presumption that when decisions are made in small groups, everyone will (or must) agree with the decision. This trap has been extensively studied, and can result in bad decisions.

The Turkey trap – see **Section 8.1**.

### *6.2 How Humans Think*

Psychologists, notably Daniel Kahneman [14], have written extensively about how humans think. Two distinct styles are recognized:

- Type 1 – Fast, intuitive, efficient. Not consciously thinking. Zoned out. Prone to bias and irrationality. If you've ever driven to work and don't remember doing it, you've experienced Type 1 thinking. Actually, humans spend the vast majority of their day in Type 1 thinking, while performing tasks, etc. most of the time very reliably.
- Type 2 – Slow and inefficient. Hard. Careful and rational. Conscious thought. Taking exams, writing procedures, performing calculations, require Type 2 thinking. Try to maintain Type 2 thinking *continuously* for more than a few minutes. It can become exhausting.

Type 2 thinking is not error free (see the definition of mistake in **Section 3.4.2**) however, in certain contexts it can be superior to Type 1 thinking. Obviously, because your brain is turned ON, which we assume is better for most people when performing tasks and making decisions.

It is impractical to expect any one to spend their day in Type 2 thinking. The so far elusive goal of Human factors Engineering is this:

> To jolt the operator out of Type 1 thinking into Type 2 thinking, just before he is about to make a catastrophic mistake.

For example, say an operator is staring blankly at a sign that states 'Reactor Valve A' getting ready to open that valve, but that the operator intended to be on 'Reactor Valve B'. As the operator reaches for the handle a small jolt of electricity or heat or something shocks him, and an audio output declares "You are on the wrong valve!"

To achieve this takes some serious human factors engineering and technologies that probably do not exist yet, or not available to general consumer.

# 7 Tool 5: Improving Barrier Effectiveness using Bow-tie

Bow-tie drawings (the left side) may be used as an effective means to communicate how barriers may fail due to human or organizational factors. Task analysis would be used to identify escalation factors (something that can cause a barrier to fail) as well as escalation factor controls (that prevent interference of the escalation factor). The same analysis can identify failures that could impact multiple barriers.

These escalation factors on the bow-tie drawings represent leading indicators of process safety and could be monitored for failure. The bow-tie can be quantified to determine barrier probability of failure. The bow-tie is the best way to show how barriers can fail and the controls to prevent failure.

Common escalation factors related to Barrier failure:

- Errors made by Stores
- Lack of technical competence
- Contractors (where they touch the barrier)
- Poor Maintenance procedures and practices
- Improper restoration following maintenance
- Lack of Change Management
- Poor Work processes
- Lack of Communications
- Wrong specification

# 8 Handling Uncertainty

Typically we treat uncertainty in a quantitative way by performing something like a sensitivity analysis (what-if this input value is changed to this) or by acknowledging data uncertainty intervals (i.e., confidence limits) for the data we use. Typically this handles uncertainty associated with random hardware failure.

There are at least two other types of uncertainty (see **Section 2.4.2** for others),

- That which we know about but do not account for, i.e., systematic human error, the subject of this paper
- That which we don't even know about (i.e., referred to as a Black Swan) addressed below

### *8.1 Black Swan Theory*

I won't attempt to explain or even summarize Black Swan theory [11]. I can't do it justice. Further, I would not listen to or read any second hand explanation of it, other than from the book itself, reference [11]. It is an important book and should be read by every practitioner of process safety risk management.

Psychologists have recognized a fundamental flaw (a bias) in humans thinking that we can project (predict) the future based on the past. This is obviously a trap that PHA teams still fall into. Many of the most infamous accidents in Industry only occurred after years of performing the same task or operation "safely." These incidents could look something like **Figure 1**, adopted from Ref [11]. From the perspective of the turkey, the probability of a catastrophic incident is 1e-3 (looking backwards) the day before it strikes. In hindsight, the catastrophic event seems readily predictable (Americans eat turkeys on Thanksgiving), and the events leading up to it can be described in a nice linear fashion (feed, water, repeat, later kill). This is hindsight bias at work.

The best we can do to handle this kind of uncertainty is to maintain a sense of vulnerability and continue to use work processes (e.g., hazard identification, change management, preventive maintenance, etc.) and tools (e.g., databases, procedures, task analysis, reliability analysis, etc.), to manage what we can predict, in hopes of effecting what we cannot predict.

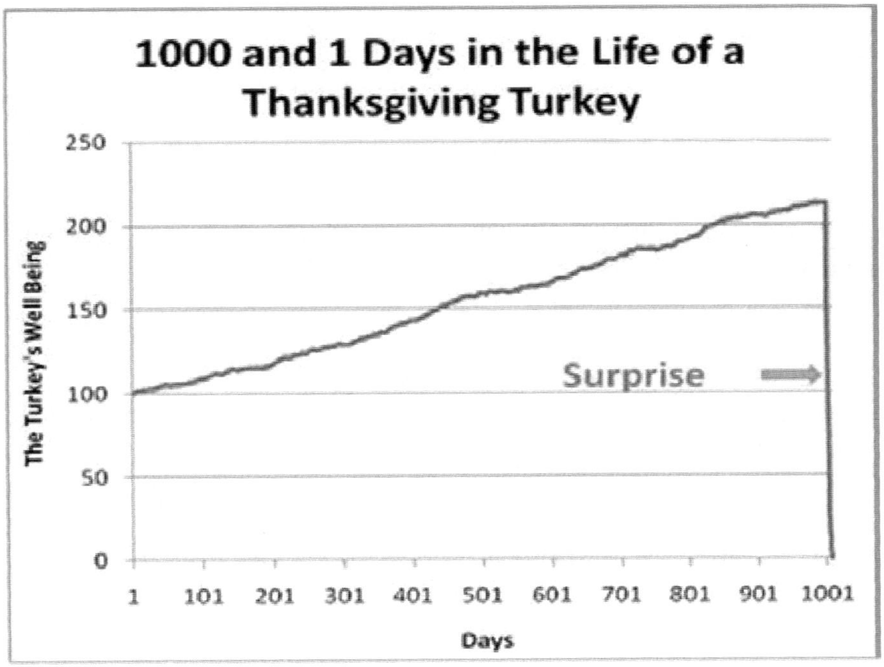


**Figure 1**. 1e-3 Probability of Catastrophe based on Past History

# 9 A Short Task Analysis Checklist

The following checklists represents a sample of the types of questions asked during a Task Analysis. A follow-up walk-thru to actually observe the task is always necessary. It is useful to observe how different the below checklists are compared to the standard human factors checklists used in PHA.

For any scenario with a human error initiating event that goes to LOPA, ask the team to answer the following questions:

- What is the cause of the error ("The operator." Why? "Wasn't paying attention." Why? "Gets pulled away during task." Why? etc.)
- What are the administrative controls in place (immediate feedback, procedure, permit, tagging, LOTO, human redundancy, housekeeping audits, etc.) that could allow recovery from the error?
- Is the task skill/ rule/ or knowledge based? Often tasks will be a combinations of the 3.
- How frequent is task performed? How long does the task take to complete?
- What is the qualification for the job?
- What Tools are required for the job?
- Can the task cross shift-change?
- What role do Contractors play in the task? Ask the same set of questions focusing on the Contractor.

For any scenario with a human response to abnormal condition (e.g., critical alarm barrier) that goes to LOPA, ask the team to answer the following questions (some of these can be observed directly during the walk-through):

Focus on the big 4: perceive, interpret, diagnose, respond (the first 3 represent the degree of situational awareness)

- Perceive: How does the Operator become aware of the alarm? Is the alarm prioritized? Is a dedicated alarm screen used? Are gray-scale graphics used?
- Interpret: What are the alarm metrics (how many alarms per hour, false alarms, etc.)? Do the metrics conform to good engineering practice? Are instrument signals trended? Is there a policy that every critical alarm that comes in must be treated as real? Are there periodic (e.g., quarterly) talk-throughs or drills for the alarm?
- Diagnose: Is the scenario known or novel? How many potential causes? How long does each cause take to diagnose? What actions are required?
- Respond: see checklist above.

For any scenario with a human barrier used in LOPA, ask the team to answer the following questions (some of these can be observed directly during the walk-through):

- What are all the human touch points and how can failure occur? E.g.,
    - Supplier
    - Purchasing

  - Stores
  - Maintenance
  - Quality Control
  - Engineering
  - Operations
  - Management
- What are the Contractor roles in the barrier?
- Has the barrier Specification (sizing, set-point, response time, etc.) been independently validated?

For any barrier or IPL that requires good situation awareness (SA*) (e.g., response to abnormal situation - AS), the following items should be looked at:

- Individual factors
  - How good is the operator's 'mental model' of the plant related to the AS (process conditions, equipment status, maintenance activities, etc.)?
  - What is the level of 'Automaticity' (how much cognition does the operator have to engage to resolve the AS)?
  - How well are the goals defined to resolve the AS?
- System factors
  - Interface design (alarm rationalization, use of gray-scale graphics, etc.)
  - System complexity (How many potential causes of the AS are there?)
  - Level of automation (Is there so much automation the operator has lost their hands-on operating skills?)

*SA is the difference between what is actually happening in the plant and what the operator thinks is happening.

These short checklists give you an idea of what task analysis is for each.

## 10 Summary

Improving Human factors in PHA-LOPA requires a diverse toolkit of methods and theory. Identification, prediction, and control of the human factors role in major accident hazards is currently soft. There is not a quick easy answer, however, improvement especially related to low hanging fruit is readily achievable.

PHA-LOPA facilitators need to begin practicing the methods of human factors and human reliability to improve said practice as it relates to major accident hazards. The skills a facilitator gains by practicing human factors and reliability will directly transfer to all of PHA-LOPA to improve decision making, the ability to write human factors scenarios, and better awareness of when further study is needed.

## 11 References

[1] Norman, Don. *The Design of Everyday Things*, Revised and Expanded Edition, Basic Books, New York, 2013.

[2] Chemical Safety Board Website. http://www.csb.gov/videos/csb-video-excerpts-from-dr-trevor-kletz/

[3] J. Robert Taylor. *Human Error in Process Plant Design and Operations – A Practitioner's Guide*, CRC Press, Taylor & Francis Group, Boca Raton, 2016.

[4] Kirwan, Barry. *A Guide to Practical Human Reliability Assessment*, CRC Press, Taylor & Francis Group, Boca Raton, 1994.

[5] CCPS. *Guidelines for Engineering Design for Process Safety* 2nd Edition, 2012.

[6] HSE. Review of Human Reliability Assessment Methods, 2009.

[7] Gawande, Atul. *The Checklist Manifesto: How to Get Things Right*, Metropolitan Books, 2009.

[8] Kletz, Trevor. *Hazard Analysis – A Quantitative Approach to Safety*, I.Chem.E. Symposium Series, No. 34, 1971.

[9] McLeod, Ronald W. *Designing for Human Reliability – Human Factors Engineering in the Oil, Gas, and Process Industries*, Gulf Professional Publishing, 2015.

[10] Hopkins, Andrew. *Disastrous Decisions – The Human and Organizational Causes of the Gulf of Mexico Blowout*, CCH, Australia, 2012.

[11] Taleb, Nassim Nicholas. *The Black Swan* 2nd Edition, Random House, New York, 2010.

[12] Swain and Guttmann. NUREG/ CR – 1278. *Handbook of Human Reliability Analysis*. 1983.

[13] Reason, James. A Life in Error – From Little Slips to Big Disasters, CRC Press, Taylor & Francis Group, Boca Raton, 2013.

[14] Kahneman, Daniel. *Thinking Fast and Slow*, Farrar, Straus & Giroux, New York, 2011.

13TH GLOBAL CONGRESS ON PROCESS SAFETY

# Waste Solvents to Trash Haulers: Lessons Learned from Hazardous Waste Accidents

**Delmar "Trey" Morrison, Ph.D., P.E.**
**Exponent, Inc.**
**4580 Weaver Parkway, Suite 100, Warrenville, IL 60555, USA**
**tmorrison@exponent.com**

**Michael Stern, Ph.D., P.E.**
**Exponent, Inc.**
**Natick, MA, USA**
**mstern@exponent.com**

**Carmen H. Osorio-Amado, Ph.D.**
**Exponent, Inc.**
**Los Angeles, CA, USA**
**cosorio@exponent.com**

Prepared for Presentation at
American Institute of Chemical Engineers
2017 Spring Meeting and 13th Global Congress on Process Safety
San Antonio, TX
March 26 – 29, 2017

# Waste Solvents to Trash Haulers: Lessons Learned from Hazardous Waste Accidents

**Delmar "Trey" Morrison, Ph.D., P.E.**
**Exponent, Inc.**
**4580 Weaver Parkway, Suite 100, Warrenville, IL 60555, USA**
**tmorrison@exponent.com**

**Michael Stern, Ph.D., P.E.**
**Exponent, Inc.**

**Carmen H. Osorio, Ph.D.**
**Exponent, Inc.**

**Keywords:** Hazardous waste, incident investigation, chemical reactivity

## Abstract

Process safety management (PSM) of hazardous chemicals is a core tenet of the chemical, oil, and gas industries. Organizations in these industries have increasingly woven the elements of PSM into their strategic vision, day-to-day operations, and interactions with contractors and vendors. This cultural evolution has, in part, been driven by major accidents in these fields. In these industries, the chemicals, their hazardous properties, and their potential interactions, are usually either well-defined or capable of being well-defined. Control and mitigation strategies for those chemicals can be developed based on this available knowledge. In contrast, companies that handle wastes have experienced catastrophic accidents due to the ill-defined nature of the hazards. This industry handles the cast-offs, dregs, and refuse, which makes identification of the hazards and prevention of incidents more challenging. How do you implement a PSM program when your input streams can vary due to processes at a plant you do not control or monitor? Does the EPA (Environmental Protection Agency) Resource Conservation and Recovery Act (RCRA) provide sufficient guidance of safely handling of wastes? The overriding theme in the incidents is that PSM is more complex to apply to unknown, mixed, and uncontrolled chemicals, both solid and liquid. To assist companies that handle wastes, this paper provides several case studies of fires and explosions involving multiple waste material handling processes at different facilities. The incidents involved numerous matrices from solid industrial wastes, EPA RCRA hazardous wastes, to waste solvents. An analysis of the lessons learned will be presented and used to provide realistic guidelines to reduce safety risks based on a pragmatic implementation of process safety management principles. The paper will also compare OSHA (Occupation Safety and Health Administration) PSM with EPA RCRA approaches to characterizing wastes versus managing the hazards, and will emphasize those elements that played critical roles in the accident case studies.

# 1 Introduction

This paper discusses several case histories of process incidents occurring at waste handling or processing facilities and compiles lessons learned from those accidents. Depending upon the source, the industry, and the regulatory classification of the waste, industrial waste may pass through a municipal waste transfer station to a landfill or be collected, aggregated and then incinerated by a RCRA (Resource Conservation and Recovery Act) [1] permitted hazardous waste Treatment, Storage, and Disposal Facility (TSDF). Managing waste is inherently more challenging than managing typical chemical process hazards where the identity and properties of the process chemicals can be effectively defined. While many types of hazardous waste facilities test for the properties of incoming waste streams, these tests can be challenging if the waste is phase separated, if the waste is still reacting and changing over time, or if the waste is incompletely identified by the generator. Similarly, investigations of accidents are complicated by the fact that wastes involved in the accident could have contained unknown chemicals with unknown properties.

Several case studies are presented in this paper. The accident locations range from waste transfer stations, waste storage facilities, waste treatment facilities, and a cement manufacturing facility using waste as a fuel source. The wastes involved in the accidents ranged from municipal waste, unregulated solid and liquid wastes, to regulated hazardous solid and liquid wastes. Significant fires and explosions occurred in these incidents. The overarching causes of these incidents were due to the gap between the true properties and the recognized properties of the wastes. While the task of characterizing waste streams may be challenging, the overall process safety and incident investigation principles remain similar. In this paper, we start with a typical decision tree analysis for screening chemical reactivity hazards and modify it for use for waste processing facilities. We will then discuss several case studies involving fires and explosions in waste handling and treatment and provide lessons learned.

# 2 Background

Application of Process Safety Management (PSM) principles in the chemical industry has been codified for highly hazardous chemicals above threshold quantities through Title 29 Part 1910.119 of the Code of Federal Regulation since 1992 [2]. The Center for Chemical Process Safety (CCPS) subsequently expanded on the process safety model presented by the regulations to develop twenty elements of risk-based process safety [3]. The overlap between the CCPS process safety model and the OSHA PSM regulatory elements (bolded) is represented in the list below (Table 1).

**Table 1. Overlap between OSHA (bold) and CCPS PSM elements**

| | |
|---|---|
| 1. Process Safety Culture | 2. Contractor Management<br>**(h) Contractors** |
| 3. Compliance with Standards | 4. Training and Performance Assurance<br>**(g) Training** |
| 5. Process Safety Competency | 6. Management of Change<br>**(l) Management of change** |
| 7. Workforce Involvement<br>**(c) Employee participation** | 8. Operational Readiness |
| 9. Stakeholder Outreach | 10. Conduct of Operations |
| 11. Process Knowledge Management<br>**(d) Process safety information** | 12. Emergency Management<br>**(n) Emergency planning and response** |
| 13. Hazard Identification and Risk Analysis<br>**(e) Process hazard analysis** | 14. Incident Investigation<br>**(m) Incident investigation** |
| 15. Operating Procedures<br>**(f) Operating procedures** | 16. Measurement and Metrics |
| 17. Safe Work Practices<br>**(k) Hot work permit** | 18. Auditing |
| 19. Asset Integrity and Reliability<br>**(j) Mechanical integrity** | 20. Management Review & Continuous Improvement |

Deficiencies in many of these elements can be traced back to the root causes of hazardous materials incidents, as will be illustrated in the case studies in this paper. However, it is uncommon for TSDFs and other waste handling facilities to formally apply PSM to their processes. Usually, facilities follow a prescriptive approach to waste characterization and handling that is based upon the EPA (Environmental Protection Agency) regulations and relies heavily upon the waste generator's characterization of the waste properties [4]. This reliance is not unusual when considering the quantity of distinct and unique waste streams that an individual municipal facility or TSDF may handle and dispose.

The primary hazard identified in the case studies is delayed reaction leading to pressure generation, heat generation, and unintended ignition of packaging or the waste material itself. This is commonly referred to as spontaneous ignition or spontaneous combustion since a piloted ignition source is not necessary. Trash and municipal waste handling facilities are frequently susceptible to delayed fires due to spontaneous combustion. Such ignition events may be caused by unintentional mixing of incompatible chemicals or exothermic oxidation of self-heating materials, such as vegetable-derived oils. For example, the National Fire Protection Association (NFPA) estimated that 4,460 annual fires involving spontaneous combustion of trash occurred between 2005 and 2009 [5].

Recent studies have identified several trends in hazardous waste TSDF fires and explosions [6], [7], [8]. Incidents can occur in any stage of the process from initial storage, waste repackaging, or transfer to the final treatment stage (i.e., incinerator, calciner, or kiln). The consequences of a fire or explosion can accelerate rapidly, depending upon the nature of the initial waste and its proximity to surrounding waste materials, surrounding waste materials' properties, and the adequacy of fire protection and emergency response actions. The causes of fires and explosions typically derive from unknown or unrecognized chemical reactivity issues with the waste material in question.

### *2.1 Waste generation in the United States*

According to the US EPA, 26,237 generators produced 33.6 million tons of hazardous waste, of which over 6 million tons were non-wastewater [9]. Of the 6 million tons, 26% is combusted for energy recovery, 19% is put into landfills, and 13% is incinerated. Over 76% of the non-wastewater hazardous waste is treated off-site by a TSDF.

The EPA statistics also show that 258.5 million tons of non-hazardous Municipal Solid Waste (MSW) was generated in the United States in 2013. While the per capita waste has slightly decreased since the year 2000, the total annual generation continues to climb due to population growth [10]. Approximately 89.4 million tons (35%) of the MSW were recycled in 2013. Another 13% of the total MSW was combusted with some amount of energy recovery. The remaining 52.6% was placed into landfills.

### *2.2 Resource Conservation and Recovery Act*

A hazardous waste can be any waste material with a potential to harm human health or the environment, either by itself or through interaction with other factors. However, for regulatory purposes in the U.S., a waste material must meet specific criteria to be considered as a hazardous waste. These conditions, summarized in Figure 1, are defined by the EPA under the RCRA in Title 40 of the Code of Federal Regulations [1].

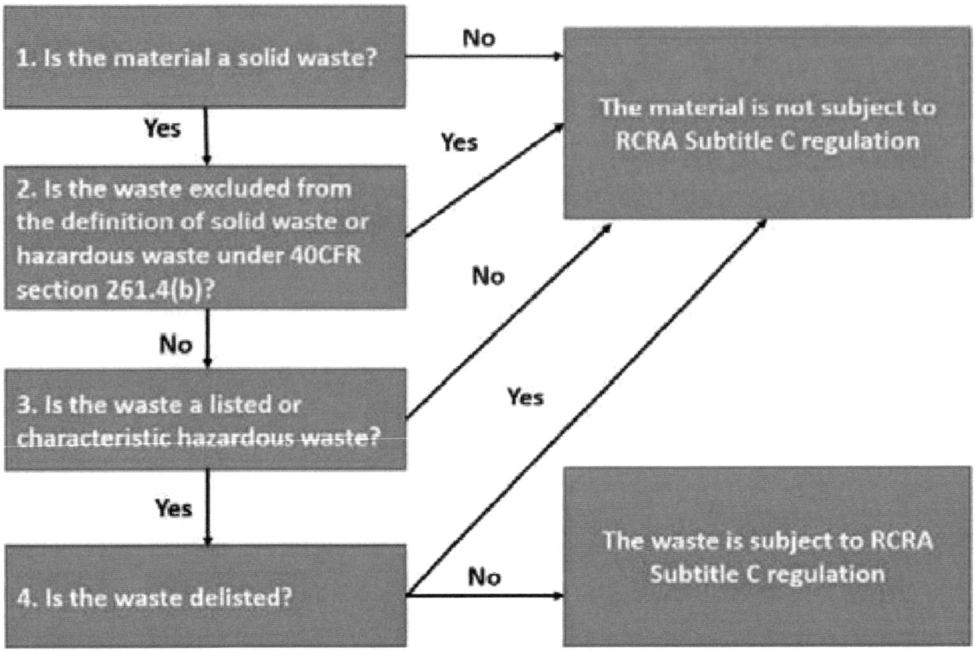


**Figure 1. EPA's Hazardous waste identification process [11]**

Per the EPA, the first condition of a hazardous waste is to be a solid waste, which paradoxically includes not only wastes that are physically solid, but also wastes that are liquid, semi-solid, or contained gaseous material. A hazardous waste is also required to meet certain characteristics based on its toxicity, ignitability, reactivity, and corrosivity, and needs to be listed in either the F, K, P, or U lists found at the 40CFR section 261 [12].

Some solid wastes are specifically excluded from RCRA regulatory compliance due to different reasons such as impracticability, economy impact, lack of information to justify applying the regulation, regulation by other laws, or they were removed (delisted) from this category. These exclusions refer to conditions (2) and (4) from the hazardous waste identification process shown in Figure 1. For example 40 CFR 261.4(b) excludes household waste and solid waste from commercial or industrial sources (that don't contain solid hazardous waste). It also excludes fly ash waste. These types of waste material may pose physical and chemical hazards, but they are not regulated under RCRA.

## *2.3 RCRA chemical compatibility testing*

As part of the RCRA regulations, hazardous waste generators are required to classify their waste. When TSDFs receive waste, they must perform pre-acceptance analysis to confirm the waste's characteristics and identify incompatible wastes [13]. The EPA states "Compatibility testing may include determining water reactivity (e.g., foaming, heat generating, explosivity), oxidation-reduction potential, spot tests for cyanide and sulfide, among others" [14]. The EPA provides a list of some common potentially incompatible wastes in Appendix V to Part 264. When analysis is required, the EPA has published guidance on how to qualitatively evaluate the compatibility of wastes [15]. The guidance document uses a two-part approach: step-by-step compatibility analysis procedures and then a compatibility chart that considers 41 reactivity groupings. To simplify this analysis when chemical testing is not required, free computer software packages are available for performing quick assessments of chemical compatibility [16].

### *2.4 Guidelines for evaluating chemical reactivity hazards*

According to the U.S. Chemical Safety Board (CSB), a chemical reactive hazard can be defined as "*a sudden event involving an uncontrolled chemical reaction with significant increases in temperature, pressure, or gas evolution that has the potential to, or has caused serious harm to people, property or the environment*" [17]. In order to decide whether a chemical can be considered a reactive hazard or not, the CCPS published a preliminary screening test based on a series of questions summarized in Figure 2 [18].

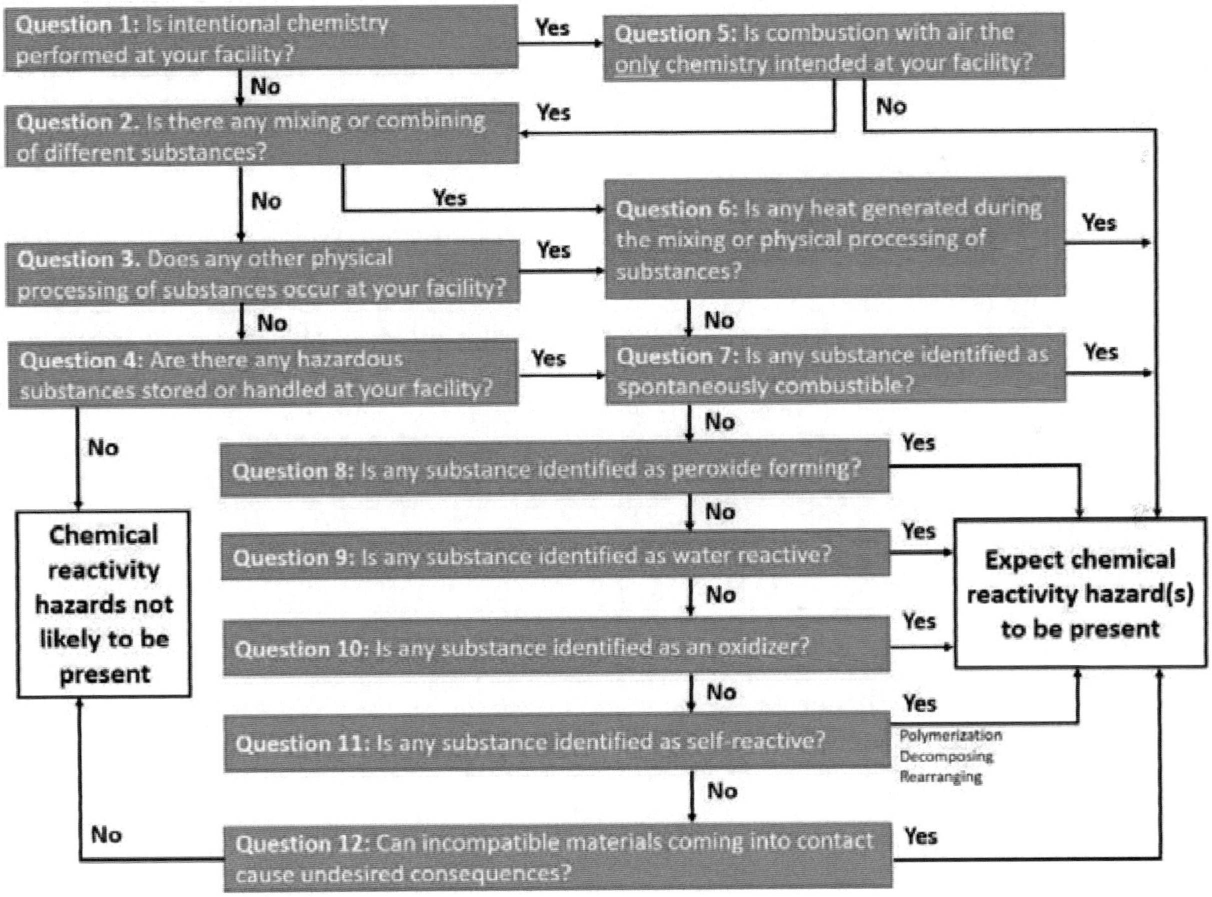


**Figure 2. Preliminary screening for chemical reactivity hazards [18].**

This test should be conducted on an ongoing basis, in particular when there have been changes in the process/conditions or a new chemical is introduced into the facility. If a chemical reactive hazard is expected to be present at the facility, the CCPS also suggests the implementation of a chemical reactivity management program structured consistent in Figure 3.

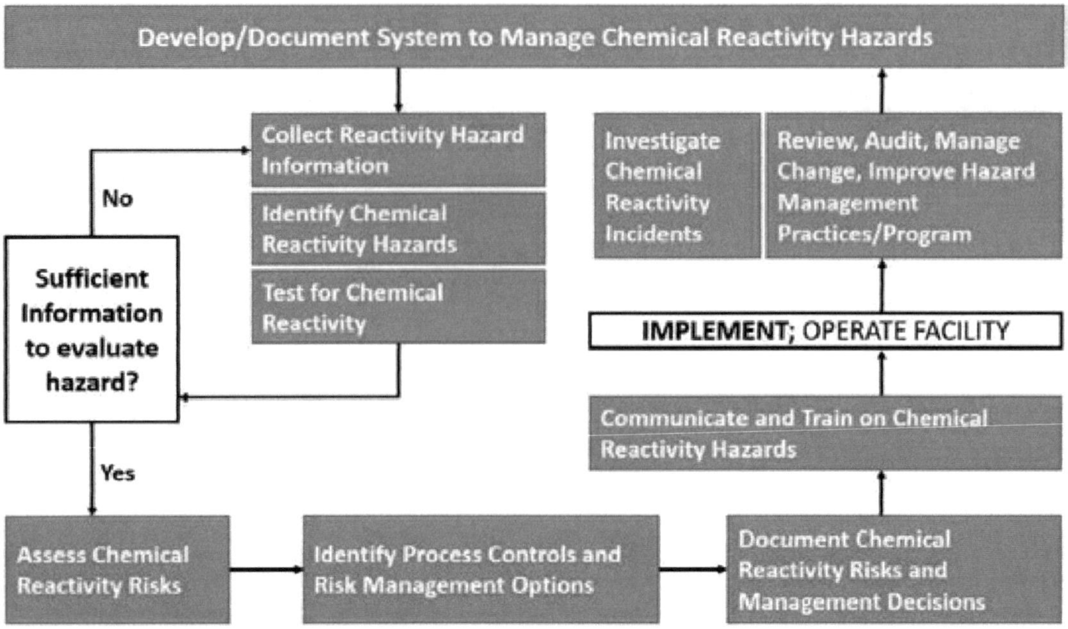


**Figure 3. Flowchart for implementing chemical reactivity hazard management [18].**

### *2.5 Why and how an event happened? Protocols for investigating reactive chemical incidents*

Given the nature of fires and explosions, especially when involving hazardous waste materials, necessary evidence may be destroyed or lost during the incident or subsequent emergency response. Thus, conclusive determination of the chemistry and proximate causes may be impossible. The investigation process must rely upon the scientific method to develop hypothetical causes, test those causes, and determine the most likely causes and contributing factors.

Two commonly relied upon resources for performing incident investigations per the scientific method are NFPA 921 *Guide for Fire and Explosion* Investigations [19] and the CCPS publication Guidelines for Investigating Chemical Process Incidents [20]. Figure 4 shows the progression of an investigation as described in Chapter 4 of NFPA 921. Two essential concepts, inductive reasoning and deductive reasoning, are used to develop and test hypotheses, respectively. Inductive reasoning is using the analyzed data to develop a consistent hypothesis. Deductive reasoning is used to test the hypothesis by comparing it to all known facts of the accident and scientific literature. NFPA 921 cautions that "[t]he investigator does not have a valid or reliable conclusion unless the hypothesis can stand the test of a careful and serious challenge." Recent editions of NFPA 921 have also added cautionary language regarding the use of *negative corpus* also known as the process of elimination. Investigators should not confuse the lack of an alternative hypothesis for affirmative proof of a hypothesis with no supporting evidence (but that cannot be ruled out with the available information).

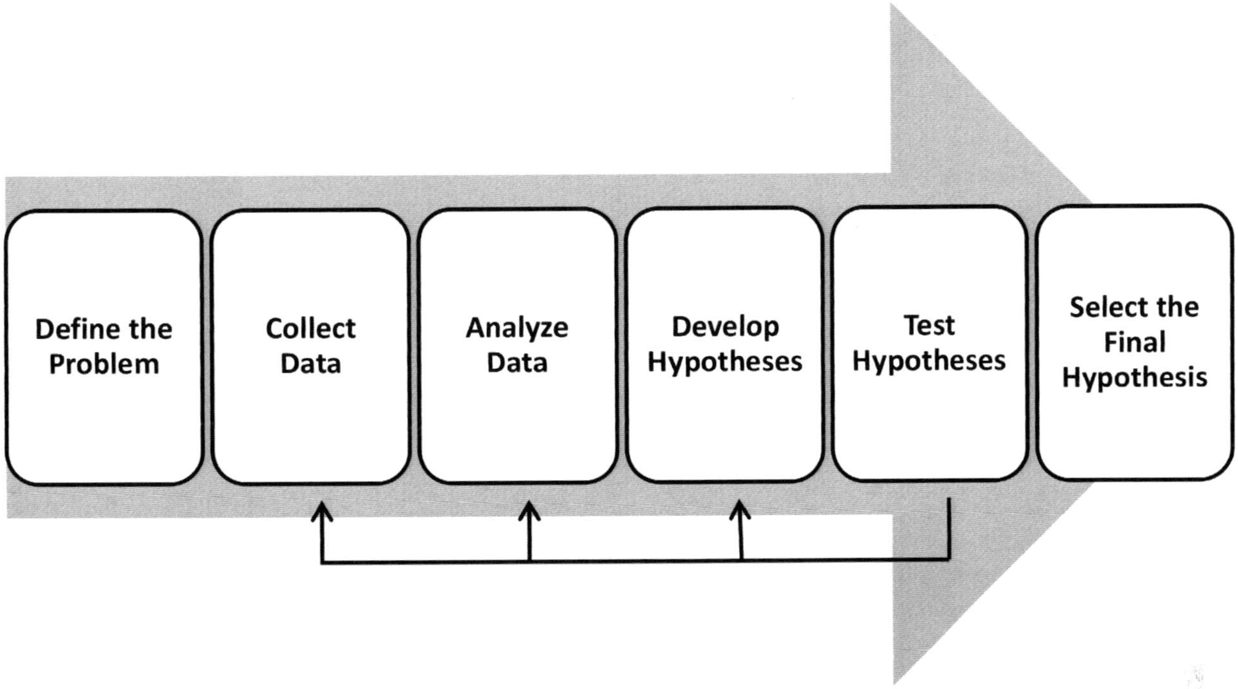


**Figure 4. The scientific method of investigation.**

The CCPS Guidelines text provides a similar scientific methodology for accident investigation but also offers information specific to industrial accident investigation. For example, information for designing incident management systems within a corporation is provided. Specific investigation and root cause analysis tools commonly employed for investigations, organized by the level of structure, are shown in Figure 5.

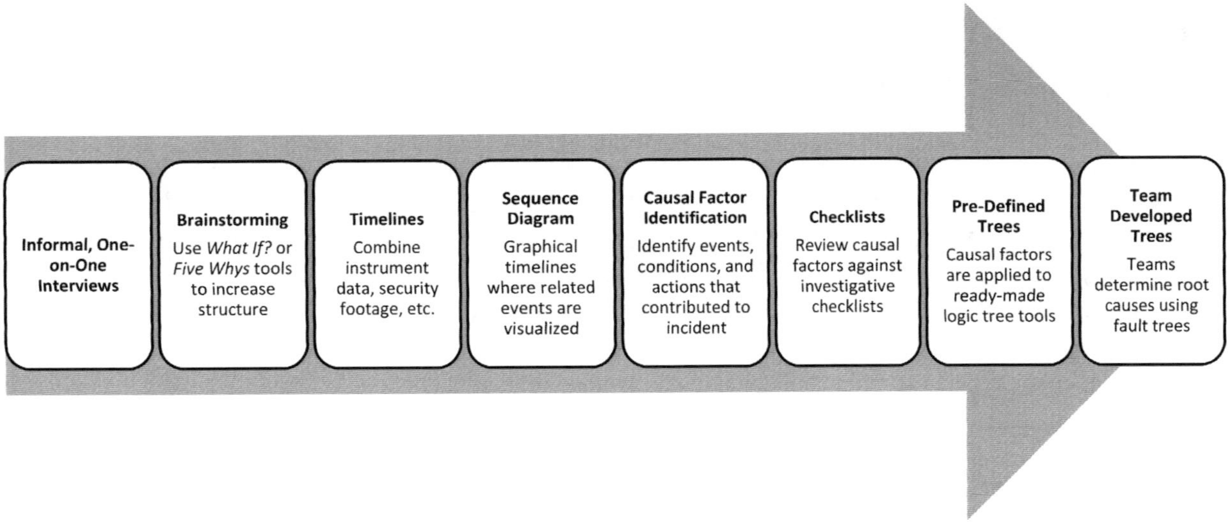


**Figure 5. Investigation tools for process incident investigations.**

The case studies below describe the causes and lessons learned from MSW handling, solvent reclamation, and TSDF incidents. All of the incidents were characterized by fires or explosions, which caused significant damage and posed potential for catastrophic injuries or fatalities to onsite personnel. Many of the incidents were investigated by external public agencies in addition to the company investigations. The authors also participated firsthand in the investigations of the incidents. Where possible, lessons learned from the investigation process as one element of PSM are shared.

# 3 MSW Transfer Facility Fires

Two fires, separated by three days, occurred at a MSW transfer station. The first fire involved a trailer of waste awaiting transfer to the landfill, and the second fire involved a different trailer and a large portion of the building and waste contained within. The facility was unsophisticated when considering the element of process safety management, which is understandable because it did not handle reactive materials. As a consequence, early investigation of the causes of the fires relied upon supposition and poor science to conclude that one waste stream was the cause of the fire. The incorrect conclusion adopted by the facility meant that the real cause went unaddressed until a later, thorough engineering investigation occurred.

The waste transfer station was a large, open warehouse structure with two sunken loading bays. The facility handled trash classified as municipal solid waste (MSW) and construction and demolition (C&D) debris. Waste materials were piled on the concrete and moved around using a front end loader. Different materials were commingled by the facility operators, both in piles and in the open trailers located in the loading bays. The commingled waste was periodically hauled to a municipal landfill. The facility did not accept RCRA hazardous waste or liquid wastes. The local regulations and the permit for the facility both acknowledged the reasonable possibility of "hot loads" - in other words, the receipt of loads of waste material that are undergoing self-heating or are already smoldering.

The first fire involved the contents and the rear of a trailer (Figure 6), which was located in a loading bay overnight. The trailer contained a mixture of bottom ash from a fluidized bed coal combustor and other unknown MSW and C&D materials. No formal investigation was conducted to attempt to determine the cause of the trailer fire. The waste material in the trailer was not fully extinguished or overhauled by the fire department, and the facility located the trailer away from the building so that it could eventually burn itself out. The fire witness statements indicated that the contents of the trailer continued to burn for the next several days. The fire damage patterns to the trailer are consistent with a fire originating at the rear wheels, perhaps due to overheated brakes. Three days after the initial fire, a second fire was discovered after a different trailer was loaded with similar waste materials from within the building and held overnight in the undamaged loading bay. The second fire burned to completion and destroyed the building.

**Figure 6. Photograph of the trailer damage resulting from the first fire.**

An early investigation of the second fire concluded that the combustor bottom ash was the equivalent of pure lime (CaO), caused both fires due to exothermic heat of hydration, and should have been classified as a RCRA hazardous waste. Pure CaO is referred to as "lime" or "quick lime." The reaction of CaO with water vapor results in solid calcium hydroxide, $Ca(OH)_2$, while mixing with excess water creates an aqueous solution, according to the chemical reaction:

$$CaO_{(s)} + H_2O \rightarrow Ca(OH)_2$$

This reaction is exothermic, meaning that it generates heat, with a heat of reaction of -65.27 kJ/mole of CaO reacted [21]. The quantity of heat released is directly proportional to the quantity of CaO and $H_2O$ undergoing the reaction.

The transfer station's incorrect initial conclusion was based on the description of the waste by the generator as "agricultural lime," which is commonly used to describe calcium carbonate ($CaCO_3$), hydroxide ($Ca(OH)_2$), and oxide (CaO) used in farming. Further, the early investigation concluded that the mischaracterization of the waste was a direct cause of the fires. A subsequent engineering investigation analyzed that conclusion and also identified several plausible causes including an electrical fire, spontaneous combustion of MSW or C&D, or re-kindling of the smoldering remains of the first fire after being transported across the facility.

Chemical analysis of the combustor bottom ash and a review of the literature indicated that the ash did not contain sufficient CaO to cause spontaneous combustion through hydration and was excluded under 40 CFR 261.4(b). The residual CaO in combustor tram will undergo hydration by reacting with water ($H_2O$) in either a vapor (e.g., atmospheric) or liquid form. The hydration reaction for residual CaO would run largely to completion within a few days when exposed to the outdoor atmosphere [22]. Comparison samples of fresh combustor bottom ash revealed that it contained approximately 10% CaO by weight.

If concentrated CaO is mixed with liquid water, the hydration reaction can lead to very high temperatures capable of igniting combustible materials. However, it is unlikely that these conditions existed in the current matter. The maximum temperature reached by the combustor ash via this reaction, if any unreacted CaO is present, will be tempered by the presence of liquid water and surrounding solid material (i.e., $CaCO_3$, $Ca(OH)_2$, $CaSO_4$, sand, and onsite waste materials). The water will absorb the released heat and act to cool the material as the water evaporates; the other solid species will act to absorb the liberated heat and increase in temperature.

The false conclusion that the ash waste was CaO, thus causing the fires, highlights the importance of precise descriptions for waste streams even if they are not bound to the RCRA regulation. Effective description can help investigators, who may be unaware of the precise waste compositions, more effectively assess the important properties of the waste stream after an accident. Similarly, investigators must be careful when examining the properties of non-RCRA waste streams as their descriptions may be more colloquial. Conclusions should not be drawn until a more detailed composition of the waste can be obtained through chemical analysis or discussion with the generator.

# 4 Solvent Recovery Plant Vapor Cloud Explosion

In 2009, a release of ignitable liquid and vapors, followed by catastrophic vapor cloud explosion and subsequent fires, took place in a solvent recovery plant. The explosion severely injured two employees, destroyed part of the plant (see Figure 7), and broke windows in neighboring facilities and residences. The explosion was investigated by the Chemical Safety and Hazard Investigation Board (CSB) [23].

The solvent recovery plant (Figure 8) was located at a facility that handled hazardous and non-hazardous waste materials that were received via tanker truck, drums, totes, and lab packs. The facility performed fuel blending, solvent recovery, and waste consolidation. Incoming dirty solvents were unloaded into tanks in the dirty tank farm. Clean solvent was recovered from the dirty solvents via one of three distillation units (Unit 1, Unit 3, and Unit 4). The recovery process involved the circulation of distillation bottoms back into the source tank in the dirty tank farm, while purified "clean" solvent was condensed and transferred to the clean solvent tank farm. Final waste liquids and sludge were shipped offsite to a waste incinerator as blended fuels.

At the time of the incident, the three Processing Units were being run by a single outside operator. Unit 1 was used to recycle a paint thinner material whose primary compound was acetone, while Unit 3 was processing a tetrahydrofuran (THF)/water mixture. Unit 4 worked as a closed circuit reflux-distillation unit where a mixture of THF and water near the azeotropic concentration was processed.

**Figure 7. Aerial view of the solvent recovery plant after the incident from the CSB report [23].**

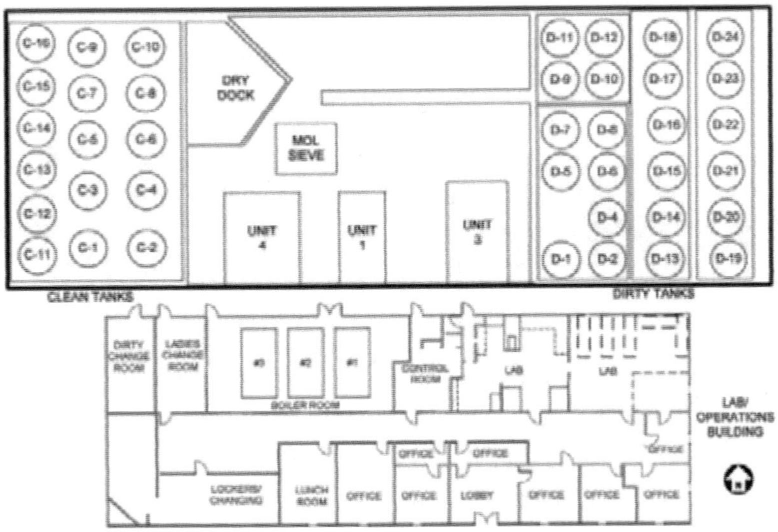


**Figure 8. Facility layout (not to scale) [23].**

Three of the six workers onsite reported hearing a loud noise like a pressurized release, accompanied by a strong solvent odor and a visible aerosol/or vapor cloud coming from the Dirty Tank Farm prior to ignition. Security camera footage from onsite and a nearby business supported the conclusion of a pressurized release, likely of Tetrahydrofuran (THF) solvent mixture; however, the exact composition and source of the flammable material released could not be determined.

The incident investigation concluded that the vapor cloud spread out from the Dirty Tank Farm, entered the laboratory/Operations Building, and ignited inside the Boiler Room causing, an initial explosion. Consequently, additional flammable material within the Dirty Tank Farm was also ignited causing a blast pressure wave (vapor cloud explosion) and subsequent pool fire that burned for several hours.

The CSB's findings from this incident included the following facility siting concerns:

- Lack of hazard analysis to determine the risks associated to the proximity of occupied buildings to processes: The most critical injured employees were at the Laboratory/operations building at the time of the incident. The nearby building was occupied with non-essential personnel during the day shift.
- Inappropriate area classification: Two boilers were in service in the Laboratory/Operation Building, where the ignition source was identified. The flammable vapor cloud was ingested and ignited by one of the boilers.

Although the facility was covered by the OSHA PSM standard, the CSB found that the TSDFs did not have industry-specific guidance for fire safety and facility siting.

# 5 Fires and Explosions in Storage Operations at TSDFs

Several fire incidents have occurred during storage of hazardous materials at TSDFs. The ultimate fate of hazardous materials is often destruction through incineration. The type of incinerator and local environmental emission limits dictate handling activities with the hazardous waste. To manage the limits on the heat of combustion for the incinerator, heavy metal emissions, and other factors, TSDFs commonly repackage wastes by either combining similar waste streams in the same container or by breaking up larger containers into smaller container sizes. For example, if the bulk heat of combustion of the 55-gallon drum of waste received by a TSDF is too high for the incinerator's operating limit, then the 55-gallon drum must either be metered (if a liquid) into the incinerator, or the drum must be subdivided into smaller containers (e.g., 30-gallon drums).

Thus, hazardous materials often pass through the following stages at a TSDF:

1. Receipt and characterization
2. Temporary storage
3. Repackaging – either bulk up or break down
4. Storage awaiting disposal
5. Final disposal

The waste generator must provide specific information on the hazardous waste that they are providing to the TSDF. The TSDF oftentimes works with the generator to help characterize the waste stream as part of the initial contracting process. The TSDF typically samples and performs basic characterization of individual wastes as they are received. However, this process is imperfect because the TSDF cannot be guaranteed that a set of samples from distinct locations in a waste container represent the bulk of the waste in that container. The potential deviation of actual properties from the results of this sample may grow significantly for larger-sized containers (e.g., a liquid tote or a 1 $m^3$ box), wastes that are heterogeneous, wastes with layers or pockets of different composition, or large volume wastes that are not completely sampled (e.g., sampling a subset of containers).

After initial contracting and waste specification, a TSDF will develop a plan for handling that waste stream, which will include sampling, characterization, and verification that as-received waste meets the contract specification. This process falls under Process Knowledge Management in the 20 elements of CCPS PSM. Deviations of as-received waste from the specification should be addressed under a Management of Change element, but the TSDF may not have a robust change management policy. Additionally, changes from the specification may not be identified during the characterization process as discussed above.

Incident investigation after a hazardous materials incident at a TSDF can be very challenging. A TSDF storage area may contain hundreds or even thousands of individual waste containers (see Figure 9). A fire or explosion in the storage area will destroy or damage the waste containers, displace waste containers, and mix the residue from involved containers. All these aspects make determination of the cause of the incident difficult. The likelihood of determination is usually directly related to whether the incipient stages of the fire or explosion were witnessed. The following three case studies provide incident summary and investigation results for a progressively larger TSDF fire incidents in order to characterize the difficulty in cause determination and lessons learned.

**Figure 9. Example of TSDF storage warehouse.**

### *5.1 Delayed Ignition - No Repackaging – Cause Undetermined*

An investigation was performed on a fire that occurred in a small TSDF's storage bay. The facility handled a wide variety of wastes including flammables materials and acids. The facility included storage bays, laboratories, treatment equipment and office areas. In the storage bay where the fire occurred, several different types of containers were present including:

- Plastic buckets
- Plastic, steel, and paper fiber drums
- Steel encaged plastic totes known as intermediate bulk containers (IBCs)

The 55-gallon drums were placed on wooden pallets in sets of four. The pallets of 55-gallon drums and IBCs were stacked two or three levels high. A photograph of the bay after the fire is shown in Figure 10.

**Figure 10. Photograph of storage area after fire.**

Based on witness statements, the fire in the storage bay was preceded by a loud hissing sound and a pressurized release of a reddish dust/mist from a container near the center row of the image. The witness did not see the exact container where the release originated, but was able to place it within a few pallet stacks in the area. Immediately after the pressurized release, a fire ensued in the area. Due to fire suppression efforts, the fire damage was contained to the storage bay where it originated. The investigation focused on the adjacent rows containing the waste drums that were damaged. The exact location of each waste container in the neighboring pallet stacks was reconstructed based upon storage records, remaining containers, damage patterns, and witness statements about container placement. The origin was confirmed to be one pallet stack in the center row.

The containers in the center row had not been repackaged or disturbed from the time of receipt until the incident; therefore, commingling of incompatible wastes or mixing of a reactive pocket within the source drum were refuted as potential causes. Delayed decomposition of waste material in one of the 55-gallon drums is the likely source of the pressurization. Delayed decomposition may be the result of chemical incompatibility or thermal instability leading to a temperature rise, gas generating reaction, or both. Eventually, the pressure increased to the level that the source container ruptured. It was most likely that this release occurred from either a steel drum or a plastic drum given their ability to contain some degree of pressurization compared to a fiber drum. After the initial pressurized release and fire, several other containers were damaged and ruptured as a result of the growing fire.

Based on the photo-documentation of the scene, it was possible to identify several potential drums that showed evidence of pressurization and/or rupture. However, due to the extent of the damage, it was not possible to identify the serial number on all of the containers, and therefore, their origin and contents. Examination of the fire damage showed evidence of burned away paint, burned pallets, and melting of non-metal containers. Based on the damage patterns, it was possible to rule out some waste containers including a set of drums containing phosphoric acid. A ground-level area containing four drums was identified as the likely area of origin, but further specificity was not possible due to the extent of the damage and several other drums that could not be conclusively eliminated. In a regular facility, the containers of all the stored containers would be known, providing valuable information to incident investigators. In a waste handling facility, significantly less certainty is available about the exact contents of each container. The ability for many of the waste container to have been the source of the pressure release significantly complicated this specific investigation. The final determination as to which container was the source could not be determined given the degree of damage.

The extent of damage does not always preclude determination of the likely cause of a fire in storage. A similar TSDF facility experienced a fire in a storage bay, which consumed all of the hazardous waste and combustible containers in the area (see Figure 11). The point of origin of the initial fire could be more precisely placed than the incident just discussed because a nearby operator observed the location of the incipient fire at ground level at a specific pallet stack. The source of the fire was most likely a leak of fuming red nitric acid that spilled onto a wood pallet holding the drums. The drums were stacked on combustible wood pallets, and fuming nitric acid is a strong acid capable of igniting such materials. The facility did not identify the potential consequences of a spill of the acid; thus, a noncombustible pallet (i.e., steel) was not selected. Use of a what-if strategy for packaging and storage selection may have prevented the fire (but not necessarily the leak).

**Figure 11. Another TSDF storage bay after extensive fire damage.**

### *5.2 EQ Industrial Services Warehouse*

A hazardous waste TSDF was complete destroyed due to a fire for which the initial cause was undetermined (see Figure 12). The magnitude of this incident forced thousands of local residents to be evacuated in two days. The CSB investigated the cause of the incident, and identified several hypothetical causes. Their recommendations focused on eliminating those hypothetical causes.

**Figure 12. Facility after the incident [24].**

Based on witness statements, a strong chlorine odor and plume of smoke came from the TSDF on the night of the incident. Initial firefighters responding to the scene reported that a small fire started in one of the hazardous waste storage bays and rapidly spread to the other locations of the facility, including a bay containing many 55-gallon drums of flammable solvents. The incident investigation revealed that the fire originated in a storage bay where containers of solid chlorine-based pool chemicals and unspent aircraft oxygen generators awaiting for final disposal.

Many incidents have occurred involving storage of chlorine-based pool chemicals (see for example [25]-[28]), and around the time of the incident, the NFPA was pursuing research into fire protection considerations for pool chemicals in warehouse settings [29]. Concentrated calcium hypochlorite or sodium dichloroisocyanurate containing pool chemicals pose potential thermal runaway concerns if contaminated with organic material or if stored at elevated temperature. Given their nature as solid oxidizers, pool chemicals would also serve to accelerate fire growth. The witness observations of a chlorine-odor associated with the smoke likely indicate that these chemicals were present near the fire origin, but the investigators could not conclude that these were cause of the fire.

Although the first material involved could not be determined due to the extent of damage, the CSB investigation findings focused on chemical oxygen generators [30]. The CSB concluded that chemical oxygen generators most likely contributed to the intensification of the fire, which quickly spread to an adjacent bay where flammable materials were stored. The investigation found that the unspent oxygen generators, coming from an aircraft maintenance facility, were not properly packaged by the aircraft maintenance personnel, and were misidentified as general oxidizer waste on shipping paperwork. A lesson learned from this incident was to consider oxygen generators that have exceed their service life as still posing a risk of heat and oxygen generation if accidentally actuated. Therefore, the CSB recommended that unspent oxygen generators should be actuated before they are transported [30].

### *5.3 Teris El Dorado Warehouse*

This incident occurred in a warehouse at a TSDF that contained 4000 to 5000 drums of different wastes. The facility operated two rotary kiln incinerators for solid disposal and thermal oxidation for liquid incineration [31]. The incident was witnessed by an operator who saw a drum rupture, which quickly spread fire throughout the warehouse contents (see Figure 13). The entire contents of the warehouse were consumed, damaged, and/or released during the incident. No injuries or fatalities were reported; however, nearby residents were evacuated during the emergency response.

The investigation found that the fire was caused by spontaneous combustion of sodium chlorate solid waste recovered from a rail car accident site. This material was brought to the facility in $1m^3$ boxes to be incinerated shortly before the incident. However, the waste generator incorrectly profiled this material as containing only sodium chlorate in soil; the waste actually contained vegetative matter, rail car components, and portions of railroad ties. The organic contaminants are incompatible with the strong oxidizer and would act as fuel sources, allowing spontaneous ignition of the mixture. The material was repackaged into smaller drums and a portion was moved into the warehouse prior to the fire. The remainder of the material had not been relocated at the time of the fire. This residual material was isolated in a trailer onsite. The trailer containing the residual material underwent spontaneous ignition several days after the original fire incident, which supported the conclusion that the sodium chlorate waste material was the cause of the fire.

**Figure 13. Aerial image during fire event [31].**

Given the non-homogeneous nature of the waste material received, grab sampling of a single container was insufficient to fully characterize the mixed material. In contrast to liquid sampling, a thorough visual examination of the mixed, non-homogeneous waste material may have identified the presence of organic materials in the oxidizer and soil waste. Further, the use of "soil" in the waste analysis plan submitted by the generator may have triggered a further review by the TSDF since soil typically contains a significant organic component. It is likely that the repackaging activity mixed reactive sodium chlorate with fresh organic fuel materials that led to the delayed spontaneous combustion event. After the incident, the facility enhanced their material analysis and handling plans for wastes containing solid oxidizers.

# 6 Fires & Explosions during Incineration Processes

## *6.1 Explosion during Solid Waste Homogenization*

An explosion occurred at an incineration facility that processed hazardous and non-hazardous solid and liquid wastes through thermal destruction in a rotary kiln. Drums of waste were homogenized in a nitrogen-inerted shredder. Drums were scheduled to be incinerated and assembled onto pallets based upon the waste characteristics. The system automatically moved individual pallets into an airlock, purged the airlock with nitrogen, then dumped the pallets into the already inerted shredder chamber. The shredder was used to shred steel drums as well as plastic drums. After shredding, the homogenized waste was conveyed to the rotary kiln via a screw conveyor.

The pallets, packaging materials, and waste materials are combustible materials, and flammable materials may also be processed based on the incinerator waste schedule. Due to the potentially flammable nature of the waste and the likelihood of mechanical sparks and heat generation from the shredder, the system was inerted with nitrogen gas. The nitrogen inerting function was performed by using a fixed timing of nitrogen addition. Although the system did not use an oxygen measurement control loop to dictate the duration of purging, it did use oxygen measurement as a permissive control for the automated conveyor movements and door opening between chambers in the system.

A central incinerator control room at the facility monitored several operating parameters, including the oxygen concentration and temperature within the inerted system. Since one of the primary safety mechanisms for the unit was a reduced oxygen atmosphere, the unit was not intended for processing oxidizing materials including solid oxidizers, oxygen containers/cylinders, explosives, and/or water reactive materials. During one waste load, an explosion occurred in the shredder chamber, which damaged the unit and caused explosion panels to open on the shredder chamber and airlock.

An approximation of the oxygen concentration data and shredder chamber temperature is provided in Figure 14. There was a lag noted in the oxygen sensor response due to the delay for gas to traverse the sampling leg to the sensor. The temperature measurement had less of a response lag; thus, the sudden temperature rise in the shredder chamber actually occurred coincidentally with the explosion and the opening of the explosion vents. On the night of the incident, an additional waste load was fed into the unit, and shredding commenced less than one minute after it was added. Approximately one minute after shredding began, the oxygen concentration started to increase from 2% to over 13% over a period of 20 seconds. According to the recorded data, the temperature in the shredder chamber suddenly increased at approximately the same time as the oxygen concentration increased. This temperature increase is consistent with the heat generated by the explosion within the shredder chamber. An engineering analysis concluded that the most likely source of the rapid oxygen influx was from an item in the waste stream that released oxygen when shredded. All potential sources of air via external sources were examined and rejected.

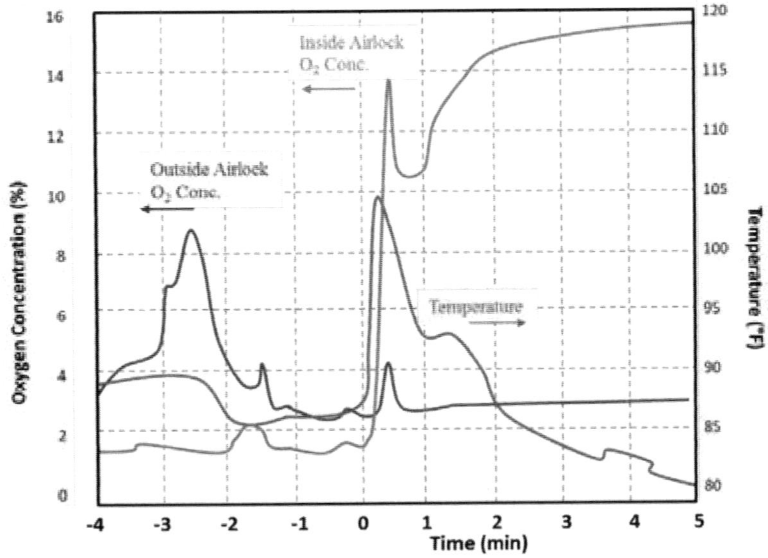


**Figure 14. Process oxygen concentration and temperature during incident.**

The explosion in the shredding system demonstrated the serious consequences of processing waste with unintended properties. In this case, a system that was intended to operate while inerted inadvertently contained compressed oxygen, resulting in a damaging explosion. While the waste was screened prior being fed into the shredder, the screening process did not identify and divert the oxygen-containing waste. In this case, the explosion mitigation system worked as intended by relieving the explosion pressure through explosion vents. The system was damaged, but the damage was not catastrophic, and there were no injuries.

### *6.2 Spontaneous Combustion at a Calciner*

A fire occurred at a cement production facility that used an engineered solid fuel composited from MSW and other industrial wastes to co-fire in the calciner. The engineered fuel was a blended mixture of solid wastes including paper, plastic, and sawdust as well as non-hazardous waste liquids, slurries, and sludge. Because the waste streams possessed variable energy content, a qualifications process (including laboratory analysis) was used to determine the appropriate mixture for creating a consistent engineered feed. A block flow diagram of the shredding, blending, and mixing process is shown in Figure 15. The engineered fuel had the potential for self-heating due to containing oxidizing materials such as polymerizing agents and vegetable-derived oils. Material handlers inspected the stockpiles for external signs of high temperature due to self-heating and excavated the piles to identify internal high temperatures. Infrared pyrometers were used for surface temperature measurements. The temperatures were monitored with the handheld pyrometers and logged prior to the loading the material into trucks for transport to the cement production facility.

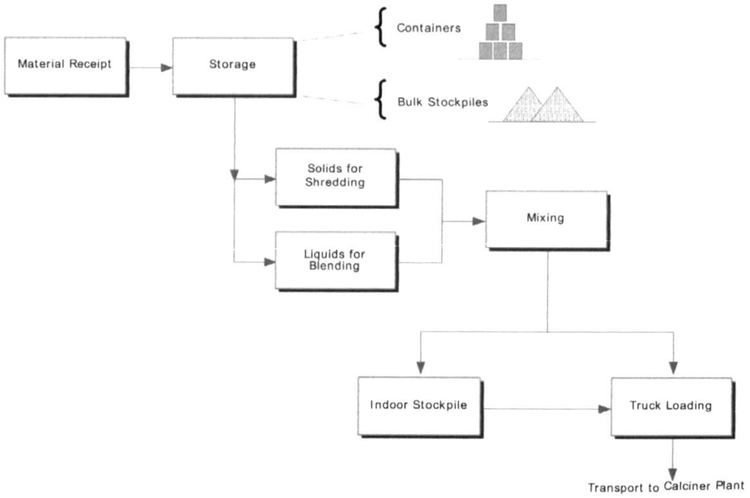


**Figure 15. Block flow diagram of the engineered solids feed preparation process.**

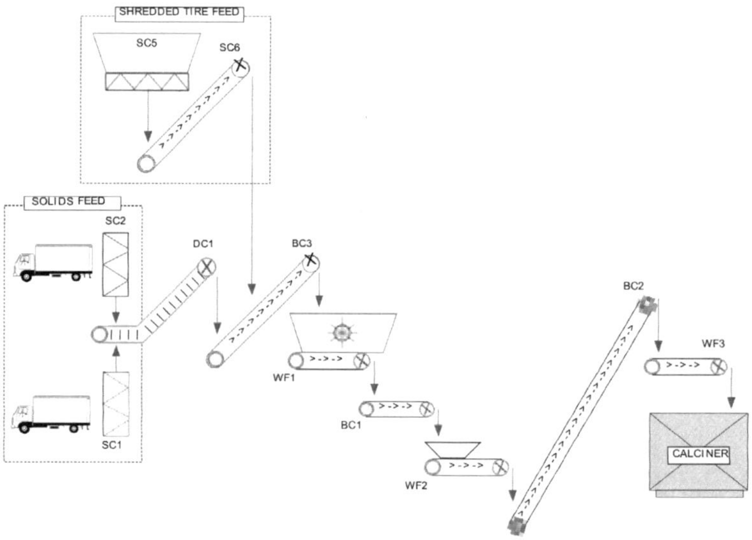


**Figure 16. Conveyor flow diagram between truck delivery and the calciner.**

The engineered fuel was transported to the cement production facility in tractor trailers for eventual feed into the calciner unit. A conveyor flow diagram is provided in Figure 16. The engineered fuel was unloaded into enclosed hoppers that discharged into horizontal screw conveyors. The horizontal screw conveyors discharged into a central drag conveyor that delivers the fuel to a belt conveyor which also received a separate shredded tire feed. The combined feed then traveled through an additional series of hoppers and conveyors before being deposited on a 1,600 feet long enclosed belt conveyor that climbed from ground level to the upper levels of the calciner structure. The belt of the long conveyor was originally a flame-resistant belt material but had been subsequently patched with non-flame resistant material. The total travel time of the engineered fuel was estimated to be 20-30 minutes by the cement facility's personnel.

A fire was discovered overnight soon after the conveyor system was restarted after a two-day outage. During that two-day period, the belts were left loaded with the engineered fuel and combined feed. When the kiln was restarted, the system was operated for approximately one hour then shut down due to an upstream clog during which a fire was discovered near the head section of the long conveyor. The belt failed and broke as a result of the fire.

An investigation into the fire cause determined that, while various friction-based ignition sources (belts, bearing, etc.) did exist within the conveyor system, the most likely cause of the fire was spontaneous ignition of the engineered feed. This was substantiated because three days after the fire, a stockpile of the engineered feed that had been fed to conveyor prior to the fire enflamed at the storage warehouse. Although staff had monitored stockpiles at the feed mixing facility for external signs of ignition, external methods can be insufficient to detect self-heating inside the pile. After exposing the self-heating pile to oxygen during transport and handling, self-heating may accelerate and lead to spontaneous ignition. The facility implemented a strategy that avoided leaving materials on conveyors for long periods of time. Through the use of conveyor interlock strategies and external dump locations, the system can be operated to empty conveyors in the event of an unintended shutdown.

### *6.3 TDI Waste Rail Tank Car*

An explosion occurred at another TSDF cement plant in the liquid fuels rail tank car unloading area [6], [32]. As consequence, the tank ruptured and was propelled about 750 feet away. Other tank cars, as well as storage tanks located nearby, were also heavily damaged (Figure 17). No injuries or fatalities were reported; however, total property losses were estimated at nearly $8.2 million.

This facility incinerated liquid hazardous waste streams as supplementary fuel in rotary kilns, and used to receive several tank cars containing column bottoms waste from a toluene diisocyanate (TDI) manufacturing process. TDI is a monomer feedstock used in the manufacture of polyurethane polymers. The waste generator had stored tank cars containing this type of waste material for several years, and reprocessed them as the opportunity arose. Prior to the incident, several tank cars were shipped to the TSDF for incineration, and had been successfully emptied, with contents incinerated in the kilns. Despite the TDI waste was partially polymerized in most tank cars, the car's contents were non-homogeneous and did not have identical properties between individual tank cars.

**Figure 17. Locations (1) and (2) indicate positions of the tank car's vessel before and after the explosion, respectively [33].**

In order to decrease the viscosity of the polymerized TDI, so that it was pumpable to the kilns, the facility had successfully used external steam heating through the built-in steam jackets on the tank cars.. A schematic of the approximately 30,000 gallon tank car is provided in Figure 18. While attempting to unload the incident TDI material, the tank car vessel was steam-heated in two opportunities to ensure the waste mixture was sufficiently fluid to be pumped. After three unsuccessful unloading attempts, the shipment was rejected, but the tank car was left in the unloading location. The explosion incident took place two days after the last heating attempt.

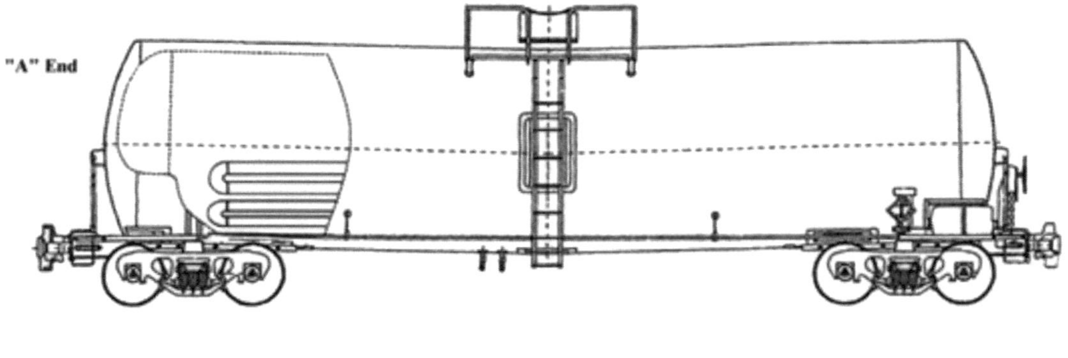


**Figure 18. TDI tank car schematic with cutaway showing the external steam heating coils. [20]**

Retained samples of the original material and residue ejected during the incident were analyzed, and the presence of self-polymerized TDI, TDI monomer, and ureas were determined. Thermal analysis of the materials indicated that the polymerized TDI and TDI monomer exhibited an onset temperature for thermal runaway that was much higher than possible with the steam heating system. A chemical hazard assessment revealed that the cause of the explosion was the runaway thermal decomposition of a contaminant mixed with the TDI waste. The TSDF operator relied on the waste generator for information regarding the waste composition and safe heating procedures. In this case, the waste generator supplied incomplete waste characterization data and inadequate instructions and warnings for heating and handling the TDI waste.

Four critical factors and lessons learned were identified from this incident [6] [20]:

(1) Use of inappropriate tank car design that was unsuitable for this service, creating difficulties for unloading of TDI solid waste.

(2) Storing of TDI waste for several years in a warm climate

(3) Providing inadequate instructions and warnings for unloading the TDI waste.

(4) Commingling an unknown contaminant with the TDI waste; this last factor was the most serious error in this incident.

# 7 Discussion and Conclusions

In the case studies presented above, hazardous properties that were not known to exist led to accidents. In several cases, the accidents were the result of a chemical reaction, leading to a release of heat or pressure that caused a storage vessel to fail. Chemical reactions due to the mixing of incompatible chemicals, storage of unstable chemicals, or heating of thermally unstable chemicals occurred in several of the case studies described. While characterization is a significant preventative safeguard for these types of failures, it may not always be possible to determine if a waste stream is completely stable based on testing; therefore, facilities may still need to consider the unlikely event of an exothermic or pressure-generating reaction within their waste streams. This may be accomplished by using separation distance or barriers between certain waste streams, fire protection systems, or fast turnaround from receipt through final processing.

In several incidents, oxidizers including oxygen tank, oxygen generators/tanks and sodium chlorate, are involved in the initiation of the incident or contribute to the severity of the incident. Because many waste streams contain flammable or combustible materials, enriched oxygen atmospheres or the presence of chemical oxidizers increases the likelihood and severity of a fire or explosion. While in many process operations, inerting atmospheres may be an effective and robust protective measure, due to the possibility of oxidizer ingress in mis-identified waste, additional protections such as explosion vents may be required. The following flowchart was created based on the cases presented in this paper, and may serve as a guideline for investigating suspected waste streams and their role on initiating a fire or explosion event.

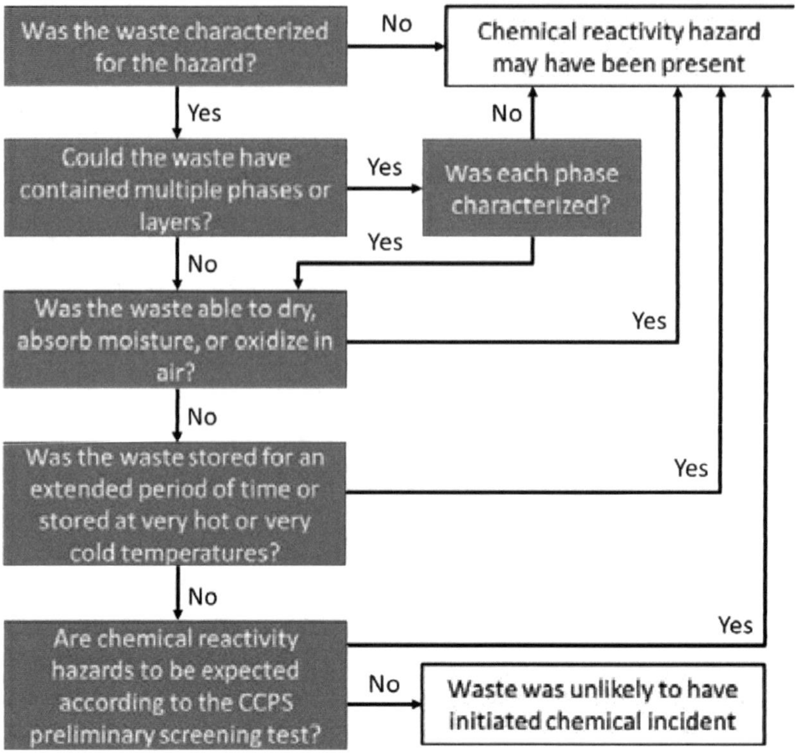


**Figure 19. Role of suspected waste stream on initiating a fire or explosion.**

## 8 References

[1] U.S. Environmental Protection Agency (EPA) Resource Conservation and Recovery Act, Title 40 of the Code of Federal Regulations.

[2] 29 CFR §§1910.119

[3] Center for Chemical Process Safety (CCPS). Guidelines for Risk Based Process Safety. Center for Chemical Process Safety, American Institute of Chemical Engineers, New York, NY, 2007.

[4] U.S. Environmental Protection Agency: Resource Conservation and Recovery Act (RCRA) Laws and Regulations, available at https://www.epa.gov/rcra. Accessed on February 15, 2017.

[5] Evarts, B. "Fires Caused by Spontaneous Combustion or Chemical Reactions," National Fire Protection Association, November 2011.

[6] Cox BL, Carpenter AR, Ogle RA. Lessons learned from case studies of hazardous waste/chemical reactivity incidents. Process Safety Progress 2014; 33(4):395–398.

[7] E.C. Buc. "Hazardous Waste Treatment, Storage, and Disposal Facility Fire Code Gap Analysis, Final Report," The Fire Protection Research Foundation, 2015.

[8] Environmental Protection Agency. Report on Emergency Incidents at Hazardous Waste Combustion Facilities and Other Treatment, Storage, and Disposal Facilities (TSDFs). EPA 1999

[9] U.S. Environmental Protection Agency (EPA): 2015 Biennial Report Summary, available at: https://rcrainfo.epa.gov/rcrainfoweb/action/modules/br/national/view. Accessed January 15, 2017.

[10] U.S. Environmental Protection Agency: Advancing Sustainable Materials Management: 2014 Fact Sheet. US EPA, November 2016.

[11] U.S. Environmental Protection Agency, Available at https://www.epa.gov/hw. Accessed on February 15, 2017.

[12] 40 CFR §§261

[13] 40 CFR §§264/265.17

[14] U.S. Environmental Protection Agency. Waste Analysis at Facilities that Generate, Treat, Store, and Dispose of Hazardous Wastes – Final: Guidance Manual (EPA 530-R-12-001). April 2015.

[15] Hatayama H.K., J. J. Chen, E. R. de Vera, R. D. Stephens, and D. L. Storm. A Method for Determining the Compatibility of Hazardous Wastes. U.S. Environmental Protection Agency, Washington, D.C., EPA/600/2-80/076 (NTIS PB80221005), 1980.

[16] National Oceanic and Atmospheric Administration (NOAA), "Chemical Reactive Worksheet," NOAA, available at http://www.aiche.org/ccps/resources/chemical-reactivity-worksheet-40. Accessed January 15, 2017.

[17] Davis E.M., D. Silva, and J. Murphy. The CCPS Chemical Reactivity Evaluation Tool. 8th Global Congress on Process Safety. April 2012.

[18] Center for Chemical Process Safety (CCPS). Essential Practices for Managing Chemical Reactivity Hazards. Center for Chemical Process Safety, American Institute of Chemical Engineers, New York, NY, 2013.

[19] National Fire Protection Agency. NFPA 921: Guide for Fire and Explosion Investigations

[20] Center for Chemical Process Safety (CCPS). Guidelines for Investigating Chemical Process Incidents. 2nd Ed. 2003

[21] Anthony, EJ, L Jia, M Caris, F Preto, and S Burwell, "An examination of the exothermic nature of fluidized bed combustion (FBC) residues," Waste Management Journal, 19 (1999).

[22] Serris, E, L Favergeon, M Pijolat, M Soustelle, P. Nortier, RS Gartner, T Chopin. "Study of the hydration of CaO powder by gas–solid reaction," Cement and Concrete Research 41(2011) 1078-1084.

[23] U.S. Chemical Safety Board (CSB). Veolia ES Technical Solutions, L.L.C. Case Study. Report 2009-10-I-OH. Washington, D.C. 2010

[24] U.S. Chemical Safety Board (CSB). EQ North Carolina Case Study. Report 2007-01-I-NC. Washington, D. C. 2008.

[25] Gray B.F. and B.W. Halliburton. 2000. "The thermal decomposition of hydrated calcium hypochlorite (UN 2880)." *Fire Safety Journal*. 35.3: 223-239.

[26] Gray B.F. 2002. Spontaneous Combustion and Self-Heating. In SFPE Handbook of Fire Protection Engineering, eds. P.J. DiNenno, D. Drysdale, C.L. Beyler, W.D. Walton, R.L.P. Custer, J.R. Hall and J.M. Watts, Quincy, Massachusetts: National Fire Protection Association, 211-228.

[27] Ogle R. A, Carpenter A. R. Toxic gas release caused by the thermal decomposition of a bulk powder blend containing sodium dichloroisocyanurate. Process Safety Progress 2003; 22:75–82.

[28] Ogle RA, Morrison DR, Carpenter AR, Su YS. Missed opportunities in reactive chemical hazard evaluations. Process Safety Progress, March, 2006, 25(1):2-7.

[29] Buc E.C. and D.J. Hoffman, "Development of an Enhanced Hazard Classification System for Oxidizers Research Project," The Fire Protection Research Foundation, One Batterymarch Park, Quincy, Massachusetts, April, 2006

[30] U.S. Chemical Safety Board (CSB). Safety Advisory Dangers of Unspent Aircraft Oxygen Generators (2007-I-NC-01-SA). 2007

[31] U.S. EPA, "Teris El Dorado Explosion," downloaded on March 13, 2017, https://response.epa.gov/site/site_profile.aspx?site_id=NRC745957.

[32] Ogle R.A, Carpenter AR, Morrison DR. Explosion of a railcar containing toluene diisocyanate waste. Process Safety Progress 2004 Dec; 23(4):316-320.

[33] National Transportation Safety Board. Rupture of a Railroad Tank Car Containing Hazardous Waste Near Clymers, Indiana. Hazardous Materials Accident Report NTSB/ HZM-01/01. Washington, DC. 2001

# 479039 ExxonMobil Torrance Refinery FCC ESP Explosion


**Abstract:**

On February 18, 2015, an explosion occurred in the electrostatic precipitator (ESP) at the ExxonMobil Refinery in Torrance, California near Los Angeles. The explosion injured four workers caused significant property damage to multiple process units within the refinery, and resulted in an off-site accidental release of catalyst dust. As a result of the incident, the FCC unit was down for over one year. The explosion occurred when hydrocarbons from the FCC unit reached the energized ESP while the unit was idling to perform maintenance on equipment that had failed. Multiple failures contributed to the incident. During the ESP explosion, there was also a near miss release of hydrofluoric acid when a large piece of ESP debris fell within feet of a storage vessel storing thousands of gallons of modified hydrofluoric acid in the neighboring alkylation unit. If the storage vessel had failed due to impact from ESP debris following the explosion, hydrofluoric acid would have been released. This presentation will detail key findings from the CSB investigation into the incident.

# Overfill to Overpressure – Investigation Findings and Contributing Events to the CAPECO Explosion

**Scott Davis**
**GexCon US**
**Bethesda, MD**
**sgdavis@gexcon.com**

**Tom DeBold**
**Claudio Marsegan**
**Bryant Hendrickson**
**GexCon US**
**Bethesda, MD**

Prepared for Presentation at
American Institute of Chemical Engineers
2017 Spring Meeting
13th Global Congress on Process Safety
San Antonio, Texas
March 26-29, 2017

# Overfill to Overpressure – Investigation Findings and Contributing Events to the CAPECO Explosion

**Scott Davis**
**GexCon US**
**Bethesda, MD**
**sgdavis@gexcon.com**

**Tom DeBold**
**Claudio Marsegan**
**Bryant Hendrickson**
**GexCon US**
**Bethesda, MD**

**Keywords:** CAPECO, petroleum, overfill, bulk storage, explosion

## Abstract

On October 23, 2009, a vapor cloud explosion occurred after an above ground gasoline storage tank was overfilled at the Caribbean Petroleum Corporation (CAPECO) tank farm facility in Bayamón, Puerto Rico. The blast registered a 2.9 on the Richter scale and sparked several fires throughout the facility that burned for another three days and damaged a total of 21 tanks before emergency responders were able to extinguish the flames.

Operations on the day of the incident included a ship transfer of unleaded gasoline to the CAPECO facility and the transfer was scheduled for four smaller capacity tanks. At approximately 10:00 pm, all of the unleaded gasoline was being transferred to tank 409 and a visible vapor cloud was noted during the walk around at 12:00 am on October 23, 2009. It was determined that tank 409 was overfilled by approximately 5,000 barrels (210,000 gallons) of gasoline, which caused gasoline to flow out of the tank vents near the top of the tank and cascade down the side forming a vapor cloud and liquid pool. This is equivalent to overfilling the tank by over 20 gasoline tanker trucks, which if 10% of the liquid formed the vapor cloud it would cover an area larger than 30 American football fields of a rich flammable mixture. The gasoline vapors mixed with the air forming an ignitable vapor cloud spanning a vast portion of the facility, which subsequently ignited neighboring tanks that burned for three days.

US regulations do not consider CAPECO and other bulk petroleum storage terminals as high hazard facilities, yet a large vapor cloud formed. This paper will present investigation findings and key lessons learned into the massive explosion and fire. This includes: effects of confinement and congestion in a seemingly open space that contributed to the incident; ship offloading and tank filling procedures that revealed significant operational deficiencies; failures to secure the confined drainage system; and level controls.

# 1 Introduction

On October 23, 2009, a vapor cloud explosion occurred after an above ground gasoline storage tank was overfilled at the Caribbean Petroleum Corporation (CAPECO) tank farm facility in Bayamón, Puerto Rico. The blast registered a 2.9 on the Richter scale and sparked several fires throughout the facility that burned for another three days and damaged a total of 21 tanks before emergency responders were able to extinguish the flames. The explosion and ensuing fires did not cause any fatalities; only a few minor injuries were reported. However, approximately 300 homes and business within 1.25 miles of the site reported damages [1].

GexCon was retained to perform an independent and objective investigation into the cause and origin of the explosion. Due to the nature of the investigation, some information provided to GexCon was confidential and could not be provided. Therefore, information provided in the reports of other agencies were used in the present analysis, and include: the ATF Investigation 764045-10-0011 Report [1]; U.S. Chemical Safety and Hazard Investigation Board Final Investigation Report – Caribbean Petroleum Tank Terminal Explosion and Multiple Tank Fires [2]; and the Puerto Rico Fire Department Fire Marshal Report [3] . This paper will present investigation findings and key lessons learned into the massive explosion and fire that occurred at the CAPECO facility. This includes: effects of confinement and congestion in a seemingly open space that contributed to the incident; ship offloading and tank filling procedures that revealed significant operational deficiencies; failures to secure the confined drainage system; and level controls.

We will first present a brief description of the facility, events leading up to the incident, and details of the explosion. Next, we will evaluate the conditions and contributing factors that lead to the explosion.

### *CAPECO Facility*

The Caribbean Petroleum Corporation (CAPECO) facility was a petroleum storage terminal located in Bayamón, Puerto Rico. At the time of the incident, the petroleum storage terminal employed approximately 65 people. Prior to the gasoline overfill and ensuing explosion, the CAPECO facility (Figure 1.1) consisted of a tank farm of above ground storage tanks, offices, decommissioned refinery, waste water treatment plant and distribution piping that serviced the San Juan Bay. The CAPECO facility stored gasoline, diesel, fuel oil, and jet fuel at the time of the incident. The tank locations along with what fuel was stored in each can be found in Figure 1.2.

The tank farm was split by roads into three areas; the northern, central, and southern tank farms. The tanks were separated by dikes or berms designed to hold 110% of the largest tank capacity in that berm (see Figure 1.2) [3]. The southern tank farm contained eleven tanks, which included diesel and fuel oil tanks, divided by six berms. Jet fuel, diesel, and fuel oil were contained in the central tank farm. The northern tank farm held gasoline, diesel, and fuel oil. CAPECO estimated that 613,816 barrels (25,780,272 gallons) of gasoline, 428,260 barrels (17,986,920 gallons) of diesel, 360,496 barrels (15,140,832 gallons) of fuel oil, and 132,724 barrels (5,574,408) of jet fuel were present at the time of the incident [4].

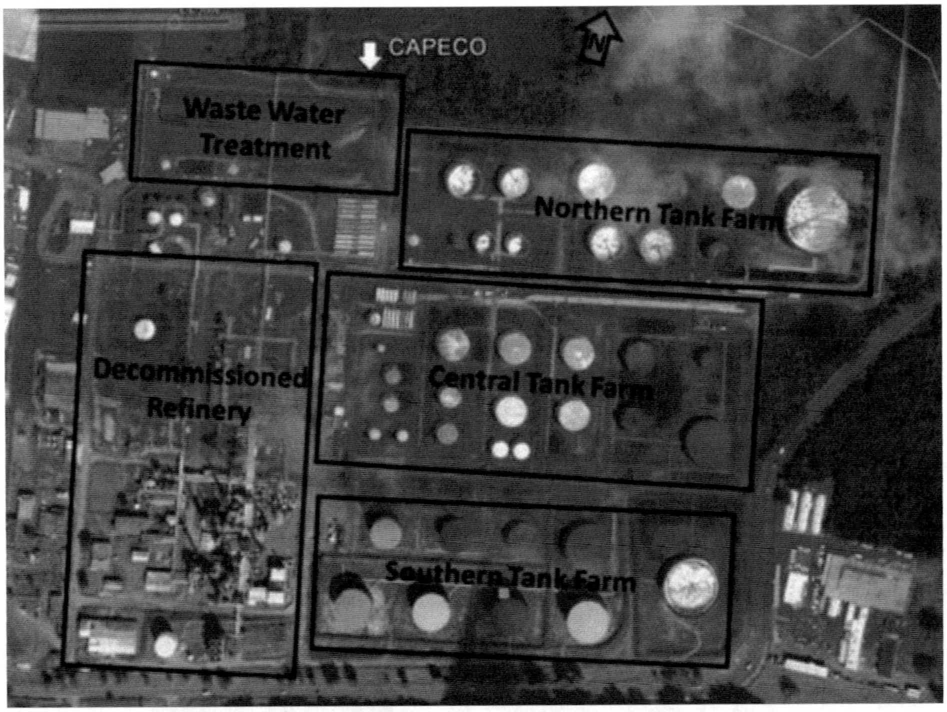


**Figure 1.1: Facility Layout.**

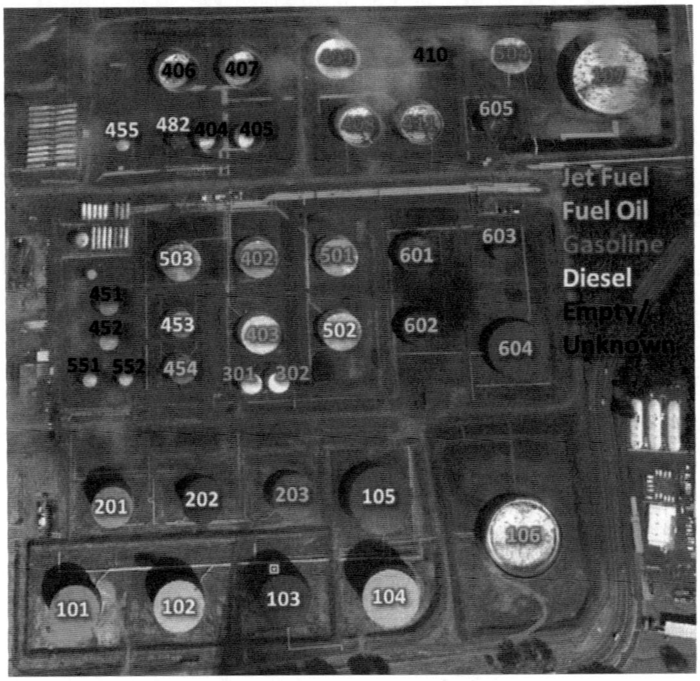


**Figure 1.2: CAPECO tank location and inventory.**

During fuel transfer operations, vessels would load/unload through the facility's 2 ½ mile pipeline which ended at a loading dock located in the San Juan Bay [2]. The terminal dock could accommodate one tanker of up to 75,000 tons. The transfers were made through six pipelines that connected to the facility transferring crude oil, FCC stock, fuel oil, gasoline, diesel, jet fuel and kerosene. The pipelines were equipped with remote operating valves that shutdown through: an electrical actuator; manually closing the valve; thermal shut-off switch set to 221°F (105°C); five push buttons in the operator's office, guards office, stairway to the main platform, fire pump building and Chiksan platform [5].

***Events Prior to the Incident***

Operations the day of the incident included: 1) A transfer of diesel fuel to the Puerto Rico Electric Power Authority Plant in Puerto Nuevo from tank 202; 2) A transfer of fuel oil to the Puerto Rico Electric Power Authority Plant in Palo Seco from tank 105; and 3) A transfer of gasoline from the Cape Bruny vessel to the CAPECO facility. Due to CAPECO's largest tank 107 (capacity 500,000 bbls – 21,000,000 gal) being used for a different quality gasoline, the transfer was scheduled for four smaller capacity tanks (see Figure 1.3) [6].

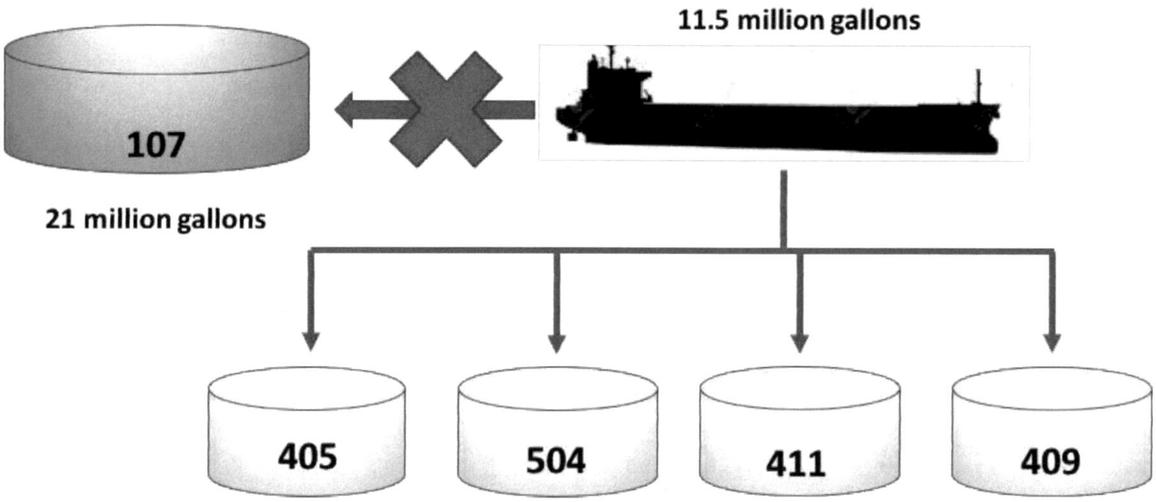


**Figure 1.3: Cape Bruny transfer plan.**

Pumping from Cape Bruny to the CAPECO facility began on October 21, 2009. Four tanks (Tank 405, 504. 409, and 411) in the northern tank farm received gasoline during the transfer. Initial pumping started at approximately 8:47 p.m. A sequence of operations can be found in Figure 1.4 [2].

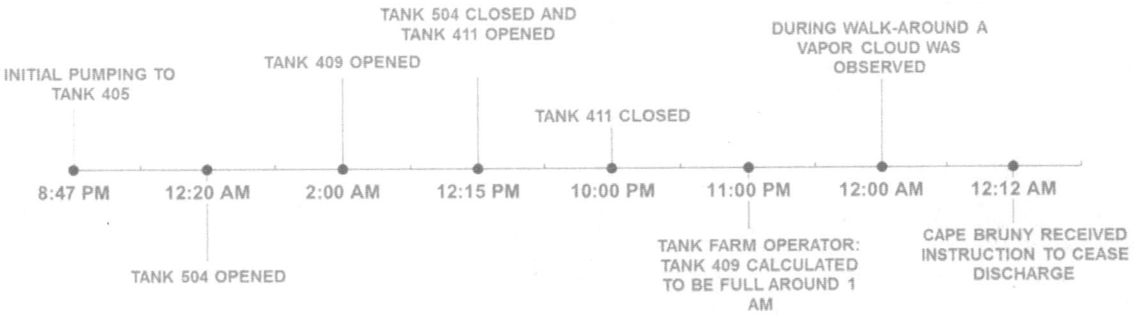


**Figure 1.4: Transfer Operations [6].**

Shortly after 11:00 p.m., the tank farm operator advised the supervisor in charge of the operations that night that the inspector found a difference between the side gauge, monitor console reading and the physical status of Tank 411 (already closed), and advised that the discrepancy was to be investigated [1].

ATF interviews determined that numerous operators and employees observed a vapor cloud starting at midnight. Tank 409 began to overfill between the 11:00 p.m. and 12:00 a.m. tank checks while the average pumping rate was approximately 11,134 bbls/hr (467,628 gal/hr) [7]. The overfilling of tank 409 caused gasoline to flow out of the tank vents near the top of the tank and cascade down the side forming a vapor cloud and liquid pool (as will be discussed in detail later). Tank 409 was contained in a berm with Tank 410 having a capacity of approximately 132,000 barrels (5,544,000 gallons) based on the 110% capacity requirement [3] . As the overfilling continued, gasoline (liquid and vapor) began to fill the containment area and expand towards the waste water treatment facility resulting in the white mist or vapor cloud that was seen in the facility by employees [6]. It was determined that approximately 5,082 barrels (213,444 gallons) of gasoline overflowed from tank 409 [7]. This is equivalent to overfilling the tank by over 20 gasoline tanker trucks, which if 10% of the liquid formed the vapor cloud it would cover an area larger than 30 American football fields of a rich flammable mixture.

The employees saw a white cloud extending approximately two to three feet above the ground by tank 404, and the white vapor cloud could be seen throughout the area between tanks 453 and 301, and near tanks 201 and 409 (Figure 1.5). The vapor cloud could only be seen in areas that were illuminated and extended approximately 200 yards wide by 400 yards long (183 m by 366 m) [1].

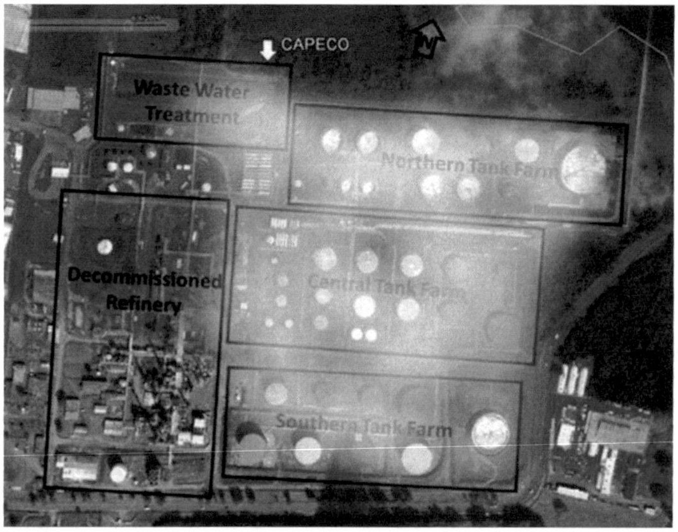


**Figure 1.5: Approximate migration area of the vapor cloud.**

### *Incident*

At approximately 12:23 a.m. on October 23, 2009, the resulting vapor cloud from the overfill of gasoline ignited on the western side of the facility, away from the tank farm area. As the flame propagated (Figure 1.6 and Figure 1.8) towards the above ground storage tanks, a large explosion (Figure 1.8) occurred in the tank farm area that registered 2.9 on the Richter scale. Due to ignition of the large vapor cloud and ensuing explosion, many of the facility storage tanks were consumed by fire, which burned for another three days before emergency responders extinguished the flames [3].

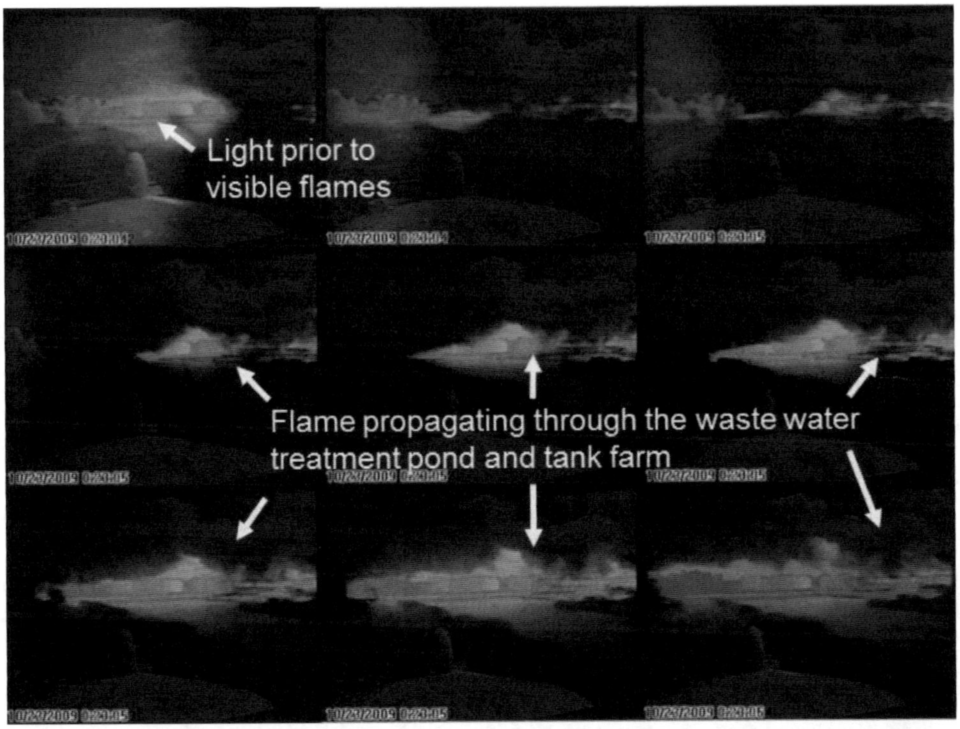


**Figure 1.6: CAPECO security footage from camera 4. Visible light is seen prior to the flame propagation through the waste water treatment pond and tank farm.**

**Figure 1.7: Approximate propagation of the flame front prior to the explosion in the underground drainage as indicated by arrows. Camera locations and angles (red-camera 3 and blue-camera 4) at the first sign of ignition.**

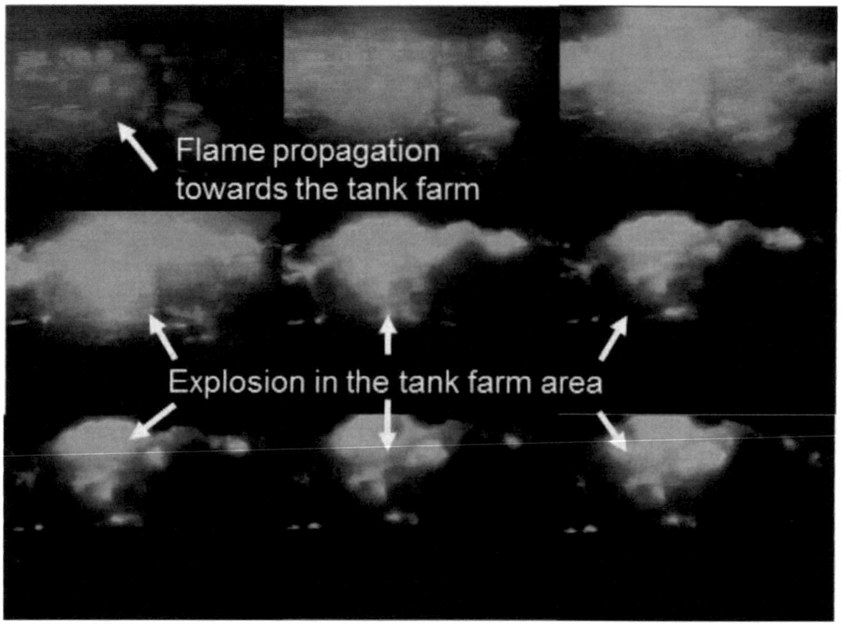


**Figure 1.8: CAPECO security footage from camera 3 as the flash fire approached the tank farm from the west, leading to an explosion within the tank farm area (top-left to bottom-right).**

### *Post Incident and Blast Damage*

Emergency crews were alerted to the incident when a nearby citizen heard a large explosion and saw flames at the CAPECO facility [1]. At approximately 12:30 a.m. on October 23, 2009 the Bayamón and Cataño Fire Departments arrived at the CAPECO facility where they waited for the site fire chief to arrive approximately 45 minutes later. Eleven of the CAPECO tanks were damaged by the initial fire and explosion.

Firefighting efforts, as shown in Figure 1.9, lasted almost three days until approximately October 25, 2009 at 11:40 a.m. when the fires were extinguished [6]. Additional tanks were damaged by ensuing fires (see Figure 1.10 and Figure 1.11) [3] and when the fire was completely extinguished, the damage caused by the large gasoline vapor cloud explosion and fire included 21 tanks onsite. Both blast and thermal damage were observed on the tanks at the CAPECO facility (see Figure 1.12), consistent with the 11 tanks initially damaged by the large vapor cloud explosion and ensuing gasoline fire, and the others falling victim to the resultant fire [1].

**Figure 1.9: Firefighting efforts after the incident [8].**

**Figure 1.10: Aerial view of the CAPECO facility after the initial gasoline vapor cloud explosion.**

**Figure 1.11: Secondary tank fires caused by the initial vapor cloud explosion and fire from the gasoline vapors.**

**Figure 1.12: Post incident damage to the tank farm.**

## 2 Physics of vapor cloud explosions

Prior to the explosion at the CAPECO facility, tank 409 was overfilled and gasoline was being discharged from the overflow vents at the top of the tank. Recall that ignition and burning of liquid hydrocarbons occurs when the hydrocarbon is in the vapor state. A large vapor cloud developed resulting in the 'white cloud' described by CAPECO employees. As the vapor cloud grew from the continuous overflow, the gasoline vapors continued to mix with air forming a flammable mixture throughout the facility. This cloud found a competent ignition source in the western side of the facility and flames propagated towards the tank farm leading up to the main explosion event. This phenomenon is referred to as a vapor cloud explosion (VCE). The specifics on how the vapor cloud was formed and on the VCE will be discussed next.

### *Formation of vapor clouds*

When evaluating the potential for large vapor cloud formation, which may lead to a massive explosion event, it is important to consider the mechanism by which a vapor and vapor clouds are formed. Recent experimental and modeling research has been performed to explore the generation of vapor clouds from overfilling gasoline storage tanks [9], [10]. The research determined that during a tank overfill, the gasoline would flow out of the ventilation holes at the top of the tank and over the deflector plate. The deflector plate channeled some of the escaped fuel onto the tank wall, but the rest ran over the top of the plate, fragmenting into droplets that cascaded through the air. Most of the fuel running down the wall hit a wind girder (a structural stiffening ring) and detached from the tank wall, creating a second cascade of droplets.

These conditions promoted the evaporation of the lighter components of gasoline (e.g., butanes, pentanes and hexanes). The free-fall of droplets leads to entrainment of air and mixing between the air and fuel vapor, and the formation of a rich fuel/air mixture. In addition, the wind conditions at the time of the overfill were calm, and had the wind speeds been higher, the evaporating gasoline vapor would be quickly diluted as it escaped the berm. Therefore, the overfill led to conditions that increased the break-up of gasoline into droplets, which significantly increased the evaporation rate of the lighter components of gasoline as compared to that from the pooled gasoline. Therefore, the overfill rate coupled with low wind speeds generated very large vapor clouds capable of filling a significant region of the CAPECO tank farm as seen in modeling shown in Figure 2.1.

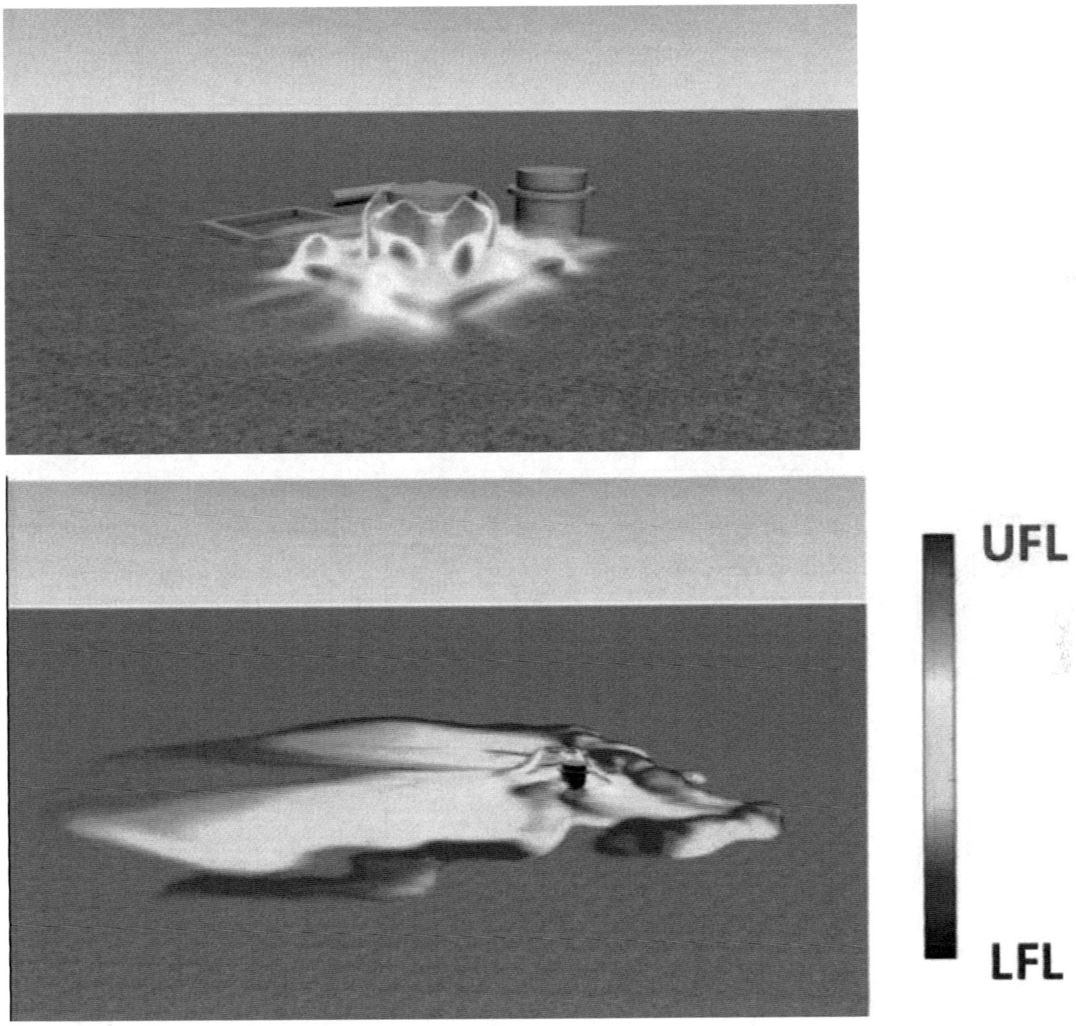


**Figure 2.1: Vapor cloud formation and growth as gasoline initially cascades down the side of a storage tank and is dispersed.**

### *Vapor cloud explosions*

Vapor cloud explosions (VCEs) are oftentimes understood as "confined explosions", where a flammable mixture of gas (or vapor) and air is ignited within a confined structure or vessel. The heat released from the burning mixture causes the gases to expand and the pressure within the confinement increases above normal atmospheric pressure; this is referred to as "overpressure". This occurs until the envelope of the structure ultimately fails, similar to an ignition of a flammable mixture within an empty tank causing the roof to blow off. The sudden failure of the structure/vessel causes a shock or blast wave to form. However, conditions at CAPECO were mostly unconfined and open, except for the underground drainage system. An open and unconfined environment presents little to no resistance of an expanding flame front when a mixture is ignited; therefore, an overpressure is not expected if the gases are allowed to freely expand. The formation of a shock or blast wave in an open area cannot be explained in the same manner, and in fact, requires a unique set of circumstances.

Overpressures can indeed be created for unconfined vapor cloud explosions and require mechanisms that accelerate the combustion or burning process. The combustion rate for burning vapor/air mixtures is measured by how fast the flame propagates (flame speed) relative to the unburned mixture. For quiescent fuel/air mixtures the flame propagates at approximately 0.5 m/s and minimal overpressure is observed as the hot gases can expand in the unconfined environment. However, the flame propagation rate is extremely sensitive to turbulence in the flammable mixture into which the flame is burning. This turbulence can be caused by repetitive obstacles, (e.g. pipes, equipment, trees) within the vapor cloud itself. This results in a feedback loop, which has been extensively studied by Gexcon (see Figure 2.2), whereby the turbulence causes the flame to burn faster, which in turn pushes the unburned gases past the obstacles even faster creating higher turbulence levels, which in turn causes the flame to burn even faster. This results in the flame accelerating and can lead to very high propagation rates.

**Figure 2.2: Flammable fuel air experiments for "unconfined" vapor clouds with no obstacles (top) and with trees (bottom). Clouds ignited on the left and maximum flame speeds with trees are six times larger than without.**

The flame may propagate at such high speeds that the expanding gas ahead of the flame is actually compressed, which can ultimately result in a shock or blast wave. The strength of the blast wave will depend on the flame speeds achieved during the acceleration process.

If the congestion in the cloud is low, the flame speeds will remain low and will not generate a shock or blast wave (i.e., a "flash" or "whoosh" type sound), while relatively higher flame speeds can occur in high congestion areas, resulting in shock wave and "boom" type sound. Flame speeds between 100-150 m/s can generate significant overpressures, and those above 400 m/s can generate pressures well in excess of 1 barg, and can even transition to detonations. This phenomenon was precisely what occurred at the Buncefield explosion in 2005, whereby the flame accelerated into highly congested tree lines that extended over 100 meters. No confinement was found at the Buncefield site and the flame accelerations, and extremely high overpressures, were exclusively caused by the tree density [11].

In order to generate the massive explosion observed at the CAPECO facility, the resulting large vapor cloud needs to migrate into a congested (as described above) or confined region to accelerate the flame. Similar to Buncefield, the CAPECO facility had congestion that included numerous low-lying pipelines connecting the storage tanks, and vegetation to the north and east of the facility. During the post-incident scene investigation, thermal damage was found to extend into the tree line approximately 50 yards past the east property fence [1]. Unlike Buncefield, the CAPECO facility had additional long channels of confined drainage pipes that were not properly closed. Recall that the vapor cloud was ignited western side of the plant propagating towards the waste water treatment (see Figure 1.7) and it propagated east towards the tank farm and vegetation. Video analysis confirmed (see Figure 2.3) that while the main flame propagated above ground through the tank farm, a faster flame front traveled via the underground drainage pipes where a fireball exited the sump pump encasement at the northeast corner of the central tank farm well before the above ground flame.

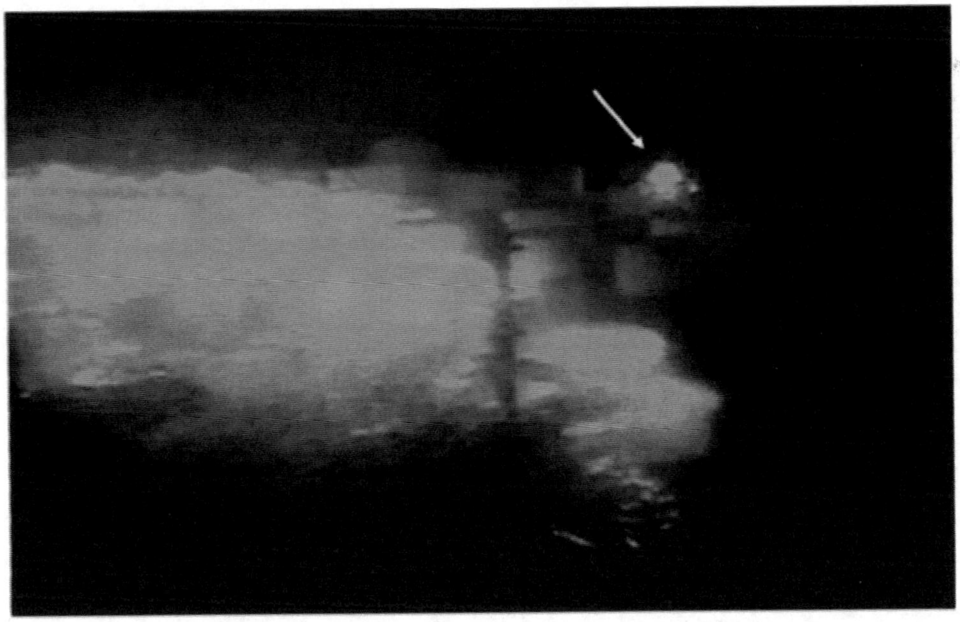

**Figure 2.3: View from camera 3 showing the flame front in the underground drainage pipes reaching the northeast corner of the central tank farm before the above ground flame front. Ignition occurred on the western side of the facility resulting in flame propagation from left to right in the image above.**

In fact, the 10 ft x 10 ft concrete lid on the sump pump, weighing about 12,000 pounds, was found over five feet from its original location [1]. Furthermore, blast damage to the above ground storage tanks was generally in the direction of the sump pump encasement and eastern vegetation and not the western side of the facility where the cloud was first ignited [1]. This confirms that the explosion pressures were exacerbated due to the confinement caused by the underground drainage pipes, which resulted in higher overpressures and faster accelerating flames that also entered the congested region. To evaluate the effect of the underground drainage system, FLACS simulation were performed to evaluate the resulting overpressures and blast damage both with and without the presence of the drainage system.

The CAPECO tank farm area (both with and without the drainage system) was modeled in FLACS, a computational fluid dynamics (CFD) tool predominantly used in the Oil & Gas and chemical industries to predict the consequences of hydrocarbon deflagrations. A large vapor cloud (e.g., stoichiometric cloud 4 meters tall) was placed throughout the tank farm and water treatment areas, and ignited in the water treatment area. Figure 2.4 shows the tank farm with the large vapor cloud and drainage system.

The flame propagation rates were evaluated both through the above ground cloud and through the drainage system. Figure 2.5 shows that the vapor cloud entering the drainage system resulted in flame speeds of over 400 m/s, significantly higher than above ground flame propagation. In fact, the confinement and acceleration of the flame in the drainage system resulted in overpressures that exceeded 2 barg (see Figure 2.6), which could possibly result in condition where the exiting deflagration transitions to a detonation. In contrast, overpressures generated by the flame burning through the above ground portion of the cloud did not exceed 0.03 barg. Therefore, the confined underground drainage system resulted in flames burning faster than the speed of sound and with pressures exceeding 2 barg, which ignited the remaining unburned vapor cloud in the tank farm. This high-pressure, supersonic flame front resulted in a significantly more devastating explosion than the “flash” or “whoosh” that would have been observed had the drainage system been properly closed prior to the overfill.

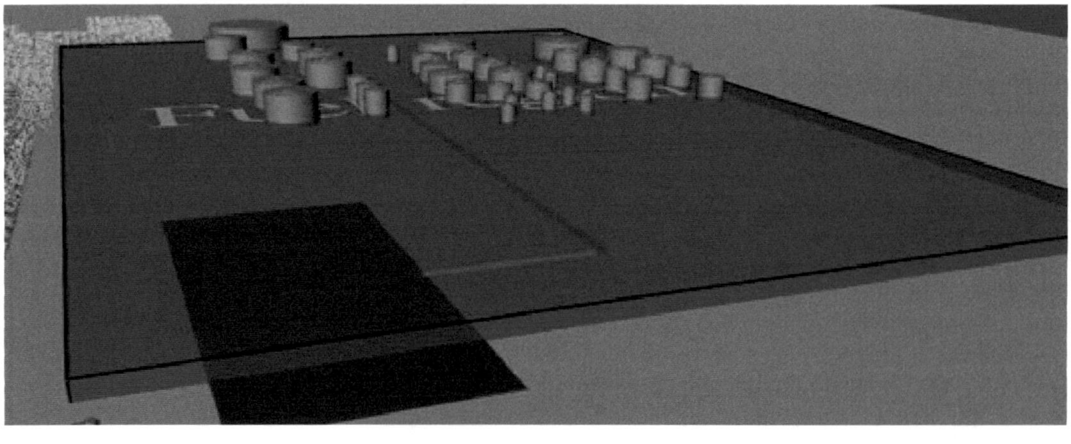

**Figure 2.4: FLACS model of the CAPECO underground drainage system.**

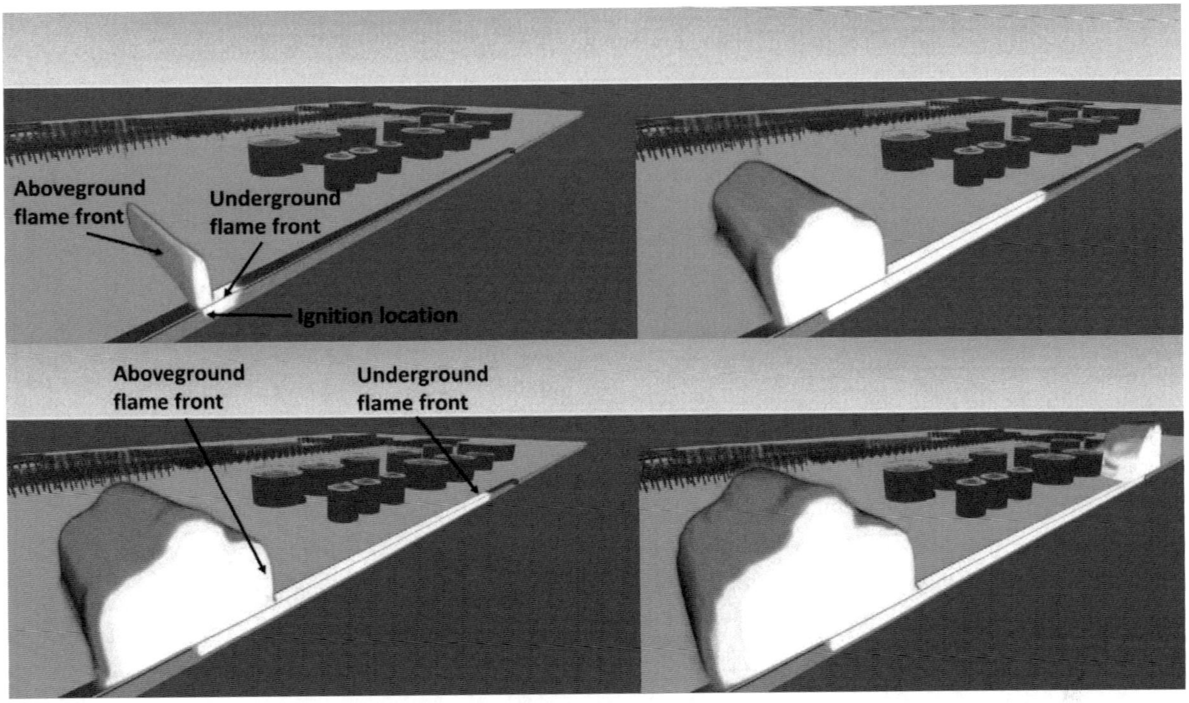


**Figure 2.5: Flame front above and underground.**

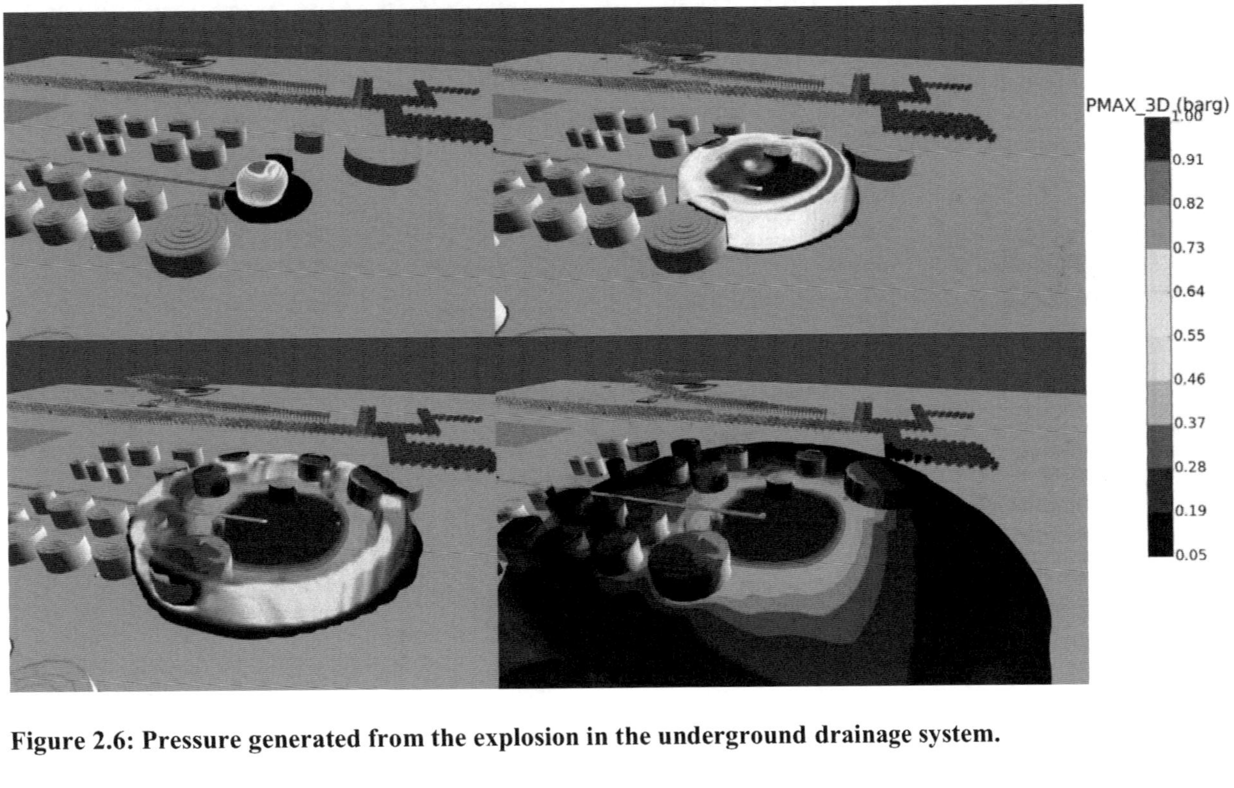


**Figure 2.6: Pressure generated from the explosion in the underground drainage system.**

# 3 Investigation findings and key lessons learned

Many of the contributors to the explosive atmosphere present at the CAPECO facility during the early morning hours on October 23, 2009 could have been avoided if the proper precautions were followed. While previous investigation reports have outlined certain contributing factors related to the incident, other major findings and lessons learned were not identified. For example, other groups have determined that the facility did not have automatic level alarms and that the automatic tank gauge system at the facility was not operational, which contributed to the event by not providing key indicators automatically and relying solely on manual measurements. However, key contributors to the incident (in addition to the automatic gauge system) were not identified and included: a failure to accurately monitor and verify the rate at which the gasoline storage tanks were being filled during the vessel offloading; and a failure to ensure "inherent safety" at the facility by not securing the dike drainage systems, which resulted high-pressure, supersonic flame fronts capable of causing the massive explosion as opposed to a "flash" or "whoosh" had the vapor cloud remained above ground. These factors will be discussed next.

***Operational deficiencies that led to the Tank 409 Overfill***

Recall that automatic controls were down and operations were manually monitored. Using the vessel discharge rates and tank strapping tables, it was estimated that when tanks 504 and 409 were being filled from 4:00 a.m. to approximately 12:00 noon on October 23, 2009, the transfer to each tank was approximately equal (see Figure 3.1). However, when 411 was brought online from 12:00 noon to 10:00 p.m. on October 23, 2009, it was determined that approximately 66% of the fuel was being transferred to tank 411 (see Figure 3.2), with the remaining fuel going to tank 409. This is consistent with operators' observations when preparing to take tank 411 offline: the valve to tank 409 was noted to be only open a little, confirming that tank 411 was receiving a much higher flow rate than tank 409 while they were being filled simultaneously [1].

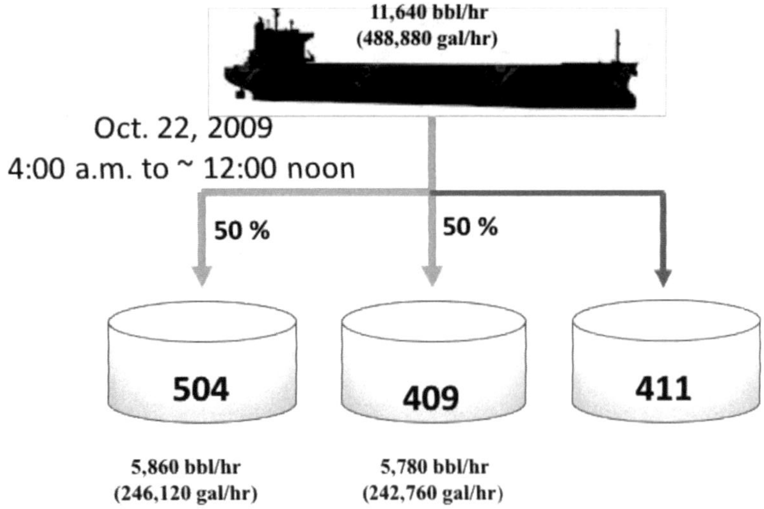


**Figure 3.1: Transfer from the Cape Bruny to tanks 504 and 409 from 4:00 a.m. to approximately 12:00 noon on October 22, 2009.**

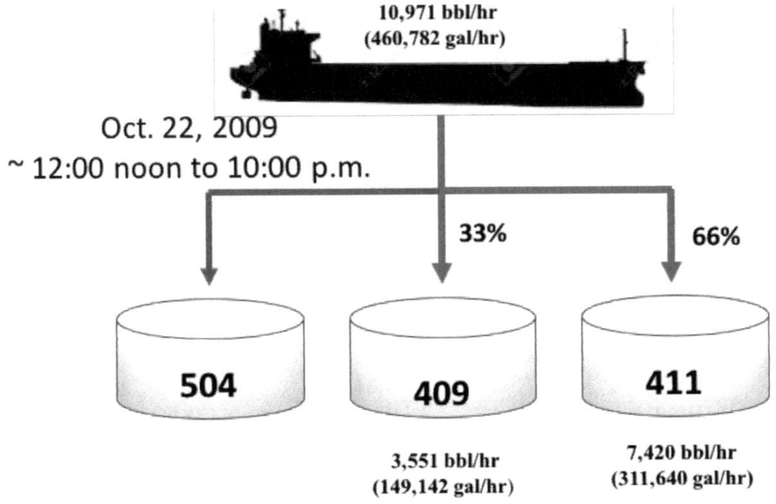


**Figure 3.2: .Transfer from the Cape Bruny to tanks 411 and 409 from approximately 12:00 noon to 10:00 p.m. on October 22, 2009.**

At 10:00 p.m. on October 22nd, tank 411 was closed and tank 409 was fully opened. Tank 409 contained approximately 98,716 gross barrels (4,146,072 gallons) of gasoline (overfill occurs at approximately 121,062 barrels).[1] The supervisor incorrectly estimated that the tank would be full at 1:00 a.m. by assuming only 33% of the discharge rate was going to tank 409. However, with the full vessel discharge of approximately 11,134 bbl/hr [7] going into tank 409, approximately 5,082 barrels (213,444 gallons) of gasoline overflowed from tank 409 contributing to the generation of a large flammable vapor cloud. This means that overflow occurred for approximately 27 minutes, beginning at approximately 11:45 p.m. on October 22nd until 12:12 a.m. the next morning.

**Table 3.1: Estimated amounts of gasoline added to each tank and overflown at the CAPECO facility.**

| Tank | Estimated gasoline added Gross Barrels (Gallons) |
|---|---|
| **405** | 4,411 (185,262) |
| **504** | 62,984 (2,645,328) |
| **411** | 74,198 (3,116,316) |
| **409** | 114,325 (4,801,650) |
| **Total** | 255,918 (255,918) |
| **Total from Cape Bruny** | 261,000 (10,962,000) |
| **Overfill** | **5,082 (213,444)** |

Therefore, one major cause of the incident was due to an incorrectly estimated fill time of 1:00 a.m. for tank 409, likely caused by ignorance of the increased flow rate to tank 409 after closing tank 411. Therefore, failure to compare vessel discharge rates to tank fill rates and verify that tank 409 had sufficient space to receive the scheduled amount of product resulted in tank overflow which generated a large flammable vapor cloud [12].

***Drainage system***

In the post incident investigation, the drainage valve in the berm surrounding tank 409 was found to be in the open position. The open valve drained to the waste water treatment facility, creating a pathway for the liquid gasoline to spread, increasing vapor generation and allowing for flammable gasoline vapor to fill the underground drainage system. The dike drainage valves should be closed during transfer operations so that in the event of an overfill, pool spread is limited. In this case, failure to close the drainage valve led to flammable vapors building up within the underground drainage system, and after ignition, resulted in a deflagration propagating within a completely confined drain and causing significant flame accelerations and overpressures. CFD analyses confirmed that these flame accelerations led to high pressure, supersonic flame fronts capable of causing the massive explosion as opposed to a "flash" or "whoosh" had the vapor cloud remained above ground. Keeping the drainage valves closed during a fuel transfer would have vastly limited the consequences and ensured inherent safety at the facility.

***Level controls***

The CAPECO facility has an automatic tank gauge system that transmitted the tank liquid depth to the operator's office. It would calculate the fill rate, as well as plot the flow rate and tank liquid depth versus time. Due to issues with the automatic tank gauge system (typical float and tape gauge), manual inspection of the tanks and calculation of the fill time had to occur through hourly readings.

# 4 Conclusions

Although US regulations do not consider bulk petroleum storage terminals as high hazards, an explosion at the CAPECO facility and ensuing fire destroyed a vast majority of the tank farm facility. This explosion was caused by a vapor cloud resulting from the overfill of a gasoline bulk storage tank, and this vapor cloud entering an improperly closed drainage system. The two main lessons learned from this incident (and previous incidents like Buncefield) include: (1) having systems and procedures in place to limit the likelihood of having such an overfill condition, and (2) providing inherent safety to such facilities to vastly limit the explosion consequences in the event such an overfill occurs (e.g. limiting confinement and congestion).

Regarding reducing the likelihood of such an event, the CAPECO facility was operating as a fully-attended facility and was in compliance with federal regulations at the time of the incident. As of today, fully-attended tank farm facilities do not require independent liquid level alarms and automatic tank gauging which allows other facilities to operate in the same fashion as CAPECO when transferring and storing flammable liquids such as gasoline. Therefore, such facilities should consider redundant controls and procedures to ensure: fill and transfer rates are constantly updated and verified throughout the transfer; and that the automatic tank gauge system is operational and complements manual measurements made at the facilities; and areas of high confinement like dike drainage systems are secured and closed prior to transfer of volatile hydrocarbons.

Inherent safety should also be evaluated at such facilities to ensure significantly lower consequences, even in the event of a significant release. CFD simulations have confirmed that physical layout of both Buncefield and CAPECO resulted in significantly higher consequences that could have resulted for identical overfill conditions. In fact, the densely coppiced tree lines at Buncefield and high confinement of the improperly opened drainage system at CAPECO resulted in explosion pressures and consequences orders of magnitude higher had such congestion/confinement not been present. Inherent safety of such facilities can be enhanced if areas of confinement and congestion are identified and mitigated in standard risk assessments. In fact, the CAPECO and Buncefield incidents would have likely resulted in large flash fires instead of the more devastating VCEs had these congestion and confinement areas been mitigated.

## References

[1] ATF Investigation 764045-10-0011 Report.

[2] U.S. Chemical Safety and Hazard Investigation Board Final Investigation Report – Caribbean Petroleum Tank Terminal Explosion and Multiple Tank Fires.

[3] Puerto Rico Fire Department Fire Marshal Report.

[4] Inventory Logs.

[5] Dock Operating Manual.

[6] ATF Investigation 764045-10-0011 Interviews.

[7] Ship Transfer Logs.

[8] Caribnewspr https://www.flickr.com/photos/43578995@N03/.

[9] Davis, SG, Hinze, PC, Hansen, OR, van Wingerden, K (2010). 2005 Buncefield Vapor Cloud Explosion: Unraveling the Mystery of the Blast. Interflame 2010, Nottingham, UK..

[10] Coldrick, S., Atkinson, G.T., and Gant, S.E., Health and Safety Laboratory, Hazards XXII, 2011, pp. 560- 569..

[11] Van Wingerden, M. and Wilkins, B. (2008) Experimental investigation of the effect of flexible obstructions on flame propagation in vapour cloud explosions, GexCon report no. GexCon-08-F44110-RA-1, rev 01.

[12] API RP 2350, Appdx B.3.3 (2005).

# Produced Water Tank Explosion Case Study

Presented by:
J. Kelly Thomas, Ph.D.
Baker Engineering and Risk Consultants, Inc.

2017 AIChE
Global Congress on Process Safety

BAKERRISK®



BAKERRISK®

## Outline

- Purpose
- Background
- Facility Description
- Incident
  - Overview
  - Fuel
  - Oxidant
  - Ignition
- Related Incidents
- Lessons Learned
- Potential Prevention and Mitigation Approaches

2

BAKERRISK

## Purpose

- Highlight commonly encountered hazard
  - Produced water tank with flammable atmosphere
  - Similar hazard associated with waste (slop) water tanks
    - Incidents involving waste water tanks in industry have occurred
- Identify lessons learned
- Identify potential prevention and mitigation approaches
  - Examples, not exhaustive compilation
  - Other approaches may provide similar level of risk reduction

3

BAKERRISK

## Background

- Produced water (brine water)
  - Water separated from oil / condensate / gas streams
  - Collected in tanks and treated as waste
  - Head space may be in flammable range depending on operating parameters and separator performance (e.g., condensate carryover)
- Not typically viewed as posing significant explosion risk
  - Reasonable on a relative basis if other explosion hazards present which pose much more severe consequences and/or risk
  - Reasonable from a building siting perspective
  - Hazard is local, although debris may be thrown considerable distance
  - However, produced water tanks can, and do, explode
- Should be considered in terms of hazard/risk to personnel

4

## Facility Description (1 of 6)

- Incident occurred at gas production site (West Virginia, 2013)
- 6 gas wells (no oil or condensate produced), each with own separator
  - Separators are 2.5 ft diameter by 8 ft long, and operate at 675 psig
  - Gas composition (97.2% $CH_4$, 2.1% $C_2H_6$, 0.1% $C_3H_8$, 0.3 $CO_2$, 0.3% $N_2$)
- Produced water tanks
  - Tanks in two groups of 3 (i.e., 6 tanks)
  - Group of 3 tanks share common overflow (3 in.) and vent (4 in.) lines
    - Common vent line not supplied by tank manufacturer
    - No flame arresters on these lines
  - Fixed roof API 12F style, 12 ft. (3.7 m) diameter, 20 feet (6.1 m) tall
  - 400 barrel (16,800 gallons, 64 $m^3$) & hydrotested at 1.5 psig (0.10 barg)
  - 2 access hatches (Enardo ES-660 and Jayco JT-8); relieve at 0.5 psig

5

## Facility Description (2 of 6)

6

## Facility Description (3 of 6)

- Fill line and access hatches

7

## Facility Description (4 of 6)

- Common vent and overflow lines

- Common vent line terminates with open end at opposite end of 3-tank group

## Facility Description (5 of 6)

- Sign at entrance to pad

9

## Facility Description (6 of 6)

- Warning sign on each produced water tank
- Flammability rating of 3 would be indicative of gasoline

10

BAKERRISK

## Incident (Overview, 1 of 7)

- Approx. 100 bbl removed from each tank every 3 or 4 days
  - Driver goes up to top of tank via catwalk, opens hatch, and gauges using wooden rod before pumping out desired liquid volume
- Driver arrived between 5:30 and 5:45 am, hooked truck up to tank manifold, and (at some point) ascended catwalk to tanks
- Explosion occurs prior to 6:45 am (February, sunrise at 7:17)
- One tank fails at shell-floor joint, ejected, and thrown 110 feet (34 m)
- One tank fails at shell-roof joint, roof thrown 165 feet (50 m)
- Catwalk damaged and displaced, throwing driver from catwalk and outside tank battery, approximately 30 feet (9 m); fatality.

11

BAKERRISK

## Incident (Overview, 2 of 7)

12

## Incident (Overview, 3 of 7)

13

## Incident (Overview, 4 of 7)

- Ejected tank

14

## Incident (Overview, 5 of 7)

- Failed tank roof
- Shell-floor joint shows uplift (was approaching joint failure)

15

## Incident (Overview, 6 of 7)

- Catwalk displacement (catwalk supported by tank attachment)

16

## Incident (Overview, 7 of 7)

- Tank structural capacity roughly 10 to 15 psig (gradually applied internal pressurization) based on past experience
- Trajectory analysis indicates in range of 5 to 15 psig required to achieve observed tank and tank roof throw distances
- Indicates weakly flammable mixture was present in two tanks which failed, and mixture in "middle" tank was either not flammable or barely flammable
- Post-incident flammable gas measurements
  - Tank in adjacent set ≈ 65% LFL days after incident
  - Truck ≈ 15% LFL 2 months after incident
  - Consistent with elevated fuel concentration in tanks at time of incident

17

## Incident (Fuel, 1 of 3)

- Separator does not remove "all gas" from the fluid, even when functioning properly and as designed
- Produced water contains gas since at elevated pressure (675 psig), which is released when pressure dropped to atmospheric
  - Gas content function of water pressure, temperature and composition
- Can have gas "carry" over or with stuck/jammed valve in separator
  - Separators were tested post-event and found to work properly
  - Concluded gas source was solely content of produced water transferred to storage tanks from separators

18

## Incident (Fuel, 2 of 3)

- Methane solubility data from Duan & Mao (2006) Geochemica, 70: 2269-86.
  - Approximately 0.043 moles/kg at relevant conditions
- See vapor space fuel concentration plot on following slide:
  - Low level to high level of 50 to 150 barrels (1/4 tank capacity)
  - Conditions present (separator pressure, fluid composition, etc.)
  - No air in-leakage (only draw in air when drop liquid level)
- Can easily exceed UFL
  - Fuel concentration would be lower at lower separator pressure
  - Minimum fuel concentration would decrease if either high level or liquid volume extracted were increased (vs. values noted above)
  - > UFL is not a "universal condition" for produced water tanks

19

## Incident (Fuel, 3 of 3)

- Fuel concentration for series of 100 bbl fill / extract cycles

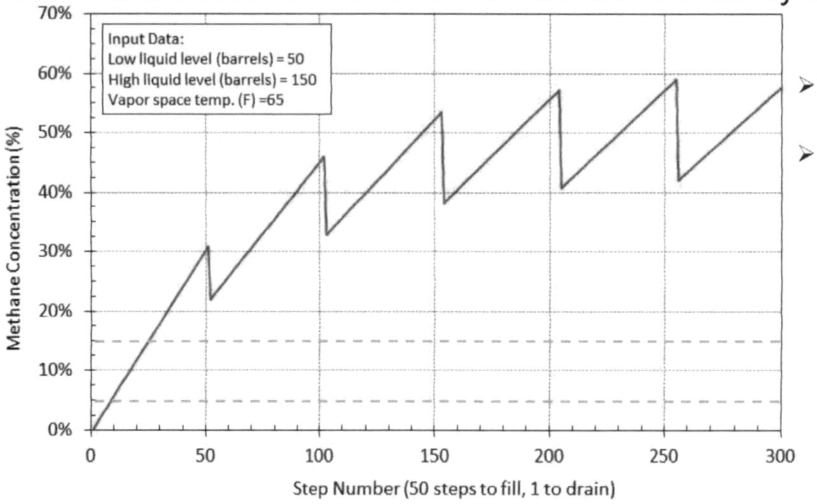


- Can easily exceed the UFL
- Issue is how get enough oxidant to be in flammable range

20

## Incident (Oxidant, 1 of 4)

- Need additional air to enter flammable range (i.e., < UFL)
- Additional air can be drawn into tank via natural convection
  - Gas within tank lighter than outside air (i.e., unless outside air is hotter)
  - Light gas can flow out common vent line, pulling air in through leak points (e.g., access hatch seals)
    - Analysis subject to uncertainty since inlet flow path not well defined, but leakage rate in range of 125 scfd (2.5 lpm, 22 bbl/day) reasonable
    - Acts to limit maximum fuel concentration
    - Steady decrease in fuel concentration when tank is "stagnant" (i.e., no produced water additions or removals)
  - Much higher air in-leakage rate when hatch is open (2 to 5 scfm)
- Tanks stagnant for period before explosion & hatched opened
- See illustrate plots on following page

21

## Incident (Oxidant, 2 of 4)

- Fuel concentration for series of 100 bbl fill / extract cycles, followed by stagnant period (125 scfd air in-leakage)

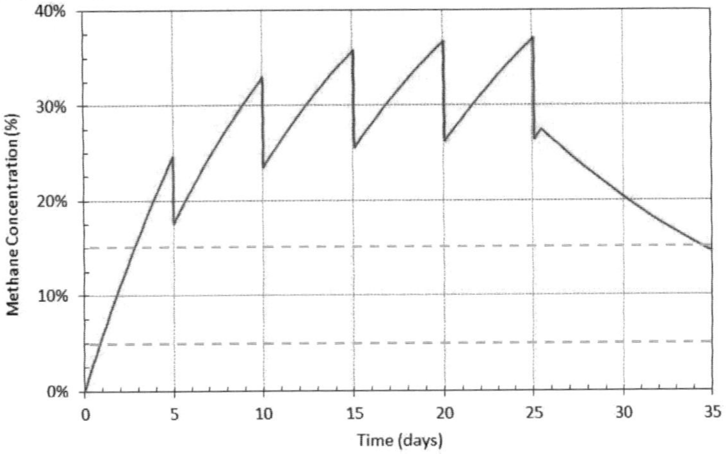


- Above UFL until enter stagnant period
- Air in-leakage during stagnant period allows fuel concentration to drop below UFL and enter flammable range
- Reach flammable condition in about 2 weeks

22

## Incident (Oxidant, 3 of 4)

- Fuel concentration for series of 100 bbl fill / extract cycles, followed by stagnant period with hatch open (2 scfm air in-leakage)

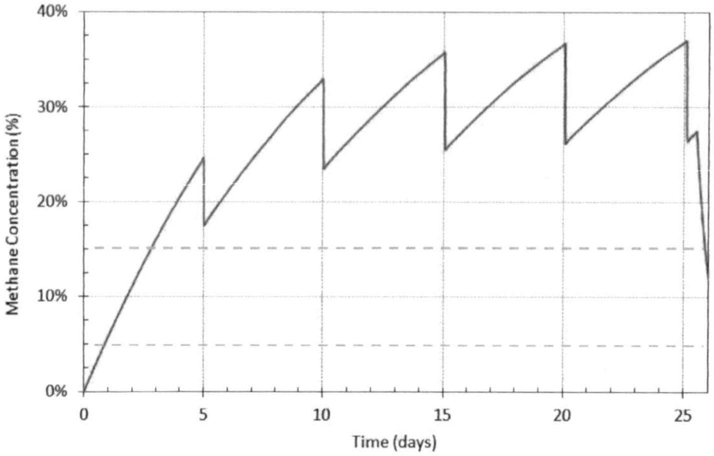


- Above UFL until hatch open for extended period
- Air in-leakage with hatch open for extended period allows fuel concentration to drop below UFL and enter flammable range
- Reach flammable condition in about 8 hours

23

## Incident (Oxidant, 4 of 4)

- Natural convection limits maximum fuel concentration and can bring tank vapor space into flammable range, particularly with a tank which is stagnant for an extended period or with a hatch left open
- Note should not conclude that require natural convection in order to form a flammable mixture with a generic produced water tank
  - Lower separator pressure would mean less gas and hence tank could be flammable for portion of time regardless of natural convection
  - Larger volume of produced water withdrawn from tank or higher tank levels could put the tank vapor space into the flammable range
- Should assume a flammable mixture can result if the fuel concentration can exceed the LFL

24

BAKERRISK

## Incident (Ignition Source)

- Driver was performing operation before sunrise
  - No headlamp or flashlight recovered from scene
  - Lighter found 25 feet from body (in same direction as thrown)
  - Driver was a smoker and one glove was off
  - Opened cigarette pack present in truck
  - Reasonable to assume driver used lighter to check liquid level
  - Driver not equipped with an LFL meter
- Static charge cannot be completely ruled out
- Flame propagation to adjacent tanks via connecting lines
- Ignition potential recognized (e.g., API RP 505)

25

BAKERRISK

## Related Incidents

- Produced water tank explosions have happened in the past
- Cursory literature search
  - Texas (2013): Set of six 200 bbl produced water tanks, 2 people smoking on top of tank, tank exploded
  - Canada: Welding on tank which had contained produced water resulted in explosion
- CSB investigation report (2011-H-1):
  - "Public Safety at Oil & Gas Storage Facilities"
  - 26 incidents over 27 year period (44 fatalities and 25 injuries)
  - Includes at least one produced water tank
- Produced water SDS packages vary in flammability warnings (as does the signage on the produced water trucks)

26

BAKERRISK

## Lessons Learned

- Produced water tank explosions can and do happen
  - Lesson already learned in prior incidents .....
- Hazard awareness may not be adequate among all personnel working at oil and gas production sites
- Development of flammable atmosphere in tank depends on separator pressure, tank arrangement and tank ventilation
  - Should assume tank vapor space may be flammable
  - Can also have separator malfunction leading to increased fuel concentration

27

BAKERRISK

## Potential Prevention and Mitigation Approaches

- Increase hazard awareness for all personnel working at site
- Measure level at grade to eliminate routine access to most hazardous location (i.e., tank roof)
- Inert tanks using natural gas
  - Take feed from product gas line
  - Eliminate open vent paths and institute pressure control system (flow as required to maintain positive pressure and preclude air in-leakage),
  - Will result in increased emission of natural gas from facility
- Use a "lead tank" as a secondary separator
  - Inert only this tank with natural gas (limit use of natural gas)
  - Pump or gravity flow from "lead tank" to other produced water tanks
- Monitor tank vapor space

28

BAKERRISK®

# Questions

?

29

# 481704 Milestone Company Incidents and Their Impact on Air Products Process Safety Journey

**Abstract:**

**Air Products has experienced several significant Process Safety events in our history that have helped define the Process Safety culture that exists at the company today. This presentation describes these events and their root causes, as well as the changes that they launched in the company's approach to Process Safety. The first incident was an explosion at a Nitrous Oxide plant in 1976 that resulted in four on-site fatalities. In the aftermath of this event the need for a formal Hazard Review Process was identified. The second event occurred in 1978 at an air separation plant in which a cryogenic Liquid Oxygen storage tank was ruptured and 800 tons of Liquid Oxygen was released. The Liquid Oxygen vaporized and drifted off-site, resulting in five off-site fatalities. This incident highlighted the necessity of Risk Quantification. While these milestone events served as the impetus for Air Products Process Safety journey, that journey continues to current day, and the presentation will close with a brief update on future direction.**

# 481925 Plymouth Incident

**Abstract:**

The Williams Company LNG facility on their Northwest Pipeline system located in the state of Washington experience an explosion March 30th 2014. The incident occurred during a routine start-up of the liquefaction process. Directly after the facility was deemed safe the site preservation process was initiated and the investigation began. The various regulatory agencies were involved during the investigation to help determine the cause and to provide recommendations to prevent like kind occurrences elsewhere within the industry. Additional 3rd party firms were called in to review potential structural deficiencies, design flaws, metallurgy testing, and auto ignition testing.

# AUTHOR INDEX

# AUTHOR INDEX